Lecture Notes in Physics

Springer
Berlin
Heidelberg
New York
Barcelona
Hong Kong
London
Milan
Paris
Singapore
Tokyo

Physics and Astronomy   ONLINE LIBRARY

http://www.springer.de/phys/

The Editorial Policy for Proceedings

The series Lecture Notes in Physics reports new developments in physical research and teaching – quickly, informally, and at a high level. The proceedings to be considered for publication in this series should be limited to only a few areas of research, and these should be closely related to each other. The contributions should be of a high standard and should avoid lengthy redraftings of papers already published or about to be published elsewhere. As a whole, the proceedings should aim for a balanced presentation of the theme of the conference including a description of the techniques used and enough motivation for a broad readership. It should not be assumed that the published proceedings must reflect the conference in its entirety. (A listing or abstracts of papers presented at the meeting but not included in the proceedings could be added as an appendix.)

When applying for publication in the series Lecture Notes in Physics the volume's editor(s) should submit sufficient material to enable the series editors and their referees to make a fairly accurate evaluation (e.g. a complete list of speakers and titles of papers to be presented and abstracts). If, based on this information, the proceedings are (tentatively) accepted, the volume's editor(s), whose name(s) will appear on the title pages, should select the papers suitable for publication and have them refereed (as for a journal) when appropriate. As a rule discussions will not be accepted. The series editors and Springer-Verlag will normally not interfere with the detailed editing except in fairly obvious cases or on technical matters.

Final acceptance is expressed by the series editor in charge, in consultation with Springer-Verlag only after receiving the complete manuscript. It might help to send a copy of the authors' manuscripts in advance to the editor in charge to discuss possible revisions with him. As a general rule, the series editor will confirm his tentative acceptance if the final manuscript corresponds to the original concept discussed, if the quality of the contribution meets the requirements of the series, and if the final size of the manuscript does not greatly exceed the number of pages originally agreed upon. The manuscript should be forwarded to Springer-Verlag shortly after the meeting. In cases of extreme delay (more than six months after the conference) the series editors will check once more the timeliness of the papers. Therefore, the volume's editor(s) should establish strict deadlines, or collect the articles during the conference and have them revised on the spot. If a delay is unavoidable, one should encourage the authors to update their contributions if appropriate. The editors of proceedings are strongly advised to inform contributors about these points at an early stage.

The final manuscript should contain a table of contents and an informative introduction accessible also to readers not particularly familiar with the topic of the conference. The contributions should be in English. The volume's editor(s) should check the contributions for the correct use of language. At Springer-Verlag only the prefaces will be checked by a copy-editor for language and style. Grave linguistic or technical shortcomings may lead to the rejection of contributions by the series editors. A conference report should not exceed a total of 500 pages. Keeping the size within this bound should be achieved by a stricter selection of articles and not by imposing an upper limit to the length of the individual papers. Editors receive jointly 30 complimentary copies of their book. They are entitled to purchase further copies of their book at a reduced rate. As a rule no reprints of individual contributions can be supplied. No royalty is paid on Lecture Notes in Physics volumes. Commitment to publish is made by letter of interest rather than by signing a formal contract. Springer-Verlag secures the copyright for each volume.

The Production Process

The books are hardbound, and the publisher will select quality paper appropriate to the needs of the author(s). Publication time is about ten weeks. More than twenty years of experience guarantee authors the best possible service. To reach the goal of rapid publication at a low price the technique of photographic reproduction from a camera-ready manuscript was chosen. This process shifts the main responsibility for the technical quality considerably from the publisher to the authors. We therefore urge all authors and editors of proceedings to observe very carefully the essentials for the preparation of camera-ready manuscripts, which we will supply on request. This applies especially to the quality of figures and halftones submitted for publication. In addition, it might be useful to look at some of the volumes already published. As a special service, we offer free of charge LATEX and TEX macro packages to format the text according to Springer-Verlag's quality requirements. We strongly recommend that you make use of this offer, since the result will be a book of considerably improved technical quality. To avoid mistakes and time-consuming correspondence during the production period the conference editors should request special instructions from the publisher well before the beginning of the conference. Manuscripts not meeting the technical standard of the series will have to be returned for improvement.

For further information please contact Springer-Verlag, Physics Editorial Department II, Tiergartenstrasse 17, D-69121 Heidelberg, Germany

Series homepage – http://www.springer.de/phys/books/lnpp

Klaus R. Mecke Dietrich Stoyan (Eds.)

Statistical Physics and Spatial Statistics

The Art of Analyzing and Modeling Spatial Structures and Pattern Formation

Springer

Editors

Klaus R. Mecke
Fachbereich Physik
Bergische Universität Wuppertal
42097 Wuppertal, Germany

Dietrich Stoyan
Institut für Stochastik
TU Bergakademie Freiberg
09596 Freiberg, Germany

Library of Congress Cataloging-in-Publication Data applied for.

Die Deutsche Bibliothek - CIP-Einheitsaufnahme

Statistical physics and spatial statistics : the art of analyzing and
modeling spatial structures and pattern formation / Klaus R. Mecke ;
Dietrich Stoyan (ed.). - Berlin ; Heidelberg ; New York ; Barcelona ;
Hong Kong ; London ; Milan ; Paris ; Singapore ; Tokyo : Springer,
2000
  (Lecture notes in physics ; Vol. 554)
  (Physics and astronomy online library)

ISSN 0075-8450
ISBN 978-3-642-08725-7   e-ISBN 978-3-540-45043-6

Springer-Verlag Berlin Heidelberg New York
a member of BertelsmannSpringer Science+Business Media GmbH

© Springer-Verlag Berlin Heidelberg 2010
Printed in Germany

Cover design: *design & production*, Heidelberg

Printed on acid-free paper

Preface

Modern physics is confronted with a large variety of complex spatial structures; almost every research group in physics is working with spatial data. Pattern formation in chemical reactions, mesoscopic phases of complex fluids such as liquid crystals or microemulsions, fluid structures on planar substrates (well-known as water droplets on a window glass), or the large-scale distribution of galaxies in the universe are only a few prominent examples where spatial structures are relevant for the understanding of physical phenomena. Numerous research areas in physics are concerned with spatial data. For example, in high energy physics tracks in cloud chambers are analyzed, while in gamma ray astronomy observational information is extracted from point patterns of Cherenkov photons hitting a large scale detector field. A development of importance to physics in general is the use of imaging techniques in real space. Methods such as scanning microscopy and computer tomography produce images which enable detailed studies of spatial structures.

Many research groups study non-linear dynamics in order to understand the time evolution of complex patterns. Moreover, computer simulations yield detailed spatial information, for instance, in condensed matter physics on configurations of millions of particles. Spatial structures also derive from fracture and crack distributions in solids studied in solid state physics. Furthermore, many physicists and engineers study transport properties of disordered materials such as porous media.

Because of the enormous amount of information in patterns, it is difficult to describe spatial structures through a finite number of parameters. However, statistical physicists need the compact description of spatial structures to find dynamical equations, to compare experiments with theory, or to classify patterns, for instance. Thus they should be interested in spatial statistics, which provides the tools to develop and estimate statistically such characteristics. Nevertheless, until now, the use of the powerful methods provided by spatial statistics such as mathematical morphology and stereology have been restricted to medicine and biology. But since the volume of spatial information is growing fast also in physics and material science, physicists can only gain by using the techniques developed in spatial statistics.

The traditional approach to obtain structure information in physics is Fourier transformation and calculation of wave-vector dependent structure functions. Surely, as long as scattering techniques were the major experimental set-up in order to study spatial structures on a microscopic level, the two-point correlation function was exactly what one needed in order to compare experiment and theory. Nowadays, since spatial information is ever more accessible through digitized images, the need for similarly powerful techniques in real space is obvious.

In the recent decades spatial statistics has developed practically independently of physics as a new branch in statistics. It is based on stochastic geometry and the traditional field of statistics for stochastic processes. Statistical physics and spatial statistics have many methods and models in common which should facilitate an exchange of ideas and results. One may expect a close cooperation between the two branches of science as each could learn from the other. For instance, correlation functions are used frequently in physics with vague knowledge only of how to estimate them statistically and how to carry out edge corrections. On the other hand, spatial statistics uses Monte Carlo simulations and random fields as models in geology and biology, but without referring to the helpful and deep results already obtained during the long history of these models in statistical physics. Since their research problems are close and often even overlap, a fruitful collaboration between physicists and statisticians should not only be possible but also very valuable. Physicists typically define models, calculate their physical properties and characterize the corresponding spatial structures. But they also have to face the 'inverse problem' of finding an appropriate model for a given spatial structure measured by an experiment. For example, if in a given situation an Ising model is appropriate, then the interaction parameters need to be determined (or, in terms of statistics, 'estimated') from a given spatial configuration. Furthermore, the goodness-of-fit of the Ising model for the given data should be tested. Fortunately, these are standard problems of spatial statistics, for which adequate methods are available.

The gain from an exchange between physics and spatial statistics is two-sided; spatial statistics is not only useful to physicists, it can also learn from physics. The Gibbs models used so extensively today in spatial statistics have their origin in physics; thus a thorough study of the physical literature could lead to a deeper understanding of these models and their further development. Similarly, Monte Carlo simulation methods invented by physicists are now used to a large extent in statistics. There is a lot of experience held by physicists which statisticians should be aware of and exploit; otherwise they will find themselves step by step rediscovering the ideas of physicists.

Unfortunately, contact between physicists and statisticians is not free of conflicts. Language and notation in both fields are rather different. For many statisticians it is frustrating to read a book on physics, and the same is true for statistical books read by physicists. Both sides speak about a strange language and notation in the other discipline. Even more problems arise from different traditions and different ways of thinking in these two scientific areas. A typical example, which is discussed in this volume, is the use of the term 'stationary' and the meaning of 'stationary' models in spatial statistics. This can lead to serious misunderstandings. Furthermore, for statisticians it is often shocking to see how carelessly statistical concepts are used, and physicists cannot understand the ignorance of statisticians on physical facts and well-known results of physical research.

The workshop 'Statistical Physics and Spatial Statistics' took place at the University of Wuppertal between 22 and 24 February 1999 as a purely German event. The aim was simply to take a first step to overcome the above mentioned difficulties. Moreover, it tried to provide a forum for the exchange of fundamental ideas between physicists and spatial statisticians, both working in a wide spectrum of science related to stochastic geometry. This volume comprises the majority of the papers presented orally at the workshop as plenary lectures, plus two further invited papers. Although the contributions presented in this volume are very diverse and methodically different they have one feature in common: all of them present and use geometric concepts in order to study spatial configurations which are random.

To achieve the aim of the workshop, the invited talks not only present recent research results, but also tried to emphasize fundamental aspects which may be interesting for the researcher from the other side. Thus many talks focused on methodological approaches and fundamental results by means of a tutorial review. Basic definitions and notions were explained and discussed to clarify different notations and terms and thus overcome language barriers and understand different ways of thinking.

Part 1 focuses on the statistical characterization of random spatial configurations. Here mostly point patterns serve as examples for spatial structures. General principles of spatial statistics are explained in the first paper of this volume. Also the second paper 'Stationary Models in Stochastic Geometry - Palm Distributions as Distributions of Typical Elements. An Approach Without Limits' by Werner Nagel discusses key notions in the field of stochastic geometry and spatial statistics: stationarity (homogeneity) and Palm distributions. While a given spatial structure cannot be stationary, a stationary *model* is often adequate for the description of real geometric structures. Stationary models are very useful, not least because they allow the application of Campbell's theorem (used as Monte Carlo integration in many physical applications) and other valuable tools. The Palm distribution is introduced in order to remove the ambiguous notion of a 'randomly chosen' or 'typical' object from an infinite system.

In the two following contributions by Martin Kerscher and Karin Jacobs et al. spatial statistics is used to analyze data occurring in two prominent physical systems: the distribution of galaxies in the universe and the distribution of holes in thin liquid films. In both cases a thorough statistical analysis not only reveals quantitative features of the spatial structure enabling comparisons of experiments with theory, but also enables conclusions to be drawn about the physical mechanisms and dynamical laws governing the spatial structure.

In Part 2 geometric measures are introduced and applied to various examples. These measures describe the morphology of random spatial configurations and thus are important for the physical properties of materials like complex fluids and porous media. Ideas from integral geometry such as mixed measures or Minkowski functionals are related to curvature integrals, which characterize connectivity as well as content and shape of spatial patterns. Since many physical phenomena depend crucially on the geometry of spatial structures, integral geometry may provide useful tools to study such systems, in particular, in combination with the Boolean model. This model, which is well-known in stochastic geometry and spatial statistics, generates random structures through overlapping random 'grains' (spheres, sticks) each with an arbitrary random location and orientation. Wolfgang Weil focuses in his contribution on recent developments for inhomogeneous distributions of grains. Physical applications of Minkowski functionals are discussed in the paper by Klaus Mecke. They range from curvature energies of biological membranes to the phase behavior of fluids in porous media and the spectral density of the Laplace operator. An important application is the morphological characterization of spatial structures: Minkowski functionals lead to order parameters, to dynamical variables or to statistical methods which are valuable alternatives to second-order characteristics such as correlation functions.

A main goal of stereology, a well-known method in statistical image analysis and spatial statistics, is the estimation of size distributions of particles in patterns where only lower-dimensional intersections can be measured. Joachim Ohser and Konrad Sandau discuss in their contribution to this volume the estimation of the diameter distribution of spherical objects which are observed in a planar or thin section. Rüdiger Hilfer describes ideas of modeling porous media and their statistical analysis. In addition to traditional characteristics of spatial statistics, he also discusses characteristics related to percolation. The models include random packings of spheres and structures obtained by simulated annealing. The contribution of Helmut Hermann describes various models for structures resulting from crystal growth; his main tool is the Boolean model.

Part 3 considers one of the most prominent physical phenomena of random spatial configurations, namely phase transitions. Geometric spatial properties of a system, for instance, the existence of infinite connected clusters, are intimately related to physical phenomena and phase transitions as shown by Hans-Otto Georgii in his contribution 'Phase Transition and Percolation in Gibbsian Particle Models'. Gibbsian distributions of hard particles such as spheres or discs are often used to model configurations in spatial statistics and statistical physics. Suspensions of sterically-stabilized colloids represent excellent physical realizations of the hard sphere model exhibiting freezing as an entropically driven phase transition. Hartmut Löwen gives in his contribution 'Fun with Hard Spheres' an overview on these problems, focusing on thermostatistical properties.

In many physical applications one is not interested in equilibrium configurations of Gibbsian hard particles but in an ordered packing of finite size. The question of whether the densest packing of identical coins on a table (or of balls in space) is either a spherical cluster or a sausage-like string may have far-reaching physical consequences. The general mathematical theory of finite packings presented by Jörg M. Wills in his contribution 'Finite Packings and Parametric Density' to this volume may lead to answers by means of a 'parametric density' which allows, for instance, a description of crystal growth and possible crystal shapes.

The last three contributions focus on recent developments of simulation techniques at the interface of spatial statistics and statistical physics. The main reason for performing simulations of spatial systems is to obtain insight into the physical behaviour of systems which cannot be treated analytically. For example, phase transitions in hard sphere systems were first discovered by Monte Carlo simulations before a considerable amount of rigorous analytical work was performed (see the papers by H. Löwen and H.-O. Georgii). But also statisticians extensively use simulation methods, in particular MCMC (Markov Chain Monte Carlo), which has been one of the most lively fields of statistics in the last decade of 20th century. The standard simulation algorithms in statistical physics are molecular dynamics and Monte Carlo simulations, in particular the Metropolis algorithm, where a Markov chain starts in some initial state and then 'converges' towards an equilibrium state which has to be investigated statistically. Unfortunately, whether or not such an equilibrium configuration is reached after some simulation time cannot be decided rigorously in most of the simulations. But Elke Thönnes presents in her contribution 'A Primer on Perfect Simulation' a technique which ensures sampling from the equilibrium configuration, for instance, of the Ising model or the continuum Widomn-Rowlinson model.

Monte Carlo simulation with a fixed number of objects is an important tool in the study of hard-sphere systems. However, in many cases grand canonical simulations with fluctuating particle numbers are needed, but are generally considered impossible for hard-particle systems at high densities. A novel method called 'simulated tempering' is presented by Gunter Döge as an efficient alternative to Metropolis algorithms for hard core systems. Its efficiency makes even grand canonical simulations feasible. Further applications of the simulated tempering technique may help to overcome the difficulties of simulating the phase transition in hard-disk systems discussed in the contribution by H. Löwen.

The Metropolis algorithm and molecular dynamics consider each element (particle or grain) separately. If the number of elements is large, handling of them and detecting neighbourhood relations becomes a problem which is approached by Jean-Albert Ferrez, Thomas M. Liebling, and Didier Müller. These authors describe a dynamic Delaunay triangulation of the spatial configurations based on the Laguerre complex (which is a generalization of the well-known Voronoi tessellation). Their method reduces the computational cost associated with the implementation of the physical laws governing the interactions between the particles. An important application of this geometric technique is the simulation of granular media such as the flow of grains in an hourglass or the impact of a rock on an embankment. Such geometry-based methods offer the potential of performing larger and longer simulations. However, due to the increased complexity of the applied concepts and resulting algorithms, they require a tight collaboration between statistical physicists and mathematicians.

It is a pleasure to thank all participants of the workshop for their valuable contributions, their openness to share their experience and knowledge, and for the numerous discussions which made the workshop so lively and fruitful. The editors are also grateful to all authors of this volume for their additional work; the authors from the physical world were so kind to give their references in the extended system used in the mathematical literature. The organizers also thank the 'Ministerium für Schule und Weiterbildung, Wissenschaft und Forschung des Landes Nordrhein-Westfalen' for the financial support which made it possible to invite undergraduate and PhD students to participate.

Wuppertal   *Klaus Mecke*
Freiberg   *Dietrich Stoyan*
June 2000

Contents

Part I

Spatial Statistics and Point Processes

Basic Ideas of Spatial Statistics

Dietrich Stoyan

Institut für Stochastik, TU Bergakademie Freiberg
D-09596 Freiberg

Abstract. Basic ideas of spatial statistics are described for physicists. First an overview of various branches of spatial statistics is given. Then the notions of stationarity or homogeneity and isotropy are discussed and three stationary models of stochastic geometry are explained. Edge problems both in simulation and statistical estimation are explained including unbiased estimation of the pair correlation function. Furthermore, the application of Gibbs processes in spatial statistics is described, and finally simulation tests are explained.

1 Introduction

The aim of this paper is to describe basic ideas of spatial statistics for physicists. As the author believes, methods of spatial statistics may be useful for many physicists, in particular for those who study real irregular or 'random' spatial geometrical structures. Stochastic geometry and spatial statistics offer many useful models for such structures and powerful methods for their statistical analysis.

Spatial statistics consists of various subfields with different histories. The book [4] is perhaps that book which describes the most branches of spatial statistics and gives so the most complete impression. The perhaps largest field, *geostatistics*, studies random fields, i.e. random structures where in every point of space a numerical value is given as, for example, a mass density or an air pollution parameter. There are many special books on geostatistics, e.g. [3] and [45]. Other branches of spatial statistics are described also in the books [2,29,37] and [40]. An area with a rather long history is *point process* statistics, i.e. the statistical analysis of irregular point patterns of, for example, positions of galaxies or centres of pores in materials. Note that statisticians use the word 'process' where physicists would prefer to speak of 'fields'; typically, there is no time-dependence considered.

There are attempts to analyse statistically also *fibre processes* and *surface processes*. A fibre process (or field) is a random collection of fibres or curves in space as, for example, dislocation lines ([39]). Also the random system of segments in the last figure of the paper by H.-O. Georgii in this volume can be interpreted as a fibre process. A surface process is a stochastic model for a random system of two-dimensional objects, modelling perhaps boundaries of particles in space or cracks in soil or rocks.

Point processes, fibre and surface processes are particular cases of *random sets*. Here for every deterministic point x the event that it belongs to the set depends on chance. It is possible to interpret a random set as a particular random field having only the values 0 and 1, but the theory of random sets contains also ideas which do not make sense for random fields in general; an example are random chord lengths generated by intersection with test lines. A very valuable tool in the statistics of random sets (but also for filtration and image analysis) is mathematical morphology, see the classical book [33], and the more recent books [16] and [35].

There are widely scattered papers on the statistics of *fractals*, i.e. on the statistical determination of the fractal dimension for given planar or spatial samples. A recent reference to the particular case of rough surfaces is [5].

In the last five years several books have been published on *shape statistics*, see [7,34] and also [40]. The aim is here the statistical analysis of objects like particles or biological objects like bones, the description of statistical fluctuations both of shape and size. Until now, mainly that case is studied (which is typical for biology) where the usually planar objects are described by characteristic points on their outline, called 'landmarks'. But there are also attempts to create a statistical theory for 'particles' (such as sand grains), where usually such landmarks do not make sense. The simplest approach is via shape rations or indices ([40]) or 'shape finders' as in Sect. 3.3.7 of M. Kerscher's contribution in this volume.

A special subfield of random set statistics is *stereology*. The aim of classical stereology is the investigation of spatial structures by planar sections, to analyse statistically the structures visible on the section planes and to transform then the results into characteristics of the spatial structure. This is a very elegant procedure, and the most famous stereological result is perhaps the solution of the Wicksell problem, which yields the diameter distribution of spheres in space as well as the mean number of spheres per volume unit based on measurement of section circle diameters. The paper by J. Ohser and K. Sandau in this volume describes modern stereological methods in the spirit of the classical approach. The experience that important spatial characteristics cannot be estimated stereologically and new microscopical techniques (e.g. confocal microscopy) have led to new statistical methods which also go under the name stereology though they use three-dimensional measurement. But also there difficult problems remain such as, for example, spatial measurement of particles. Local stereology (see [19]) shows e.g. how mean particle volumes can be estimated by length measurement.

Spatial statisticians try to develop statistical procedures for determining general characteristics of structures such as

- *intensity* ρ (mean number of points of a point process per volume unit; in spatial statistics frequently the character λ is used, and in stereological context N_V; N_V = number per volume);
- *volume fraction* η (mean fraction of space occupied by a random set; in spatial statistics frequently the character p is used and in stereological context V_V; V_V = volume per volume);

- specific surface content (mean surface area of a surface process per volume unit; in stereological context the character S_V is used; S_V = surface per volume);
- pair correlation function $g(r)$, see Sect. 4;
- covariance (often not called 'covariance function'; $C(r)$ = probability that the members of a point pair of distance r both belong to a given random set).

Statistical research leads to so-called 'non-parametric estimators' for these and other characteristics. The aim is to obtain unbiased estimators, which are free of systematic errors. Furthermore, a small estimation variance or squared deviation is wanted.

An important role play *stochastic models*, both in statistical physics and spatial statistics. In the world of mathematics such models are developed and investigated in stochastic geometry. As expressed already in the foreword, both sides, physicists and statisticians could learn a lot from the other side, since the methods and results are rather different. Two statistical problems arise in the context of models: estimation of model parameters and testing the goodness-of-fit of models, see Sect. 6.

In the last years a further topic of statistical research has appeared: the problem of *efficient simulation* of stochastic models. Starting from ideas which came originally from physicists, simulation algorithms have been developed and investigated systematically which improve the original Metropolis algorithm. The aim is to save computation time and to obtain precise results. The papers by G. Döge, J.-A. Ferrez, Th. M. Liebling and D. Müller, H.-O. Georgii and E. Thönnes in this volume describe some of these ideas.

Mathematically, two general ideas play a key role in spatial statistics: *random sets* and *random measures*. With the exception of random fields all the geometrical structures of spatial statistics can be interpreted as random sets. Fundamental problems can be solved by means of the corresponding theory created by G. Matheron and D.G. Kendall, which is described in texts such as [21] and [27]; physicists may begin with the simplified descriptions in [33,36] and [37].

A measure is a function Φ which assigns to a set A a number $\Phi(A)$, satisfying some natural conditions such as that the measure of a union of disjoint sets is equal to the sum of the measures of the components. A well-known measure is the volume or, in mathematical terms, the Lebesgue measure denoted here by ν; generalizations are the Minkowski measures. Any random set is accompanied by random measures. If the random set is a fibre process then e.g. the following two random measures may be of interest, the total fibre length or the number of fibre centres. In the first case, $\Phi(A)$ is the total fibre length in A. Here A is a deterministic set (sometimes called 'test set' or 'sampling window'), and the value $\Phi(A)$ is a random variable. Characteristics such as intensity and pair correlation function have their generalized counterparts in the theory of random measures; in particular, η is the intensity and $C(r)/\eta^2$ is the pair correlation function of the volume measure associated with the random set. The idea of using random

measures in the context of stochastic geometry and spatial statistics goes back
to G. Matheron and J. Mecke.

2   Stationarity and Isotropy

A frequently used basic assumption in spatial statistics is that the structures
analysed are stationary. Similarly as with the use of the word 'process', the
physicist should be aware that 'stationary' means in spatial statistics typically
'homogenous'. It means that the distribution of the structure analysed is translation invariant. Mathematically, this is described as follows.

Let Φ be the random structure. The probability that Φ has some propoerty,
can be written as

$$P(\Phi \in Y), \qquad (1)$$

where Y is a subset of a suitable phase space $\mathcal{N}$ and P denotes probability.

Example. Let Φ be a point process and Y be the set of all point patterns in
space which do not have any point within the ball $b(o, r)$ of radius r centred at
the origin o. Then $P(\Phi \in Y)$ is the probability that the point of Φ closest to o
has a distance larger than r from o. As a function of r, this probability is often
denoted as $1 - H_s(r)$, and $H_s(r)$ is called *spherical contact distribution function.*

The structure Φ is called *stationary* if for all $r \in R^d$ and all $Y \in \mathcal{N}$

$$P(\Phi \in Y) = P(\Phi_r \in Y), \qquad (2)$$

where Φ_r is the structure translated by the vector r.

This can be rewritten as

$$P(\Phi \in Y) = P(\Phi \in Y_r), \qquad (3)$$

where Y_r is the shifted set Y in the phase space.

Example. In the case of a stationary point process Φ it is

$$
\begin{aligned}
&P(\Phi \text{ does not have any point in } b(o, r)) = \\
&P(\Phi \text{ does not have any point in } b(r, r))
\end{aligned}
\qquad (4)
$$

for all r and r, i. e., the position of the test sphere is unimportant.

The definition of stationarity makes only sense for infinite structures, since
a bounded structure can be never stationary.

Isotropy is analogously defined. The structure Φ is called *isotropic* if for all
rotations r around the origin o and all $Y \in \mathcal{N}$

$$P(\Phi \in Y) = P(\mathbf{r}\Phi \in Y), \qquad (5)$$

where $\mathbf{r}\Phi$ is the structure rotated by $\mathbf{r}$. A structure which is both stationary and
isotropic is called motion-invariant.

Mathematicians know that there are strange stationary sets such as the empty set or the infinite set of lines $y = n + u$ in the (x, y)-plane, $n = 0, \pm 1, \ldots$, where u is a random variable with uniform distribution on the interval $[0, 1]$. A stronger property is *ergodicity*, which ensures that spatial averages taken over one sample equal local averages over the random fluctuations. Implicitly ergodicity is quite often assumed in spatial statistics, where frequently only a unique sample is analysed, for example a particular mineral deposit or forest. The difficult philosophical problems in this context are discussed in [22].

The properties of stationarity and ergodicity can never be tested statistically in their full generality. They can be proved mathematically for the stochastic models below, but in applications the decision is leaved to the statistician. She or he can test aspects of the invariance properties, can visually inspect the sample(s), look for trends or use a priori knowledge on the structure investigated.

Note that stationarity is defined without limit procedures, and the same is true for characteristics related to stationary structures such as volume fraction η. For a stationary random set X, η is simply the volume of X in any test set of volume 1. It is a mathematical theorem that for an ergodic X, η is obtained as a limit for large windows. This limit-free approach is discussed in the paper by W. Nagel in this volume. The following section describes three stochastic models of spatial statistics as models in the whole space. In Sect. 5 a similarly defined stationary Gibbs process is discussed.

Mathematicians consider their approach as natural and are perhaps not quite happy with texts such as passages in Löwen's paper in this volume (around formulas (2) or (20)). So to say, they start in the thermodynamical limit, and consider ρ, η and $g(r)$ as quantities corresponding only to the stationary case.

3   Three Stationary Stochastic Models

The Homogeneous Poisson Process

For spatial statisticians, the homogeneous (or stationary) Poisson process is the most important point process model. It is the model for a completely random distribution of points in space, without any interaction. Its distribution is given by one parameter λ, the intensity, the mean number of points per volume unit. The process has two properties which determine its distribution:

(a) For any bounded set B, the random number of points in $A, \Phi(A)$, has a Poisson distribution with parameter $\lambda\nu(A)$, where $\nu(A)$ is the volume of A. That means,

$$P(\Phi(A) = i) = \frac{[\lambda\nu(A)]^i}{i!} \exp(-\lambda\nu(A)), i = 0, 1, \ldots \qquad (6)$$

(b) For any integer k and any pairwise disjoint sets $B_1, \ldots, B_k$ the random point numbers in the sets, $\Phi(B_1), \ldots, \Phi(B_k)$, are independent.

These properties imply stationarity and isotropy because of the translation and rotation invariance of volume. A further implication is that under the assumption that in a given set A there are just n points, the point positions are independent and uniformly distributed within A. This property is important for the simulation of a Poisson process in A: first a Poisson random number n for parameter $\lambda\nu(A)$ is determined and then n independent uniform positions within A. Figure 1 shows a simulated sample of a Poisson process.

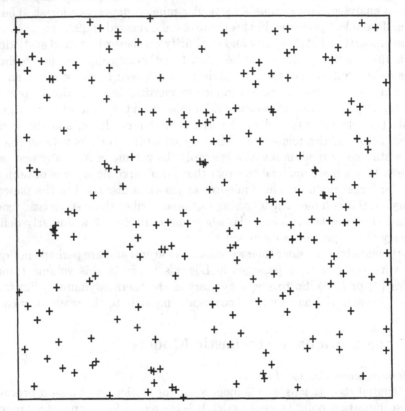

Fig. 1. A simulated sample of a homogeneous Poisson process.

The Boolean Model

Also the Boolean model is defined from the very beginning as a model in the whole space. It is a mathematically rigorous formulation of the idea of an 'infinite system of randomly scattered particles'. So it is a fundamental model for geometrical probability, stochastic geometry and spatial statistics. The Boolean model has a long history. The first papers on the Boolean model appeared in the beginning of the 20th century, see the references in [37], which include also papers of various branches of physics. The name "Boolean model" was coined in G. Matheron's school in Fontainebleau to discriminate this set-theoretic model from (other) random fields appearing in geostatistical applications.

The Boolean model is constructed from two components: a system of grains and a system of germs. The germs are the points $r_1, r_2, \ldots$ of a homogeneous Poisson process of intensity ρ. (The paper by W. Weil in this volume considers the inhomogeneous case.) The grains form a sequence of independent identically distributed random compact sets K_n. Typical examples are spheres (the most popular case in physics), discs, segments, and Poisson polyhedra. A further random compact set K_0 having the same distribution as the K_n is sometimes called the 'typical grain'.

The Boolean model $\varXi$ is the union of all grains shifted to the germs,

$$\varXi = \bigcup_{n=1}^{\infty} (K_n + r_n),$$

see Fig. 2, which shows the case of circular grains.

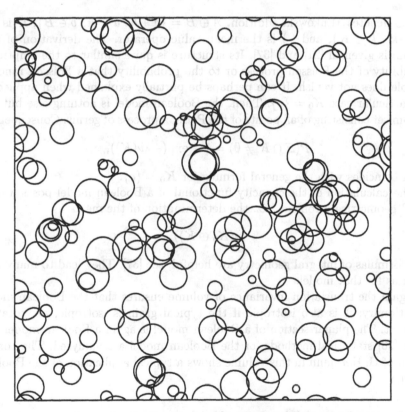

Fig. 2. A simulated sample of a Boolean model with random circular grains, which is the set-theoretic union of all disks shown. The disk centres coincide with the points in Fig. 1.

Very often it is assumed that the typical grain K_0 is convex; only Boolean models with convex grains are discussed henceforth. But this does not mean that non-convex grains are unimportant. For example, the case where K_0 is a finite point set corresponds to Poisson cluster point processes.

The parameters of a Boolean model are intensity ρ and parameters characterizing the typical grain K_0. While for simulations the complete distribution of K_0 is necessary, for a statistical description it often suffices to know that the basic assumption of a Boolean model is acceptable and to have some parameters such as mean area $\overline{A}$, mean perimeter $\overline{U}$ or, if a set-theoretic characterization is needed, the so-called Aumann mean of K_0.

The distribution of the Boolean model $\varXi$ is, as for any random set, determined by its capacity functional, $P(\varXi \cap K \neq \emptyset)$, the probability that the test set K does not intersect $\varXi$. It is given by the simple formula

$$P(\varXi \cap K \neq \emptyset) = 1 - \exp(-\lambda \langle \nu(K_0 \oplus \check{K}) \rangle) \qquad \text{for } K \in \mathbb{K}.$$

Here $\oplus$ denotes Minkowski addition, $A \oplus B = \{a + b : a \in A, \ b \in B\}$, $\check{K}$ is the set $\{-k : k \in K\}$, and $\langle \ \rangle$ is the mean value operator. The derivation of this formula is given in [21] and [37]. Its structure is quite similar to the emptiness probability of the Poisson process or to the probability that a Poisson random variable does not vanish. It can perhaps be partially explained when applied to the particular case $K_0 = \{o\}$. Then, the Boolean model is nothing else but the random set consisting of all points of the Poisson process of germs. Consequently,

$$P(\varXi \cap K \neq \emptyset) = 1 - \exp(-\lambda \nu(K)),$$

which coincides with the general formula for $K_0 = \{o\}$.

The calculation of the capacity functional of a Boolean model poses a non-trivial geometrical problem, viz. the determination of the mean

$$\langle \nu(K_0 \oplus \check{K}) \rangle.$$

Here formulas of integral geometry are helpful, see [37]. They lead to many nice formulas for that model.

Again the translation invariance of volume ensures that the Boolean model is stationary; it is also isotropic if the typical grain is isotropic, i.e. rotation invariant. The planar section of a Boolean model is again a Boolean model.

In [26] statistical methods for the Boolean model are analysed. The contribution of H. Hermann in this volume shows a typical application of the Boolean model.

The Random Sequential Adsorption Model

The RSA model is a famous model of hard spheres in space, which is called SSI model in spatial statistics (simple sequential inhibition). In the physical literature (see, for example, [17]) it is often defined for a bounded region B as follows. Spheres of equal diameter $\sigma = 2R$ are placed sequentially and randomly in B. If a new sphere is placed so in B that it intersects a sphere already existing then the new sphere is rejected. The process of placing spheres is stopped when it is impossible to place any new sphere. Clearly the distribution of the spheres in B depends heavily on shape and size of B. But very often it is obvious that physicists have in mind a homogeneous or stationary structure in the whole space which is observed only in B, see [8]. Figure 3 shows a simulated sample in a square.

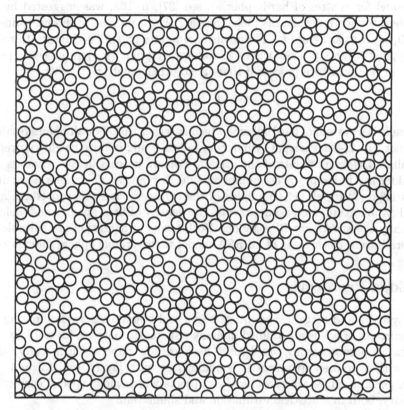

Fig. 3. A simulated sample of a planar RSA model in a quadratic region.

There are two ways to define the RSA model as a stationary and isotropic structure. One was suggested by J. Møller. It uses a random birth process as in [30] and [37], p. 185 (a birth-and-death process with vanishing death rate). Such a process starts from a homogeneous Poisson process on the product space of $\mathbb{R}^d \times [0, \infty)$, where the latter factor is interpreted as 'time'. With each 'arrival' (x_i, t_i) of the process a sphere of radius R is associated. It is assumed that an arrival is deleted when its sphere overlaps the sphere of any other arrival (x_j, t_j) with $t_i > t_j$. Then as time tends to infinity the retained spheres give a packing of the space, the RSA model; no further sphere of radius R can be placed without intersecting one of the existing spheres. The corresponding birth rate at r for the point configuration of sphere centres φ is

$$b(x, \varphi) = 1 - 1_{\varphi \oplus b(o,R)}(x),$$

where $1_A(x) = 1$ if $x \in A$ and 0 otherwise.

The second form of modeling, which is related to the idea of the dependent thinning procedure which leads to Matérn's second hard core process (a particular model for centres of hard spheres), see [37], p. 163, was suggested by M. Schlather, see for more details [38]. Take a Poisson process of intensity one in $\mathbb{R}^d \times [0, \infty)$ consisting of $(d+1)$-dimensional points (r, t). The Matern thinning rule applied to this process works as follows: A point (r, t) produces a point $r \in \mathbb{R}^d$ of the hard core process if there is no other point (r', t') with

$$\|r - r'\| < h \quad \text{and} \quad t' < t. \tag{7}$$

The result is a system of hard spheres which is rather thin. For all points retained construct $(d+1)$-dimensional cylinders of radius σ and infinite height centred at the points. Delete all points of the original Poisson process in the cylinders and reconsider the Poisson process points outside. A point (r, t) of them is retained if it satisfies (7) for all (r', t') outside of the cylinders, and this procedure is repeated ad infinitum, increasing stepwise the density of hard spheres and yielding eventually the stationary RSA model. Both forms of definition are suitable for a generalisation to the case of an RSA model with variable sphere diameters.

4   Edge Problems

For physicists and materials scientists edges and boundaries are fascinating objects; surfaces of solids are studied in many papers. In contrast, for a statistician boundaries pose annoying problems. There are few papers which study structures with a gradient (towards a boundary) or with layers, see [13] and [14]. But in general, edges are considered as objects which make special corrections necessary, both in statistical estimation and simulation.

Edge-Correction in Simulation
The simulation of stationary structures is an important task. Clearly, it is only possible to simulate them in bounded windows and it is the aim to simulate typical pieces which include also interaction to structure elements outside of the window.

Often it is sufficient for obtaining an 'exact sample' to simulate the structure in an enlarged window. However, this is not recommendable for hard-core Gibbs processes. It cost the author a lost bet for a crate of beer to be paid to H. Löwen to learn this. He tried to simulate a planar hard disk Gibbs process with free boundary and disk diameter 1 in a square window of side length 20 in order to obtain a stationary sample of about 180 points in the central square of side length 14 and had to learn that the area fraction obtained was considerably smaller (0.696) than the result with periodic boundary conditions in the smaller square (0.738).

It seems that the method of periodic boundary conditions (or simulation on a torus if the window is a planar rectangle) is a good ad-hoc method. More sophisticated methods are finite-size scaling (see [46]) and perfect simulation in space, see E. Thönnes' and H.-O. Georgii's papers in this volume.

Statistical Edge-Correction

The spatial statistician wants to avoid systematic errors or biases in estimation procedures. This aim implies in many cases edge-correction. It is explained here for a particular problem, the estimation of the pair correlation function of a stationary and isotropic point process of intensity ρ based on the points observed in a bounded window of observation W, see for details [42]. The pair correlation (or distribution) function $g(r)$ can be defined heuristically as follows. (See also H. Löwen's paper in this volume.) Consider two infinitesimal balls of volumes dV_1 and dV_2 of distance r. The probability to find in both balls each a point is

$$\lambda^2 g(r) dV_1 dV_2.$$

The statistical estimation follows this definition. A naive estimator, which is quite good for very large samples and small fluctuations of local point density, is

$$\hat{g}_n(r) = \sum_{i,j}^{\neq} \frac{k(\|R_i - R_j\| - r)}{\nu(W)} \Big/ \hat{\rho}^2.$$

The summation goes here over all pairs of different points $((R_i, R_j)$ as well as $(R_i, R_j))$ of a distance between $r - h$ and $r + h$. The sampling window is denoted by W, its volume (or area) by $\nu(W)$, and

$$k(z) = 1_{[-h,h]}(z)/2h$$

where h is called bandwidth. Finally, $\hat{\rho}$ is the intensity estimator

$$\hat{\rho} = \text{number of points in } W/\nu(W) = N/\Omega.$$

For large r and small W, the estimator $\hat{g}_n(r)$ has a considerable bias (= difference of estimator minus true value), since for a point close to the boundary ∂W of W some of the partner points of distance r are outside of W. Therefore the bias will be negative.

A naive way to improve this situation could be to include in $\hat{g}_n(r)$ only point pairs for which at least one member has a distance r from ∂W. This method is called 'minus-sampling' and means of course a big loss of information. 'Plus-sampling' would mean that for all points in W additionally the neighbours outside of W within a distance r are known. One can usually not hope to be able to apply plus-sampling, but sometimes (for estimating other characteristics) there is no better idea than to use minus-sampling.

A much better idea of edge-correction, which can be applied in pair correlation estimation, is to use a Horvitz-Thompson estimator, see [1]. The idea is here to weight the point pairs according to their frequency of observation. The observation of a point pair of large distance r is less likely than that of a small distance. Therefore pairs with a large distance get a big weight and so on. One can show that the weight

$$(\nu(W \cap W_{R_i - R_j}))^{-1}$$

where $W_r = W + r = \{y : y = w + r, \; w \in W\}$, is just the right weight, yielding an unbiased estimator of $p(r) = \lambda^2 g(r)$,

$$\hat{p}(r) = \sum_{i,j}^{\neq} \frac{k(\|R_i - R_j\| - r)}{\nu(W \cap W_{R_i - R_j})}. \tag{8}$$

Then $g(r)$ is estimated by division by the squared intensity ρ^2. It is not the best solution to use simply $\hat{\rho}^2$, to square $\hat{\rho} = N/\Omega$. The mean of the unbiased estimator $\hat{\rho}$ is not $\hat{\rho}^2$. It is better to use an adapted estimator $\rho_S(r)$, which depends on r and, particularly for large r, to replace $\hat{\rho}_S(r)^2$ by a better estimator of ρ^2, see [42].

5   Gibbs Point Processes

Some statisticians say that the 19th century was the century of the Gaussian distribution, while the 20th century was that of Gibbs distributions ; probably many physicists will agree. In many situations distributions appear which are of the form

probability of configuration = exp{− energy of configuration}.

Even the Gaussian distribution can be seen as a particular case. Physicists know that such distributions result from a maximization problem and use the idea of maximum entropy in some statistical problems (see [9,18] and [23]).

Until now in this approach the configurations have been mainly point configurations, and here only this case is considered. The 'points' may be ideal points or centres of objects such as hard spheres or locations of trees. Some exceptions are structures studied in [24] and their statistical counterparts (see [20] and [25]) and the fibre model in [12], p. 109.

For physicists Gibbs point processes (or ensembles) are models of their own interest. Typically they start with a potential function and then study the statistical properties of the ensemble in the belief so to study physically relevant structures. This is very well demonstrated in the paper by H. Löwen in this volume. One of the most important questions in this context is that of phase transition, the existence of different distributions for the same model parameters.

The approach of statisticians is quite different. For them the point pattern is given, they assume that it follows the Gibbs process model and want to estimate its parameters. Typically, they look for simple models and therefore prefer in the Gibbs approach models which are based on pair potentials. If they are successful, they have then the problem of interpretation of the estimated pair potential in terms of the data, for example biologically, which is not always simple, see [44]. Finally, they have then to simulate the process for carrying out Monte Carlo tests (see Sect. 6) and visualisation.

Statisticians study Gibbs processes (or Markov point processes) both in bounded regions and in the whole space, see also the text by H.-O. Georgii in this volume. Before the latter case, which is for physicists perhaps not so natural, will be discussed, the case of a bounded region is considered. Here both the canonical and the grand canonical ensemble are used, where the grand canonical case poses existence problems, particularly in the case of clustering or a pair potential with attraction. In the case with fixed point number n and pair potential V the joint density of the n points in W is

$$f(R_1, \ldots, R_n) = \exp\{-\sum_{i<j}^{N} V(|R_i - R_j|)\}/Z. \tag{9}$$

Here the normalizing constant Z is the classical canonical partition function which is very difficult to determine. Usually, V depends on some parameters, which have to be estimated and which also have influence on Z. If the statistician has n points $R_1, \ldots, R_n$ in W she or he could start the estimation of the parameters using the maximum likelihood method. It consists just in the task to determine that parameters which maximize $f(R_1, \ldots, R_n)$ for the given $R_1, \ldots, R_n$. But this is very difficult since Z is unknown. Ogata and Tanemura (1981) used approximations of Z derived by statistical physicists. An alternative solution is based on simulation, and the method is then called 'Monte Carlo likelihood inference', see [11]. Many point patterns (mainly of forestry) have been analysed by these methods. By the way, for modelling inhomogeneous tree distribution even Gibbs process with external potential are applied, see [41]. Other approaches for this problem consist in thinning homogeneous Gibbs processes or in transforming them.

Stationary Gibbs Point Processes
Statisticians have developed a theory of stationary Gibbs point processes, which are as the models in Sect. 3 defined in the whole space, without any limiting procedure. A sketch of the theory is given in [37], where also the relevant references are given, to which [10] should be added.

A stationary Gibbs point process Φ satisfies the Georgii-Nguyen-Zessin equation

$$\rho\langle(f(\Phi \setminus \{o\}))\rangle_o = \langle(f(\Phi)\exp\{-E(o, \Phi)\})\rangle \tag{10}$$

Here ρ is the intensity of the process and f is any non-negative function which assignes a number to the whole point process. $\langle \ \rangle_o$ means expectation with respect to the Palm distribution; this is a conditional mean under the condition that in the origin o there is a point of Φ, see Nagel's paper in this volume for an exact definition. The term '$\setminus\{o\}$' means that o is not included in the left-hand mean. $E(o, \Phi)$ is the 'local energy', the energy needed to add the point o to the configuration Φ; if the energy of the Gibbs point process is given by a pair potential V, then

$$E(o,\Phi) = -\mu + \sum_{x\in\Phi} V(\|x\|),$$

where μ is a further parameter called chemical activity. Analogs of (10) are known in statistical physics under the name 'excess particle equations' (or similar) and have been used since 1960 in the 'scaled particle theory', see Sect. 3.5 of H. Löwen's contribution to this volume and [15], Sect. 2.4 and formula (2.4.30). The author confesses that he (as a trained mathematician) prefers clearly the stationary formalism rather than that in [15], which uses N points. By the way, Formula (10) is given in a bastard notation; formula (5.5.18) in [37] is perhaps nicer.

An important particular case is the hard core Gibbs process, with the simple pair potential

$$V(r) = \begin{cases} \infty & \text{for } r < \sigma \\ 0 & \text{otherwise}. \end{cases}$$

The corresponding stationary Gibbs point process is an important stochastic model for a system of hard spheres of diameter σ or rods of length σ in the one-dimensional case. For this process, (10) simplifies as

$$\rho\langle f(\Phi\setminus\{o\})\rangle_o = e^\mu\langle f(\Phi)\mathbf{1}_{Y_\sigma}(\Phi)\rangle \tag{11}$$

where Y_r is the set of all point patterns with no point in the ball $b(o,r)$. The particular choice of

$$f(\Phi) = \mathbf{1}_{Y_r}(\Phi), \quad r \geq \sigma$$

yields

$$\lambda P_o(Y_r) = e^\mu P(Y_r) \quad \text{for} \quad r \geq \sigma, \tag{12}$$

where P_o is the Palm distribution of Φ and P the usual stationary distribution of Φ. Formula (12) can be rewritten as

$$\rho[1 - D(r)] = [1 - H_s(r)]e^\mu \quad \text{for} \quad r \geq \sigma. \tag{13}$$

Here $D(r)$ is the nearest neighbour distance distribution function (the d.f. of the random distance from a randomly chosen point to its nearest neighbour in Φ) and $H_s(r)$ is the spherical contact distribution function (the d.f. of the random distance from the origin o to its nearest neighbour in Φ, see also section 3.3.2 of M. Kerscher's paper in this volume). This equation already appears in other notation in [31]. It yields a very rough approximation of the intensity ρ, which is for fixed σ a function of μ,

$$\frac{e^\mu}{1 + b_d\sigma^d e^\mu} \leq \rho(\mu) \leq \frac{e^\mu}{1 + b_d(\frac{\sigma}{2})^d e^\mu},$$

where b_d is the volume of the unit sphere in $\mathbb{R}^d$. The function $\rho(\mu)$ is increasing in μ, see section 4.3 of H.-O. Georgii's paper in this volume. The paper by G. Döge in this volume describes how $\rho(\mu)$ can be investigated by simulated tempering, a particular simulation method.

Note that all results for the one-dimensional stationary hard core Gibbs point process (see Section 2.2 of Löwen's paper) in the physical literature can be obtained (without any limiting procedure) by using (10) for suitable functions f. It can be shown that the point process Φ of rod centres is a renewal process, i.e. the random distances between subsequent points are completely independent. The distribution function F of the distance between any two subsequent points is given by

$$F(r) = 0 \qquad \text{for} \qquad r \leq \sigma$$

and

$$F(r) = 1 - e^{-\beta(r-\sigma)} \qquad \text{for} \qquad r > \sigma$$

with β satisfying

$$\ln \beta + \beta \sigma = \mu.$$

The mean inter-point distance is $m = \sigma + \beta^{-1}$.

Spatial statisticians use the stationary Gibbs point process for statistical analyses for point patterns which were considered as samples of stationary point process. The aim is then to estimate the chemical activity μ and the pair potential V. As an approximation the methods for finite Gibbs point process can be used, while a true stationary approach is the Takacs-Fiksel method. (Both approaches were compared in [6].) The idea of the Takacs-Fiksel method is to determine empirical analogues of both sides of equation (10) for a series of 'test functions' f and to chose μ and V so that the sum of squared differences becomes a minimum. Details of the method are described in [40], an example is discussed in [37], p. 183. There the points are positions of 60 years old spruces in a German forest. The pair potential was estimated as

$$\hat{V}(r) = \begin{cases} \infty & \text{for } r < 1m \\ 6.0e^{-1.0r} & \text{for } 1 \leq r < 3.5m \\ 0 & \text{otherwise}. \end{cases}$$

6   Statistical Tests

A honest spatial statistician does not stop her/his work when she/he has found a stochastic model for her/his data. No, she/he will also test the goodness-of-fit of the model. But this is a difficult task, since the classical goodness-of-fit tests such as χ^2 or Kolmogorov-Smirnov are usually not applicable in spatial statistics. These tests are designed to test that an unknown distribution function is equal to a theoretical function, perhaps a distribution function of a normal distribution and it is assumed that the data come from independent observations. In contrast, the data of spatial statistics are typically highly correlated and often the distribution cannot be characterized by a distribution function only.

A fundamental hypothesis of spatial statistics is that of complete spatial randomness (CSR) of a given point pattern. If a homogeneous pattern is assumed, the CSR hypothesis is the same as that the point pattern is a sample of a homogeneous Poisson process. For this hypothesis there exist various tests. Their application depends on the data. If they are counts in cells then the dispersion index test is used. This is in essence a χ^2 goodness-of-fit test of the hypothesis that the counts follow a uniform distribution. (Here property (b) and the conditional uniformity property of the homogeneous Poisson process are used.) A more powerful test is the L test. It is explained here to give a typical example for a test in spatial statistics.

The L function of a point process Φ is defined as follows. Consider the mean number of points in a sphere of radius r centred at a randomly chosen point of Φ. (A 'randomly chosen point' or a 'typical point' is a point of Φ obtained in a sampling scheme where every point has the same chance to be chosen; an exact definition needs the use of Palm distributions, see W. Nagel's text in this volume.) Denote this mean, which depends on r, by $\rho K(r)$, where ρ is the intensity of Φ. The function $K(r)$ appearing here is called Ripley's K function. It is related to the pair correlation function $g(r)$ by

$$g(r) = \frac{d}{dr} K(r) \Big/ d b_d r^{d-1}.$$

For a Poisson process it is

$$K(r) = b_d r^d.$$

It makes sense to transform the K function to obtain the L function given by

$$L(r) = \sqrt[d]{K(r)/b_d}.$$

It satisfies in the Poisson case

$$L(r) = r.$$

For a given point pattern, the L function can be estimated statistically (see [32,40] and [42]); the root transform stabilizes estimation variances. If the empirical L function, $\hat{L}(r)$, shows large deviations from r, then the statistician may conclude that the data do not come from a Poisson process. The deviation measure used is

$$\delta = \max_{(r)} |r - \hat{L}(r)|.$$

Since usually the maximum deviations appear for smaller r, no upper bound on r appears here. The distribution of δ under the Poisson hypothesis depends on window size and process intensity. Ripley [32] (see also [40], p. 225) gives for the planar case the critical value

$$\tau_{0.95} = 1.45 \frac{\sqrt{\text{window area}}}{\text{point number}}$$

for the error probability of $\alpha = 0.05$, a value which was found by simulation. (The factor is 1.68 for $\alpha = 0.01$.) If δ for a given pattern is larger than $\tau_{0.95}$ then at the level of 0.95 the CSR hypothesis is rejected.

An analogous test is possible for any other point process, to test the hypothesis that a given point pattern can be considered as a sample of certain point process. Assume that the L function of the process under the hypothesis is known, perhaps only after simulation. Then

$$\delta = \max_{(r)} |L(r) - \hat{L}(r)|$$

is calculated for the given sample, yielding the value δ^*. Furthermore the theoretical point process is simulated 999 times in just the same window. For each simulation the δ value is determined. Then the obtained 1000 δ values are ordered in ascending order. If δ^*, the empirical value, is too big, i. e. belongs to the 50 biggest values, the hypothesis is rejected at the level 0.95.

Such a test is called a Monte Carlo test, and tests of this type are frequently used in spatial statistics. There not only the L function is used, but also the D function and H_s function mentioned in section 5, or the J function defined by $J(r) = (1 - D(r))/(1 - H_s(r))$, see [43]. Also quite other characteristics can be used, for example, intensity ρ or volume fraction η. Then again the deviation of the empirical and theoretical values can be investigated, usually by simulation.

It is also possible (and perhaps more convenient) to consider fluctuations of real-valued charateristics without regard to theoretical values. Such a value can be $L(r)$ for a particularly important distance r. Then there is given an empirical value $\hat{L}(r)$ and 999 values obtained by simulation. If then $\hat{L}(r)$ belongs to the 25 smallest or 25 largest $L(r)$ values, the model hypothesis is rejected. Often this test is made sharper by considering the graph of $\hat{L}(r)$ and the 2.5 % and 97.5 % envelope of the simulated L functions, see section 15.8 in Stoyan and Stoyan (1994). The level of this test is smaller than 0.95 but the true level is unknown.

Often not 999 simulations are carried out. Instead, sometimes only 19 simulations are made and the hypothesis is rejected if the empirical value is outside of the interval formed by the extreme simulated values or the empirical function is outside of the band formed by the 19 simulated functions. The statistician then believes to work close to a level of 0.95.

Note a subtle difficulty. Typically, the model parameters are estimated from the same data which are later used for determining the deviation measure δ. Then of course δ should tend to be smaller than for the simulated samples, and the test is a bit favourable to the hypothesis. To determine then the true test level is difficult or at least time-consuming.

Acknowledgment

The author thanks H.-O. Georgii and K. Mecke for helpful discussions on Sect. 5. He is also grateful to D. J. Daley for a discussion of the distribution of the one-dimensional hard core Gibbs process and to H. Löwen, K. Mecke and M. Schmidt for comments on an earlier version of this paper.

References

1. Baddeley, A. (1999): 'Spatial sampling and censoring'. In: [2], pp. 37–78
2. Barndorff-Nielsen, O.E., W.S. Kendall, M.N.M. van Lieshout (1999): *Stochastic Geometry. Likelihood and Computation.* (Chapman & Hall / CRC, Boca Raton, London, New York, Washington)
3. Chilès, J.-P., P. Delfiner (1999): *Geostatistics. Modeling Spatial Uncertainity* (J. Wiley & Sons, New York)
4. Cressie, N. (1993): *Statistics of Spatial.* Data. (J. Wiley & Sons, New York)
5. Davies, S., P. Hall (1999): 'Fractal analysis of surface roughness by using spatial data', *J. Roy. Statist. Soc.* B **61**, pp. 3–37
6. Diggle, P.J., T. Fiksel, P. Grabarnik, Y. Ogata, D. Stoyan, D., M. Tanemura (1994): 'On parameter estimation for pairwise interaction point processes'. *Int. Statist. Rev.* **62**, pp. 99-117
7. Dryden, I.L., K.V. Mardia (1998): *Statistical Shape Analysis* (J. Wiley & Sons, Chichester)
8. Evans, J.W. (1993): 'Random and cooperative adsorption'. *Rev. Modern Phys* **65**, pp. 1281–1329
9. Frieden, B.R. (1983): *Probability, Statistical Optics, and Data Testing* (Springer-Verlag, Berlin, Heidelberg, New York)
10. Georgii, H.-O. (1976): 'Canonical and grand canonical Gibbs states for continuum systems'. *Comm. Math. Phys.* **48**, pp. 31–51
11. Geyer, C. (1999): 'Likelihood inference for spatial point processes'. In: [2], pp. 79–140.
12. Grenander, U. (1996): *Elements of Pattern Theory* (Johns Hopkins University Press, Baltimore and London)
13. Hahn, U., A. Micheletti, R. Pohlink, D. Stoyan, H. Wendrock (1999): 'Stereological analysis and modelling of gradient structures''. *J. Microsc.* **195**, pp. 113-124
14. Hahn, U., D. Stoyan (1999): 'Unbiased stereological estimation of the surface area of gradient surface processes'. *Adv. Appl. Prob.* **30** pp. 904–920
15. Hansen, J.-P., I.R. McDonald (1986): *Theory of Simple Liquids* (Academic Press, London, Orlands, New York)
16. Heijmans, H.J.A.M. (1994): *Morphological Image Operators* (Academic Press, New York, London)
17. Hinrichsen, E.L., J. Feder, T. Jøssang (1986): 'Geometry of random sequentail adsorption'. *J. Statist. Phys.* **44**, pp. 793–827
18. Jaynes, E.T. (1994): *Probability Theory: The Logic of Science.* Fragmentary edition of 1994. See http://omega.albany.edu:8008/JaynesBook.
19. Jensen, E.B.V. (1998): *Local Stereology* (World Scientific, Singapore)
20. Kendall, W.S., M.N.M. van Lieshout, A.J. Baddeley (1999): 'Quermass-interaction processes: Conditions for stability'. *Adv. Appl. Prob.* **31**, pp. 315–342
21. Matheron, G. (1975): *Random Sets and Integral Geometry* (J. Wiley & Sons, New York)
22. Matheron, G. (1989): *Estimating and Choosing* (Springer-Verlag, Berlin, Heidelberg, New York)
23. McGibbon, A.J., S.J. Pennycook, D.E. Jesson (1999): 'Crystal structure retrieval by maximum entropy analysis of atomic resolution incoherent images'. *J. Microsc.* **195**, pp. 44–57
24. Mecke, K. (1996): 'A morphological model for complex fluids'. *J. Phys. Condens. Matter* **8**, pp. 9663–9667

25. Møller, J. (1999): 'Markov chain Monte Carlo and spatial point processes'. In: [2], pp. 141-172
26. Molchanov, I. (1997): *Statistics of the Boolean Model for Practitioners and Mathematicians* (J. Wiley & Sons, Chichester)
27. Molchanov, I. (1999): 'Random closed sets'. In: [2], pp. 285-331
28. Ogata, Y., M. Tanemura (1981): 'Estimation of interaction potentials of spatial point-patterns through the maximum-likelihood procedure'. *Ann. Inst. Statist. Math.* **33**, pp. 315-338
29. Ohser, J., F. Mücklich (2000): *Statistical Analysis of Materials Structures* (J. Wiley & Sons, Chichester)
30. Preston, C.J. (1976): *Random Fields.* Springer Lecture Notes in Mathematics 534. (Springer-Verlag Berlin, Heidelberg, New York)
31. Reiss, H., H.L. Frisch, J.L. Lebowitz (1959): 'Statistical mechanics of rigid spheres'. *J. Chem. Phys.* **31**, pp. 369-380
32. Ripley, B.D. (1988): *Statistical Inference for Spatial Processes* (Cambridge University Press, Cambridge, New York)
33. Serra, J. (1982): *Image Analysis and Mathematical Morphology* (Academic Press, London, New York)
34. Small, C.G. (1996): *The Statistical Theory of Shape* (Springer-Verlag, New York, Berlin, Heidelberg)
35. Soille, P. (1999): *Morphological Image Analysis. Principles and Applications.* (Springer-Verlag, Berlin, Heidelberg, New York)
36. Stoyan, D. (1998): 'Random sets: models and statistics'. *Int. Stat. Rev.* **66**, pp. 1-27
37. Stoyan, D., W.S. Kendall, J. Mecke (1995): *Stochastic Geometry and its Applications* (J. Wiley & Sons, Chichester)
38. Stoyan, D., M. Schlather (2000): 'On the random sequential adsorption model with disks of different sizes', submitted
39. Stoyan, D., H. Stoyan (1986): 'Simple stochastic models for the analysis of dislocation distributions'. *Physica Stat. Sol.* (a) **97**, pp. 163-172
40. Stoyan, D., H. Stoyan (1994): *Fractals, Random Shapes and Point Fields* (J. Wiley & Sons, Chichester)
41. Stoyan, D., H. Stoyan (1998): 'Non-homogeneous Gibbs process models for forestry – a case study'. *Biometrical J.* **40**, pp. 521-531
42. Stoyan, D., H. Stoyan (2000): 'Improving ratio estimators of second order point process characteristics'. *Scand. J. Statist.*
43. Thönnes, E., M.C. van Lieshout (1999): 'A comparative study on the power of Van Lieshout and Baddeley's *J*-function'. *Biometrical J.* **41**, pp. 721-734
44. Tomppo, E. (1986): 'Models and methods for analysing spatial patterns of trees'. *Comm. Inst. Forest. Fennicae* **138**, Helsinki
45. Wackernagel, H. (1998): *Multivariate Geostatistics.* 2nd edition (Springer-Verlag, Berlin-Heidelberg-New York)
46. Weber, H., D. Marx, K. Binder (1995): 'Melting transition in two dimensions: A finite-size scaling analysis of bond-orientational order in hard disks'. *Phys. Rev. B* **51**, pp. 14636-14651

Stationary Models in Stochastic Geometry – Palm Distributions as Distributions of Typical Elements. An Approach Without Limits

Werner Nagel

Fakultät für Mathematik und Informatik, Friedrich-Schiller-Universität Jena
D-07740 Jena, Germany

Abstract. The text introduces basic notions for stationary models in Stochastic Geometry. Those are the models themselves, the intensity and the Palm distribution as the distribution of the typical object in a random geometric structure. The Campbell theorem is presented as the main tool to investigate relations between quantitative characteristics without considerations of limits in space. The application to random tessellations is demonstrated as an example.

1 Introduction

The present introductory text is written to encourage the reader in taking note of models, tools and results provided in the field of Stochastic Geometry. There has been considerable progress during the last decades, mainly due to a successful application of the theory of point processes.

Two key notions are those of stationarity or - synonymously - homogeneity and of the Palm distribution. With a minimum of formalism and appealing to the intuition the basic concepts are introduced here.

Possible applications are illustrated in some examples. For detailed and comprehensive presentations we refer to monographs, see [1,3,5,6].

In Stochastic Geometry one deals with random geometric structures in the Euclidean plane $\mathbb{R}^2$, in the space $\mathbb{R}^3$ or more general in $\mathbb{R}^d$, $d \geq 1$. The spatial structures are considered as random variables Φ with domain $\mathcal{N}$, the set of all possible realisations φ of Φ. Thus $\mathcal{N}$ is the set of all geometric patterns which may occur in the model of interest. Usually, these realisations φ are understood as geometric structures which extend over the whole $\mathbb{R}^d$. The probability distribution of Φ is denoted by P_Φ, and this is a distribution on the set $\mathcal{N}$.

From a mathematical point of view most of the spaces and sets ($\mathbb{R}^d$, $\mathcal{N}$ and so on) have to be endowed with an appropriate σ-algebra, and all the occurring functions have to be measurable functions. In order to facilitate reading we will not mention these assumptions in the following. Notice that an exact formulation of definitions and theorems would require to explicate them.

2   Quantitative Analysis
   of Irregular Geometric Structures

Its a key task in Stochastic Geometry to choose and to define quantities or parameters which reflect certain features of a structure. Their choice depends on the application. For precise definitions of such quantities there are two principal ways: the ergodic approach or the method based on Palm distributions.

In order to explain these two approaches we start with two simple examples.

Example 1: Point pattern in the Euclidean plane $\mathbb{R}^2$.
Quantitative features are

- the mean number of points per unit area,
- the mean number per unit area of points with a distance d to the next neighbour smaller than a given r,
- the mean number of points in a certain sector of a disk, given that the centre of the disk lies in a point of the pattern.

Example 2: Marked point pattern.
Starting from a point pattern, a geometric figure - e.g. a segment, a circle, a polygon - is attached to each point. These figures are referred to as marks. For such structures the quantitative analysis of the underlying point pattern can also be of interest but, furthermore, the probability distribution of the marks has to be taken into consideration. For the example of segments as marks one can investigate

- the mean total length of segments per unit area,
- the mean number per unit area of points marked with segments of a direction in a certain sector of angles,
- the mean total length per unit area of segments with its direction in a certain sector of angles.

A further class of models is formed by random tessellations ('mosaics') of $\mathbb{R}^d$. These are treated in section 7.

In the ergodic approach limits are used in order to define the notions rigorously. For example, the mean number of points per unit area is introduced as the limit, for an infinitely expanding window W, of the ratio of the number of points in that window and the area of the window. This limit is considered for a fixed realisation φ of the point process, and the ergodic assumption says that this limit is the same for (almost) all realisations of a given process. This means that each of the realisations φ bears the whole information about the process Φ. Notice that in a bounded window some of the quantitative features can be determined only up to edge effects. For example, if the distance to the next neighbour is the parameter of interest then it can happen that a point in W has its next neighbour outside W. Such edge effects have to be taken into consideration when the ergodic limit is determined.

In the alternative approach which uses Palm distributions no limit in space is used. It is assumed that in an arbitrarily chosen window all essential information can be gained, if sufficiently many realisations or samples are considered. That means for the example of the mean number of points that in a fixed window W several realisations φ are observed and the mean number of points is divided by the area of W.

This is based on the concepts of spatial stationarity and the Palm distribution. It is the purpose of the present text to explain this concept and to illustrate its use. It has to be mentioned that also here the problem of edge effects occurs.

3   Stationary Models for Random Geometric Structures

At the beginning, it should be emphasised, that 'randomness' in Stochastic Geometry does *not* mean 'Poisson distribution' or 'independence'. Sometimes in applied papers the term 'purely random' is used as a synonym for 'Poisson process'. In random structures spatial interdependencies may occur between elements or particles such as clustering or repulsion.

Stationarity or homogeneity means the invariance with respect to translations *of the distribution* of a random geometric structure Φ. This means that the distribution of the part of the structure which is visible inside a window W does not change if the window is shifted. Of course, this invariance is not valid for single realisations φ of Φ. For a fixed realisation the observed sample depends on the location of the window.

In order to formalise the notion of stationarity consider the group of translations of the Euclidean space $\mathbb{R}^d$:

$$\{T_x : x \in \mathbb{R}^d\}, \quad T_x = \mathbb{R}^d \rightarrow \mathbb{R}^d, \text{ with } T_x y = y - x \text{ for all } y \in \mathbb{R}^d.$$

Thus T_x can be interpreted as the shift of the coordinate system such that its origin o is translated to the point x. It is clear how T_x acts on subsets of $\mathbb{R}^d$ and hence how it acts on the set $\mathcal{N}$ of all realisations φ of a random geometric structure Φ. We write $T_x \Phi$ for the shift of Φ by $-x$. If P_Φ is the distribution of Φ then the distribution of the shifted structure $T_x \Phi$ is denoted by $P_\Phi \circ T_x^{-1}$; it is the image of the probability measure P_Φ under the operation T_x.

Definition 3.1: *The random geometric structure Φ in $\mathbb{R}^d$ is stationary (homogeneous), iff*

$$P_\Phi = P_\Phi \circ T_x^{-1} \text{ for all } x \in \mathbb{R}^d. \tag{1}$$

Formula (1) means that for any set $A \subseteq \mathcal{N}$ holds

$$P_\Phi(A) = P_\Phi(T_x^{-1} A),$$

i.e. the set $T_x^{-1} A$ of shifted realisations has the same probability as the set A. The translation T_x^{-1} is the inverse to T_x, i.e. $T_x^{-1} = T_{-x}$, $x \in \mathbb{R}^d$.

When a model is chosen to describe a real structure the question has to be answered whether a stationary model is adequate. Basically, each image of a structure *can* be considered as an cutting of a realisation φ of a stationary random structure Φ. Hence, one should not ask whether a real geometric structure *is* stationary, but whether a stationary model is useful to treat the problem of interest and what the consequences of such a model choice are. It is clear that in a stationary model any spatial stochastic trend is a-priori neglected.

4   The Intensity and the Campbell Theorem

From now on, let Φ be a *stationary point process* in $\mathbb{R}^d$. For a bounded observation window $W \subset \mathbb{R}^d$ - e.g. a rectangle or a parallelepiped, respectively - denote by $\Phi(W)$ the random number of points of Φ in W. The stationarity of Φ implies that the expectation $\mathbb{E}\Phi(W)$ is invariant with respect to translations of W, i.e.

$$\mathbb{E}\Phi(W) = \mathbb{E}\Phi(T_x W) \text{ for all } x \in \mathbb{R}^d.$$

Thus $\mathbb{E}\Phi(W)$ is a translation invariant measure on $\mathbb{R}^d$. This yields that there is a non-negative real number λ_Φ such that

$$\mathbb{E}\Phi(W) = \lambda_\Phi \cdot \nu^d(W) \tag{2}$$

where ν^d denotes the d-dimensional volume (corresponding to the Lebesgue measure). This is a consequence of the assertion that the volume measure is, up to a constant, the only translation invariant measure in the Euclidean space.

Definition 4.1: *For a stationary point process Φ in $\mathbb{R}^d$ the constant*

$$\lambda_\Phi = frac\mathbb{E}\Phi(W)\nu^d(W) \tag{3}$$

is called the intensity *or density of Φ.*

The value of λ_Φ does neither depend on the location of W nor on its shape or size even if W occurs on the right hand side of (3).

The interpretation is simple: λ_Φ is the mean number of points of Φ per unit d-dimensional volume. In the literature, the density is also denoted by N_A or N_V in the planar or spatial cases respectively.

Remark: Also for other stationary structures Φ than point processes and other functionals $F(\Phi \cap W)$ instead of the number $\Phi(W)$ the intensity

$$\lambda_{F,\Phi} = \frac{\mathbb{E}F(\Phi \cap W)}{\nu^d(W)}$$

can be introduced. To do this, the functional must be appropriately chosen, i.e. F must be translation invariant itself and σ-additive. Such considerations lead to several mean values per unit volume or more abstractly to the notion of the intensity of stationary random measures, in particular of Minkowski measures, see [3,5]. An example is L_A, the mean total length per unit volume of a stationary planar segment process, which was described above as a marked point process.

From formula (2) the Campbell theorem can be derived relatively easy. Denote the indicator function of W by 1_W, i.e.

$$1_W(x) = \begin{cases} 1 \text{ if } x \in W, \\ 0 \text{ if } x \notin W, \end{cases} x \in \mathbb{R}^d.$$

Then (2) can be rewritten as

$$\mathbb{E}\left(\sum_{x \in \Phi} 1_W(x)\right) = \lambda_\Phi \int 1_W(x)\mathrm{d}x. \tag{4}$$

Now it can be shown by standard methods of integration theory that (4) remains true if 1_W is replaced by more general functions h. This yields the Campbell Theorem which is basic in point process theory, and it is impossible to trace back where it appeared for the first time.

Theorem 4.1: [Campbell Theorem]
Let Φ be a stationary point process in $\mathbb{R}^d$ with intensity $0 < \lambda_\Phi < \infty$ and let $h : \mathbb{R}^d \to [0, \infty)$ be a nonnegative function. Then

$$\mathbb{E}\left(\sum_{x \in \Phi} h(x)\right) = \lambda_\Phi \int h(x)\mathrm{d}x. \tag{5}$$

Notice that h is a function depending on the location of a single point x of the point pattern but *not* on the whole realisation φ of Φ, i.e. relations of $x \in \Phi$ to other points of the process, e.g. the distance to the next neighbour, cannot be expressed in this way.

An application of formula (5) is Monte-Carlo-integration of a function h. The integral on the right hand side can be approximated by an averaged sum of h-values in simulated random points $x \in \mathbb{R}^d$ of a point process.

5 The Palm Distribution and the Refined Campbell Theorem

The key notion of the Palm distribution of a stationary (point) process formalises the idea of 'a typical point' or 'a typical object' in a random geometric structure.

It may be ambiguous to speak of a 'randomly chosen' element (point, particle, ...) in a structure. Therefore the way of such a choice has to be described and formalised precisely. Here this is done for point processes but the method can be extended to other structures and to attached random measures.

Definition 5.1: *Let Φ be a stationary point process with intensity $0 < \lambda_\Phi < \infty$ and distribution P_Φ. The Palm distribution P_0 of Φ is defined by*

$$P_0(A) = \frac{1}{\lambda_\Phi \cdot \nu^d(W)} \mathbb{E} \left(\sum_{x \in \Phi} 1_A(T_x \Phi) \cdot 1_W(x) \right) \tag{6}$$

for all events $A \subseteq \mathcal{N}$ and windows $W \subset \mathbb{R}^d$ with $0 < \nu^d(W) < \infty$.

It can be proven that - due to stationarity - this definition does *not* depend on the particular choice of the window W.

The idea behind (6) is the desire to consider the conditional probability '$P_\Phi(A|o \in \Phi)$', i.e. the probability that a realisation has the property expressed by A under the condition that the process Φ has a point in the origin o. The problem is that for stationary processes $P_\Phi(0 \in \Phi) = 0$. Therefore, as done in (6), one considers all points $x \in \Phi \cap W$. In a window W of finite volume there are (almost surely) finitely many points x of Φ. Each of these points is 'visited' by the origin of the coordinate system - this is expressed by $T_x\Phi$. It is checked for each of the shifted realisations whether they belong to $A \subseteq \mathcal{N}$ or not. Therefore the Palm distribution is also interpreted as the distribution of the point process when the process is considered from its 'typical' point. Formula (6) yields the expression

$$P_0(A) = \frac{\text{mean number of points } x \in \Phi \cap W \text{ with } T_x\Phi \in A}{\text{mean number of points } x \in \Phi \cap W} \tag{7}$$

This formula also suggests the way of generalising this concept to other geometric structures than point processes: Replace 'number of points x...' by 'functional of the set of all x...'

An example of an event $A \in \mathcal{N}$ which is considered in the quantitative analysis of point patterns is, for given $B \subset \mathbb{R}^d$ and $k = 0, 1, 2, ...,$
$A_k(B) = \{\varphi \in \mathcal{N} : o \in \varphi, \ \varphi(B \setminus \{o\}) = k\}$
i.e. the set of all realisations φ with one point in the origin and exactly k further points within the set B. This set B may or may not contain the origin. Important examples for B are circles, circular sectors or circular rings, respectively.
Such events $A_k(B)$ are suitable to express the so-called second reduced moment measure $\mathcal{K}$ of Φ which is

$$\lambda \mathcal{K}(B) = \sum_{k=0}^{\infty} k \cdot P_0(A_k(B)) = \int \varphi(B \setminus \{o\}) P_0(d\varphi) = \mathbb{E}_0 \Phi_0(B \setminus \{o\}),$$

see [5], chapter 4. Thus $\lambda \mathcal{K}(B)$ is the mean number of points in $B \setminus \{o\}$ with respect to the Palm distribution. It can be estimated by visiting all points $x \in \varphi$, counting the number of points different from x in the set $B + x$, and then averaging this number. Notice that an edge effect correction is necessary in order to get unbiased estimates of $\lambda \mathcal{K}(B)$ form observations in a window W.

The second reduced moment measure $\mathcal{K}$ is closely related to the pair correlation function g which is used in physics. This function is given by

$$g(r) = \frac{r}{d \cdot \nu^d(b(o,r))} \frac{\mathrm{d}}{\mathrm{d}r} \mathcal{K}(b(o,r))$$

where $b(o,r)$ is the ball with radius r and center o.

Replacing the function $\mathbf{1}_A \cdot \mathbf{1}_W$ in the definition (6) by a more general function yields

Theorem 5.1: [Refined Campbell Theorem (J. Mecke, 1967)]
Let Φ be a stationary point process in $\mathbb{R}^d$ with intensity $0 < \lambda_\Phi < \infty$ and $f : \mathbb{R}^d \times \mathcal{N} \to [0,\infty)$ a nonnegative function. Then

$$\mathbb{E}\left(\sum_{x \in \Phi} f(x, T_x \Phi) \right) = \lambda_\Phi \mathbb{E}_0 \left(\int f(x, \Phi_0) \mathrm{d}x \right) \qquad (8)$$

where Φ_0 is a point process with distribution P_0, i.e. the Palm distribution of Φ, and $\mathbb{E}_0$ is the expectation with respect to P_0.

Notice that Φ_0 is a point process which has a point in the origin o with probability 1.

Another version of (8) can be derived by substituting $g(x, \Phi) = f(x, T_x \Phi)$.

$$\mathbb{E}\left(\sum_{x \in \Phi} g(x, \Phi) \right) = \lambda_\Phi \mathbb{E}_0 \left(\int g(x, T_{-x}\Phi_0) \mathrm{d}x \right) \qquad (9)$$

In contrast to the Campbell Theorem 4.1, in the refined version, functions f or g respectively, are considered which may depend on single points x of the process *and* also on the whole realisation φ to which x belongs. Applications of the Refined Campbell Theorem will be given in section 7 on random tessellations.

6   Marked Point Processes

A realisation φ of a point process Φ is an enumerable subset $\{x_1, x_2, ...\} \subset \mathbb{R}^d$. In order to describe tessellations or ensembles of particles it is often appropriate to use marked point processes as models. A mark $m_i \in M$ is attached to each point $x_i \in \varphi$. The set M is the set of all possible marks. Thus the realisations of a marked point process are of the form

$$\{(x_1, m_1), (x_2, m_2), ...\} \subset \mathbb{R}^d \times M.$$

Examples are

Example 3: Process of balls in the space with x_i as the centre and positive random radius m_i, $i = 1, 2, ...$,

Example 4: Process of segments in the plane given by the midpoints x_i and length and direction $m_i = (l_i, \alpha_i) \in (0, \infty) \times [0, \pi)$, $i = 1, 2, ...$

For a window $W \subset \mathbb{R}^d$ and a set of marks $B \subseteq M$

$$\Phi(W \times B) = \sum_{(x,m) \in \Phi} \mathbf{1}_W(x) \cdot \mathbf{1}_B(m) \tag{10}$$

denotes the random number of points located in W which have marks in B.

Notice that formally a marked point process is a usual point process on the space $\mathbb{R}^d \times M$. But often, from a methodological angle it is more convenient to distinguish between points and marks.

In order to deal with stationary marked point processes, translations have to be defined adequately. Examples (3) and (4) suggest that a translation of a marked point has to be defined as

$$T_x(x_i, m_i) = (x_i - x, m_i), \quad x \in \mathbb{R}^d. \tag{11}$$

The translation of the point by minus x means that in particular x itself goes to o. The marks remain unchanged. This can be applied to a whole marked point process Φ, and the translated version of it is denoted by $T_x\Phi$. As before, the distribution of Φ is P_Φ and that of $T_x\Phi$ is written as $P_\Phi \circ T_x^{-1}$, respectively.

Definition 6.1: *The marked point process Φ on $\mathbb{R}^d$ with mark space M is stationary if*

$$P_\Phi \circ T_x^{-1} = P_\Phi \text{ for all } x \in \mathbb{R}^d$$

with T_x according to (11).

For a stationary marked point process Φ consider the expectation of the random number given in (10). Stationarity implies that

$$\mathbb{E}\Phi(W \times B) = \mathbb{E}\Phi((T_xW) \times B) \text{ for all } x \in \mathbb{R}^d.$$

Thus for all $B \subseteq M$ the measure $\mathbb{E}\Phi(\cdot \times B)$ as a functional of $W \subseteq \mathbb{R}^d$ is invariant with respect to translations and hence equals the d-dimensional volume measure up to a factor. This yields the following factorisation.

Theorem 6.1: *Let Φ be a stationary marked point process in $\mathbb{R}^d$ with the mark space M. Then there are a number λ_Φ and a probability measure μ on M such that for all $W \subset \mathbb{R}^d$ with $0 < \nu^d(W) < \infty$ and $B \subseteq M$*

$$\lambda_\Phi = \frac{\mathbb{E}\Phi(W \times M)}{\nu^d(W)} \tag{12}$$

and

$$\mathbb{E}\Phi(W \times B) = \lambda_\Phi \cdot \nu^d(W) \cdot \mu(B). \tag{13}$$

The probability measure μ on M is referred to as *mark distribution*. With respect to the Palm distribution, μ can also be interpreted as the distribution of the mark attached to the 'typical point' of the process.

Rewrite (13) as

$$\mu(B) = \frac{\mathbb{E}\Phi(W \times B)}{\lambda_\Phi \nu^d(W)} = \frac{\mathbb{E}\Phi(W \times B)}{\mathbb{E}\Phi(W \times M)}, \tag{14}$$

i.e. $\mu(B)$ is the ratio of the mean number of points in W with mark in B to the mean total number of points in W.

For stationary marked point processes the above mentioned factorisation of the intensity yields the following version of the Refined Campbell Theorem.

Theorem 6.2: *Let Φ be a stationary marked point process in $\mathbb{R}^d$ with mark space M, intensity $0 < \lambda_\Phi < \infty$ and mark distribution μ. Then for all nonnegative functions $f : \mathbb{R}^d \times M \times \mathcal{N} \to [0, \infty)$*

$$\mathbb{E}\left(\sum_{(x,m)\in\Phi} f(x, m, T_x\Phi) \right) = \lambda_\Phi \mathbb{E}_0 \left(\int \int f(x, m, \Phi_0)\mathrm{d}x \ \mu(\mathrm{d}m) \right). \tag{15}$$

Notice that here $\mathcal{N}$ denotes the set of all realisations of a marked point process.

7 Application to Random Tessellations of the Plane

A random tessellation of the Euclidean plane $\mathbb{R}^2$ can be modelled as a random variable Φ with values in

$\mathcal{N}$ - the set of all tessellations of $\mathbb{R}^2$ into convex and bounded polygonal cells, such that any circle in $\mathbb{R}^2$ intersects only finitely many of these cells.

In the following we assume that Φ is *stationary*, i.e. its distribution is invariant with respect to translations of the tessellations.

In order to apply point process methods it is useful to endow the tessellation with point processes:

α_0 - the point process of nodes,
α_1 - the point process of centres of the cell edges,
α_2 - the point process of centroids of the cells,

and

$\beta = \alpha_0 \cup \alpha_1 \cup \alpha_2$ - the union of these point processes.

There are several ways to define the centroid of a polygonal cell. For our purposes it must not be specified. Of course, one has to settle the same type of centroid for all cells of the tessellation.

The stationarity of Φ implies the stationarity of all these point processes. Their intensities, as they have been defined in (3), are denoted by $\lambda_0, \lambda_1, \lambda_2, \lambda_\beta$, respectively.

These point processes can be marked in several ways. Examples of marks are:

for α_0: the number of edges emanating from a node,
for α_1: the length of the edge, the direction of the edge,
for α_2: the area, the perimeter of the cell, the number of edges of the cell.

The choice of appropriate marks depends on the questions to be answered.

Here, we will consider some basic mean values and their interdependencies inherent in all stationary tessellations with convex polygonal cells.

As an example for the derivation of such a relation consider the point process $\alpha_2(\Phi)$ of cell centroids of Φ. The Refined Campbell Theorem 5.1 yields

$$\mathbb{E}_\Phi \left(\sum_{x \in \alpha_2(\Phi)} f(x, T_x \Phi) \right) = \lambda_2 \mathbb{E}_{0,2} \left(\int f(x, \Phi_{0,2}) \mathrm{d}x \right) \tag{16}$$

for all $f : \mathbb{R}^2 \times \mathcal{N} \to [0, \infty)$. Here $\mathbb{E}_\Phi$ denotes the expectation with respect to the distribution P_Φ of Φ, and $\mathbb{E}_{0,2}$ the expectation with respect to the Palm distribution $P_{0,2}$ of the point process α_2, and $\Phi_{0,2}$ is a random tessellation with distribution $P_{0,2}$. Notice that, with probability 1, the random tessellation $\Phi_{0,2}$ has a cell centroid in the origin o.

Insert $f(x, \Phi) = \mathbf{1}_{C_0\Phi}(x)$ into (16), where $C_0\Phi$ denotes that cell of Φ which contains the origin o. Thus we obtain

$$\mathbb{E}_\Phi \sum_{x \in \alpha_2(\Phi)} \mathbf{1}_{C_0\Phi}(x) = 1 = \lambda_2 \mathbb{E}_{0,2} \nu^2(C_0\Phi_{0,2}) \tag{17}$$

The left equality is due to the fact that $C_0\Phi$ contains exactly one centroid point. On the right hand side occurs the mean area ν^2 of the typical cell of Φ, since it is the mean area with respect to the Palm distribution. Notice that in contrast to this, $\mathbb{E}_\Phi \nu^2(C_0\Phi)$ would be the mean area of the cell of Φ containing the origin. This is a random cell which is chosen from the tessellation by a mechanism which prefers larger cells proportional to their area. Therefore, the corresponding distribution of the cell of Φ which contains the origin is referred to as the area weighted cell distribution.

Formula (17) provides

$$\mathbb{E}_{0,2} \nu^2(C_0\Phi_{0,2}) = \frac{1}{\lambda_2} \tag{18}$$

This means that the mean area of the typical cell equals the reciprocal of the mean number of cell centroids per unit area. This is plausible but a strict formal proof of it without the tool of Palm distributions would by no means be as short as that one given above. For example, the determination of ergodic limits requires subtle considerations concerning edge effects.

Reconsider (16) and put $f(x, \Phi) = [\nu^2(C_0\Phi)]^{-1}1_{C_0\Phi}(x)$. This yields

$$\underbrace{\mathbb{E}_\Phi \frac{1}{\nu^2(C_0\Phi)} \sum_{x \in \alpha_2(\Phi)} 1_{C_0\Phi}(x)}_{1} = \lambda_2 \mathbb{E}_{0,2} \underbrace{\left(\frac{1}{\nu^2(C_0\Phi)} \int 1_{C_0\Phi}(x) \mathrm{d}x \right)}_{1}. \qquad (19)$$

Thus

$$\mathbb{E}_\Phi \frac{1}{\nu^2(C_0\Phi)} = \lambda_2. \qquad (20)$$

This formula says that the mean reciprocal area of the area weighted cell equals the mean number of cell centroids per unit area. This is not obvious. The connection between (18) and (20) can be stated by explicating the area weighted distribution of cells.

Now consider a more complicated application of the Refined Campbell Theorem. Denote
$\overline{n_{02}}$ - the mean number of cells meeting in the typical node,
$\overline{n_{20}}$ - the mean number of nodes on the boundary of the typical cell.
Then

$$\lambda_2 \overline{n_{20}} = \lambda_0 \overline{n_{02}} \qquad (21)$$

Even if this relation appears plausible it is not trivial. Notice that there arise products of mean values of random variables which do not necessarily coincide with the mean of the product. The formula can be proven by using the following symmetry property of the Palm distribution.

Theorem 7.1: (J. Mecke, 1967)
Let Φ be a stationary point process in $\mathbb{R}^d$ with intensity $0 < \lambda_\Phi < \infty$ and $f : \mathbb{R}^d \times \mathcal{N} \to [0, \infty)$ a nonnegative function. Then

$$\mathbb{E}_0 \left(\sum_{x \in \Phi_0} f(x, \Phi) \right) = \mathbb{E}_0 \left(\sum_{x \in \Phi_0} f(-x, T_x\Phi) \right) \qquad (22)$$

Denote by $\mathbb{E}_{0,\beta}$ and $\mathbb{E}_{0,0}$ the expectations with respect to the Palm distributions of the point processes β and α_0 respectively. Random tessellations with these Palm distributions are denoted by $\Phi_{0,\beta}$ and $\Phi_{0,0}$, respectively. Further, $\mathcal{C}(\Phi)$ denotes the set of all cells of Φ.

Then the consecutive application of the definition of $\overline{n_{20}}$, of (16) and the analogue to (16) for $\mathbb{E}_{0,\beta}$ respectively, yield

$$\lambda_2\,\overline{n_{20}}$$

$$= \lambda_2\,\mathbb{E}_{0,2}\sum_y \mathbf{1}_{\alpha_0(\Phi_{0,2})\cap(C_0\Phi_{0,2})^{cl}}(y)$$

$$= \lambda_2\,\mathbb{E}_{0,2}\sum_y \mathbf{1}_{\alpha_0(\Phi_{0,2})\cap(C_0\Phi_{0,2})^{cl}}(y)\int \mathbf{1}_{[0,1]^2}(x)\mathrm{d}x$$

$$= \mathbb{E}_\Phi\sum_{x\in\alpha_2(\Phi)} \mathbf{1}_{[0,1]^2}(x)\sum_y \mathbf{1}_{\alpha_0(T_x\Phi)}(y)\,\mathbf{1}_{(C_0T_x\Phi)^{cl}}(y)$$

$$= \mathbb{E}_\Phi\sum_{x\in\beta(\Phi)} \mathbf{1}_{\alpha_2(\Phi)}(x)\mathbf{1}_{[0,1]^2}(x)\sum_y \mathbf{1}_{\alpha_0(T_x\Phi)}(y)\,\mathbf{1}_{(C_0T_x\Phi)^{cl}}(y)$$

$$= \lambda_\beta\,\mathbb{E}_{0,\beta}\sum_y \mathbf{1}_{\alpha_0(\Phi_0)}(y)\,\mathbf{1}_{(C_0\Phi_0)^{cl}}(y)\int \mathbf{1}_{[0,1]^2}(x)\mathbf{1}_{\alpha_2(T_{-x}\Phi_0)}(x)\mathrm{d}x$$

$$= \lambda_\beta\,\mathbb{E}_{0,\beta}\sum_y \mathbf{1}_{\alpha_0(\Phi_0)}(y)\,\mathbf{1}_{(C_0\Phi_0)^{cl}}(y)\,\mathbf{1}_{\alpha_2(\Phi_0)}(o)$$

Now the application of Theorem 7.1 and of the relation between $\lambda_\beta\,\mathbb{E}_{0,\beta}$ and $\lambda_0\,\mathbb{E}_{0,0}$ which can be derived in analogous steps as above provide

$$\lambda_\beta\,\mathbb{E}_{0,\beta}\sum_y \mathbf{1}_{\alpha_0(\Phi_0)}(y)\,\mathbf{1}_{(C_0\Phi_0)^{cl}}(y)\,\mathbf{1}_{\alpha_2(\Phi_0)}(o)$$

$$= \lambda_\beta\,\mathbb{E}_{0,\beta}\sum_y \mathbf{1}_{\alpha_0(\Phi_0)}(y)\sum_{C\in\mathfrak{C}(\Phi_0)} \mathbf{1}_{C^{cl}}(o)\mathbf{1}_{C^{cl}}(y)\,\mathbf{1}_{\alpha_2(\Phi_0)}(o)$$

$$= \lambda_\beta\,\mathbb{E}_{0,\beta}\sum_y \mathbf{1}_{\alpha_0(T_y\Phi_0)}(-y)\sum_{C\in\mathfrak{C}(T_y\Phi_0)} \mathbf{1}_{C^{cl}}(o)\mathbf{1}_{C^{cl}}(-y)\,\mathbf{1}_{\alpha_2(T_y\Phi_0)}(o)$$

$$= \lambda_\beta\,\mathbb{E}_{0,\beta}\mathbf{1}_{\alpha_0(\Phi_0)}(o)\sum_y \mathbf{1}_{\alpha_2(\Phi_0)}(y)\sum_{C+y\in\mathfrak{C}(\Phi_0)} \mathbf{1}_{C^{cl}+y}(y)\mathbf{1}_{C^{cl}+y}(o)$$

$$= \lambda_\beta\,\mathbb{E}_{0,\beta}\mathbf{1}_{\alpha_0(\Phi_0)}(o)\sum_y \mathbf{1}_{\alpha_2(\Phi_0)}(y)\sum_{C\in\mathfrak{C}(\Phi_0)} \mathbf{1}_{C^{cl}}(y)\mathbf{1}_{C^{cl}}(o)$$

$$= \lambda_0\,\mathbb{E}_{0,0}('\ number\ of\ cells\ which\ contain\ the\ origin')$$

$$= \lambda_0\,\overline{n_{02}}$$

The upper index 'cl' indicates the topological closure of a set in $\mathbb{R}^2$, i.e. the set including its boundary. Thus formula (21) is shown.

We conclude with a survey of relations for mean values of stationary planar tessellations, see [2] or [5]. Supplementary notations are

L_A - mean total edge length per unit area,
$\overline{l_0}$ - mean total length of edges emanating from the typical node,
$\overline{l_1}$ - mean length of the typical edge,
$\overline{l_2}$ - mean perimeter of the typical cell,
$\overline{a_2}$ - mean area of the typical cell.

Then

$$\lambda_1 = \lambda_0 + \lambda_2,$$

$$\overline{a_2} = \tfrac{1}{\lambda_2},$$

$$\overline{l_1} = \tfrac{L_A}{\lambda_0 + \lambda_2},$$

$$\overline{n_{02}} = 2 + 2\tfrac{\lambda_2}{\lambda_0}, \quad \overline{n_{20}} = 2 + 2\tfrac{\lambda_0}{\lambda_2},$$

$$\overline{l_0} = 2\tfrac{L_A}{\lambda_0}, \qquad \overline{l_2} = 2\tfrac{L_A}{\lambda_2}.$$

Observe the duality of the formulae when the indices 0 and 2 are interchanged. The first formula is the classical Euler formula in terms of intensities.

This system of equations shows that for *planar* stationary tessellations the system of ten mean values $\{\lambda_0, \lambda_1, \lambda_2, L_A, \overline{n_{02}}, \overline{n_{20}}, \overline{l_0}, \overline{l_1}, \overline{l_2}, \overline{a_2}\}$ can be reduced to three parameters, e.g. $\{L_A, \lambda_0, \lambda_2\}$ without loss of information, since all the others can be expressed by them. This reflects interdependence of parameters inherent to all planar tessellations.

An example of a corollary from this equations is

$$\frac{1}{\overline{n_{02}}} + \frac{1}{\overline{n_{20}}} = \frac{1}{2}.$$

Since each node in $\mathbb{R}^2$ has at least three emanating edges, also the mean $\overline{n_{02}} \geq 3$ and consequently,

$$\overline{n_{20}} \leq 6.$$

This means that there exists no stationary planar tessellation where the mean number of edges of the typical polygon is greater than six. Notice that $\overline{n_{20}} = 6$ if and only if $\overline{n_{02}} = 3$, i.e. if all nodes have exactly three emanating edges.

The results in the present section have been given as examples for the application of Palm distributions and of the Refined Campbell Theorem. Much more comprehensive presentations of random tessellations can be found in [2,4,5].

References

1. Kallenberg, O. (1986): *Random Measures* (Akademie-Verlag and Academic Press, Berlin and Orlando)
2. Mecke, J. (1984): 'Parametric representation of mean values for stationary random mosaics', *Math. Operationsf. Statist., Ser. Statistics* **15**, pp. 437–442
3. Mecke, J., R. Schneider, D. Stoyan, W. Weil (1990): *Stochastische Geometrie* (Birkhäuser, Basel)
4. Møller, J. (1989): 'Random tessellations in $\mathbb{R}^d$', *Adv. Appl. Prob.* **21**, pp. 37-73
5. Stoyan, D., W.S. Kendall, J. Mecke (1995): *Stochastic Geometry and its Applications* (John Wiley & Sons, Chichester)
6. Stoyan, D., H. Stoyan (1994): *Fractals, Random Shapes and Point Fields* (John Wiley & Sons, Chichester)

Statistical Analysis of Large-Scale Structure in the Universe

Martin Kerscher

Ludwig–Maximilians–Universität, Sektion Physik, Theresienstraße 37
80333 München, Germany

Abstract. Methods for the statistical characterization of the large–scale structure in the Universe will be the main topic of the present text. The focus is on geometrical methods, mainly Minkowski functionals and the J function. Their relations to standard methods used in cosmology and spatial statistics and their application to cosmological datasets will be discussed. A short introduction to the standard picture of cosmology is given.

1 Introduction

A fundamental problem in cosmology is to understand the formation of the large–scale structure in the Universe. Normally theoretical models of large–scale structure, whether involving analytical predictions or numerical simulations, are based on some form of random or stochastic initial conditions. This means that a statistical interpretation of clustering data is required, and that statistical tools must be deployed in order to discriminate between different cosmological models. Moreover the identification and characterization of specific geometric features in the galaxy distribution like walls, filaments, and clusters will deepen our understanding of structure formation, assist in the construction of approximations and also help to constrain cosmological models.

During the past two decades enormous progress has been made in the mapping of the distribution of galaxies in the Universe. Using the measured redshifts of galaxies as distance indicators, and knowing their angular positions on the sky, we can obtain a three–dimensional view of the distribution of luminous matter in the Universe. Presently available redshift surveys already permit the detailed study of the statistical properties of the spatial distribution of galaxies. Surveys of galaxy redshifts that cover reasonable solid angles and are significantly deeper than those presently available present important challenges, and not just for the observers. A precise definition of the statistical methods is needed to extract most out of the costly data, and this is an important goal for theorists.

A complete review of the variety of statistical methods used in cosmology is not attempted. The focus of this overview will be on methods of point process statistics using geometrical ideas like Minkowski functionals and the J function; moment based methods will also be mentioned. For reviews with a different emphasis see e.g.[14,15,30,68,84] and [86].

This text is organized as follows:
In Sect. 2 we will give a short introduction to the common theoretical "prejudice"

in cosmology and describe some observational issues. We briefly comment on two–point correlations (Sect. 3.1) and moment based methods (Sect. 3.2), and focus on Minkowski functionals (Sect. 3.3) and the J function, as well as its extensions the J_n functions (Sect. 3.4). In Sect. 4 we summarize and provide an outlook.

2   Cosmological Models and Observations

Most cosmological models studied today are based on the assumption of homogeneity and isotropy (see however [20] and [21]). Observationally one can find evidence that supports these assumptions on very large scales, the strongest being the almost perfect isotropy of the cosmic microwave background radiation (after assigning the whole dipole to our proper motion relative to this background). The relative temperature fluctuations over the sky are of the order of 10^{-5} as shown in Fig. 1. This tells us that the Universe *was* nearly isotropic and, with some additional assumptions, homogeneous at the time of decoupling of approximately 13Gy (Giga years) ago.

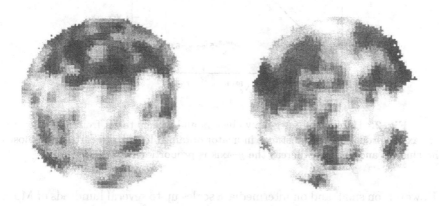

Fig. 1. Projection of the temperature fluctuations in the microwave background radiation as observed by the COBE satellite (from [100]). The relative fluctuations are of the order of 10^{-5}.

For such a highly symmetric situation the universal expansion may be described by a position vector $\mathbf{x}_H(t)$ at time t that can be calculated from the initial position $\mathbf{x}_i$

$$\mathbf{x}_H(t) = a(t)\,\mathbf{x}_i \tag{1}$$

using the scale factor $a(t)$ with $a(t_i) = 1$. The dynamical evolution of $a(t)$ is determined by the Friedmann equations (see e.g. [82]). As a direct consequence the velocities may be approximated by the Hubble law,

$$\mathbf{v}_H(t) = H(t)\,\mathbf{x}_H(t) \tag{2}$$

relating the distance vector $\mathbf{x}_H(t)$ with the velocity $\mathbf{v}_H(t)$ by the Hubble parameter $H(t) = \dot{a}(t)/a(t)$. Indeed such a mainly linear relationship is observed for galaxies (see Fig. 2). The deviations visible may be assigned to peculiar motions, as caused by mass density perturbations.

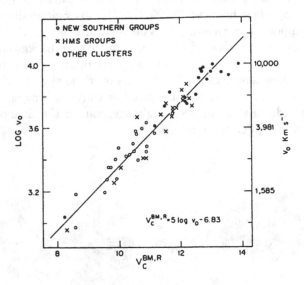

Fig. 2. Hubble law for galaxy clusters and groups taken from [93].
The x–axis is proportional to distance indicator obtained from the a certain luminosity of the clusters and groups, whereas the y–axis is proportional to the redshift.

However, on small and on intermediate scales up to several hundreds of Mpcs, there are significant deviations from homogeneity and isotropy as visible in the spatial distribution of galaxies. (Mega parsec (Mpc) is the common unit of length in cosmological applications with 1pc=3.26 light years.) Large holes, filamentary as well as wall–like structures are observed (Fig. 3, see also sect. 3.3.4).

One of the goals in cosmology is to understand how these large scale structures form, given a nearly homogeneous and isotropic matter distribution at some early time. In the Newtonian approximation the process of structure formation is modeled using a self gravitating pressure–less fluid, with the mass

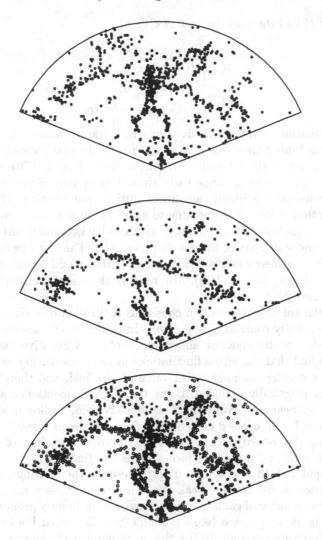

Fig. 3. In the upper two panels, the position of the galaxies in two neighboring slices with an angular extent of 135×5 deg^2, and a maximum distance of $120h^{-1}$Mpc from our galaxy which is located at the tip of the cone. The galaxies are shown projected along the angular coordinate spanning only 5deg. In the lower plot both slices are shown projected on top of each other (data from [48] and [49]).

density $\varrho(\mathbf{x}, t)$ and the velocity field $\mathbf{v}(\mathbf{x}, t)$:

$$\partial_t \varrho + \nabla(\varrho \mathbf{v}) = 0,$$
$$\partial_t \mathbf{v} + (\mathbf{v} \cdot \nabla)\mathbf{v} = \mathbf{g},$$
$$\nabla \times \mathbf{g} = 0, \tag{3}$$
$$\nabla \cdot \mathbf{g} = -4\pi G \varrho.$$

The first equation is the continuity equation, stating mass–conservation, the second comes from momentum conservation with the gravitational acceleration $\mathbf{g}(\mathbf{x}, t)$ self–consistently determined from the mass density. With small fluctuations in ϱ and $\mathbf{v}$ given at some early time, this system of partial differential equations constitutes a highly non–linear initial value problem. Up to now no general solution is known. Approximate solutions may be constructed using a perturbative expansion around the homogeneous background solutions either for the fields ϱ and $\mathbf{v}$ directly or for the characteristics. The first one is called Eulerian perturbation theory (see e.g. [86]), whereas the second is named Lagrangian perturbation theory (see e.g. [19]). Also numerical integration with N–body simulations is used.

The initial conditions are often chosen as realization of a Gaussian random field for the density contrast $(\varrho - \varrho_H)/\varrho_H$. In principle a Gaussian random field model for the density contrast allows for unphysical negative mass densities, however we find that the initial fluctuations in the mass density are by a factor of 10^5–times smaller than the mean value of the field, and therefore negative densities are practically excluded. Using the methods mentioned above we can follow the nonlinear time evolution of the density field, leading to a highly non–Gaussian field. In this evolved mass density field galaxies are identified sometimes also utilizing the velocity field. Moreover, our understanding of the physical processes determining the galaxy formation is still limited.

Two popular stochastic models used to describe the distribution of galaxies are the Poisson model and the peak selection. In the Poisson model we assume that the mean number of galaxies inside a region C is directly proportional to the total mass inside this region (see e.g. [86], often also called Poisson sampling). Hence the intensity measure $\Lambda(C)$ – the mean number of galaxies inside C – is

$$\Lambda(C) \propto \int_C d\mathbf{x}\, \varrho(\mathbf{x}). \tag{4}$$

If the mass density ϱ is modeled as a random field the Poisson model results in a double–stochastic point process, i.e. a Cox process [109].

Within the peak selection model, galaxies appear only at the peaks of the density field above some given threshold [7]. This model is an example for an "interrupted point process" [109]. In Fig. 4 we illustrate both models in the one–dimensional case. There are also dynamically and micro–physically motivated models for the identification of galaxies in simulations we do not cover here ([52,53,126]).

As we have seen several "parameters" enter these partly deterministic, partly stochastic models for the galaxy distribution. Before describing the statistical

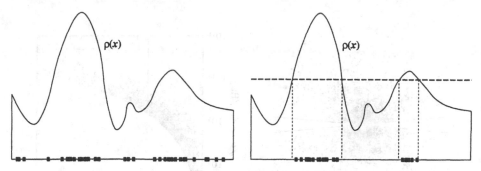

Fig. 4. The left figure illustrates the Poisson model, whereas the right figure shows the peak selection for the same density field.

methods used to constrain these parameters, typical observational problems entering the construction of galaxy catalogues will be mentioned.

The starting point is the two–dimensional distribution of galaxies on the celestial sphere. Their angular positions are known to a high precision compared to their radial distance r. In most galaxy catalogues the radial distance is estimated utilizing the redshift:

$$z = \frac{\lambda_{obs} - \lambda_{lab}}{\lambda_{lab}}, \tag{5}$$

with the observed wavelength of a spectral line λ_{obs} and with the wavelength of the same spectral line measured in a laboratory λ_{lab}. Out to several hundreds of Mpc's the relation between the radial distance r and the redshift z is to a good approximation

$$cz \approx |v_H| + u = H_0 r + u, \tag{6}$$

with the velocity of light c, and the Hubble parameter H_0 at present time (see (2)). u is the radial component of the peculiar velocity, i.e. the local deviation from the global expansion due to inhomogeneities. Galaxy catalogues sampled homogeneously and with r determined independently from the redshift are still rare. Therefore the distance is simply estimated by

$$r = \frac{cz}{H_0}, \tag{7}$$

neglecting the peculiar velocities u. This is often called "working in redshift space". There is still some controversy about the actual value of the Hubble parameter which is parameterized by the number h: $H_0 = h\,100\mathrm{km\,sec^{-1}\,Mpc^{-1}}$. Likely values are in the range $h = 0.5 - 0.8$.

Furthermore we have to face another problem. The majority of galaxy catalogues is flux (i.e. magnitude) limited. This means that the catalogue is complete for galaxies with a flux higher than some minimum flux f_{min}. As a first approximation the absolute luminosity L of a galaxy with observed flux f_{obs} at distance r may be calculated by $L = 4\pi r^2\,f_{obs}$. Hence at larger distances we observe

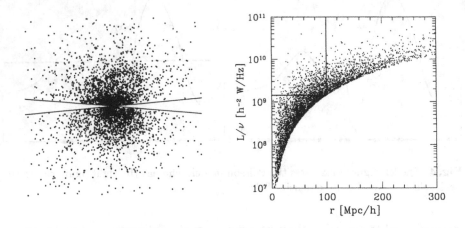

Fig. 5. In the left figure the spatial distribution of the galaxies taken from the IRAS 1.2Jy galaxy catalogue [35], projected along one axis. The horizontally cones indicate the region where the observation was obscured due to the absorption in our own galaxy. In the right plot the absolute luminosity of a galaxy against its radial distance is shown, each point represents one galaxy. The volume limited subsample with limiting distance of $100h^{-1}$Mpc includes only the galaxies in the marked upper left corner of the figure.

only the brightest galaxies as can be seen in Fig. 5, resulting in a systematically in–homogeneously sampled point–set in three dimensions. To construct a homogeneously sampled point set from such a galaxy catalogue we may restrict ourselves to galaxies closer than $r_{\lim}$ with a absolute luminosity higher than $L_{\lim} = 4\pi r^2 f_{\min}$. This procedure is called "volume limitation". Such a set of galaxies for $r_{\lim} = 100h^{-1}$Mpc is marked in Fig. 5 and the spatial distribution is shown in Fig. 6. Especially in the direction of the disc of our galaxy, in the galactic plane, we suffer from extinction mainly due to dust. To take care of this we use a cut of 5 to 30 degrees (depending on the catalogue under consideration) around the galactic plane, resulting in a deformed sampling window as it can be seen in Fig. 6.

The following discussion will refer to a set of points $X = \{x_i\}_{i=1}^N$. The objects located at these points are either galaxies, or galaxy clusters, and also super–clusters (clusters of galaxy clusters). Galaxies are well defined objects in space, with an extent of typically $0.03h^{-1}$Mpc. Similarly, galaxy clusters are well defined objects, clearly visible in the two–dimensional distribution of galaxies, with a typical extent of $1\text{-}3h^{-1}$Mpc. Whether the combination of galaxy clusters to super–clusters is a reasonable concept is still some matter of debate [55].

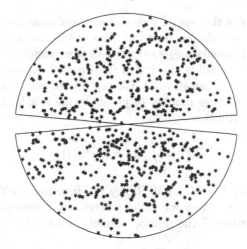

Fig. 6. The spatial distribution of IRAS galaxies in a volume limited sample with a depth of $100h^{-1}$Mpc, projected along one coordinate axis. This volume limited sample is formed by the galaxies shown in the upper left corner of the plot with luminosity against radial distance (Fig. 5).

3   Statistics of Large Scale Structure

New observations of our Universe will give us an increasingly precise mapping of the galaxy distribution around us ([39,66]). But we will have only *one* realization. This makes a statistical analysis problematic, especially model assumptions like stationarity (homogeneity) and isotropy may be tested locally only. For an interesting discussion of such problems see [69]. Still, global methods like the Minkowski functionals give us information on the shape and topology of this point set.

A pragmatic interpretation is that with a statistical analysis of a galaxy catalogue, one wants to constrain parameters of the cosmological models. These models incorporate some randomness, quantifying our ignorance of the initial conditions, or our limited understanding of the exact physical processes leading to the formation of galaxies.

3.1   Two–point Statistics

Second–order statistics, also called two–point statistics, are still among the major tools to characterize the spatial distribution of galaxies. With the mean number density, or intensity, denoted by ρ, the product density

$$\rho_2(\mathbf{x}_1, \mathbf{x}_2)dV(\mathbf{x}_1)dV(\mathbf{x}_2) = \rho^2 g(r)\ dV(\mathbf{x}_1)dV(\mathbf{x}_2) \tag{8}$$

describes the probability to find a point in the volume element $dV(\mathbf{x}_1)$ and another point in $dV(\mathbf{x}_2)$, at the distance $r = |\mathbf{x}_1 - \mathbf{x}_2|$; $|\cdot|$ is the Euclidean norm (we assume stationarity and isotropy). The product density $\rho_2(\mathbf{x}_1, \mathbf{x}_2)$ is the Lebesgue density of the second factorial moment measure (e.g. [109]). Often the (full) two–point correlation function, also called pair correlation function, $g(r)$ and the normed cumulant $\xi_2(r) = g(r) - 1$ are considered. Throughout the cosmological literature $\xi_2(r)$ is also called (two–point) correlation function [86]. For a Poisson process one has $g(r) = 1$. Closely related is the correlation integral $C(r)$ (e.g. [38]), the average number of points inside a ball of radius r centred on a point of the distribution

$$C(r) = \int_0^r ds \, \rho \, 4\pi s^2 g(s), \tag{9}$$

which is related by $K(r) = C(r)/\rho$ to Ripley's K function, see Stoyans's paper in this volume. Another common way to characterize the second–order properties is the excess fluctuation of the number density inside of C with respect to a Poisson process:

$$\sigma^2(C) = \frac{1}{|C|^2} \int_C dx \int_C dy \, \xi_2(|x - y|). \tag{10}$$

Often the power spectrum $P(k)$ is used to quantify the second order statistical properties of the point distribution [86]. $P(k)$ may be defined as the Fourier transform of $\xi_2(r) = g(r) - 1$:

$$P(k) = \frac{1}{(2\pi)^3} \int dx \, e^{-ik\cdot x} \xi_2(|x|), \tag{11}$$

with $k = |k|$.

Observed Two–Point Correlations The first analysis of a galaxy cata-logue using the two–point correlation function was presented by Totsuji and Kihara [121]. Following the work of Peebles [86], today the two–point correla-tion function has become *the* standard tool, applied to nearly every cosmological dataset. The need for boundary corrected estimators was recognized early. Sev-eral estimators have been introduced, with differing claims on their applicability ([43,56,65,89,111]). A clarification for cosmological applications is attempted in [62].

Figure 7 shows the (full) correlation function $g(r)$ and the normed cumulant $\xi_2(r)$ determined from a volume limited sample of the Southern Sky Redshift Survey 2 (SSRS2; [26]) with 1179 galaxies. The strong clustering of galaxies, due to their gravitational interaction, is shown by large values of $g(r)$ and $\xi_2(r)$ for small r.

Of special physical interest is, whether the two–point correlations are scale–invariant. A scale–invariant $g(r) \propto r^{D-3}$ is an indication for a fractal distribution of the galaxies ([67,113]). A scale–invariant $\xi_2(r) \propto r^{-\gamma}$ is expected in critical phenomena (see [36,37]).

Now lets look at the log–log plot in Fig. 7. Willmer et al. [128] give a scale–invariant fit of $\xi_2(r) \propto r^{-\gamma}$ with a scaling exponent $\gamma = 1.81$ in the range of

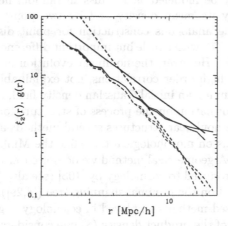

Fig. 7. Estimated two–point correlation function $g(r)$ (solid) and the normed cumulant $\xi_2(r) = g(r) - 1$ (dashed) in a double logarithmic plot for the volume limited sample from the SSRS2 with $100h^{-1}$Mpc depth. The results of the minus (reduced–sample) estimator and the Fiksel [34] estimator are shown, illustrating that only on large scales differences occur. The straight lines correspond to $g(r) \propto r^{-1}$ (solid) and $\xi_2(r) \propto r^{-1.81}$ (dashed).

3-12h^{-1}Mpc for the volume limited sample with $100h^{-1}$Mpc. However on smaller scales the slope of ξ_2 is flattening, suggesting that a scale–invariant function $\propto r^{-\gamma}$ gives only a poor description of the observed $\xi_2(r)$ in this SSRS2–sample. If we look at the correlation function $g(r)$ in Fig. 7, the observed data may be approximated by $g(r) \propto r^{3-D}$ with $D = 2$ over the larger range from 0.5-20h^{-1}Mpc. However the scale–invariance of $g(r)$ is observed over less than 2 decades only, and therefore an estimate of a fractal dimension D from the scaling exponent of $g(r)$ may be misleading ([56,71,72,108]). On large scales the observed $g(r)$ also deviates from a purely scale invariant model, and shows a tendency towards unity. This however depends on the estimator chosen. In this specific sample, a scale–invariant $g(r)$ seems to be suitable, but this is not so clear from other data sets. Also the result on small scales might be unreliable due to the small number of pairs with a short separation. For a comprehensive analysis of the SSRS2 catalogue focusing on two–point properties and scaling see [22].

Hence, currently we cannot exclude a scale–invariant $g(r)$, a scale–invariant $\xi_2(r)$, or no scale–invariance at all, with the limited observational range provided by the available three–dimensional catalogues. Hopefully this controversial issue will be clarified in the near future by the advent of deeper galaxy catalogues ([39,66]).

3.2 Higher Moments

The two–point correlation function plays an important role in cosmology, since the inflationary paradigm suggests that the initial deviations from the homogeneous density field may be modeled as a Gaussian random field, stochastically completely specified by its mean density and its two–point correlation function (see e.g. [16]). The analogous construction for point distributions is the Gauss–Poisson process [77], with subtle but important differences from the Gaussian random field model. However, the nonlinear evolution of the mass density given by (3) generates high order correlations, not explainable within a Gaussian model. Hence, assuming an initial Gaussian density field, these higher order correlations give us information on the process of structure formation.

To investigate these nonlinear structures several methods are used. In Sects. 3.3 and 3.4 we will focus on morphological tools like the Minkowski functionals and on the J function. A geometrical method we do not cover in this text is percolation analysis as introduced to cosmology by [105] (see also [91]). Yet another one is the analysis based on the minimal spanning tree ([8,29]). A description of the direct, moment based methods employed in cosmology is given now [86]:

As a generalization of the product density (8) one considers n–th order product densities

$$\rho_n(\mathbf{x}_1, \ldots, \mathbf{x}_n) \, dV_1 \ldots dV_n. \tag{12}$$

giving the probability of finding n points in the volume elements dV_1 to dV_n, respectively. Again ρ_n is the Lebesgue densities of the n–th factorial moment measures [109]. In physical applications the (normalized) cumulants are often considered. As an example we look at the three–point correlations:

$$\rho_3(\mathbf{x}_1, \mathbf{x}_2, \mathbf{x}_3) =$$
$$\rho^3 \Big(1 + \xi_2(|\mathbf{x}_1 - \mathbf{x}_2|) + \xi_2(|\mathbf{x}_2 - \mathbf{x}_3|) + \xi_2(|\mathbf{x}_1 - \mathbf{x}_3|) + \xi_3(\mathbf{x}_1, \mathbf{x}_2, \mathbf{x}_3) \Big). \tag{13}$$

The three–point correlation function , i.e. the cumulant ξ_3, describes the correlation of three points in addition to their correlations determined from the pairs. For a Poisson process all ξ_n with $n \geq 2$ equal zero. A general definition of the n–point correlation functions ξ_n is possible using generating functions (e.g. [15,27]). Although the interpretation is straightforward, the application is problematic, because a large number of triples etc. are needed to get a stable estimate. Therefore, one looks for ξ_n, $n = 3, 4, \ldots$ mainly in angular, two–dimensional, surveys (e.g. [117]); for a recent three dimensional analysis see [50].

More stable estimates of n–point properties, but with reduced informational content, may be obtained using counts–in–cells [86]. For a test volume C, typically chosen as a sphere, we are interested in the probability $P_N(C)$ of finding exactly N points in C. These $P_N(C)$ determine the one–dimensional (marginal) distributions considered in spatial statistics [109]. For a Poisson process we have

$$P_N(C) = \frac{(\rho|C|)^N}{N!} \exp(-\rho|C|), \tag{14}$$

with the volume $|C|$ of the set C. Of special interest is the "void probability" $P_0(C)$, which serves as a generating functional for all the $P_N(C)$, and relates the $P_N(C)$ with the n–point correlation functions discussed above (see [27, 112,

127] and [6]). For a sphere B_r we have $P_0(B_r) = 1 - F(r) = 1 - H_s(r)$, with the spherical contact distribution $F(r)$, also denoted by $H_s(r)$ (see Sect. 3.4).

To facilitate the interpretation of the counts–in–cells one considers their n–th moments:

$$\sum_{N=0}^{\infty} N^n P_N(C). \tag{15}$$

They can be expressed by the n–th moment measures μ_n (for their definition see e.g. [109]):

$$\mu_n(C, \dots, C) = \sum_{N=0}^{\infty} N^n P_N(C). \tag{16}$$

Especially the centered moments can be related easily to the n–point correlation functions. As an example consider the third centered moment with $\overline{N} = \rho|C|$ (e.g.[24]):

$$\sum_{N=0}^{\infty} (N - \overline{N})^3 P_N(C) = \overline{N} + 3\overline{N}^2 \sigma^2(C) + \rho^3 \int_C dx \int_C dy \int_C dz \, \xi_3(x, y, z) \tag{17}$$

where $|C|$ is the volume of C, and $\sigma^2(C)$ given by (10). This centered moment incorporates information from the two–point and three–point correlations integrated over the domain C. One may go one step further. The factorial moments

$$\sum_{N=0}^{\infty} N(N - 1) \cdots (N - n + 1) P_N(C). \tag{18}$$

attracted more attention recently, since they may be estimated easier with a small variance ([115,118]), and offer a concise way to correct for typical observational problems ([25,116]). The factorial moments may be expressed by the n–th factorial moment measures α_n [109] or the n–th order product densities:

$$\sum_{N=0}^{\infty} N(N - 1) \cdots (N - n + 1) P_N(C) = \alpha_n(C, \dots, C)$$

$$= \int_C dx_1 \dots \int_C dx_n \, \rho_n(x_1, \dots, x_n), \tag{19}$$

yielding a simple relation with the integrated n–point correlation functions by (13) and its generalizations for higher n.

The moments and the factorial moments are well defined quantities for a stationary point process. Especially the relation of the (factorial) moments to the n–point correlation functions in (17) and (19) is valid for any stationary point process. It is worth to note that this does not depend on Poisson sampling from a density field (4). A lot of work is devoted to relate the properties of the counts in cells with the dynamics of the underlying matter field (see e.g. [13,17,51,83]). However, this relation is depending on the galaxy identification scheme. Typically the Poisson model is assumed (4).

3.3 Minkowski Functionals

Minkowski functionals, also called Quermaß integrals are well known in stochastic and integral geometry (see e.g. [41,63,104,124]). Quantities like volume, surface area, and sometimes also integrated mean curvature and Euler characteristic were used to describe physical processes and to construct models. Such models and significant extensions of them were put into the context of integral geometry just recently ([73,75]), see also the article by K. Mecke in this volume. The first cosmological application of all Minkowski functionals is due to [74], marking the advent of Minkowski functionals as analysis tools for point processes. In the following years Minkowski functionals became more and more common in cosmology. The interested reader may consider the articles [12,28,47,60,61,81,87,92,95-102,129]. In the next section a short introduction to Minkowski functionals will be given. See also the articles by K. Mecke and W. Weil in this volume.

A Short Introduction Usually we are dealing with d–dimensional Euclidean space $\mathbb{R}^d$ with the group of transformations G containing as subgroups rotations and translations. One can then consider the set of convex bodies embedded in this space and, as an extension, the so called convex ring $\mathcal{R}$ of all finite unions of convex bodies. In order to characterize a body B from the convex ring, also called a poly-convex body, one looks for scalar functionals M that satisfy the following requirements:

- *Motion Invariance:* The functional should be independent of the body's position and orientation in space,

$$M(gB) = M(B) \text{ for any } g \in G, \text{ and } B \in \mathcal{R}. \tag{20}$$

- *Additivity:* Uniting two bodies, one has to add their functionals and subtract the functional of the intersection,

$$M(B_1 \cup B_2) = M(B_1) + M(B_2) - M(B_1 \cap B_2) \text{ for any } B_1, \text{ and } B_2 \in \mathcal{R}. \tag{21}$$

- *Conditional (or convex) continuity :* The functionals of convex approximations to a convex body converge to the functionals of the body,

$$M(K_i) \to M(K) \text{ as } K_i \to K \text{ for } K, K_i \in \mathcal{K}. \tag{22}$$

This applies to convex bodies only, *not* to the whole convex ring. The convergence for bodies is with respect to the Hausdorff–metric.

One might think that these fairly general requirements leave a vast choice of such functionals. Surprisingly, a theorem by Hadwiger states that in fact there are only $d+1$ independent such functionals in $\mathbb{R}^d$. To be more precise:
Hadwiger's theorem [41]: There exist $d+1$ functionals M_μ on the convex ring $\mathcal{R}$ such that any functional M on $\mathcal{R}$ that is motion invariant, additive and conditionally continuous can be expressed as a linear combination of them:

$$M = \sum_{\mu=0}^{d} c_\mu M_\mu, \text{ with numbers } c_\mu. \tag{23}$$

In this sense the $d + 1$ Minkowski functionals give a complete and up to a constant factor unique characterization of a poly-convex body $B \in \mathcal{R}$. The four most common normalizations are M_μ, V_μ, W_μ, and the intrinsic volumes $\overline{V}_\mu$ defined as follows (ω_μ is the volume of the μ–dimensional unit ball):

$$V_\mu = \frac{\omega_{d-\mu}}{\omega_d} M_\mu, \quad \overline{V}_{d-\mu} = \frac{\omega_{d-\mu}}{\omega_d} \binom{d}{\mu} M_\mu,$$

$$W_\mu = \frac{\omega_\mu \omega_d}{\omega_{d-\mu}} M_\mu, \quad \text{with} \quad \omega_\mu = \frac{\pi^{\mu/2}}{\Gamma(1 + d/2)}.$$

Table 1. The most common notations for Minkowski functionals in three–dimensional space expressed in terms of the corresponding geometric quantities.

geometric quantity	μ	M_μ	V_μ	W_μ	$\overline{V}_{3-\mu}$	ω_μ
V volume	0	V	V	V	V	1
A surface	1	$A/8$	$A/6$	$A/3$	$A/2$	2
H int. mean curvature	2	$H/2\pi^2$	$H/3\pi$	$H/3$	H/π	π
χ Euler characteristic	3	$3\chi/4\pi$	χ	$4\pi\chi/3$	χ	$4\pi/3$

In three–dimensional Euclidean space, these functionals have a direct geometric interpretation as listed in Table 1.

The Germ–Grain Model Now the Minkowski functionals are used to describe the geometry and topology of a point set $X = \{\mathbf{x}_i\}_{i=1}^N$. Direct application gives rather boring results, $V_\mu(X) = 0$ for $\mu = 0, 1, 2$ and $V_3(X) = N$. However, one may think of X as a skeleton of more complicated spatial structures in the universe (see e.g. Fig. 3). Decorating X with balls of radius r puts "flesh" on the skeleton in a well defined way. Also non–spherical grains may be used.

The Minkowski functionals for the union set of these balls $A_r = \bigcup_{i=1}^N B_r(\mathbf{x}_i)$ give non–trivial results, depending on the point distribution considered. We will use r as a diagnostic parameter specifying a neighborhood relations, to explore the connectivity and shape of A_r.

Let X be a finite subset of a realization of a Poisson process inside some finite domain W. Then A_r is a part of a realization of the *Boolean grain model*, illustrated in Fig. 8. For these randomly placed balls the mean volume densities m_μ of the Minkowski functionals are known (e.g. [75,104], also called intensities of Minkowski functionals).

$$m_0(A_r) = 1 - e^{-\rho M_0}, \quad m_2(A_r) = e^{-\rho M_0} (M_2\rho - M_1^2\rho^2),$$

$$m_1(A_r) = e^{-\rho M_0} M_1\rho, \quad m_3(A_r) = e^{-\rho M_0} (M_3\rho - 3M_1 M_2\rho^2 + M_1^3\rho^3),$$

(24)

Fig. 8. Randomly distributed points decorated with balls of varying radius r - a realization of the Booelean grain model.

with the number density ρ and

$$M_0 = \frac{4\pi}{3}r^3, \quad M_1 = \frac{\pi}{2}r^2, \quad M_2 = \frac{4}{\pi}r, \quad M_3 = \frac{3}{4\pi}. \tag{25}$$

Starting from a general point process, decorating it with spheres, we arrive at the *germ–grain model* (see also [109]). The Minkowski functionals or their volume densities calculated for the set A_r may be use as tools to describe the underlying point distribution, directly comparable to standard point process statistics like the two–point correlation function (Sect. 3.1) or the nearest neighbor distribution (Sect. 3.4). Indeed, the volume density $m_0(A_r)$ equals the spherical contact distribution or equivalently the void probability minus one: $m_0(A_r) = F(r) = H_s(r) = P_0(B_r) - 1$ (see also Sect. 3.2). Expressions relating the Minkowski functionals of such a set A_r, with the n–point correlation functions of the underlying point-process may be found in [73,74] and [101] and the contribution by K. Mecke in this volume.

Already for moderate radii r nearly the whole space is filled with up by A_r, leading to $m_0(A_r) \approx 1$ and $m_\mu(A_r) \approx 0$, with $\mu > 0$. This illustrates the different role the radius r plays for the Minkowski functionals compared to the distance r as used in the two–point correlation function $g(r)$. Already for a fixed radius, the Minkowski functionals of A_r are sensitive to the global geometry and topology of A_r and, hence, of the decorated point set (see also Sect. 3.3.5). Indeed point sets with an identical two–point correlation function, but with clearly different large scale morphology may be generated easily (see e.g. [5,114]).

All galaxy catalogues are spatially limited. To estimate the volume densities of Minkowski functionals for such a realization of the germ–grain model given by the coordinates of galaxies, we use boundary corrections based on principal kinematical formula (see [75,102,109]):

$$m_\mu(A_r) = \frac{M_\mu(A_r \cap W)}{M_0(W)} - \sum_{\nu=0}^{\mu-1} \binom{\mu}{\nu} m_\nu(A_r) \frac{M_{\mu-\nu}(W)}{M_0(W)}, \tag{26}$$

We use the convention $\sum_{n=i}^{j} x_n = 0$ for $j < i$. An example illustrating these boundary corrections is given in [58].

In the following an application of these methods to a catalogue of galaxy clusters [61] (an earlier analysis of a smaller cluster catalogue was already given by [74]) and to a galaxy catalogue will illustrate the qualitative and quantitative results obtainable with global Minkowski functionals.

Cluster Catalogues The spatial distribution of centers of galaxy clusters, using the Abell/ACO cluster sample of [88], was analyzed with Minkowski functionals applied to the germ–grain model [61]. At first a qualitative discussion of the observed features is presented, followed by a comparison with models for the cluster distribution.

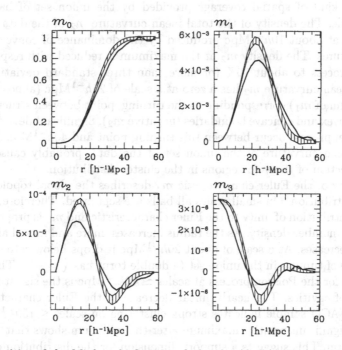

Fig. 9. Densities of the Minkowski functionals for the Abell/ACO (solid line) and a Poisson process (shaded area) with the same number density. The shaded area gives the statistical variance of the Poisson process calculated from 100 different realizations.

The most prominent feature of the volume densities of all four Minkowski functionals are the broader extrema for the Abell/ACO data as compared to the results for the Poisson process (see Fig. 9). This is a first indication for enhanced clustering. Let us now look at each functional in detail:

The density of the Minkowski functional m_0 measures the density of the covered volume. On scales between $25h^{-1}$Mpc and $40h^{-1}$Mpc, m_0 as a function of r

lies slightly below the Poisson data. The volume density is lower because of the clumping of clusters on those scales.

The density of the Minkowski functional m_1 measures the surface density of the coverage. It has a maximum at about $20h^{-1}$Mpc both for the Poisson process and for the cluster data. This maximum is due to the granular structure of the union set on the relevant scales. At the same scales, we find the maximum deviation from the characteristics for the Poisson process. The lower values of m_1 for the cluster data with respect to the Poisson are again an indication of a significant clumping of clusters at these scales. The functional m_1 shows also a positive deviation from the Poisson on scales of $(35\ldots50)h^{-1}$Mpc where more coherent structures form in the union set than in the Poisson process, keeping the surface density larger.

The densities of the Minkowski functionals m_2 and m_3 characterize in more detail the kind of spatial coverage provided by the union set of balls in the data sample. The density of the total mean curvature m_2 of the data reaches a maximum at about $10h^{-1}$Mpc produced by the dominance of convex (positive m_2) structures. The density m_2 at the maximum is reduced with respect to the Poisson process to about 70% (or more than three standard deviations). The integral mean curvature m_2 has a zero at a scale of $25h^{-1}$Mpc (almost the scale of maximum of m_1) corresponding to the turning–point between structures with mainly convex and concave boundaries (negative m_2). Significant deviations from the Poisson process occur between this turning point and $40h^{-1}$Mpc due to the smaller mean curvature of the union set of the data, probably caused by the interconnection of the void regions in the cluster distribution.

The density of the Euler characteristic m_3 describes the global topology of the cluster distribution. On small scales all balls are separated. Therefore, each ball gives a contribution of unity to the Euler characteristic and m_3 is proportional to the cluster number density. As the radius increases, more and more balls overlap and m_3 decreases. At a scale of about $20h^{-1}$Mpc it drops below zero due to the emergence of tunnels in the union set (a double torus has $\chi = -1$). The positive maximum for the Poisson process at scales $\simeq 40h^{-1}$Mpc is the signature for the presence of cavities. The nearly linear decrease of the Euler characteristic for the Abell/ACO sample indicates strong clustering on scales $\leq 15h^{-1}$Mpc. The lack of a significant positive maximum after the minimum shows that only a few cavities form. This suggests a support dimension for the distribution of clusters of less than three. The presence of voids on scales of 30 to $45h^{-1}$Mpc is shown by the enhanced surface area m_1 and the reduced integral mean curvature m_2, while on these scales the Euler characteristic m_3 is approximately zero.

The emphasis of [61] was on the comparison with cosmological model predictions. For this purpose artificial cluster distributions were constructed, from the density field of N–body simulations. Such simulations are still quite costly, and therefore only four specific models were investigated. In Fig. 10 the comparison of the observations with the Standard Cold–Dark–Matter (SCDM) model is shown. This model shows too little clustering on small scales, as it is clearly seen by the enhanced maxima of the surface area m_1 and the integral mean curvature

m_2, as well as in the flatter decrease of the Euler characteristic m_3. Additionally, the higher volume m_0 indicates weak clumping and to few coherent structures also on large scales. These deviations may be quantified using some norm for

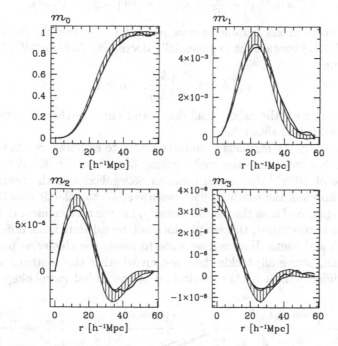

Fig. 10. Densities of the Minkowski functionals for the Abell/ACO (solid line in both panels) compared to the SCDM (shaded area in top panel). The shaded area gives 1σ-error bars of the variance among different realizations.

the comparison of the observational data with the model prediction (for details see [61]). A comparison of the clusters distribution with CDM–models using the power spectrum (11) lead to similar conclusions [90].

Large Fluctuations A physically interesting point is how well defined are the statistical properties of the galaxy or cluster distribution, determined from one spatially limited realization only. Or in other words, how large are the fluctuations of the morphology for a domain of given size? Kerscher et al. [60] investigate this using Minkowski functionals, the J function (see Sect. 3.4), and the two–point statistic σ^2 defined by (10).

By normalizing with the functional $M_\mu(B_r)$ of a single ball we can introduce normalized, dimensionless Minkowski functionals $\Phi_\mu(A_r)$,

$$\Phi_\mu(A_r) = \frac{m_\mu(A_r)}{\rho M_\mu(B_r)}, \tag{27}$$

where ρ is the number density. In the case of a Poisson process the exact mean values are known (24). For decorating spheres with radius r one obtains:

$$\Phi_0^P = (1 - e^{-\eta})\ \eta^{-1},\ \Phi_1^P = e^{-\eta},$$
$$\Phi_2^P = e^{-\eta}\ (1 - \tfrac{3\pi^2}{32}\eta),\ \Phi_3^P = e^{-\eta}\ (1 - 3\eta + \tfrac{3\pi^2}{32}\eta^2),\tag{28}$$

with the dimensionless parameter $\eta = \rho M_0(B_r) = \rho\ 4\pi r^3/3$. For $\mu \geq 1$ the measures $\Phi_\mu(A_r)$ contain the exponentially decreasing factor $e^{-\eta(r)}$. We employ the reduction

$$\phi_\mu(A_r) = \frac{\Phi_\mu(A_r)}{\Phi_1^P(A_r)},\quad \mu \geq 1,\tag{29}$$

and thereby remove the exponential decay and enhance the visibility of differences in the displays shown below.

We now apply the methods introduced above to explore a redshift catalogue of 5313 IRAS selected galaxies with limiting flux of 1.2 Jy [35]. A volume limited sample of $100h^{-1}$Mpc depth contains 352 galaxies in the northern part, and 358 galaxies in the southern part (with respect to galactic coordinates), as shown in Fig. 6. As far as the number density, i.e. the first moment of the galaxy distribution is concerned, the sample does not reveal significant differences between north and south. However, we want to assess the clustering properties of the data and, above all, tackle the question whether the southern and northern parts differ or not. A characterization of the global morphology using the

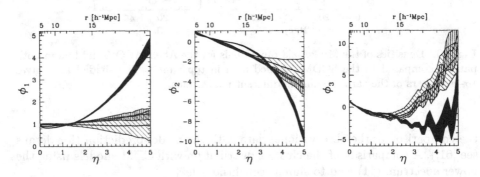

Fig. 11. Minkowski functionals ϕ_μ of a volume limited sample with $100h^{-1}$Mpc depth extracted from the IRAS 1.2 Jy catalogue; the dark shaded areas represent the southern part, the medium shaded the northern part, and the dotted a Poisson process with the same number density. The shaded areas are the 1σ errors estimated from twenty realizations for the Poisson process and from twenty errors using a Jackknife procedure with 90% sub-sampling, for the data.

Minkowski functionals (Fig. 11) shows that in both parts of the 1.2 Jy catalogue the clustering of galaxies on scales up to $10h^{-1}$Mpc is clearly stronger than in the case of a Poisson process, as inferred from the lower values of the functionals for the surface area, ϕ_1, the integral mean curvature, ϕ_2, and the Euler characteristic, ϕ_3. Moreover, the northern and southern parts differ significantly, with the northern part being less clumpy. The most conspicuous features are the enhanced surface area ϕ_1 in the southern part on scales from 12 to $20h^{-1}$Mpc and the kink in the integral mean curvature ϕ_2 at $14h^{-1}$Mpc. This behavior indicates that dense substructures in the southern part are filled up at this scale (i.e. the balls in these substructures overlap without leaving holes).

These strongly fluctuating clustering properties are also visible in the J function (Sect. 3.4), and the $\sigma^2(B_r)$ (see (10)). An analysis of possible contaminations and systematic selection effects showed that these fluctuations are real structural differences in the galaxy distribution on scales of $100h^{-1}$Mpc even extending to $200h^{-1}$Mpc (see also [59]). It is interesting to note that an N–body simulation in a periodic box with side–length of $250h^{-1}$Mpc [64] was not able to reproduce these large–scale fluctuations.

Minkowski Functionals of Excursion Sets In the preceding section the Minkowski functionals were used to characterize the union set of balls, the body A_r. Consider now a smooth density or temperature field $u(\mathbf{x})$. We wish to calculate the Minkowski functionals of an excursion set Q_ν over a given threshold ν (see Fig. 12), defined by

$$Q_\nu = \{\mathbf{x} \mid u(\mathbf{x}) \geq \nu\}. \tag{30}$$

This threshold ν will be used as a diagnostic parameter. The geometry and

Fig. 12. The black set marks the excursion set Q_ν of a Gaussian density field with increasing ν from left to right. Only the highest peaks remain for large ν.

topology of random fields $u(\mathbf{x})$ and their excursion sets was studied extensively by Adler [2]. Two complementary calculation methods for the Minkowski functionals of the excursion set Q_ν were presented by Schmalzing and Buchert [97].

Starting with a given point distribution a density field may be constructed with a folding employing some kernel $k_\epsilon(\mathbf{z})$ of width ϵ

$$u(\mathbf{y}) = \sum_{i=1}^{N} k_\epsilon(\mathbf{x}_i - \mathbf{y}). \tag{31}$$

Often a triangular or a Gaussian kernel sometimes with an adaptive smoothing scale $\epsilon(\mathbf{y})$ are used. A discussion of smoothing techniques may be found in [107].

The Euler characteristic χ of the excursion set is directly related to the genus G of the iso–density surface separating low from high density regions:

$$G(\partial Q_\nu) = 1 - 2\chi(Q_\nu). \tag{32}$$

The analysis of cosmological density field using the genus of iso–density surfaces is a well accepted tool in cosmology (see [23,76,125 and refs. therein), now incorporated in the more general analysis using Minkowski functionals. Especially the Euler characteristic of excursion sets has also applications in other fields like medical image processing [130].

Gaussianity of the Cosmic Microwave Background As already mentioned in Sect. 3.2 it is physically very interesting, whether the observed fluctuations in the temperature field of the cosmic microwave background radiation (CMB), as shown in Fig. 1, are compatible with a Gaussian random field model. For a Gaussian random field Tomita [120] obtained analytical expressions for the Minkowski functionals of Q_ν in arbitrary dimensions. Since the temperature fluctuations are given on the celestial sphere, an adopted integral geometry for spaces with constant curvature must be used ([94]). Schmalzing and Górski [100] took this geometric constraint and further complications due to boundary and binning effects, as well as noise contributions into account. They find no significant deviation from a Gaussian random field for the resolution of the COBE data set.

Other methods to test for Gaussianity are based on a wavelet analysis [47] on high–order correlation functions [45] or on the two–point correlation function of peaks in the temperature fluctuations [46].

Geometry of Single Objects – Shape–finders Looking at high thresholds ν, the excursion set is mainly composed out of separated regions (see Fig. 12). The morphology of these regions may be characterized using Minkowski functionals and the derived shape–finders [92]. Employing the following ratios of the Minkowski functionals $H_1 = V_0/(2V_1)$, $H_2 = 2V_1/(\pi V_2)$ and $H_3 = 3V_2/(4V_3)$ one may construct the dimensionless shape–finders *planarity* P and *filamentarity* F

$$P = \frac{H_2 - H_1}{H_2 + H_1} \quad \text{and} \quad F = \frac{H_3 - H_2}{H_3 + H_2}. \tag{33}$$

A simple example [98] is provided by a cylinder of radius r and height λr with the Minkowski functionals

$$V_0 = \pi r^3 \lambda, \quad V_1 = \tfrac{\pi}{3}r^2(1+\lambda), \quad V_2 = \tfrac{1}{3}r(\pi+\lambda), \quad V_3 = 1. \tag{34}$$

The shape–finders planarity P and filamentarity F for this specific example are plotted against each other in Fig. 13. Indeed this is nothing else but an inverted

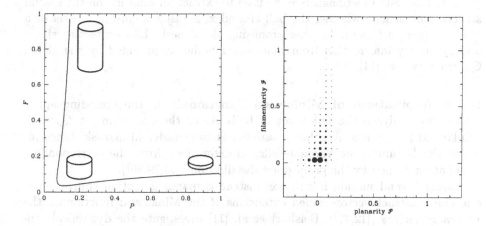

Fig. 13. On the left side a plot of the shape–finders for the cylinder with varying λ is shown, illustrating the turnover from $\lambda \approx 0$, a plane geometry ($P \approx 1$, $F \ll 1$), through a roughly spherical ($P \approx 0$, $F \approx 0$) to a mainly line like geometry ($P \ll 1$, $F \approx 1$) for $\lambda \gg 1$. On the right side a frequency histogram of the shape–finders determined from the excursion sets of an N–body simulation is shown. Larger circles correspond to more objects within the shape–finder bin (from [98]).

Blaschke diagram for the form factors ([40,103]). Following [98] the shape–finders may be written in terms of the form factors. With

$$x = \frac{\pi V_0 V_2}{4 V_1^2}, \quad y = \frac{8 V_1 V_3}{3 \pi V_2^2} \tag{35}$$

one obtains

$$P = \frac{1-x}{1+x} \quad \text{and} \quad F = \frac{1-y}{1+y}. \tag{36}$$

The isoperimetric inequalities [103] assure that $0 \leq P, F \leq 1$ for convex bodies. For a sphere one gets $P = 0 = F$.

One of the results obtained with the shape–finders applied to single objects in the excursion sets of N–body simulations [98] is given in Fig. 13. This histogram shows that the majority of the regions inside the excursion set has $P \approx 0 \approx F$, and a smaller fraction has $P \approx 0$, $F > 0$, whereas only a few of the regions have $F \approx 0$ $P > 0$. Interpreting regions with e.g. $P \approx 0$, $F > 0$ as filamentary or line–like structures is tempting but dangerous, since also non–convex regions are considered. Also, the histogram was constructed from the excursion sets of all thresholds under consideration.

It does not seem to be possible to construct shape–finders based on the global scalar Minkowski functionals facilitating a unique interpretation for non–convex sets. Abandoning the density field approach, and going back to the germ–grain model, and the Minkowski functional of a union set of balls $A_r = $

$\bigcup_{i=1}^{N} B_r(\mathbf{x}_i)$, one may assign a partial Minkowski functional to each ball. These *partial* Minkowski functionals may be used to extract information on the spatial structure elements – whether the ball around $\mathbf{x}_i$ is inside a cluster, a sheet or a filament (see [73,87,99]). Another promising *global* method for extracting shape and symmetry information from *non-convex* bodies is provided by the global Quermaß vectors [11].

Other Applications of Minkowski Functionals In the preceding applications we analyzed the union set of balls A_r or the excursion set Q_ν with Minkowski functionals. Another possibility is to consider Minkowski functionals of the Delauney– or Voronoi–cells, as determined from the corresponding tessellation defined by the given point distribution ([54,79,80]).

Going beyond motion invariance, instead demanding motion equivariance, one can construct vector–valued extensions of the Minkowski functionals, the *Quermaß vectors* ([12,42]). Beisbart et al. [11] investigate the dynamical evolution of the substructure in galaxy clusters using Quermaß vectors (see also [10]).

3.4 The J Function

Other methods to characterize the spatial distribution of points, well known in spatial statistics, are the spherical contact distribution $F(r)$ (also denoted by $H_s(r)$), i.e. the distribution function of the distances r between an arbitrary point and the nearest object in the point set X, and the nearest neighbor distance distribution $G(r)$, that is defined as the distribution function of distance r of an object in X to the nearest other object in X. $F(r)$ is related to the void probability $P_0(B_r)$ by $F(r) = 1 - P_0(B_r)$. For a Poisson distribution it is simply

$$G(r) = F(r) = 1 - \exp\left(-\rho\frac{4\pi}{3}r^3\right). \tag{37}$$

Recently, van Lieshout and Baddeley [122] suggested to use the ratio

$$J(r) = \frac{1 - G(r)}{1 - F(r)} \tag{38}$$

as a further distributional characteristic. For a Poisson distribution $J(r) = 1$ follows directly from (37). As shown by [122], a clustered point distribution implies $J(r) \leq 1$, whereas regular structures are indicated by $J(r) \geq 1$. However, Bedford and van den Berg [9] showed that $J = 1$ does not imply a Poisson process. For several point process models $J(r)$, or at least limiting values for $J(r)$, are known ([122]). The J function was considered by White [127] as the "first conditional correlation function" and used by Sharp [106] to test hierarchical models. The relation between $J(r)$ and the cumulants $\xi_n(r)$ was used by Kerscher [55]. An empirical study of the performance of the J function for several point process models is given by [119]. A refined definition of the J function "without edge correction" may be especially useful for a test on spatial randomness [4].

Clustering of Galaxies The J function may be used to characterize the distribution of galaxies or galaxy clusters and for the comparison with the results from simulations, similar to the application of the Minkowski functionals in sect. 3.3.3. This approach was pursued by [57]. The Perseus–Pisces redshift survey ([123] and refs. therein) was compared with galaxy samples constructed from a mixed dark matter simulation. The observed $J(r)$ determined from a volume limited sample with $79h^{-1}$Mpc depth differs significantly from the results of the simulations (Fig. 14). Especially on small scales the galaxy distribution shows a stronger clustering, as seen by steeper decreasing $J(r)$. We also could show that modeling the galaxy distribution with a simple Poisson cluster process is not appropriate.

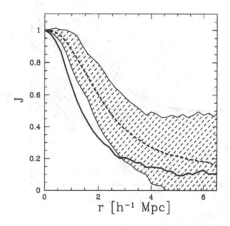

Fig. 14. $J(r)$ for the volume limited sample from Perseus–Pisces redshift survey (solid line) and the 1σ range determined from galaxy samples generated by a mixed dark matter simulation.

Regularity in the Distribution of Super–Clusters? Einasto et al. [32] report a peak in the 3D–power spectrum (the Fourier transform of ξ_2) of a catalogue of clusters on a scale of $120h^{-1}$Mpc. Broadhurst et al. [18] observed periodicity on approximately the same scale in an analysis of 1D–data from a pencil–beam redshift survey. As is well known from the theory of fluids, the regular distribution (e.g. of molecules in a hard–core fluid) reveals itself in an oscillating two–point correlation function and a peak in the structure function respectively (see e.g. [44], and the contribution of H. Löwen in this volume). In accordance with this an oscillating two–point correlation function $\xi_2(r)$ or at least a first peak was reported on approximately the same scale (e.g. [31] and [78]). The existence of regularity on large scales implies a preferred scale in the initial conditions, which would be of major physical interest.

Using the $J(r)$ function [55] investigates the super–cluster catalogue [33] constructed from an earlier version of the cluster catalogue by [3] using a friend–of–friends procedure. (The friend–of–friends procedure is called single linkage clustering in the mathematical literature). Comparing with Poisson distributed points one clearly recognizes that the super–cluster catalogue is a regular point distribution (Fig. 15). However, a similar signal for $J(r)$ may be obtained by starting with a Poisson process followed by a friend–of–friends procedure with the same linking length as used in the construction of the super–cluster catalogue. Only some indication for a regular distribution on large scales remains, showing that this super–cluster catalogue is seriously affected by the construction method.

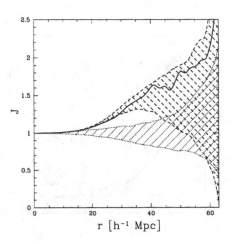

Fig. 15. $J(r)$ determined from the super–cluster sample (solid line) is shown together with the 1–σ range determined from a pure Poisson process (dotted area) and a Poisson process followed by a similar friend–of–friends procedure (dashed area) as used to construct the super–cluster catalogue.

G_n and F_n As a direct generalization of the nearest neighbor distance distribution one may consider the n–th neighbor distance distributions $G_n(r)$, the distribution of the distance r to the n–th nearest point (e.g. [110]). For a Poisson process in three dimensions we have

$$G_n(r) = 1 - \frac{\Gamma\left(n, \rho\frac{4\pi}{3}r^3\right)}{\Gamma(n)}, \tag{39}$$

shown in Fig. 16. $\Gamma(n,x) = \int_x^\infty ds\, s^{n-1} e^{-s}$ is the incomplete Gamma–function, $\Gamma(n) = \Gamma(n,0)$ the complete. Clearly $G_1(r) = G(r)$. In Fig. 16 the curves for the first five $G_n(r)$ for a Poisson process are shown, together with their densities

$p_n(r)$ defined by

$$G_n(r) = \int_0^r ds\, p_n(s). \tag{40}$$

The sum of these densities is directly related to the two–point correlation function [70]

$$g(r)\, \rho\, 4\pi r^2 = \sum_{n=1}^{\infty} p_n(r). \tag{41}$$

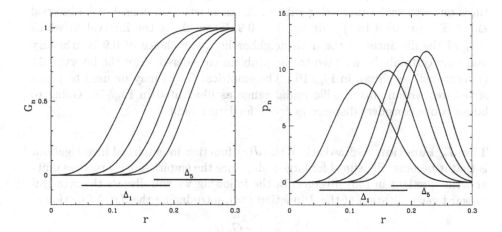

Fig. 16. In the left plot you see the $G_n(r)$ with $n = 1, \ldots, 5$ for a Poisson process with $\rho = 100$. In the right plot the corresponding densities $p_n(r)$ are shown.

The n–th spherical contact distribution $F_n(r)$ is the distribution function of the distances r between an arbitrary point and the n–th closest object in the point set X (we assume that the n–th closest point is unique). Clearly $F_1(r) = F(r)$. For stationary and isotropic point processes $F_n(r)$ is the probability to find at least n points inside a sphere B_r with radius r, and therefore

$$F_n(r) = \sum_{i=n}^{\infty} P_i(B_r) = 1 - \sum_{i=0}^{n-1} P_i(B_r), \tag{42}$$

where $P_i(B_r)$ are the counts–in–cells as discussed in Sect. 3.2.

For a Poisson process with number density ρ we obtain directly from (14)

$$F_n(r) = 1 - \exp(-\rho|B_r|) \sum_{i=0}^{n-1} \frac{(\rho|B_r|)^i}{i!}, \tag{43}$$

which is essentially the series expansion of the incomplete gamma function (see e.g. [1]). Therefore,

$$F_n(r) = 1 - \frac{\Gamma\left(n, \rho\frac{4\pi}{3}r^3\right)}{\Gamma(n)}, \tag{44}$$

and we explicitly see that for a Poisson process

$$F_n(r) = G_n(r) \tag{45}$$

This is a special case of the "Slivnyak's theorem" [109].

A very interesting feature of the $G_n(r)$ and $F_n(r)$ is their sensitivity to structures on large scales increasing with n. As an illustration consider the interval $\Delta_n \subset \mathbb{R}^+$ specified by $\int_{\Delta_n} ds \, p_n(s) = 0.9$. Then Δ_n is the interval in which 90% of the distances to the n–th neighbor lie. (The choice of 0.9 is arbitrary and may certainly be adopted to the problem considered. Also the interval Δ_n is "centered" as shown in Fig. 16.) The empirical $G_n(r)$ may be used to probe structures within this specific radial range as illustrated in Fig. 16. Going to larger n one considers distance intervals for larger radii.

The J_n Function A drawback of the $J(r)$ function in empirical investigations is that it becomes ill defined for large radii, since the empirical $F(r)$ reaches unity and the quotient in (38) diverges. In the following we will discuss the straightforward generalization of the J function (38), introducing the $J_n(r)$ functions:

$$J_n(r) = \frac{1 - G_n(r)}{1 - F_n(r)}. \tag{46}$$

From (45) we obtain directly for a Poisson process

$$J_n(r) = 1 \quad \text{for all } n. \tag{47}$$

Qualitatively we expect the same behavior of the $J_n(r)$ functions as for the $J(r)$ function, but now for a radius r in the interval Δ_n (defined at the end of Sect. 3.4.3).

- If a point distribution shows clustering on scales r in Δ_n, the $G_n(r)$ increases faster than for a Poisson process since the n–th nearest neighbor is typically closer. $F_n(r)$ increases more slowly than for a random distribution. Both effects result in a $J_n(r) \leq 1$.
- On the other hand, for a point distribution regular on the scale r in Δ_n, $G_n(r)$ increases more slowly than for a Poisson process, since the n–th neighbor is found at a finite characteristic distance. $F_n(r)$ increases stronger since the distance from a random point to the n–th closest point on the regular structure is typically smaller. This results in $J_n(r) \geq 1$.
- $J_n(r) = 1$ indicates the transition from regular to clustered structures on scales r in Δ_n.

With a simple point process model we illustrate these properties. In a Matérn cluster process a single cluster consists out of μ points in the mean, randomly distributed inside a sphere of radius R, where the number of points follows a Poisson distribution. The clusters centers (not belonging to the point process) form a Poisson process with a density of ρ/μ [109]. In Fig. 17 the strong clustering in the Matérn cluster process is visible from a decline of the $J_n(r)$. This decline becomes weaker with increasing n. For large radii r the J_n acquire a constant value. Investigating larger scales, i.e. for large n, the constant value of J_n shows a trend towards unity, i.e. we start to "see" the Poisson distribution of the clusters centers.

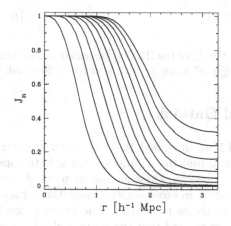

Fig. 17. The $J_n(r)$ with $n = 1, \ldots 10$ (bending up successively) for a Matérn cluster process with $\mu = 10$ and $R = 1.5h^{-1}$Mpc calculated using the reduced sample estimators.

On our Way to Large Scales A similar behavior may be identified in the galaxy distribution. We calculate the J_n functions for a volume limited sample of galaxies extracted from the IRAS 1.2 Jy catalogue with $200h^{-1}$Mpc depth using the reduced sample estimator for both F_n and G_n. For small n, i.e. small scales, the $J_n(r)$ are all smaller than unity, indicating clustering out to scales of $40h^{-1}$Mpc(see Fig. 18). For large n the J_n are consistent with no clustering, i.e. $J_n = 1$. However a trend towards a J_n larger than unity, indicating regularity is observed. Clearly, the results obtained from this sparse sample with 280 galaxies only may serve mainly as an illustration of the method – to obtain decisive results we will have to wait for deeper surveys.

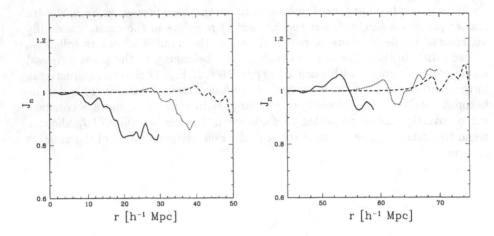

Fig. 18. In the left plot the J_n of the IRAS galaxies are shown with $n = 1, 4, 7$ (solid, dotted, dashed), in the right plot the J_n with $n = 10, 15, 20$ (solid, dotted, dashed).

4   Summary and Outlook

In Sects. 3.3.3 and 3.4.1 we discussed that advanced geometrical methods like the Minkowski functionals and the J function are able to constrain parameters of cosmological models. However, these geometric methods are not only limited to the parameter estimation in cosmological simulations, they are also valuable tools as point process statistics in general. The direct probe of galaxy surveys with geometrical methods showed that the large–scale structure exhibits strong morphological fluctuations (Sect. 3.3.4). Such fluctuations are often attributed to "cosmic variance" in an Universe homogeneous on very large scales. However the fluctuations are astonishingly large even on scales of $200h^{-1}$Mpc. A preferred scale, may be viewed as an indication for a homogeneous galaxy distribution on large scales. Especially geometric methods like the J and J_n functions may be helpful to identify a preferred scale in the galaxy distribution.

Perspectives for future research might be as follows:
Starting with the Minkowski functionals or other well founded geometrical tools, more specialized methods may be constructed to understand certain features in the galaxy distribution in detail. An example are the vector valued extension of the Minkowski functionals, the Quermaß–vector, used in the investigation of the substructure in galaxy clusters.
In empirical work, one has to determine these geometrical measures from a given point set. The construction of estimators with well understood distributional properties is crucial to be able to draw decisive conclusions from the data.
Using these geometrical methods as tools for constraining the cosmological parameters will be one way to go. Currently this is mainly performed by comparisons with N–body simulations. Clearly a more direct link between the geometry and the dynamics of matter in the Universe promoting our understanding how

structures form is desirable. Carefully constructed approximations may be the key ingredient.

Another way in trying to understand structure formation is to directly investigate the appearance of geometric features like walls, filaments, and clusters – or to identify a preferred scale showing up in a regular distribution on large scales. Such findings will guide us in the construction of approximations, which are able to reproduce such geometric features.

Acknowledgements

I would like to thank Claus Beisbart, Jörg Retzlaff, Dietrich Stoyan for comments on the manuscript and Jens Schmalzing who kindly provided the Figs. 1, 12, 13. Special thanks to Thomas Buchert and Herbert Wagner. Their constant support, the inspiring discussions, and the helpful criticism significantly influenced my understanding of physics, cosmology, and statistics; the emphasis of morphological measures was always a major concern.

References

1. Abramowitz, M.X., I.A. Stegun (1984): *Pocketbook of Mathematical Functions* (Harri Deutsch, Thun, Frankfurt/Main)
2. Adler, R.J. (1981): *The geometry of random fields* (John Wiley & Sons, Chichester)
3. Andernach, H., E. Tago (1998): 'Current status of the ACO cluster redshift compilation'. In: *Proc. of the 12th Potsdam Cosmology Workshop 1997, Large Scale Structure: Tracks and Traces*, ed. by V. Müller, S. Gottlöber, J.P. Mücket, J. Wambsgans (World Scientific, Singapore), Astro-ph/9710265
4. Baddeley, A.J., M. Kerscher, K. Schladitz, B. Scott (1999): 'Estimating the J function without edge correction', *Statist. Neerlandica*, in press
5. Baddeley, A.J., B.W. Silverman (1984): 'A cautionary example on the use of second–order methods for analyzing point patterns', *Biometrics*, **40**, pp. 1089–1093
6. Balian, R., R. Schaeffer (1989): 'Scale–invariant matter distribution in the Universe I. counts in cells', *Astron. Astrophys.*, **220**, pp. 1–29
7. Bardeen, J.M., J.R. Bond, N. Kaiser, A.S. Szalay (1986): 'The statistics of peaks of Gaussian random fields', *Ap. J.*, **304**, pp. 15–61
8. Barrow, J.D., D.H. Sonoda, S.P. Bhavsar (1985): 'Minimal spanning trees, filaments and galaxy clustering', *Mon. Not. Roy. Astron. Soc.*, **216**, pp. 17–35
9. Bedford, T., J. van den Berg (1997): 'A remark on the van Lieshout and Baddeley J–function for point processes', *Adv. Appl. Prob.*, **29**, pp. 19–25
10. Beisbart, C., T. Buchert (1998): 'Characterizing cluster morphology using vector-valued minkowski functionals'. In: *Proc. of the 12th Potsdam Cosmology Workshop 1997, Large Scale Structure: Tracks and Traces*, ed. by V. Müller, S. Gottlöber, J.P. Mücket, J. Wambsgans (World Scientific, Singapore), astro-ph/9711034
11. Beisbart, C., T. Buchert, M. Bartelmann, J.G. Colberg, H. Wagner (2000): 'Morphological evolution of galaxy clusters: First application of the quermaß vectors', in preparation
12. Beisbart, C., T. Buchert, H. Wagner (1999): 'Morphometry of galaxy clusters', submitted

13. Bernardeau, F., L. Kofman (1995): 'Properties of the cosmological density distribution function', *Ap. J.*, **443**, pp. 479–498
14. Bertschinger, E. (1992): 'Large-scale structures and motions: Linear theory and statistics'. In: *New Insights into the Universe*, ed. by V. Martinez, M. Portilla, D. Saez, number 408 in Lecture Notes in Physics (Springer Verlag, Berlin), pp. 65–126
15. Borgani, S. (1995): 'Scaling in the universe', *Physics Rep.*, **251**, pp. 1–152
16. Börner, G. (1993): *The Early Universe, Facts and Fiction*, 3rd ed. (Springer Verlag, Berlin)
17. Bouchet, F.R., R. Juszkiewicz, S. Colombi, R. Pellat (1992): 'Weakly nonlinear gravitational instability for arbitrary Omega', *Ap. J. Lett.*, **394**, pp. L5–L8
18. Broadhurst, T.J., R.S. Ellis, D.C. Koo, A.S. Szalay (1990): 'Large–scale distribution of galaxies at the galactic poles', *Nature*, **343**, pp. 726–728
19. Buchert, T. (1996): 'Lagrangian perturbation approach to the formation of large–scale structure'. In: *Proc. of the international school of physics Enrico Fermi. Course CXXXII: Dark matter in the Universe*, ed. by S. Bonometto, J. Primack, A. Provenzale (Società Italiana di Fisica, Varenna sul Lago di Como)
20. Buchert, T. (1999): 'On average properties of inhomogeneous fluids in general relativity: I. dust cosmologies', *G.R.G.* in press, gr-qc/9906015
21. Buchert, T., J. Ehlers (1997): 'Averaging inhomogeneous Newtonian cosmologies' *Astron. Astrophys.*, **320**, pp. 1–7
22. Cappi, A., C. Benoist, L. Da Costa, S. Maurogordato (1998): 'Is the universe a fractal?, results from the SSRS2', *Astron. Astrophys.*, **335**, pp. 779–788
23. Coles, P., A. Davies, R.C. Pearson (1996): 'Quantifying the topology of large–scale structure', *Mon. Not. Roy. Astron. Soc.*, **281**, pp. 1375–1384
24. Coles, P., F. Lucchin (1994): *Cosmology: The origin and evolution of cosmic structure* (John Wiley & Sons, Chichester)
25. Colombi, S., I. Szapudi, A.S. Szalay (1998): 'Effects of sampling on statistics of large scale structure', *Mon. Not. Roy. Astron. Soc.*, **296**, pp. 253–274
26. da Costa, L.N., C.N.A. Willmer, P. Pellegrini, O.L. Chaves, C. Rite, M.A.G. Maia, M.J. Geller, D.W. Latham, M.J. Kurtz, J.P. Huchra, M. Ramella, A.P. Fairall, C. Smith, S. Lipari (1998): 'The Southern Sky Redshift Survey', *A. J.*, **116**, pp. 1–7
27. Daley, D.J., D. Vere-Jones (1988): *An Introduction to the Theory of Point Processes* (Springer Verlag, Berlin)
28. Dolgov, A., A. Doroshkevich, D. Novikov, I. Novikov (1999): 'Geometry and statistics of cosmic microwave polarization', Astro-ph/9901399
29. Doroshkevich, A.G., V. Müller, J. Retzlaff, V. Turchaninov (1999): 'Superlarge–scale structure in n–body simulations', *Mon. Not. Roy. Astron. Soc.*, **306**, pp. 575–591
30. Efstathiou, G. (1996): 'Observations of large–scale structure in the Universe'. In: *LesHouches Session LX: cosmology and large scale structure, august 1993*, ed. by R. Schaeffer, J. Silk, M. Spiro, J. Zinn-Justin (Elsevier, Amsterdam), pp. 133–252
31. Einasto, J., M. Einasto, P. Frisch, S. Gottlöber, V. Müller, V. Saar, A.A. Starobinsky, E. Tago, E., D.T.H. Andernach (1997): 'The supercluster–void network. II. an oscillating cluster correlation function', *Mon. Not. Roy. Astron. Soc.*, **289**, pp. 801–812
32. Einasto, J., M. Einasto, M., S. Gottlöber, V. Müller, V. Saar, A.A. Starobinsky, E. Tago, D. Tucker, H. Andernach, P. Frisch (1997): 'A 120-Mpc periodicity in the three–dimensional distribution of galaxy superclusters', *Nature*, **385**, pp. 139–141
33. Einasto, M., E. Tago, J. Jaaniste, J. Einasto, H. Andernach, H. (1997): 'The supercluster–void network. I. the supercluster catalogue and large–scale distribution', *Astron. Astrophys. Suppl.*, **123**, pp. 119–133

34. Fiksel, T. (1988): 'Edge–corrected density estimators for point processes', *Statistics*, **19**, pp. 67–75

35. Fisher, K.B., J.P. Huchra, M.A. Strauss, M. Davis, A. Yahil, D. Schlegel (1995): 'The IRAS 1.2 Jy survey: Redshift data', *Ap. J. Suppl.*, **100**, pp. 69–103

36. Gaite, J., A. Dominguez, J. Perez-Mercader (1999): 'The fractal distribution of galaxies and the transition to homogeneity', *Ap. J.*, **522**, pp. L5–L8

37. Goldenfeld, N. (1992): *Lectures on Phase Transitions and the Renormalization Group* (Addison–Wesley, Reading, MA)

38. Grassberger, P., I. Procaccia (1984): 'Dimensions and entropies of strange attractors from fluctuating dynamics approach', *Physica D*, **13**, pp. 34–54

39. Gunn, J.E. (1995): 'The Sloan Digital Sky Survey', *Bull. American Astron. Soc.*, **186**, p. 875

40. Hadwiger, H. (1955): *Altes und Neues über konvexe Körper* (Birkhäuser, Basel)

41. Hadwiger, H. (1957): *Vorlesungen über Inhalt, Oberfläche und Isoperimetrie* (Springer Verlag, Berlin)

42. Hadwiger, H., R. Schneider (1971): 'Vektorielle Integralgeometrie', *Elemente der Mathematik*, **26**, pp. 49–72

43. Hamilton, A.J.S. (1993): 'Toward better ways to measure the galaxy correlation function', *Ap. J.*, **417**, pp. 19–35

44. Hansen, J.P., I.R. McDonnald (1986): *Theory of Simple Liquids* (Academic Press, New York and London)

45. Heavens, A.F. (1999): 'Estimating non–Gaussianity in the microwave background', *Mon. Not. Roy. Astron. Soc.*, **299**, pp. 805–808

46. Heavens, A.F., R.K. Sheth (1999): 'The correlation of peaks in the microwave background', submitted, astro-ph/9906301

47. Hobson, M.P., A.W. Jones, A.N. Lasenby (1999): 'Wavelet analysis and the detection of non–Gaussianity in the CMB', *Mon. Not. Roy. Astron. Soc.*, **309**, pp. 125–140

48. Huchra, J.P., M.J. Geller, H.G. Corwin Jr. (1995): 'The CfA redshift survey: Data for the NGP + 36 zone', *Ap. J. Suppl.*, **99**, pp. 391–403

49. Huchra, J.P., M.J. Geller, V. De Lapparent, H.G. Corwin Jr. (1990): 'The CfA redshift survey – data for the NGP + 30 zone', *Ap. J. Suppl.*, **72**, pp. 433–470

50. Jing, Y. P., G. Börner (1998): 'The three–point correlation function of galaxies', *Ap. J.*, **503**, pp. 37–47

51. Juszkiewicz, R., D.H. Weinberg, P. Amsterfamski, M. Chodorowski, F.R. Bouchet (1995): 'Weakly nonlinear Gaussian fluctuations and the Edgeworth expansion', *Ap. J.*, **442**, pp. 39–56

52. Kates, R., E. Kotok, A. Klypin (1991): 'High–resolution simulations of galaxy formation in a cold dark matter scenario', *Astron. Astrophys.*, **243**, pp. 295–308

53. Kauffmann, G., A. Nusser, M. Steinmetz (1997): 'Galaxy formation and large-scale bias', *Mon. Not. Roy. Astron. Soc.*, **286**, pp. 795–811

54. Kerscher, M. (1998a): *Morphologie großräumiger Strukturen im Universum* (GCA–Verlag, Herdecke)

55. Kerscher, M. (1998b): 'Regularity in the distribution of superclusters?' *Astron. Astrophys.*, **336**, pp. 29–34

56. Kerscher, M. (1999): 'The geometry of second–order statistics – biases in common estimators', *Astron. Astrophys.*, **343**, pp. 333–347

57. Kerscher, M., M.J. Pons-Bordería, J. Schmalzing, R. Trasarti-Battistoni, V.J. Martínez, T. Buchert, R. Valdarnini (1999): 'A global descriptor of spatial pattern interaction in the galaxy distribution', *Ap. J.*, **513**, pp. 543–548

58. Kerscher, M., J. Schmalzing, T. Buchert (1996): 'Analyzing galaxy catalogues with Minkowski functionals'. In: *Mapping, measuring and modelling the Universe*, ed. by P. Coles, V. Martínez, M.J. Pons Bordería (Astronomical Society of the Pacific, Valencia), pp. 247–252

59. Kerscher, M., J. Schmalzing, J., T. Buchert, H. Wagner (1996): ' The significance of the fluctuations in the IRAS 1.2 Jy galaxy catalogue'. In: *Proc. 2nd SFB workshop on Astro–particle physics Ringberg 1996, Report SFB375/P002*, ed. by R. Bender, T. Buchert, P. Schneider (Ringberg, Tegernsee), pp. 83–98

60. Kerscher, M., J. Schmalzing, T. Buchert, H. Wagner (1998): 'Fluctuations in the 1.2 Jy galaxy catalogue', *Astron. Astrophys.*, **333**, pp. 1–12

61. Kerscher, M., J. Schmalzing, J. Retzlaff, S. Borgani, T. Buchert, S. Gottlöber, V. Müller, M. Plionis, H. Wagner, H. (1997): 'Minkowski functionals of Abell/ACO clusters', *Mon. Not. Roy. Astron. Soc.*, **284**, pp. 73–84

62. Kerscher, M., I. Szapudi, A. Szalay (1999): 'A comparison of estimators for the two-point correlation function: dispelling the myths', submitted

63. Klain, D.A., C.C. Rota (1997): *Introduction to Geometric Probability* (Cambridge University Press, Cambridge)

64. Kolatt, T., A. Dekel, G. Ganon, J.A. Willick (1996): 'Simulating our cosmological neighborhood: Mock catalogs for velocity analysis', *Ap. J.*, **458**, pp. 419–434

65. Landy, S.D., A.S. Szalay (1993): 'Bias and variance of angular correlation functions', *Ap. J.*, **412**, pp. 64–71

66. Maddox, S. (1998): 'The 2df galaxy redshift survey: Preliminary results'. In: *Proc. of the 12th Potsdam Cosmology Workshop 1997, Large Scale Structure: Tracks and Traces*, ed. by V. Müller, S. Gottlöber, J.P. Mücket, J. Wambsgans (World Scientific, Singapore), Astro-ph/9711015

67. Mandelbrot, B. (1982): *The Fractal Geometry of Nature* (Freeman, San Francisco)

68. Martínez, V.J. (1996): 'Measures of galaxy clustering'. In: *Proc. of the international school of physics Enrico Fermi. Course CXXXII: Dark matter in the Universe*, ed. by S. Bonometto, J. Primack, A. Provenzale (Società Italiana di Fisica, Varenna sul Lago di Como)

69. Matheron, G. (1989): *Estimating and Choosing: An Essay on Probability in Practice* (Springer Verlag, Berlin)

70. Mazur, S. (1992): 'Neighborship partition of the radial distribution function for simple liquids', *J. Chem. Phys.*, **97**, pp. 9276–9282

71. McCauley, J.L. (1997): 'Are galaxy distributions scale invariant? a perspective from dynamical systems theory', submitted, astro-ph/9703046

72. McCauley, J.L. (1998): 'The galaxy distributions: Homogeneous, fractal, or neither?' *Fractals*, **6**, pp. 109–119

73. Mecke, K.R. (1994): *Integralgeometrie in der Statistischen Physik: Perkolation, komplexe Flüssigkeiten und die Struktur des Universums* (Harri Deutsch, Thun, Frankfurt/Main)

74. Mecke, K.R., T. Buchert, H. Wagner (1994): 'Robust morphological measures for large–scale structure in the Universe', *Astron. Astrophys.*, **288**, pp. 697–704

75. Mecke, K.R., H. Wagner (1991): 'Euler characteristic and related measures for random geometric sets', *J. Stat. Phys.*, **64**, pp. 843–850

76. Melott, A.L. (1990): 'The topology of large–scale structure in the Universe', *Physics Rep.*, **193**, pp. 1–39

77. Milne, R.K., M. Westcott (1972): 'Further results for Gauss–poisson processes', *Adv. Appl. Prob.*, **4**, pp. 151–176

78. Mo, H.J., Z.G. Deng, X.Y. Xia, P. Schiller, G. Börner, G. (1992): 'Typical scales in the distribution of galaxies and clusters of galaxies from unnormalized pair counts', *Astron. Astrophys.*, **257**, pp. 1–10

79. Muche, L. (1996): 'Distributional properties of the three–dimensional Poisson Delauney cell', *J. Stat. Phys.*, **84**, pp. 147–167

80. Muche, L. (1997): 'Fragmenting the universe and the Voronoi tessellation', preprint, Freiberg

81. Novikov, D., H.A. Feldman, S.F. Shandarin (1999): 'Minkowski functionals and cluster analysis for CMB maps', *Int. J. Mod. Phys.*, **D8**, pp. 291–306

82. Padmanabhan, T. (1993): *Structure formation in the Universe* (Cambridge University Press, Cambridge)

83. Padmanabhan, T., K. Subramanian (1993): 'Zel'dovich approximation and the probability distribution for the smoothed density field in the nonlinear regime', *Ap. J.*, **410**, pp. 482–487

84. Peacock, J.A. (1992): 'Satistics of cosmological density fields', In: *New Insights into the Universe*, ed. by V. Martinez, M. Portilla, D. Saez, number 408 in Lecture Notes in Physics, (Springer Verlag, Berlin), pp. 65–126

85. Peebles, P. (1973): 'Satistical analysis of catalogs of extragalactic objects. I. theory', *Ap. J.*, **185**, pp. 413–440

86. Peebles, P.J.E. (1980): *The Large Scale Structure of the Universe* (Princeton University Press, Princeton, New Jersey)

87. Platzöder, M., T. Buchert (1995): 'Applications of Minkowski functionals for the statistical analysis of dark matter models'. In: *Proc. of "1st SFB workshop on Astroparticle physics"*, *Ringberg, Tegernsee*, ed. by A. Weiss, G. Raffelt, W. Hillebrandt, F. von Feilitzsch, pp. 251–263, astro-ph/9509014

88. Plionis, M., R. Valdarnini (1991): 'Evidence for large–scale structure on scales about 300/h Mpc', *Mon. Not. Roy. Astron. Soc.*, **249**, pp. 46–62

89. Pons–Bordería, M.-J., V.J. Martínez, D. Stoyan, H. Stoyan, E. Saar (1999): 'Comparing estimators of the galaxy correlation function', *Ap. J.*, **523**, pp. 480–491

90. Retzlaff, J., S. Borgani, S. Gottlöber, A. Klypin, V. Müller (1998): 'Constraining cosmological models with cluster power spectra', *New Astronomy*, **3**, pp. 631–646

91. Sahni, V., B.S. Sathyaprakash, S.F. Shandarin (1997): 'Probing large-scale structure using percolation and genus curves', *Ap. J.*, **476**, pp. L1–L5

92. Sahni, V., B.S. Sathyaprakash, S.F. Shandarin (1998): 'Shapefinders: A new shape diagnostic for large–scale structure', *Ap. J.*, **495**, pp. L5–L8

93. Sandage, A. (1995): 'Practical cosmology: Inventing the past' In: *The Deep Universe*, ed. by A. Sandage, R. Kron, M. Longair, Saas–Fee Advanced Course Lecture Notes 1993 (Swiss society for Astrophysics and Astronomy, Springer Verlag, Berlin), pp. 1–232

94. Santaló, L. A. (1976): *Integral Geometry and Geometric Probability* (Addison–Wesley, Reading, MA)

95. Sathyaprakash, B.S., V. Sahni, S.F. Shandarin (1998): 'Morphology of clusters and superclusters in N-body simulations of cosmological gravitational clustering', *Ap. J.*, **508**, pp. 551–569

96. Sathyaprakash, B.S., V. Sahni, S.F. Shandarin, K.B. Fisher, K. B. (1998): 'Filaments and pancakes in the IRAS 1.2 Jy redshift survey', *Ap. J.*, **507**, pp. L109–L112

97. Schmalzing, J., T. Buchert (1997): 'Beyond genus statistics: A unifying approach to the morphology of cosmic structure', *Ap. J. Lett.*, **482**, pp. L1–L4

98. Schmalzing, J., T. Buchert, A. Melott, V. Sahni, B. Sathyaprakash, S. Shandarin (1999): 'Disentangling the cosmic web I: Morphology of isodensity contours', in press, astro-ph/9904384

99. Schmalzing, J., A. Diaferio (1999): 'Topology and geometry of the CfA2 redshift survey', in press, astro-ph/9910228
100. Schmalzing, J., K.M. Górski (1998): 'Minkowski functionals used in the morphological analysis of cosmic microwave background anisotropy maps', Mon. Not. Roy. Astron. Soc., **297**, pp. 355–365
101. Schmalzing, J., S. Gottlöber, A. Kravtsov, A. Klypin (1999): 'Quantifying the evolution of higher order clustering', Mon. Not. Roy. Astron. Soc., **309**, pp. 1007–1016
102. Schmalzing, J., M. Kerscher, T. Buchert (1996): 'Minkowski functionals in cosmology'. In: Proc. of the international school of physics Enrico Fermi. Course CXXXII: Dark matter in the Universe, ed. by S. Bonometto, J. Primack, A. Provenzale (Società Italiana di Fisica, Varenna sul Lago di Como), pp. 281–291
103. Schneider, R. (1993): Convex bodies: the Brunn–Minkowski theory (Cambridge University Press, Cambridge)
104. Schneider, R., W. Weil (1992): Integralgeometrie (Bernd G. Teubner, Leipzig, Berlin)
105. Shandarin, S.F. (1983): 'Percolation theory and the cell–lattice structure of the Universe', Sov. Astron. Lett., **9**, pp. 104–106
106. Sharp, N. (1981): 'Holes in the zwicky catalogue', Mon. Not. Roy. Astron. Soc., **195**, pp. 857–867
107. Silverman, B.W. (1986): Density Estimation for Statistics and Data Analysis (Chapman and Hall, London)
108. Stoyan, D. (1998): 'Caution with "fractal" point–patterns', Statistics, **25**, pp. 267–270
109. Stoyan, D., W.S. Kendall, J. Mecke (1995): Stochastic Geometry and its Applications, 2nd ed., (John Wiley & Sons, Chichester)
110. Stoyan, D., H. Stoyan (1994): Fractals, Random Shapes and Point Fields (John Wiley & Sons, Chichester)
111. Stoyan, D., H. Stoyan (2000): 'Improving ratio estimators of second order point process statistics', Scand. J. Statist., in press
112. Stratonovich, R.L. (1963): Topics in the theory of random noise, volume 1 (Gordon and Breach, New York)
113. Sylos Labini, F., M. Montuori, L. Pietronero (1998): 'Scale invariance of galaxy clustering', Physics Rep., **293**, pp. 61–226
114. Szalay, A.S. (1997): 'Walls and bumps in the Universe'. In: Proc. of the 18th Texas Symposium on Relativistic Astrophysics, ed. by A. Olinto (AIP, New York)
115. Szapudi, I. (1998): 'A new method for calculating counts in cells', Ap. J., **497**, pp. 16–20
116. Szapudi, I., S. Colombi (1996): 'Cosmic error and statistics of large scale structure', Ap. J., **470**, pp. 131–148
117. Szapudi, I., E. Gaztanaga (1998): 'Comparison of the large–scale clustering in the APM and the EDSGC galaxy surveys', Mon. Not. Roy. Astron. Soc., **300**, pp. 493–496
118. Szapudi, I., A.S. Szalay (1998): 'A new class of estimators for the n–point correlations', Ap. J., **494**, pp. L41–L44
119. Thönnes, E., M.-C. van Lieshout (1999): 'A comparative study on the power of van Lieshout and Baddeley's J–function', Biom. J., **41**, pp. 721–734
120. Tomita, H. (1986): 'Statistical properties of random interface systems', Progr. Theor. Phys., **75**, pp. 482–495
121. Totsuji, H., T. Kihara (1969): 'The correlation function for the distribution of galaxies', Publications of the Astronomical Society of Japan, **21**, pp. 221–229

122. van Lieshout, M.N.M., A.J. Baddeley (1996): 'A nonparametric measure of spatial interaction in point patterns', *Statist. Neerlandica*, **50**, pp. 344–361

123. Wegner, G., M.P. Haynes, R. Giovanelli (1993): 'A survey of the Pisces–Perseus supercluster. v – the declination strip +33.5 deg to +39.5 deg and the main supercluster ridge', *A. J.*, **105**, pp. 1251–1270

124. Weil, W. (1983): 'Stereology: A survey for geometers'. In: *Convexity and its applications*, ed. by P.M. Gruber, J.M. Wills (Birkhäuser, Basel), pp. 360–412

125. Weinberg, D.H., J.R. Gott III, A.L. Melott (1987): 'The topology of large–scale–structure. I. topology and the random phase hypothesis', *Ap. J.*, **321**, pp. 2–27

126. Weiß, A.G., T. Buchert (1993): 'High–resolution simulation of deep pencil beam surveys – analysis of quasi–periodicty', *Astron. Astrophys.*, **274**, pp. 1–11

127. White, S.D.M. (1979): 'The hierarchy of correlation functions and its relation to other measures of galaxy clustering', *Mon. Not. Roy. Astron. Soc.*, **186**, pp. 145–154

128. Willmer, C., L.N. da Costa, L. N., P. Pellegrini (1998): 'Southern sky redshift survey: Clustering of local galaxies', *A. J.*, **115**, pp. 869–884

129. Winitzki, S., A. Kosowsky (1997): 'Minkowski functional description of microwave background Gaussianity', *New Astronomy*, **3**, pp. 75–100

130. Worsley, K. (1998): 'Testing for a signal with unknown location and scale in a χ^2 random field, with an application to fMRI', *Adv. Appl. Prob.*, accepted

One may find some of the more recent articles on the preprint servers
http://xxx.lanl.gov/archive/astro-ph or
http://xxx.lanl.gov/archive/gr-qc.
Numbers like astro-ph/9710207 refer to preprints on these servers. An abstract server for articles appaering in several astrophysical journals is
http://ads.harvard.edu/.
Articles older than a few years are scanned and and may be donloaded from there.

Dynamics of Structure Formation in Thin Liquid Films: A Special Spatial Analysis

Karin Jacobs[1], Ralf Seemann[1], and Klaus Mecke[2]

[1] Abteilung Angewandte Physik, Universität Ulm
   D-89069 Ulm, Germany
[2] Fachbereich Physik, Bergische Universität Wuppertal
   D-42097 Wuppertal, Germany

Abstract. The characterization of morphologies that are not perfectly regular is a very difficult task, since there is no simple "measure" for imperfections and asymmetries. We faced this problem by trying to describe the pattern that evolves in the course of the dewetting of a thin polymer film as compared with the scenario that takes place in a thin gold film. With the help of the Minkowski functionals we found significant differences in the pattern of the two systems: We were able to distinguish a spinodal dewetting mechanism for the gold film from heterogeneous nucleation for the polymer film. Moreover, we show how a temporal development of a pattern can be characterized by Minkowski functionals.

1 Introduction

What catches our eye in Fig. 1 on page 80? The form of the artishokes and the clover, their color or the way of stacking? Where are the guiding lines for the eye? Some of the photograph's fascination might be explained by the symmetry of the arrangement of the artishokes, which is nearly a hexagonal closest packing, and some might arise from the contrast of the different morphologies: radially arranged leaves for the artishokes and a linearly layered structure for the clover. The patterns are not perfect, yet clearly visible by eye. How can one describe such patterns scientifically?

Researchers in crystallography and in solid state physics found a nomenclature which is able to characterize symmetrical, recurring patterns by sorting them into 32 possible point groups. An example for such a regular pattern is shown in Fig. 2. Still, there is no room for characterizing imperfections or asymmetries in the pattern.

In nature, symmetry and symmetry-breaking lie close together. Symmetry-breaking occurs for instance when a tree trunk ramifies or a cell divides in two. In our experiments, we probe such a symmetry-breaking process by characterizing its morphology. They are performed in an easily controllable model system, a thin liquid film lying uniformly and smoothly on a solid substrate. Here, symmetry-breaking takes place when the film does not like to wet the solid surface and beads off. This is actually an everyday phenomenon, it comprises - just to name a few - the dewetting of a water film from a freshly waxed car, the printing of ink on paper or transparencies or the trickling of water droplets down water-repellent textiles or leaves ("the lotus effect" [1]).

Thin liquid films on solid surfaces are of enormous importance in many fields of modern technology. A vast amount of research has thus been dedicated to these systems, and the interest in this field is still growing. However, most of the fundamental processes are not yet fully understood. Among them are the symmetry-breaking mechanisms in the early as well as in the late stage of dewetting. In the present study, we investigate two systems, a thin metal film and a thin polymer film, which give insight into the morphology and the dynamics of liquid structures.

2   Results and Discussion

2.1   Early Stage of Dewetting: Formation of Holes

In many experiments involving different films and substrates, it has been invariably observed that rupture proceeds by the formation of circular holes whose radius grows in time until they finally merge and lead to dewetting of the entire film [9,10,11,20,21,22,29,30,33,36] as shown in Fig. 3. The mechanism of initial hole formation is still under discussion. Let us first concentrate on this early stage of dewetting.

It is generally accepted that there are two possible mechanisms that may give rise to the formation of dry patches initiating the dewetting process. Either there is *nucleation*, e. g. from dust particles or impurities in the film, or, if long-range interactions (such as van der Waals forces) between the liquid film material and the substrate disfavor wetting, fluctuations in film thickness experience a driving force. (Fluctuations by thermal motion are always present on liquid surfaces.) Their amplitudes then grow exponentially with time, finally leading to dewetting when their size becomes equal to the film thickness. From the apparent analogy with symmetry-breaking mechanisms involved in decomposition processes [7,13], the scenario has been termed *spinodal dewetting*. In this kind of dynamical instability, there is usually a certain wavelength λ_{max} the amplitude of which grows fastest, and thus determines the dominant length scale of the emerging structure. This wavelength scales as the square of the film thickness under certain, quite general conditions, as was shown experimentally [2,35] and theoretically [3,24]. Since the number density of holes appearing during the rupture process was found in several studies to depend on the film thickness according to this scaling law, it was widely believed that spinodal dewetting is the standard mechanism leading to the rupture of polymer films. In contrast, we show in the present paper that nucleation by defects is in fact the dominating mechanism leading to the generally observed rupture scenario of polymer films.

As a convenient model system, we chose polystyrene (PS) films on silicon wafers, which is frequently used as a standard system, in particular for most of the work dedicated to the study of rupture mechanisms. The polymer films were prepared by spin casting PS from a toluene solution onto freshly cleaved mica sheets. From there they were floated onto a clean deionized water surface and picked up with silanized Si wafers. More experimental details can be found in [11]. Figure 3 shows the temporal evolution of the symmetry-breaking process and Fig. 4a displays details of the typical rupture scenario of a thin PS film on a non-wettable substrate. Here, a 47 nm PS film (molecular weight 600 kg/mol) on a silanized Si wafer is shown, annealed for 7 min at 130 °C. The average diameter of the holes, whose shape was found invariably to be circular, is 8.4 μm, the width of the distribution 0.2 μm. The narrow size distribution reflects the fact that almost all of the holes appear within a sharp time window. The areal density of the holes was found to scale in accordance with the above mentioned results obtained by other authors. Similarly, the (apparently) random spatial distribution of holes is qualitatively identical to what has been generally observed before [21,22]. It is this spatial distribution onto which we will concentrate next.

Radial Pair Correlation Analysis If hydrodynamically unstable surface ripples (spinodal dewetting) were responsible for the formation of the holes, correlations were to be expected in the geometry of the hole arrangement reflecting the presence of a critical wavelength λ_{max}. In contrast, a distribution of positions of holes following a Poisson point process would be a clear counterevidence of a spinodal process. We thus determined the two-point correlation function $g(r)$ of the point set represented by the positions of the centers of the holes and plotted it in Fig. 5b. Obviously, no feature indicative of a dominant wavelength is found. Let us consider for comparison the dewetting pattern of a liquid gold film on a quartz glass, as shown in Fig. 4b and determine $g(r)$, too (Fig. 5b). Again, as Fig. 5b demonstrates, no modulation in $g(r)$ can be detected. For this system, however, unstable surface ripples with a dominant wavelength λ_{max} had been clearly identified as the dewetting mechanism [2]. These surface ripples can even be seen with a light microscope (Fig. 6a). We have analyzed the positions of the valleys of the observed undulations (data taken from [2]), as depicted in Fig. 4b. They have been used as a point set onto which the same calculation of the correlation function $g(r)$ was applied. The result is shown in Fig. 6b. The modulation in $g(r)$ with a wavelength λ_{max} corresponding to the mean distance of the valleys in the gold film indicates correlated sites. Moreover, two valleys are at least 0.6 mean distances separated. (Experimentally, two valleys could be resolved for distances larger than about 0.1 mean distances.) The existence of such a minimum distance and of the modulation=20in $g(r)$ thus demonstrate that the dewetting structure of a thin liquid gold film on a quartz substrate is indeed strongly different from a Poisson point process. Further experiments showed that the preferred wavelength λ_{max} (also observable 'by eye' in Fig. 6a) scales as the square of the film thickness [2], as is theoretically expected for a spinodal rupture scenario.

At this point, there are two open questions: i) Can we rule out a spinodal dewetting mechanism for the PS films by looking only at $g(r)$? ii) Why is no modulation visible in $g(r)$ for the holes in the gold film?

The answer to i) and ii) is that $g(r)$ is only a two-point correlation function and therefore not sensitive for higher-order correlations. So from the point of view of $g(r)$, we can neither exclude nor corroborate spinodal dewetting as rupture mechanism for any of the systems. For the gold film it is not clear yet, whether the holes stem from valleys that reached the substrate surface or if nucleation led to holes.

Characterization by Means of Minkowski Functionals The answers to i) and ii) urge us to analyze the spatial distribution of the sites of holes (or valleys, resp.) in greater detail. We could either perform standard statistical tests like the L-test [32], or alternatively, characterize the samples with the help of Minkowski functionals.

We decided to use the latter, since the Minkowski functionals are connected to physically useful parameters like the threshold of percolation. Moreover, they are very handy to describe the spatial structure of the pattern directly and to record - for instance - structural changes as a function of time, as we will demonstrate later. Besides, characteristics based on Minkowski functionals are known to yield stable results with small statistical errors even for small samples, which is of particular importance for the analysis of experimental data. They provide statistical descriptors which contain features of n-point correlation functions at any order n and are efficient in discriminating theoretical models [15,23,27].

Let us consider the centers of the holes in a film as an ensemble of points in the plane. On each point, we put a circular disk, each with the same radius r, as depicted in Fig. 7. The set is now defined as the set union of all disks, the "coverage". The scale-dependent morphological features of this coverage are then explored by varying the disk radius and calculating the Minkowski functionals of the coverage as functions of r. In two dimensions, the Minkowski functionals are the area F, the boundary length U, and the Euler characteristic χ of the coverage. The latter is defined as the integral over the boundary curvature, extended along the entire boundary [16]. Moreover, the Euler characteristic is a measure of the connectivity of a structure.

For testing the Poisson process, we used the normalized length $x := 3Dr/L$ (L denoting the average hole distance) and the normalized functions

$$F^*(x) = 3D - \frac{|\Omega|}{\pi r^2 N} \ln\left(1 - \frac{F(x)}{|\Omega|}\right)$$

$$U^*(x) = 3D \frac{U(x)}{2\pi r N(1 - F(X)/|\Omega|)} \tag{1}$$

$$\chi^*(x) = 3D \frac{\chi(x)}{N(1 - F(X)/|\Omega|)}$$

where $|\Omega|$ denotes the area of the sampling window and N the number of holes. In the case of a Poisson process it is $F^* = 3D1$, $U^* = 3D1$, and $\chi^*(x) = 3D1 - \pi x^2$.

In Fig. 8, the behavior of the three functions is shown as open squares for the point set given by the centers of the holes displayed in Fig. 5a (left) for the PS film. The theoretical values for a Poisson point process are represented as solid lines. Excellent agreement is found for all three functionals; the small deviations for larger radii being solely due to finite size effects which occur when r comes within the range of the system size. This suggests that no lateral correlations are present in the distribution of the holes in the polymer film. We employed the same analysis also to published data of hole distributions [21,28] and found, again, no significant deviation from a Poisson point process.

Let us consider for comparison the spinodal-dewetting pattern of the liquid gold film, the positions of the holes are shown in Fig. 5a (right). The measured functions F^*, U^*, and χ^* are shown as solid circles in Fig. 8. The deviation from a Poisson process is evident; here, correlations between the sites of the holes are present. Also classical tests of point process statistics (the L-test [32] and an analogous test using Ripley's K-function [25]) reject the Poisson process hypothesis [26,31]. For the L-test, the Poisson process hypothesis was rejected for significance level $\alpha = 3D0.01$ [31].

Correlations in the distance of holes could surely also be detected with the help of a Voronoi construction [34], but this costs a little larger computational effort and is not as versatile as the Minkowski measures. Using a Fourier transformation to find out a preferred wavelength requires large point sets and a thorough consideration of the boundaries. Besides, a Fourier transformation is unsuitable to check for a Poisson distribution.

It is thus demonstrated that not only the valleys of the undulation in the gold film are correlated in their distance, but also the emerging holes. Therefore, we can conclude that the holes in the gold film stem from unstable hydrodynamic surface ripples. An involvement of a spinodal process in the rupture of the polymer films shown here, however, is very unlikely.

2.2   Late Stage of Dewetting: Growth of Holes

Up to this point, we analyzed the static features of the morphology, namely the statistical distribution of the holes. To gain further insight in the two dewetting modes, we investigated the dynamics of the rupture process, too. Here, we enter the late stage of dewetting, the growth of holes.

Characterizing the Growth of Single Holes According to the theory of spinodal dewetting, the rupture time, τ, which denotes the time after which the formation of the holes begin, should be equal for all holes and scale as $\tau \propto h^5$ [3]. For the gold films, τ is experimentally not accessible due to the specific annealing method using short laser pulses. In order to obtain the rupture time of the polystyrene films, the diameters of a number of single holes were recorded as a function of time and extrapolated to zero radius [12]. This gave a reasonable estimate of τ with a statistical scattering of less than 10%. The rupture times obtained this way are only weakly dependent on film thickness (a fit yields $\tau \propto h^{0.6\pm0.3}$). Even films with thicknesses of more than 300 nm, which is beyond the range of the van der Waals-interaction, break up within some hundred seconds. It is thus demonstrated that there is neither a lateral correlation in the hole distribution nor a temporal behavior even close to what is predicted for spinodal dewetting. This rules out completely a spinodal dewetting process to be responsible for the hole formation.

It should be noted here that a thermally activated hole formation process would proceed independently at different places on the sample, and hole formation would necessarily continue to take place as long as the sample is not fully dewetted. This contradicts observation, since almost all of the holes are present immediately after the start of dewetting, as it is clearly seen in Fig. 3. We have thus ruled out any dynamical instability as well as thermally activated hole formation as processes responsible for the generic polymer film rupture scenario. The only remaining possibility is heterogeneous nucleation from defects in the film or at the substrate surface. This is our knowledge at the moment and further experiments should clarify the kind of defect leading to rupture.

Characterizing the Entire Structure Formation Process Deeper exploration of the temporal behavior as well as the dynamics of the structure formation process by recording the radii of the holes in the case of the polymer film is limited to the roundness of these holes. As soon as the holes coalesce, or the surrounding rim of the hole gets unstable, this method is at its limits. Here, the Minkowski functionals again come into play.

By determining the three Minkowski functionals F, U and χ for every single snapshot of the time sequence, we can characterize the structure formation process from the very beginning to the final state, where the droplets of material are lying on the substrate, as shown in Fig. 3. Here, two thresholds are set such that the gray scale of the usually bright substrate is above the first threshold and the typically dark rims are below the second one. So F, U and χ of only the uniform film are analyzed, as illustrated in Fig. 9. (Note that now F characterizes the area of the uniform film, and *not* the area of the holes, in contrast to the analysis of the point distribution depicted in Fig. 7.) The Euler characteristic $\chi(t)$ is determined by calculating the mean curvature of the boundary line. The Minkowski functionals $F(t)$, $U(t)$ and $\chi(t)$ for the entire sequence are shown in Fig. 10. On one hand, $F(t)$, $U(t)$ and $\chi(t)$ serve as a kind of "fingerprint" for the temporal development of the specific morphology and, on the other hand, simultaneously, they characterize the morphology itself.

Here, the interesting question arises whether or not the growth behavior of the holes changes as soon as the holes come close to each other. Does a hole feel the neighborhood of another hole? Nearby the small liquid neck between two large holes, does the draining of the liquid change? By fitting a theoretical curve to the data in Fig. 10, we can compare the growth rate of the holes with the growth rate of the morphology. In one model, the radius of the holes, r, grows linearly with time, in the other, r grows as $r \propto t^{2/3}$. The former is applicable to simple liquids flowing with viscous energy dissipation only [4,5], the latter for complex fluids where energy is dissipated only as sliding friction between the liquid and the solid ('slippage') [6]. We expect slippage to play a role since the chain length of the polystyrene is exceeding the so-called 'entanglement length', which means that by pulling one chain, a hole bunch of chains will be moved. From the fits shown in Fig. 10 we learn that for the growth of the morphology, slippage does not play a prominent role and that purely viscous flow can be assumed to be the main mechanism of energy dissipation.

This is at variance with the results for the growth law obtained by recording the radius of single holes described above where both mechanism had to be taken into account to fit the data [12]. From these experiments we know that slippage is especially important in the early regime of hole growth. Why don't we see this with the Minkowski functionals? The explanation is the optical resolution that we have for characterizing the $r(t)$-diagram of single holes, 1 μm, in comparison with that of the dozens of holes, as shown in Fig. 3, where features smaller than roughly 10 μm cannot be resolved. This means that the magnification is too small to see deviations from a linear growth mode in the early times of hole growth with the help of the Minkowski functionals.

Another access to the growth law of the pattern is to combine the results of the "simulated" growth of holes as shown in Fig. 7, where we get the functions $F(r)$, $U(r)$ and $\chi(r)$, with the measured functionals in the form of $F(t)$, $U(t)$ and $\chi(t)$. This way, we get three independent grow laws $r(t)$, Fig. 11. For guiding the eye, a linear growth law is plotted. Substantial deviations from this law can be found for very small and very large annealing times. Here, the scatter of data is quite high indicating that at these times (or for the corresponding pictures of the holes, resp.) the Minkowski functionals are extremely sensitive to the height of the threshold within the pictures. In other words, the resolution of the images for very small and very large annealing times is too low to characterize the small features correctly with the help of the Minkowski functionals. However, from the correspondence of all three $r(t)$-curves in Fig. 11, we can derive that the holes indeed grow like the overlapping disks shown in Fig. 7 and that the growth of a hole is not changed by the vicinity of other holes. This behavior is unexpected, but is characteristic for our system and particularly convincing since it is obtained for all three functions. Further experiments should clarify whether or not this behavior changes in other systems, e.g. polystyrene films of very short chain lengths (below the entanglement length) or of films on top of other substrates.

3   Conclusions and Outlook

Before the analysis with Minkowski functionals, we knew that the surface of dewetting gold films oscillates with a certain wavelength $\lambda_{\max}$ that also scales with the film thickness as predicted for a spinodal dewetting scenario [2]. However, it was not clear whether the emerging holes in the gold film stem from these hydrodynamically unstable surface waves or were nucleated by defects. With the help of the Minkowski functionals we found out that the sites of the holes are correlated. This is a counterevidence for heterogeneous nucleation. We therefore conclude that in the case of gold films, spinodal dewetting is the dominating rupture mechanism.

For the polystyrene films it was known that the areal density of holes scales with film thickness in accordance with spinodal dewetting theories [21,22]. The rupture time τ, though, did not. In our experiment it is only weakly dependent on film thickness. Our analysis by Minkowski functionals shows that the holes are most likely randomly distributed. We conclude that in the experiments shown here, heterogeneous nucleation is responsible for symmetry-breaking. Besides, by characterizing the complete structure formation process of a polystyrene film, we found that neighboring holes do not influence each other, they rather behave like overlapping disks.

It can be summarized that Minkowski functionals are powerful tools to quantify morphological properties of patterns. Moreover, structure formation processes can be followed and characterized. In this way the "fingerprints" of different processes can be compared, independently of any statistical assumptions and without any limitation of boundary conditions or minimum number of objects. Our vision for the future is that functions based on Minkowski functionals will become standard image analysis tools in real space, like Fourier transformation is in k-space.

4   Acknowledgements

It is a pleasure to thank S. Herminghaus for inspiring discussions. We also appreciate a critical reading of the manuscript by D. Stoyan. Support from Deutsche Forschungsgemeinschaft DFG Grant No. Ja 905/1-2 is gratefully acknowledged. We also acknowledge generous supply of Si wafers by Wacker Chemitronics, Burghausen, Germany.

Figure 1: Artichokes and clover, the load of an Egyptian farmer's tractor [18].

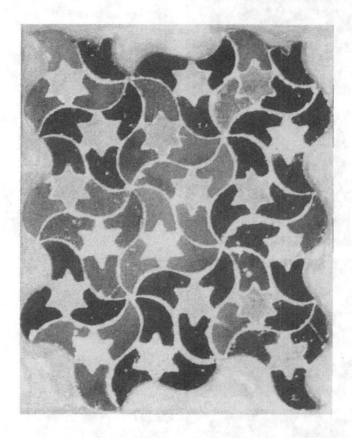

Figure 2: Example for a geometric pattern: a ceramic tile of the Alhambra Palace (Granada, Spain).

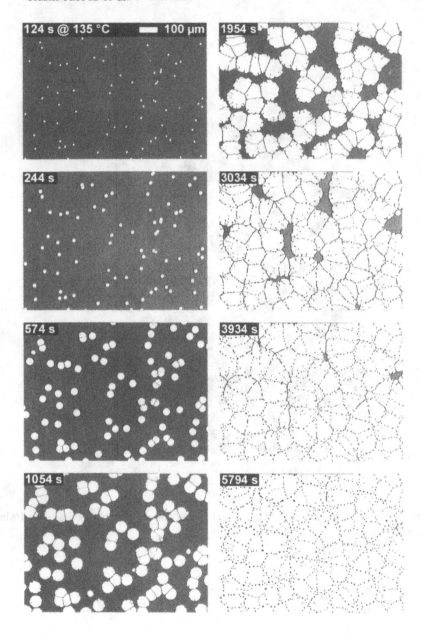

Figure 3: Series of photographs of a dewetting film, viewed through a reflection light microscope. The 80 nm thick polystyrene film on a specially treated Si wafer is liquid and beads off the non-wettable surface. Due to mass conservation, the material that formerly covered the hole has been accumulated in the form of a liquid rim surrounding the hole. The annealing time in seconds is given on each photograph.

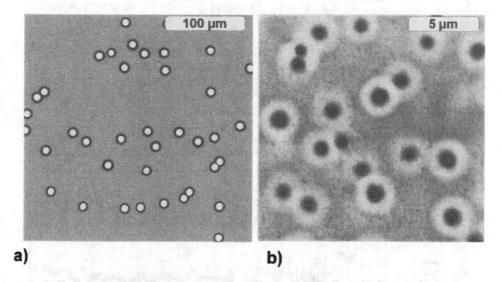

a)   **b)**

Figure 4: a) Light microscopy image of a 60 nm thick polystyrene film on a silanized Si wafer, annealed for 7 min at 133 °C, b) Atomic force microscopy (AFM) image of a 100 nm thick gold film on a quartz glass. The film was molten by a short laser pulse. The height scale is represented in shades of grey, entirely ranging 150 nm from black (deep) to white (elevated area).

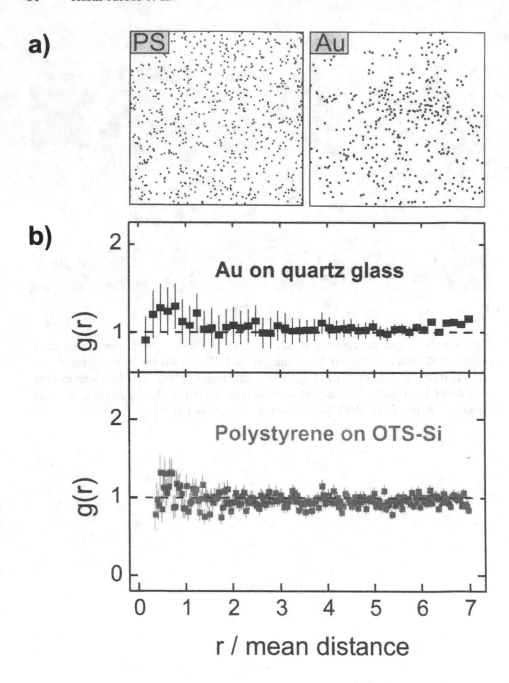

Figure 5: a) Positions of holes in the PS film (left) and in the Au film (right), as extracted from Fig. 4a) and Fig. 4b). b) $g(r)$ of the Au and the PS film for the above shown point pattern. r is given in units of the mean distance of objects.

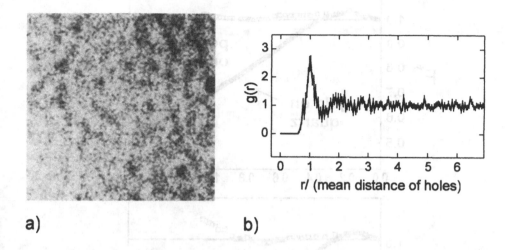

a)   b)

Figure 6: a): Detail taken by a light microscope of a dewetting Au film [19]. The snapshot characterizes the situation some fractions of nanoseconds prior to the one shown in Fig. 4: The undulations of the film are slightly visible. The dominant wavelength λ_{max} of the system is 2.4(2)μm, as determined in [19]. Width of the image: 24 μm. b) Pair correlation function, $g(r)$, for the valley positions of the dewetting Au film shown on the left. r is given in units of the mean distance of objects.

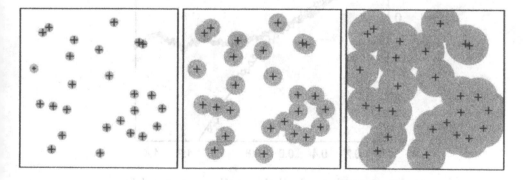

Figure 7: The positions of the holes, marked with a cross, are decorated each with a disk, whose radius increases from left to right. The Minkowski functionals in two dimensions include area F (the grey area), boundary length U between grey and white area and the Euler characteristic χ, which is a measure of the connectivity of the grey structure.

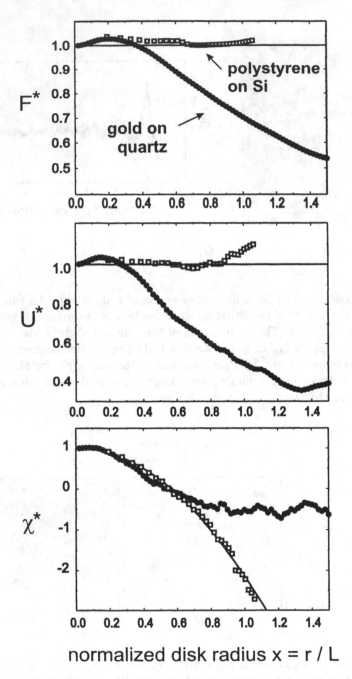

Figure 8: Normalized morphological measures F^*, U^* and χ^* of the Au (full circles) and of the PS film (open squares) as a function of the normalized radius x, $x =$3D r/L, of the disks with mean distance L. The solid lines mark the expected behavior for Poisson point process.

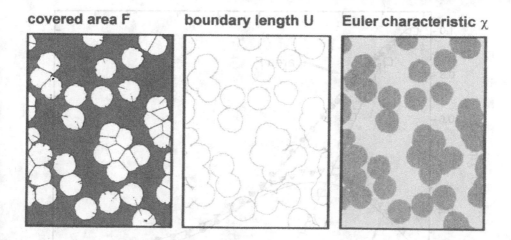

Figure 9: Analyzing structures with the Minkowski functionals - measuring area (right), boundary length (middle) and Euler characteristic of the structure (left). As an example serves a snapshot of a dewetting polystyrene film.

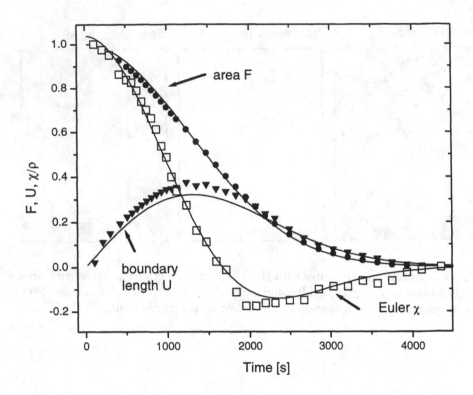

Figure 10: The functions $F(t)$, $U(t)$ and $\chi(t)$ of a temporal series of photographs, some of which are shown in Fig. 3. The diagram serves as a "fingerprint" of the entire structure formation process. The solid lines mark a fit to the data assuming a linear growth behavior of the radii of the holes. $\chi(t)$ is normalized to the total number of holes per area, ρ.

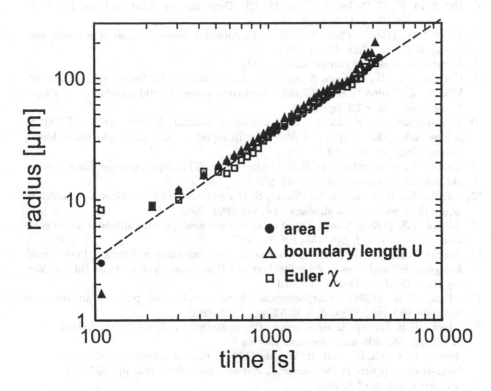

Figure 11: Growth laws $r(t)$ as derived from comparing the functions in the form of $F(r)$, $U(r)$ and $\chi(r)$ with the time dependent curves $F(t)$, $U(t)$ and $\chi(t)$. The dashed line is a guide to the eye for a linear growth of $r(t)$.

References

1. Barthlott, W., C. Neinhuis (1997): 'Purity of the sacred lotus, or escape from contamination in biological surfaces', *Planta* **202**, pp. 1–8
2. Bischof, J., D. Scherer, S. Herminghaus, P. Leiderer (1996): 'Dewetting modes of thin metallic films: Nucleation of holes and spinodal dewetting', *Phys. Rev. Lett.* **77**, pp. 1536–1539
3. Brochard-Wyart, F., J. Daillant (1990): 'Drying of solids wetted by thin liquid films', *Can. J. Phys.* **68**, pp. 1084–1088
4. Brochard-Wyart, F., P.G. De Gennes, H. Hervet, C. Redon (1994): 'Wetting and slippage of polymer melts on semi-ideal surfaces', *Langmuir* **10**, pp. 1566–1572

5. Brochard-Wyart, F., J.M. di Meglio, D. Quere (1987): 'Dewetting: Growth of dry regions from a film covering a flat solid or a fiber', *C. R. Acad. Sci. Paris II* **304**, pp. 533–558

6. Brochard, F., C. Redon, C. Sykes (1992): 'Dewetting of ultrathin liquid films', *C. R. Acad. Sci. Paris II* **314**, pp. 19–24

7. Cahn, J.W. (1965): 'Phase separation by spinodal decomposition in isotropic systems', *J. Chem. Phys.* **42**, pp. 93–99

8. Experimental details can be found in [11].

9. Herminghaus, S., A. Fery, S. Schlagowski, K. Jacobs, R. Seemann, H. Gau, W. Mönch, T. Pompe (2000): 'Liquid= microstructures at solid interfaces', *J. Phys.: Condensed Matter* **12**, pp. 57-74

10. Herminghaus, S., K. Jacobs, K.R. Mecke, J. Bischof, A. Fery, M. Ibn-Elhaj, S. Schlagowski (1998): 'Spinodal dewetting in liquid crystal and liquid metal films', *Science* **282**, pp. 916–919

11. Jacobs, K., S. Herminghaus, K.R. Mecke (1998): 'Thin liquid polymer films rupture via defects', *Langmuir* **14**, pp. 965–969

12. Jacobs, K., R. Seemann, G. Schatz, S. Herminghaus (1998): 'Growth of holes in liquid films with partial slippage', **14**, pp. 4961–4963

13. Mitlin, V.S. (1993): 'Dewetting of solid surface: analogy with spinodal decomposition', *J. Coll. Interf. Sci.* **156**, pp. 491–497

14. Mecke, K.R. (1994): *Integralgeometrie in der Statistischen Physik - Perkolation, komplexe Flüssigkeiten und die Struktur des Universum, Reihe Physik Bd. 25* (Verlag Harri Deutsch, Frankfurt a.M.)

15. Mecke, K.R. (1996): 'Morphological characterization of patterns in reaction-diffusion systems', *Phys. Rev. E* **53**, pp. 4794–4800

16. Mecke, K.R. (1998): 'Integral geometries in statistical physics', *Int. J. Mod. Phys.* **B 12**, pp. 861–899 and references therein

17. Mecke, K.R., Th. Buchert, H. Wagner (1994): 'Robust morphological measures for large-scale structure in the universe', *Astron. Astrophys.* **288**, pp. 697–704

18. Picture courtesy of J. Wolf

19. Picture is a detail of Fig. 3 in [2].

20. Redon, C., F. Brochard-Wyart, F. Rodelez (1991): 'Dynamics of dewetting', *Phys. Rev. Lett.* **66**, pp. 715–718

21. Reiter, G. (1992): 'Dewetting of thin polymer films', *Phys. Rev. Lett.* **68**, pp. 751–754

22. Reiter, G. (1993): 'Unstable thin polymer films: rupture and dewetting processes', *Langmuir* **9**, pp. 1344–1351

23. Rosenfeld, A., A.C. Kak (1976): in: *Digital picture processing* (Academic Press, New York) and references therein

24. Ruckenstein, E., R.K. Jain (1974): 'Spontaneous rupture of thin liquid films', *Faraday Trans.* **70**, pp. 132–147

25. Ripley, B. (1976): 'The second-order analysis of stationary point processes', *J. Appl. Probab.* **13**, pp. 255–266

26. Schladitz, K.: private communication

27. Serra, J. (1982): in: *Image analysis and mathematical morphology*, Vol. 1 and 2, (Academic Press, New York)

28. Sharma, A., G. Reiter (1996): 'Instability of thin polymer films on coated substrates: rupture, dewetting and drop formation', *J. Coll. Interface Sci.* **178**, pp. 383–390

29. Shull, K.R., T.E. Karis (1994): 'Dewetting dynamics for large equilibrium contact angles', *Langmuir* **10**, pp. 334–339

30. Stange, T.G., D.F. Evans, W.A. Hendrickson (1997): 'Nucleation and growth of defects leading to dewetting of thin polymer films', *Langmuir* **13**, pp. 4459–4465
31. Stoyan, D.: private communication
32. Stoyan, D., W.S. Kendall, J. Mecke (1995): in: *Stochastic geometry and its applications*, 2nd ed. (John Wiley Ltd.)
33. van der Wielen, M.W.J., M.A. Cohen-Stuart, G.J. Fleer (1998): 'Autophobicity and layering behavior of thin liquid-crystalline polymer films', **14**, pp. 7065–7071
34. Weaire, D., N. Rivier (1984): 'Soap, cells and statistics - random patterns in two dimensions', *Contem. Phys.* **25**, pp. 59–99
35. Xie, R., A. Karim, J.F. Douglas, C.C. Han, R.A. Weiss (1998): 'Spinodal dewetting of thin polymer films', *Phys. Rev. Lett.* **81**, pp. 1251–1254
36. Yerushalmi-Rozen, R., J. Klein, L.W. Fetters (1994): 'Suppression of rupture in thin, nonwetting liquid films', *Science* **263**, pp. 793–795

Part II

Integral Geometry and Morphology of Patterns

Mixed Measures
and Inhomogeneous Boolean Models

Wolfgang Weil

Mathematisches Institut II, Universität Karlsruhe
D-76128 Karlsruhe, Germany

Abstract. The Boolean model is the basic random set model for many applications. Its main advantage lies in the fact that it is determined by a single measure-valued parameter, the intensity measure. Whereas classically Boolean models were studied which are stationary and isotropic, some of the methods and results have been extended to the non-isotropic situation. More recent investigations consider inhomogeneous Boolean models, i.e. random sets without any invariance property. Density formulae for inhomogeneous Boolean models make use of local variants of the classical quermassintegrals, the surface area measures and curvature measures. Iterations of translative integral formulae for curvature measures lead to further measures of mixed type.

In this survey, we describe some of these local and mixed-type functionals from integral geometry and show how they can be used to extend density formulae for Boolean models from the stationary and isotropic case to the non-isotropic situation, and finally to inhomogeneous Boolean models.

1 Motivation

Throughout the following, we describe some recent developments in the study of random collections X of particles and their union sets Z. Random systems of particles are observed and investigated in many applied sciences and often overlappings of the particles make the direct measurement of particle characteristics difficult or even impossible. In these practical situations, the dimension is usually 2 or 3. In order to give a unified exposition of the problems and results, we work here in the space $\mathbb{R}^d$ with a general dimension d, $d \geq 2$.

In the worst case, only the union set

$$Z = \bigcup_{K \in X} K$$

of the particles is visible and sometimes only the part $Z \cap W$ of this union set Z in a sampling window W (which we assume to be a full dimensional, compact and convex set) is observable. The main goal is then to determine (or better: estimate) characteristical quantities of the particle system X (like mean values or distributions) from observations of Z, respectively $Z \cap W$. An important aspect of random sets Z which arise as the union of particle systems X is that they can be an appropriate model even in situations where there are no particles in the background of the physical situation. If a given complicated structure has to be modelled by a random closed set Z, the class of particle models is often the only practicable class to start with. Here again the statistical task is to estimate the characteristical quantities of the (artifical) particle system X from observations of Z.

Of course, this is only a realistic task if we make a suitable assumption on the distribution of the particles. A simple, yet very effective *assumption* is the following: We assume that the number N of particles of X which hit the sampling window W (hence have visible parts in W) is a random variable with the additional property that, given $N = k$, these k particles are independent and identically distributed. Although this is a slightly vague formulation, it specifies a unique model for X, namely the *Poisson process* on the space of particles. For simplicity, we assume here that the particles are compact and convex (so-called convex bodies), although some of the methods and results we describe can be extended to non-convex particles as well. Thus, we consider a Poisson process X on the space $\mathcal{K}^d$ of convex bodies in $\mathbb{R}^d$. This process is characterized by the property

$$\mathbb{P}(X(\mathcal{A}) = k) = e^{-\Theta(\mathcal{A})} \frac{\Theta(\mathcal{A})^k}{k!}, \qquad \text{for } k = 0, 1, \ldots.$$

Here, $\mathcal{A}$ is a subset of $\mathcal{K}^d$ and $X(\mathcal{A})$ denotes the number of particles of X in $\mathcal{A}$. Examples for subsets $\mathcal{A}$ of interest are

$$\mathcal{A} = \{K \in \mathcal{K}^d : K \cap W \neq \emptyset\}$$

or

$$\mathcal{A} = \{K \in \mathcal{K}^d : K \subset W\}.$$

(Mathematically, a more rigorous setting is necessary which requires to supply $\mathcal{K}^d$ with a natural distance (the Hausdorff metric) and the corresponding system of Borel sets; $\mathcal{A}$ then runs through all Borel subsets of $\mathcal{K}^d$.) $\Theta(\mathcal{A})$ is the parameter of the Poisson distribution, hence it equals the expectation of $X(\mathcal{A})$,

$$\Theta(\mathcal{A}) = \mathbb{E}X(\mathcal{A})$$

(thus, $\Theta(\mathcal{A})$ gives the mean number of particles of X which belong to the set $\mathcal{A}$). As a function of $\mathcal{A}$, Θ is a measure on $\mathcal{K}^d$ (a σ-additive set function on the Borel subsets of $\mathcal{K}^d$), it is called the *intensity measure* of X. The measure Θ need not be finite, in fact in most of the applications, which we discuss later, we have $\Theta(\mathcal{K}^d) = \infty$. However, we assume that Θ is *locally finite* in the sense that $\Theta(\{K \in \mathcal{K}^d : K \cap M \neq \emptyset\}) < \infty$, for all $M \in \mathcal{K}^d$. Another natural assumption we make is that $\Theta(\{K\}) = 0$, for all singleton sets $\{K\}, K \in \mathcal{K}^d$ (we then say that Θ is *atom-free*). The following existence and uniqueness theorem is a classical result in abstract point process theory (see, for example, [1]).

Theorem. *Each measure Θ on $\mathcal{K}^d$, which is locally finite and atom-free, is the intensity measure of some Poisson process X on $\mathcal{K}^d$. This Poisson process X is uniquely determined (in distribution).*

If X is a Poisson process on $\mathcal{K}^d$, the union set Z is a random closed set (RACS) in the sense of [10]. Z is called a *Boolean model* (with convex grains); see also

Sect. 3 of the paper by D. Stoyan in this volume. The basic *statistical problem* mentioned at the beginning can now be reformulated: Determine (or estimate) Θ on the basis of observations of Z.

We also understand now why the Boolean model is an appropriate model for many different random structures in practice. Its main advantage is that it is completely determined by the single parameter Θ. The estimation of Θ is a problem which belongs to the area of *Spatial Statistics*. In the general situation considered so far, Θ is an arbitrary (locally finite and atom-free) measure on $\mathcal{K}^d$, hence much too complicated to be accessible in practice. The situation becomes easier if further assumptions are made, for example on the shape of the particles. A simple model of that kind consists of balls with random radii which are independent of the location. In this case, Θ can be expressed by the spatial density of the midpoints and the radii distribution. We discuss corresponding formulae at the end of this paper. For the main part, we allow quite general intensity measures Θ, but concentrate on the estimation of certain mean values. We refer to [23], for general information about Boolean models, and to [14], for further results concerning the estimation of Θ.

In order to see which mean values of X can be estimated from observations of Z, we consider a geometric functional φ and assume that $\varphi(Z \cap W)$ is observable. The following *assumptions* on φ are partially enforced by the mathematical methods which we will use later, partially they are also reasonable from a practical point of view.

(a) φ is defined on the *convex ring* $\mathcal{R}^d$; this is the collection of all finite unions of convex bodies (the reason for this assumption is that, for our Boolean model Z, we have $Z \cap W \in \mathcal{R}^d$ almost surely),

(b) φ is *additive*,

$$\varphi(C \cup D) + \varphi(C \cap D) = \varphi(C) + \varphi(D), \qquad \text{for } C, D \in \mathcal{R}^d,$$

(c) φ is *translation invariant*,

$$\varphi(C + x) = \varphi(C), \qquad \text{for } C \in \mathcal{R}^d, x \in \mathbb{R}^d,$$

(d) φ is *continuous* (on $\mathcal{K}^d$),

$$\varphi(K_i) \to \varphi(K), \qquad \text{as } K_i \to K \text{ in the Hausdorff metric on } \mathcal{K}^d.$$

Volume and surface area are functionals which fulfill these conditions, further examples are discussed in Sect. 2.

Concerning such a functional φ, the mean information about Z contained in the sampling window W is $\mathbb{E}\varphi(Z \cap W)$, this is also the quantity for which $\varphi(Z \cap W)$ is an unbiased estimator. Hence, we shall try to express $\mathbb{E}\varphi(Z \cap W)$ in terms of Θ. By the independence properties of Poisson processes, the additivity of φ (in form of the inclusion-exclusion principle), and the theorems of Fubini and Campbell, we obtain for any convex body $K_0 \in \mathcal{K}^d$

$$\mathbb{E}\varphi(Z \cap K_0) = \mathbb{E}\varphi\left(\bigcup_{K \in X}(K \cap K_0)\right)$$

$$= \sum_{k=1}^{\infty}\frac{(-1)^{k+1}}{k!}\mathcal{E}\sum_{(K_1,\dots,K_k)\in X_{\neq}^k}\varphi(K_0 \cap K_1 \cap \dots \cap K_k)$$

$$= \sum_{k=1}^{\infty}\frac{(-1)^{k+1}}{k!}\int_{\mathcal{K}^d}\cdots$$

$$\int_{\mathcal{K}^d}\varphi(K_0 \cap K_1 \cap \dots \cap K_k)\Theta(dK_1)\cdots\Theta(dK_k), \qquad (1)$$

where $X_{\neq}^k$ denotes the process of k-tupels from X with pairwisely disjoint entries. The question arises how to simplify this expression.

Let us first make the *additional assumption* that the Boolean model Z is *stationary*, hence has a distribution which is invariant under translations (in other words, Z and $Z + x$ have the same distribution for all $x \in \mathbb{R}^d$). The stationarity of Z is equivalent to the corresponding property of X and thus equivalent to the translation invariance of Θ. This invariance has important consequences for the structure of Θ. In order to describe these, we decompose each particle $K \in \mathcal{K}^d$ into its "location" and its "shape". As location parameter, we choose the Steiner point $s(K)$ of K (the centroid of the Gaussian curvature measure on the boundary of K). Other centre points (like the midpoint of the circumsphere of K) are also possible, but the Steiner point is especially adapted to the later representations. We let

$$\mathcal{K}_0^d = \{K \in \mathcal{K}^d : s(K) = 0\}$$

denote the set of convex bodies with Steiner point at the origin (the space of "shapes"). Each particle $K \in \mathcal{K}^d$ has a unique representation $(z, M) \in \mathbb{R}^d \times \mathcal{K}_0^d$ with $K = M + z$, namely $z = s(K)$ and $M = K - s(K)$. Using this representation, we can interpret Θ as a measure on the product space $\mathbb{R}^d \times \mathcal{K}_0^d$. Moreover, for stationary Z, this measure is translation invariant with respect to the first component. Such a measure necessarily has a product form,

$$\Theta = \gamma(\lambda_d \otimes \mathbb{P}_0), \qquad (2)$$

where $\gamma \in (0, \infty)$ is a constant (the *intensity* of X), λ_d is the Lebesgue measure on $\mathbb{R}^d$, and $\mathbb{P}_0$ is a probability distribution on $\mathcal{K}_0^d$, the *distribution of the typical grain*. (The finiteness of γ stems from the local finiteness of Θ which we assumed; we also exclude the case $\gamma = 0$ in the following, since it belongs to the process X with no particles at all.) For stationary Z, we thus obtain from (1)

$$\mathbb{E}\varphi(Z \cap K_0)$$

$$= \sum_{k=1}^{\infty}\frac{(-1)^{k+1}}{k!}\gamma^k\int_{\mathcal{K}_0^d}\cdots\int_{\mathcal{K}_0^d}\phi(K_0, K_1, \dots, K_k)\mathbb{P}_0(dK_1)\cdots\mathbb{P}_0(dK_k)(3)$$

with

$$\phi(K_0, K_1, \ldots, K_k)$$

$$= \int_{\mathcal{R}^d} \cdots \int_{\mathbb{R}^d} \varphi(K_0 \cap (K_1 + x_1) \cap \cdots \cap (K_k + x_k)) \lambda_d(dx_1) \cdots \lambda_d(dx_k) \tag{4}$$

In order to obtain more explicit formulae here, we face two problems:

(1) Find appropriate geometric functionals φ.

(2) Find, for such φ, translative integral formulae (in particular, iterated versions).

2 Geometric Functionals

The class of functionals $\varphi : \mathcal{R}^d \to \mathbb{R}$ which fulfill conditions (a) - (d) is still rather big. By a theorem of [5] it is sufficient to consider additive, translation invariant and continuous functionals on $\mathcal{K}^d$ since any such functional φ can be extended to a functional on $\mathcal{R}^d$ which fulfills (a) - (d). Additive functionals on $\mathcal{K}^d$ are also called *valuations* and there is an extensive theory on valuations including translation invariant and continuous ones (see the surveys [11] and [12], as well as the book [19]). A complete description of continuous translation invariant valuations is however not known, in this generality.

An approach which is reasonable from a geometric point of view is to restrict the class of functionals further by replacing condition (c) by

(c′) φ is *motion invariant*,

$$\varphi(gK) = \varphi(K), \qquad \text{for } K \in \mathcal{K}^d, g \in G_d,$$

where G_d denotes the group of (rigid) motions in $\mathbb{R}^d$ (and where we already used the fact that it is sufficient to assume invariance on $\mathcal{K}^d$).

We already mentioned the most simple example of a continuous, motion invariant valuation, the volume functional $\varphi = V_d$ (the Lebesgue measure). Further examples like the surface area can be obtained from V_d by differentiation, due to the *Steiner formula* for the outer parallel body,

$$V_d((K \oplus \epsilon B^d) \setminus K) = \sum_{j=0}^{d-1} \epsilon^{d-j} \kappa_{d-j} V_j(K), \qquad \text{for } \epsilon > 0. \tag{5}$$

Here, we used $\oplus$ to denote vector addition of sets, B^d for the unit ball in $\mathbb{R}^d$, and $\kappa_{d-j} = \lambda_{d-j}(B^{d-j})$, $j = 0, \ldots, d-1$. The coefficients $V_0(K), \ldots, V_{d-1}(K)$ in the polynomial expansion (5) are called the *intrinsic volumes* of K. The functional $\varphi = V_j$ satisfies conditions (a)–(d) as well as (c′). The intrinsic volumes are proportional to the classical quermassintegrals $W_i(K)$, namely we have

$$\kappa_{d-j} V_j(K) = \binom{d}{j} W_{d-j}(K), \qquad \text{for } j = 0, \ldots, d-1.$$

The advantage of the intrinsic volumes is that they only depend on the dimension of K and not on the dimension d of the surrounding space. In particular, for a j-dimensional body K in $\mathbb{R}^d$, $j < d$, $V_j(K)$ is the j-dimensional volume of K, $V_j(K) = \lambda_j(K)$. The intrinsic volumes can be represented as curvature integrals over the boundary of K, but some of them have a more direct interpretation. Namely, $V_0(K)$ is the (Euler) characteristic, that is, $V_0(K) = 1$ if $K \neq \emptyset$ and $V_0(\emptyset) = 0$. $V_1(K)$ is proportional to the mean width of K and, at the other end, $V_{d-1}(K)$ is half the surface area of K.

The importance of the intrinsic volumes is expressed by the following fundamental theorem [9], for a more recent and simplified proof).

Theorem (Hadwiger). *For each motion invariant, continuous valuation φ on $\mathcal{K}^d$ there are constants $c_0, \ldots, c_d$ such that*

$$\varphi = \sum_{j=0}^{d} c_j V_j.$$

Intrinsic volumes do not give any information on the orientation of a set. Such direction dependent quantities can be obtained by suitable variations of (5).

The first possibility is to replace the unit ball in (5) by a more general convex body M, which is possible with the use of Minkowski's *mixed volumes*. We obtain

$$V_d((K \oplus \epsilon M) \setminus K) = \sum_{j=0}^{d-1} \epsilon^{d-j} \binom{d}{j} V(K\,[j], M\,[d-j]), \qquad \text{for } \epsilon > 0. \quad (6)$$

The mixed functionals $V(K\,[j], M\,[d-j])$ on the right-hand side may be defined by this polynomial expansion, but they are also special cases of a general expansion of the volume of sum-sets (see [19], for definitions and properties of these mixed volumes).

A second possibility is to consider local counterparts of (5). There are two approaches in this direction. For each Borel set β in $\mathbb{R}^d$ and $\epsilon > 0$, we may consider the local outer parallel set

$$A_\epsilon(K, \beta) = \{x \in (K \oplus \epsilon B^d) \setminus K : p(K, x) \in \beta\},$$

where $p(K, \cdot)$ denotes the metric projection onto K. Then,

$$\lambda_d(A_\epsilon(K, \beta)) = \sum_{j=0}^{d-1} \epsilon^{d-j} \kappa_{d-j} \Phi_j(K, \beta), \qquad \text{for } \epsilon > 0. \quad (7)$$

By this expansion, local functionals $\Phi_0(K, \cdot), \ldots, \Phi_{d-1}(K, \cdot)$ are defined which are (finite and nonnegative) measures on $\mathbb{R}^d$ and are actually supported by the boundary bd$K$ of K. These are the *curvature measures* of [4]. $\Phi_0(K, \cdot)$ measures the Gaussian curvature in the boundary points of K and $\Phi_{d-1}(K, \cdot)$ is half the

$(d-1)$-dimensional Hausdorff measure on $\mathrm{bd}K$. Alternatively, we may define a local parallel set $B_\epsilon(K,\omega)$ for each Borel set ω in the unit sphere S^{d-1} of $\mathbb{R}^d$,

$$B_\epsilon(K,\omega) = \{x \in (K \oplus \epsilon B^d) \setminus K : u(K,x) \in \omega\}.$$

Here, $u(K,x) = (x-p(K,x))/\|x-p(K,x)\|$ is the outer normal vector in $p(K,x)$, pointing in direction of x. Then, as a counterpart to (7), we get

$$\lambda_d(B_\epsilon(K,\omega)) = \sum_{j=0}^{d-1} \epsilon^{d-j} \kappa_{d-j} \Psi_j(K,\omega), \qquad \text{for } \epsilon > 0. \tag{8}$$

By (8), the *surface area measures* $\Psi_0(K,\cdot),\ldots,\Psi_{d-1}(K,\cdot)$ of K are defined. They have been introduced first by Aleksandrov, Fenchel and Jessen (see [19], for details, connections and generalizations). Here, $\Psi_0(K,\cdot)$ plays a special role since it is the spherical Lebesgue measure (and hence independent of K). For each j, the curvature measure $\Phi_j(K,\cdot)$ determines K uniquely. For the surface area measures, this is true for $j \geq 1$ and uniqueness holds only up to translations. We remark that there are common generalizations of curvature measures and surface area measures, the support measures which are supported by the generalized normal bundle of K. We also refer to [19] for the relations between these measures and mixed volumes. For the following, we mention that

$$V_j(K) = \Phi_j(K,\mathbb{R}^d) = \Psi_j(K,S^{d-1}), \qquad \text{for } j = 0,\ldots,d-1,$$

and

$$V(K\,[d-1], M\,[1]) = \frac{2}{d} \int_{S^{d-1}} h(M,u)\Psi_{d-1}(K,du), \tag{9}$$

where $h(M,\cdot) = \max_{x \in K}\langle x,\cdot\rangle$ is the *support function* of M.

Mixed volumes, curvature measures and surface area measures share a number of properties with the intrinsic volumes. In particular, they are additive and continuous (where we use the weak convergence for measures), hence they have a (unique) additive extension to the convex ring $\mathcal{R}^d$. Obviously, they are no longer motion invariant, but translation invariant (mixed volumes and surface area measures), respectively translation covariant (curvature measures).

3   Integral Geometry

In order to approach our second problem, namely to find translative integral formulae for the geometric functionals φ under consideration, we proceed similarly to the last section. Again, we first replace the translative situation by a motion invariant one, that is, we consider kinematic integral formulae. For this purpose, we make use of the (up to a normalizing constant unique) invariant measure μ on the group G_d of rigid motions (for more details on this measure and the following integral geometric formulae, see [21]). The classical result of integral geometry is the *Principal Kinematic Formula*, due to Blaschke, Santaló and Chern,

$$\int_{G_d} V_j(K \cap gM)\mu(dg) = \sum_{k=j}^{d} \alpha_{djk} V_k(K) V_{d+j-k}(M), \qquad \text{for } j = 0, \ldots, d, \quad (10)$$

which holds for $K, M \in \mathcal{R}^d$. The constants α_{djk} depend on the normalization of μ and can be given explicitly.

The following local version of (10) for curvature measures is due to Federer,

$$\int_{G_d} \Phi_j(K \cap gM, \beta \cap g\beta')\mu(dg)$$

$$= \sum_{k=j}^{d} \alpha_{djk} \Phi_k(K, \beta) \Phi_{d+j-k}(M, \beta'), \qquad (11)$$

for $j = 0, \ldots, d$,

$K, M \in \mathcal{R}^d$, and Borel sets $\beta, \beta' \subset \mathbb{R}^d$. Here, we have also used the measure $\Phi_d(K, \cdot)$ which is defined as the Lebesgue measure λ_d restricted to K. Both formulae, (10) and (12) can be iterated easily.

The corresponding translative versions look naturally more complicated. The *Principle Translative Formula* has the form

$$\int_{\mathbb{R}^d} V_j(K \cap (M + x))\lambda_d(dx) = \sum_{k=j}^{d} V_{k,d+j-k}^{(j)}(K, M), \qquad \text{for } j = 0, \ldots, d \quad (12)$$

and $K, M \in \mathcal{R}^d$, with *mixed functionals* $V_{k,d+j-k}^{(j)}$. The indices here indicate the degree of homogeneity, hence $V_{k,d+j-k}^{(j)}(K, M)$ is homogeneous of degree k in K and of degree $d + j - k$ in M. A special case is $j = 0$, since

$$V_{k,d-k}^{(0)}(K, M) = \binom{d}{k} V(K\,[k], M^*\,[d-k]), \qquad \text{for } k = 0, \ldots, d$$

and $K, M \in \mathcal{K}^d$, where M^* denotes the reflection of M in the origin. For $K, M \in \mathcal{R}^d$, the formula holds true if, in case $k = 0$, the right-hand mixed volume is interpreted as $V_0(K)V_d(M^*)$ (and symmetrically for $k = d$).

The iteration of (12) is possible but produces new mixed functionals,

$$\int_{\mathbb{R}^d} \cdots \int_{\mathbb{R}^d} V_j(K_0 \cap (K_1 + x_1) \cap \cdots \cap (K_k + x_k))\lambda_d(dx_1) \cdots \lambda_d(dx_k)$$

$$= \sum_{\substack{m_0, \ldots, m_k = j \\ m_0 + \cdots + m_k = kd+j}} V_{m_0, \ldots, m_k}^{(j)}(K_0, \ldots, K_k), \qquad (13)$$

for $j = 0, \ldots, d$.

Fortunately, the sequence of mixed functionals $V^{(j)}_{m_0,\ldots,m_k}(K_0,\ldots,K_k)$ is limited in view of the condition

$$m_0 + \cdots + m_k = kd + j, \qquad \text{for } m_0,\ldots,m_k \in \{j,\ldots,d\},$$

and the fact that

$$V^{(j)}_{m_0,\ldots,m_k}(K_0,\ldots,K_k) = \frac{1}{\kappa_{d-j}} V^{(0)}_{m_0,\ldots,m_k,d-j}(K_0,\ldots,K_k,B^d),$$

as well as

$$V^{(j)}_{m_0,\ldots,m_{k-1},d}(K_0,\ldots,K_{k-1},K_k) = V^{(j)}_{m_0,\ldots,m_{k-1}}(K_0,\ldots,K_{k-1})V_d(K_k).$$

If we replace the intrinsic volume by the corresponding curvature measure, we get a result, analogous to (12)

$$\int_{\mathbb{R}^d} \Phi_j(K \cap (M + x), \beta \cap (\beta' + x))\lambda_d(dx)$$
$$= \sum_{k=j}^{d} \Phi^{(j)}_{k,d+j-k}(K,M;\beta \times \beta'), \qquad (14)$$

for $j = 0,\ldots,d$,

with mixed measures $\Phi^{(j)}_{k,d+j-k}(K,M;\cdot)$ on $\mathbb{R}^d \times \mathbb{R}^d$. The iteration of (14) yields

$$\int_{\mathbb{R}^d} \cdots \int_{\mathbb{R}^d} \Phi_j(K_0 \cap (K_1 + x_1) \cap \cdots \cap (K_k + x_k), \beta_0 \cap (\beta_1 + x_1) \cap \cdots$$
$$\cap(\beta_k + x_k))\lambda_d(dx_1)\cdots\lambda_d(dx_k)$$
$$= \sum_{\substack{m_0,\ldots,m_k=j \\ m_0+\cdots+m_k=kd+j}}^{d} \Phi^{(j)}_{m_0,\ldots,m_k}(K_0,\ldots,K_k;\beta_0 \times \cdots \times \beta_k), \qquad (15)$$

for $j = 0,\ldots,d$,

where the mixed measure $\Phi^{(j)}_{m_0,\ldots,m_k}(K_0,\ldots,K_k;\cdot)$ is now a measure on the k-fold product space $(\mathbb{R}^d)^k = \mathbb{R}^d \times \cdots \times \mathcal{R}^d$.

4   Boolean Models

Returning to equations (3) and (4) for a stationary Boolean model Z, we first assume that Z is, in addition, isotropic. Then, the distribution of the typical grain, $\mathbb{P}_0$, is rotation invariant and hence we can replace the iterated translation integral by an iterated kinematic integral. Thus, for a stationary and isotropic Boolean model Z and the intrinsic volume V_j, we obtain

$$\mathbb{E}V_j(Z \cap K_0)$$

$$= \sum_{k=1}^{\infty} \frac{(-1)^{k+1}}{k!} \gamma^k \int_{\mathcal{K}_0^d} \cdots$$

$$\int_{\mathcal{K}_0^d} \phi(K_0, K_1, \ldots, K_k) \mathbb{P}_0(dK_1) \cdots \mathbb{P}_0(dK_k) \qquad (16)$$

with

$$\phi(K_0, K_1, \ldots, K_k)$$

$$= \int_{G_d} \cdots \int_{G_d} V_j(K_0 \cap g_1 K_1 \cap \cdots \cap g_k K_k) \mu(dg_1) \cdots \mu(dg_k). \qquad (17)$$

We define

$$\overline{V}_j(X) = \gamma \int_{\mathcal{K}_0^d} V_j(K) \mathcal{P}_0(dK)$$

and call this the j-th *quermass density* of the particle system X, $j = 0, \ldots, d$. A corresponding quantity for Z can be defined by a limit process,

$$\overline{V}_j(Z) = \lim_{r \to \infty} \frac{\mathbb{E} V_j(Z \cap rK)}{V_d(rK)},$$

for an arbitrary 'window' $K \in \mathcal{K}^d$ with inner points. By (16), (17), and the iterated version of (10), it is easily seen that this limit exists and is independent of K. We even get an explicit expression for the alternating sum in (16) and the series expansion of the exponential function leads to the following classical result.

Theorem. *For a stationary and isotropic Boolean model Z in $\mathbb{R}^d$, we have*

$$\overline{V}_d(Z) = 1 - e^{-\overline{V}_d(X)}$$

and

$$\overline{V}_j(Z) = e^{-\overline{V}_d(X)} \Big\{ \overline{V}_j(X)$$

$$+ \sum_{k=2}^{d-j} \frac{(-1)^{k+1}}{k!} \sum_{\substack{m_1, \ldots, m_k = j+1 \\ m_1 + \cdots + m_k = (k-1)d+j}}^{d-1} c_{m_1, \ldots, m_k}^{(j)} \overline{V}_{m_1}(X) \cdots \overline{V}_{m_k}(X) \Big\},$$

for $j = 0, \ldots, d-1$.

The constants $c_{m_1, \ldots, m_k}^{(j)}$ occurring here are known explicitly.

For dimension $d = 2$, the intrinsic volumes equal (or are proportional to) the *area* A, the *boundary length* L, and the *Euler characteristic* χ. The formulae then reduce to

$$\overline{A}(Z) = 1 - e^{-\overline{A}(X)},$$

$$\overline{L}(Z) = e^{-\overline{A}(X)}\overline{L}(X),$$

$$\overline{\chi}(Z) = e^{-\overline{A}(X)}\left(\gamma - \frac{1}{4\pi}\overline{L}^2(X)\right).$$

For dimension $d = 3$, we have correspondingly the *volume* V, the *surface area* S, the *integral of mean curvature* M, and the *Euler characteristic* χ. The formulae are then the following,

$$\overline{V}(Z) = 1 - e^{-\overline{V}(X)},$$

$$\overline{S}(Z) = e^{-\overline{V}(X)}\overline{S}(X),$$

$$\overline{M}(Z) = e^{-\overline{V}(X)}\left(\overline{M}(X) - \frac{\pi^2}{32}\overline{S}^2(X)\right)$$

$$\overline{\chi}(Z) = e^{-\overline{V}(X)}\left(\gamma - \frac{1}{4\pi}\overline{M}(X)\overline{S}(X) + \frac{\pi}{384}\overline{S}^3(X)\right).$$

An interesting aspect of these formulae is that they allow an estimation of the intensity γ of the particle system X solely on the basis of measurements of the union set Z. Since the formulae extend to particles in $\mathcal{R}^d$, this is even true for non-convex particles, provided they belong to the class $\mathcal{R}^d$ and are simply connected.

Now we consider the non-isotropic case. Here, we have to use (13) in (4) and obtain the following generalization of the last theorem (see [24]).

Theorem. *For a stationary Boolean model Z in $\mathbb{R}^d$, we have*

$$\overline{V}_d(Z) = 1 - e^{-\overline{V}_d(X)}$$

and

$$\overline{V}_j(Z) = e^{-\overline{V}_d(X)}\Big\{\overline{V}_j(X)$$

$$+ \sum_{k=2}^{d-j}\frac{(-1)^{k+1}}{k!}\sum_{\substack{m_1,\dots,m_k=j+1 \\ atopm_1+\cdots+m_k=(k-1)d+j}}\overline{V}^{(j)}_{m_1,\dots,m_k}(X,\dots,X)\Big\},$$

for $j = 0,\dots,d-1$.

The two-dimensional version now reads

$$\overline{A}(Z) = 1 - e^{-\overline{A}(X)},$$

$$\overline{L}(Z) = e^{-\overline{A}(X)}\overline{L}(X),$$

$$\overline{\chi}(Z) = e^{-\overline{A}(X)}\left(\gamma - \overline{A}(X,X^*)\right),$$

where we used the *mixed area* $A(K, M)$ for the mixed volume $V(K\,[1], M\,[1])$ in the plane and where M^* denotes the reflection of the set M in the origin. Since the density $\overline{A}(X, X^*)$ of the mixed area cannot be expressed in terms of $\overline{A}(X)$ or $\overline{L}(X)$, the three equations cannot be used, as in the isotropic case, to obtain an estimator for γ (by only measuring $\overline{A}(Z)$, $\overline{L}(Z)$ and $\overline{\chi}(Z)$).

The solution in this situation is to use a local counterpart of the second equation, namely

$$\overline{\Psi}_1(Z, \cdot) = e^{-\overline{A}(X)} \overline{\Psi}_1(X, \cdot). \tag{18}$$

In view of the symmetry of the mixed area $A(K, M)$ (in K and M) and (9), (18) is equivalent to a corresponding equation for densities of support functions (where we use again additive extension on the left-hand side),

$$\overline{h}(Z, \cdot) = e^{-\overline{A}(X)} \overline{h}(X, \cdot). \tag{19}$$

The measure $\overline{\Psi}_1(Z, \cdot)$ is the *mean normal measure* of the random set Z, its normalized version gives the distribution of the outer normal in a typical boundary point of Z. In the symmetric case (that is, the case where $\overline{\Psi}_1(Z, \cdot)$ is an even measure), $\overline{\Psi}_1(Z, \cdot)$ can be estimated from random lines or line segments (see [15]), an estimation procedure in the general case is given in [18]. The mean support function $\overline{h}(Z, \cdot)$ is convex (as follows from (19)), it is thus the support function of a convex body $M(Z)$, the *mean body* of Z. The mean normal measure $\overline{\Psi}_1(Z, \cdot)$ is the surface area measure of the mean body,

$$\overline{\Psi}_1(Z, \cdot) = \Psi_1(M(Z), \cdot).$$

Since

$$\begin{aligned}
\overline{A}(X, X^*) &= \gamma^2 \int_{\mathcal{K}_0^2} \int_{\mathcal{K}_0^2} A(K, M^*) \mathbb{P}_0(dK) \mathbb{P}_0(dM) \\
&= \int_{S^1} \overline{h}(X, -u) \overline{\Psi}_1(X, du),
\end{aligned}$$

we thus can use the equation for $\overline{\chi}(Z)$ to get an estimator for γ. More details can be found in [25].

A similar method still works for $d = 3$ and $d = 4$ (see [26,28]).

Finally, we consider the non-stationary case. Non-stationary random sets and, in particular, non-stationary Boolean models have been of recent interest and a number of authors have considered appropriate definitions of the volume density, the surface area density and densities of other quermass integrals for three-dimensional random set models, in particular for Boolean models of balls (see [6,7,8,13,16,17]). We describe here a general approach for non-stationary Boolean models, due to [2,3].

If X (and hence Z) is inhomogeneous (not stationary), (2) does not hold anymore, and we have to replace it by an appropriate desintegration property of the intensity measure Θ,

$$\Theta(\mathcal{A}) = \int_{\mathcal{K}_0^d} \int_{\mathbb{R}^d} 1_{\mathcal{A}}(M + x) f(x) \lambda_d(dx) \mathbb{P}_0(dM), \tag{20}$$

for all Borel sets $\mathcal{A} \subset \mathcal{K}^d$. The function f may be called the *local intensity* (or *spatial density*), $\mathbb{P}_0$ is again the *distribution of the typical particle* (in [2,3] a more general situation is discussed where the local intensity may even depend on M). The definition of the quermass densities of Z as a limit for increasing sampling windows, which worked in the stationary case, is not applicable in the non-stationary case anymore. However, the quermass density $\overline{V}_j(Z)$ of Z (in the stationary case) fulfills also the alternative representation

$$\mathbb{E}\Phi_j(Z, \cdot) = \overline{V}_j(Z)\lambda_d,$$

hence $\overline{V}_j(Z)$ is the (constant) density of the measure $\mathbb{E}\Phi_j(Z, \cdot)$ with respect to the Lebesgue measure (similar results hold for the densities and mixed densities of X). This shows how to define local densities in the inhomogeneous case. As was shown in [2,3],

(1) $\mathbb{E}\Phi_j(Z, \cdot)$ is absolutely continuous to λ_d. We denote the Radon-Nikodym derivative by $\overline{V}_j(Z; \cdot)$ and call this the *(local) quermass density* of Z.

(2) $\mathcal{E}\sum_{(K_1,\ldots,K_k)\in X_{\neq}^k} \Phi_{m_1,\ldots,m_k}^{(j)}(K_1, \ldots, K_k, \cdot)$ is absolutely continuous to the product measure $\lambda_d^k = \lambda_d \otimes \cdots \otimes \lambda_d$. We denote the corresponding Radon-Nikodym derivative by $\overline{V}_{m_1,\ldots,m_k}^{(j)}(X, \ldots, X; \cdot)$.

Note that the quermass density $\overline{V}_j(Z; \cdot)$ is now a function on $\mathbb{R}^d$ which is almost everywhere defined (accordingly, the mixed density $\overline{V}_{m_1,\ldots,m_k}^{(j)}(X, \ldots, X; \cdot)$ is a function on $(\mathbb{R}^d)^k$).

The following theorem of [2,3] is the counterpart to the results in the stationary, respectively stationary and isotropic case. Its proof is based on the iterated translative formula (15) for curvature measures.

Theorem. *For a general Boolean model Z in $\mathbb{R}^d$ fulfilling (20), we have for almost all $z \in \mathbb{R}^d$*

$$\overline{V}_d(Z; z) = 1 - e^{-\overline{V}_d(X;z)}$$

and

$$\overline{V}_j(Z; z) = e^{-\overline{V}_d(X;z)}\Big\{\overline{V}_j(X; z)$$
$$+ \sum_{k=2}^{d-j} \frac{(-1)^{k+1}}{k!} \sum_{\substack{m_1,\ldots,m_k=j+1 \\ m_1+\cdots+m_k=(k-1)d+j}}^{d-1} \overline{V}_{m_1,\ldots,m_k}^{(j)}(X, \ldots, X; z, \ldots, z)\Big\},$$

for $j = 0, \ldots, d-1$.

Generalizations to densities of mixed volumes are considered in [28].

In order to interpret this result, we concentrate on the planar case again and assume a process X of discs with random radii. Let G be the distribution function of the radii, and let us assume that G has a density g (the following formulae can be easily extended to more general situations). Let

$$\tilde{G}(x) = 1 - G(\|x\|),$$

$$\tilde{g}(x) = g(\|x\|),$$

$$\hat{g}(x) = \frac{g(\|x\|)}{\|x\|},$$

for all $x \in \mathbb{R}^2 \setminus \{0\}$. Then,

$$\overline{A}(X; \cdot) = f * \tilde{G},$$

$$\overline{L}(X; \cdot) = f * \tilde{g},$$

$$\overline{\chi}(X; \cdot) = f * \hat{g},$$

where $*$ denotes the (ordinary) convolution of functions. As a consequence, we get the following theorem.

Theorem. *Let Z be a general Boolean model of discs in $\mathcal{R}^2$ with radius distribution function G (and corresponding density function g) and local intensity function f. Then, we have for almost all $z \in \mathbb{R}^2$*

$$\overline{A}(Z; z) = 1 - e^{-(f*\tilde{G})(z)},$$

$$\overline{L}(Z; z) = e^{-(f*\tilde{G})(z)}(f * \tilde{g})(z),$$

$$\overline{\chi}(Z; z) = e^{-(f*\tilde{G})(z)}\left[(f * \hat{g})(z) - ((f \otimes f) *_{\alpha \cdot \sin \alpha} (\tilde{g} \otimes \tilde{g}))(z, z)\right],$$

with the $\alpha \cdot \sin \alpha$-convolution of the tensor products defined as

$$((f \otimes f) *_{\alpha \cdot \sin \alpha} (\tilde{g} \otimes \tilde{g}))(z, z)$$

$$= \int_{\mathbb{R}^2} \int_{\mathbb{R}^2} f(z - x) f(z - y) \alpha(x, y) \sin \alpha(x, y) \tilde{g}(x) \tilde{g}(y) \lambda_2(dx) \lambda_2(dy)$$

($\alpha(x, y)$ is the angle between x and y).

This and other results are discussed in [28]. Related formulae have also been given in dimensions 2 and 3 by [13]. In the recent paper [29], the formulae for discs are extended to more general particle shapes and it is shown that, also in the inhomogeneous planar case, the quermass densities and mixed densities of Z determine the density $\overline{\chi}(X; \cdot)$ of the Euler characteristic of X uniquely.

References

1. D. Daley, D. Vere-Jones (1988): *An Introduction to the Theory of Point Processes.* (Springer, New York)
2. Fallert, H. (1992): *Intensitätsmaße und Quermaßdichten für (nichtstationäre) zufällige Mengen und geometrische Punktprozesse.* PhD Thesis, Universität Karlsruhe.

3. Fallert, H. (1996): 'Quermaßdichten für Punktprozesse konvexer Körper und Boolesche Modelle'. *Math. Nachr.* **181**, pp. 165–184.
4. Federer, H. (1959): 'Curvature measures'. *Trans. Amer. Math. Soc.* **93**, pp. 418–491.
5. Groemer, H. (1978): 'On the extension of additive functionals on classes of convex sets'. *Pacific J. Math.* **75**, pp. 397–410.
6. Hahn, U., A. Micheletti, R. Pohlink, D. Stoyan, H. Wendrock (1999): 'Stereological analysis and modelling of gradient structures'. *J. Microscopy* **195**, pp. 113–124.
7. Hahn, U., D. Stoyan (1998): 'Unbiased stereological estimation of the surface area of gradient surface processes'. *Adv. Appl. Probab. (SGSA)* **30**, pp. 904–920.
8. Hug, D., G. Last (1999): 'On support measures in Minkowski spaces and contact distributions in stochastic geometry'. *Annals Probab.*, to appear.
9. Klain, D., G.-C. Rota (1997): *Introduction to Geometric Probability.* (Cambridge Univ. Press, Cambridge)
10. Matheron, G. (1975): *Random Sets and Integral Geometry.* (Wiley, New York)
11. McMullen, P. (1993): 'Valuations and dissections'. In: *Handbook of Convex Geometry*, ed. by P. Gruber, J.M. Wills (Elsevier Science Publ., Amsterdam), pp. 933–988.
12. McMullen, P., R. Schneider (1983): 'Valuations on convex bodies'. In: *Convexity and Its Applications.* ed. by P. Gruber, J.M. Wills (Birkhäuser, Basel), pp. 170–247.
13. Mecke, K. (1994): *Integralgeometrie in der statistischen Physik.* (Verlag Harri Deutsch, Thun)
14. Molchanov, I.S. (1997): *Statistics of the Boolean Model for Practitioners and Mathematicians.* (Wiley, New York)
15. Molchanov, I.S., D. Stoyan (1994): 'Asymptotic properties of estimators for parameters of the Boolean model'. *Adv. Appl. Probab.* **26**, pp. 301–323.
16. Quintanilla, J., S. Torquato (1997a): 'Microstructure functions for a model of statistically inhomogeneous random media'. *Phys. Rev. E* **55**, pp. 1558–1565.
17. Quintanilla, J., S. Torquato (1997b): 'Clustering in a continuum percolation model'. *Adv. Appl. Probab. (SGSA)* **29**, pp. 327–336.
18. Rataj, J. (1996): 'Estimation of oriented direction distribution of a planar body'. *Adv. Appl. Probab. (SGSA)* **28**, pp. 394–404.
19. Schneider, R. (1993): *Convex Bodies: the Brunn-Minkowski Theory.* (Cambridge Univ. Press, Cambridge)
20. Schneider, R., W. Weil (1986): 'Translative and kinematic integral formulae for curvature measures'. *Math. Nachr.* **129**, pp. 67–80.
21. Schneider, R., W. Weil (1992): *Integralgeometrie.* (Teubner, Stuttgart)
22. Schneider, R., J.A. Wieacker (1993): 'Integral geometry'. In: *Handbook of Convex Geometry*, ed. by P. Gruber, J.M. Wills (Elsevier Science Publ., Amsterdam), pp. 1349–1390.
23. Stoyan, D., W.S. Kendall, J. Mecke (1995): *Stochastic Geometry and Its Applications.* 2nd edn. (Wiley, New York)
24. Weil, W. (1990): 'Iterations of translative integral formulae and non-isotropic Poisson processes of particles'. *Math. Z.* **205**, pp. 531–549.
25. Weil, W. (1995): 'The estimation of mean shape and mean particle number in overlapping particle systems in the plane'. *Adv. Appl. Probab.* **27**, pp. 102–119.
26. Weil, W. (1999a): 'Intensity analysis of Boolean models'. *Pattern Recognition* **32**, pp. 1675–1684.
27. Weil, W. (1999b): 'Mixed measures and functionals of translative integral geometry'. *Math. Nachr.*, to appear.

110   Wolfgang Weil

28. Weil, W. (1999c): 'Densities of mixed volumes for Boolean models', in preparation.
29. Weil, W. (1999d): 'A uniqueness problem for non-stationary Boolean models'. *Suppl. Rend. Circ. Mat. Palermo, II. Ser.*, to appear.

Additivity, Convexity, and Beyond: Applications of Minkowski Functionals in Statistical Physics

Klaus R. Mecke

Fachbereich Physik, Bergische Universität
D - 42097 Wuppertal, Germany

Abstract. The aim of this paper is to point out the importance of geometric functionals in statistical physics. Integral geometry furnishes a suitable family of morphological descriptors, known as Minkowski functionals, which are related to curvature integrals and do not only characterize connectivity (topology) but also content and shape (geometry) of spatial patterns. Since many physical phenomena depend essentially on the geometry of spatial structures, integral geometry may provide useful tools to study physical systems, in particular, in combination with the Boolean model, well known in stochastic geometry. This model generates random structures by overlapping 'grains' (spheres, sticks) each with arbitrary location and orientation. The integral geometric approach to stochastic structures in physics is illustrated by applying morphological measures to such diverse topics as complex fluids, porous media and pattern formation in dissipative systems. It is not intended to cover these topics completely but to emphasize unsolved physical problems related to geometric features and to present ideas and proposals for future work in possible collaboration with spatial statisticians and statistical physicists.

1 Motivation: Complex Patterns in Statistical Physics

The spatial structure of systems becomes more and more important in statistical physics. For instance, transport properties in porous media depend on the shape and statistical distribution of the pores (see Figs. 1 and 2, [13,49], and the paper by Hilfer in this volume). Also the interest in microemulsions and colloidal suspensions rests primarily on the complex spatial structures of mesophases, e.g., the bicontinuous phases on a mesoscopic scale of an emulsion of oil and water stabilized by amphiphiles [17]. Consider, for instance, an ensemble of hard *colloidal particles* (black points in Fig. 1) surrounded by a fluid wetting layer (white). The interactions between these colloids are then given by the free energy of the spatially complex structured fluid film which may cause eventually a phase separation of the hard particles (see the paper by H. Löwen in this volume).

Dissipative structures in hydrodynamics, mesophases of liquid crystals, or Turing patterns occurring in chemical reactions are other examples where physics has to focus on the spatial patterns when trying to understand the physical properties of these systems [25,40]. For instance, inhomogeneous concentrations of molecules can occur spontaneously in chemical reactions when diffusion of the species plays a role. The regular hexagonal cells, the stripe patterns and the turbulent, time dependent structures shown in Fig. 4 are only a few examples of the many different morphologies which may occur [25]. Rapid cooling of a fluid from above the critical point deep into the two-phase coexistence of liquid and vapor is probably the easiest way to generate inhomogeneous concentrations of particles. The so-called spinodal decomposition of liquid and vapor exhibit transient spatial structures so that size and shape of homogeneous phases are changing in time [7,20].

The physical properties of such complex spatially fluctuating structures are studied in statistical physics combined with many different disciplines and methods. For instance, pattern growth is a phenomenon of nonlinear dynamics, random spatial structures are considered in stochastic geometry and integral geometry, whereas digital image analysis and mathematical morphology have been developed mainly in biology, medical and material science. Therefore, the **aim of morphology in statistical physics** is the application of mathematical models, particularly models using overlapping grains of arbitrary size and shape (germ grain models), to describe stochastic geometries which occur in physical systems and to use the methods of spatial statistics and integral geometry to analyze them [43,61]. The present paper attempts to present different applications of Minkowski functionals grouped in four main sections, where at the end of each subsection problems are mentioned which will be solved hopefully in the near future.

First, Minkowski functionals M_ν are introduced as *additive* functionals of spatial patterns in Sect. 2. Minkowski functionals play a crucial role in integral geometry, a mathematical discipline aiming for a geometric description of objects using integral quantities instead of differential expressions. In the three-dimensional Euclidean space the family of Minkowski functionals consists of the volume $V = M_0$, the surface area $S = 8M_1$ of the pattern, its integral mean curvature $H = 2\pi^2 M_2$, and integral of Gaussian curvature, i.e., the Euler characteristic $\chi = \frac{4\pi}{3}M_3$. Additivity of the functionals is the relevant condition for Hadwiger's theorem which is the backbone of integral geometry. Therefore, not only theorems and formula but also many physical applications depend on additivity. Three examples are given where such geometric measures are important in physics: the energy of a biological membrane is determined by its curvature. Phase coexistence of fluids in porous media depends on the structure function of the substrate and therefore on the geometry of the pores. And the spectral density of the Laplace operator, which is needed whenever a field in a domain is oscillating, is given in geometric terms of the domain. All three examples reveal a limited applicability of Minkowski functionals due to non-additive features of the physical system considered. The hope of future work is that the additive frame given by Minkowski functionals can be extended to additional non-additive measures without loosing the appealing features of integral geometry.

In Sect. 3 a major problem in statistical physics is considered, namely the definition of relevant order parameters of spatial structures. Since the amount of spatial information is growing fast in physics and material science due to scanning microscope techniques, digital image recording, and computer simula-

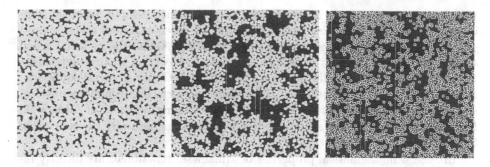

Fig. 1. *Porous media* (left) can be described by overlapping grains (spheres, discs) distributed in space. If the density of grains (white) decreases below a threshold, an infinite cluster of connected pores (black area) is spanning through the whole system. This cluster of pores enables the transport of fluids, for instance. The knowledge of the dependence of the so-called percolation threshold on the shape and distribution of the grains is essential for many applications. Inhomogeneous domains of thermo-dynamic stable phases of complex fluids may also be described by overlapping grains [9,35,38,39,43]. Such configurations resemble, for instance, the structure of *microemul-sions* (figure in the middle) or an ensemble of hard *colloidal particles* (black points in the figure on the right) surrounded by a fluid wetting layer (white). The interac-tions between these colloids, as well as the free energy of the homogeneous oil phase in a microemulsion are given by a bulk term (volume energy), a surface term (surface tension), and curvature terms (bending energies) of the white region covered by the overlapping shapes. Thus, the spatial structure of the phases, i.e., the morphology of the white regions determines the configurational energy which determines itself the spatial structure due to the Boltzmann factor in the partition function of a canonical ensemble. A main feature of complex fluids is the occurrence of different length scales: the clusters of the particles, i.e., the connected white regions are much larger than the 'microscopic' radius of the discs and the typical nearest neighbor distance within a cluster.

tions, the scientist faces the problem of reducing the information to a limited number of relevant quantities. So far powerful methods have been developed in Fourier space, namely structure functions and more recently wavelet anal-ysis. But techniques to analyze spatial information directly in real space may be very useful for physicists in order to get more relevant spatial information out of their data which may be complement to structure functions measured by scattering techniques in Fourier space. Such techniques and measures have been developed in spatial statistics and the interested reader is referred to the papers by D. Stoyan and W. Nagel in this volume. To this world also belong the additive Minkowski functionals which may offer robust morphological mea-sures as powerful tools which is illustrated by three examples: they can be used as *order parameters* characterizing pattern transitions in dissipative systems, as *dynamical quantities* characterizing spinodal decomposition, or as generalized *molecular distribution functions* characterizing the atomic structure of simple fluids. The additivity of the Minkowski functionals seems to be the relevant property which causes the interesting features of robustness and universal form of these morphological order parameters. But all three examples reveal a need of theory in statistical physics: so far there exist no Landau theory as for usual order parameters of phase transitions, no dynamic theory as for time dependent two-point correlation functions, and no perturbation theory (integral equations) as for structure functions of fluids. Since fractal patterns occur in many physical systems we consider in a fourth subsection the scaling behavior of Minkowski functionals for self-similar structures which generalizes naturally the concept of fractal dimensions.

In Sect. 4 mathematical properties of Minkowski functionals are directly related to problems occurring in statistical physics: the kinematic formula helps to calculate virial coefficients and to find accurate density functionals; the notion of an 'excluded volume of a grain' helps to predict percolation thresholds; and the completeness of additive functionals allows the formulation of a general morphological model for complex fluids. All three examples exhibit promising advances although severe drawbacks occur: third and higher virial coefficients in a cluster expansion are not yet tractable in terms of Minkowski functionals; the accuracy of threshold estimates gained by local Minkowski functionals need to be improved; and the thermodynamics of complex fluids require analytic expressions for Minkowski functionals in the Boolean model beyond mean values of additive measure.

The previous problem immediately leads to Sect. 5 where the Boolean model is used to obtain explicit analytic expressions for thermal averages needed in statistical physics: mean values (intensities) of Minkowski functionals for inhomogeneous distributions of grains which are necessary for an improved density functional theory; mean values of correlated distributions of grains which are necessary for the morphology of most spatial patterns in physics; second order moments of the Minkowski functionals which are necessary for high temperature expansions and fluctuation theory in thermal systems. Since many models in statistical physics are using lattices as a basis for spatial configurations, the concepts of integral geometry on lattices are introduced in the last subsection. Applying an adopted kinematic formula one can calculate not only exact mean values and variances of Minkowski functionals but also distribution functions of local configurations.

The aim of this paper is neither to introduce or review geometric functionals and integral geometry (recent developments are presented by W. Weil in this volume), nor to give a complete list of references. Also, it is not intended to cover every interesting phenomenon related to the physical topics considered but only to apply the concepts of integral geometry and the notion of morphological measures such as Minkowski functionals to selected problems among them and to focus on applications, problems and possible advances in the near future. The synopsis at the end of this paper may help to give an overview of *Integral Geometry in Statistical Physics*.

But let us first define the stage of most physical applications considered here: typical spatial configurations in statistical physics show striking similarities to

configurations of the so-called *Boolean model* (see also Fig. 2 in the paper by
D. Stoyan in this volume). This stochastic model generates spatial structures
by overlapping grains of arbitrary shape and size. For convenience, spheres or
discs of fixed radius are often used. Such configurations are shown in Fig. 1 of
the present paper where correlations between the grains are introduced by physically motivated interactions. But also the structures in Fig. 4 can be described
as configurations of an 'overlapping grain model', because they consist of a lattice of squared pixels on a non-visible 'microscopic' scale. Since a number of
lattice models have been defined and extensive computer simulations have been
performed for percolation phenomena [60] and microemulsions [18,46] not many
continuum models have been studied. A model based on correlated overlapping
grains (germ grain model, see Fig. 1) seems to be a promising starting point. Also
for the *morphological analysis* of point patterns - experimentally observable as
holes in thin liquid films (see Fig. 6 and the paper by K. Jacobs et al. in this volume), as positions of particles in a fluid (Fig. 9), or as distribution of the galaxies
in the universe (see the paper by M. Kerscher in this volume) - the overlapping
grain model and the Minkowski functionals may be used. For instance, if one
decorate each point of the pattern, i.e., each particle in Fig. 9 and each galaxy
in the universe with a sphere of varying radius R one obtains a scale dependent
covering of space which can be morphologically characterized in the same way as
the configurations shown in Fig. 1 and 2 for each radius r. Computer algorithms
have been developed in two and three dimensions to calculate these r-dependent
Minkowski functions and in particular the Euler characteristic of such coverings
which can be compared with analytic results of stochastic models [9,10,36].

Integral geometry turns out to be an important mathematical method to describe complex patterns in space because of a complete family of morphological
measures, the so-called *Minkowski functionals* [21,52]. These measures are increasingly used in digital picture analysis, in particular due to their stereological
properties [50,57,64]. A prominent member of this family is the *Euler characteristic* which describes the connectivity of spatial structures. Thus, these measures
do not only describe size (volume) and geometry (curvatures) of domains such
as shown in Figs. 1, 2, and 4 but also the topology of the pattern. Besides the
definition of morphological measures integral geometry provide theorems, formulae, and elegant calculus in order to derive exact results, in particular, for the
Boolean model [34], a standard model of stochastic geometry [61]. This will be
considered in detail in Sect. 5. Sections 2-4 focus solely on the measures them-self
and their occurrence in statistical physics.

2   Geometric Functionals: Curvatures in Physics

Let us first recall in this section some basic facts from integral geometry since the
methods are perhaps not widely known among physicists. We compile only some
pertinent facts in order to introduce the notations and refer to the literature
[21,52,54,64) and to the contribution by W. Weil in this volume for more details
and thorough derivations of the theorems and formulae.

Geometric functionals, known as Minkowski functionals or intrinsic volumes (quermassintegrale, curvature integrals) in integral geometry, may be introduced as integrals of curvatures using differential geometry of smooth surfaces. Let A be a compact domain in $\mathbb{R}^d$ with regular boundary $\partial A \in \mathcal{C}^2$ and $d-1$ principal radii of curvature R_i ($i = 1, \ldots d-1$). The functionals $W_\nu(A)$, with $\nu \geq 1$, can be defined by the surface integrals

$$W_{\nu+1}(A) = \frac{1}{(\nu+1)\binom{d}{\nu+1}} \int_{\partial A} S_\nu(\frac{1}{R_1}, \ldots, \frac{1}{R_{d-1}}) dS \qquad (1)$$

where S_ν denotes the ν-th elementary symmetric function and dS the $(d-1)$-dimensional surface element. Especially in three dimensions one obtains

$$W_1 = \frac{1}{3} \int dS, \quad W_2 = \frac{1}{3} \int H dS, \quad W_3 = \frac{1}{3} \int G dS \qquad (2)$$

with the Gaussian

$$G = \frac{1}{R_1 R_2} \qquad (3)$$

and the mean curvature

$$H = \frac{1}{2}\left(\frac{1}{R_1} + \frac{1}{R_2}\right). \qquad (4)$$

Although the Minkowski functionals are introduced as curvature integrals, they are well-defined for polyhedra with singular edges [52]. Therefore, one can define these functionals naturally for lattice configurations (Sect. 5.4). It is convenient to normalize the functionals

$$M_\nu(A) = \frac{\omega_{d-\nu}}{\omega_\nu \omega_d} W_\nu(A), \quad \nu = 0, \ldots, d, \qquad (5)$$

using the volume ω_d of a d-dimensional unit sphere $\omega_d = \pi^{d/2}/\Gamma(1 + d/2)$, namely $\omega_1 = 2$, $\omega_2 = \pi$, and $\omega_3 = 4\pi/3$. In three dimensions the family of Minkowski functionals consists of the volume $V = M_0$, the surface area $S = 8M_1$ of the coverage, its integral mean curvature $H = 2\pi^2 M_2$, and the Euler characteristic $\chi = \frac{4\pi}{3} M_3$. In the mathematical literature the normalization $V_\nu(A) = \binom{d}{\nu} W_{d-\nu}(A)/\omega_{d-\nu}$ is frequently used, for instance, by W. Weil in this volume, where V_ν are called the *intrinsic volumes* of A (also 'quermassintegral' or 'curvature integral' are used).

Before the main theorems and formulae of integral geometry are used in Sect. 4, let me first give some interesting properties of Minkowski functionals and examples of physical systems where they play a major role. The most important property of Minkowski functionals is additivity, i.e., the functional of the union $A \cup B$ of two domains A and B is the sum of the functional of the single domains subtracted by the intersection

$$M_\nu(A \cup B) = M_\nu(A) + M_\nu(B) - M_\nu(A \cap B). \qquad (6)$$

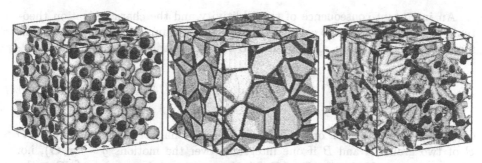

Fig. 2. Parallel sets of non-convex spatial configurations [2]: (A) the parallel body A_R of distance R of Poisson distributed points is the Boolean model of overlapping spheres of radius R; but the parallel body of facets (B) and edges (C) of a Voronoi tessellation yields completely different spatial structures and Minkowski functionals $M_\nu(A_R)$ as functions of R. The fraction of the covered volume is $v = 0.25$ in all three configurations. Parallel surface can not only be used to characterize spatial non-convex patterns such as point distributions (Sect. 3.3), foams, gels, fractals (Sect. 3.4), or chemical patterns (Sect. 3.1), but also to define appropriate stochastic models for porous media, network models for percolation, and fluid wetting layers near substrates (Sect. 2.2).

This relation generalizes the common rule for the addition of the volume of two domains to the case of a general morphological measure, i.e. the measure of the double-counted intersection has to be subtracted.

A remarkable theorem is the 'completeness' of the Minkowski functionals proven 1957 by H. Hadwiger [21]. This *characterization theorem* asserts that any additive, motion-invariant and conditionally continuous functional $\mathcal{M}$ is a linear combination of the $d + 1$ Minkowski functionals M_ν,

$$\mathcal{M}(A) = \sum_{\nu=0}^{d} c_\nu M_\nu(A) \,, \tag{7}$$

with real coefficients c_ν independent of A. Motion-invariance of the functional means that the functional $\mathcal{M}$ does not dependent on the location and orientation of the grain A. Since quite often the assumption of a homogeneous and isotropic system is made in physics, motion-invariance is not a very restrictive constraint on the functional. Nevertheless, in the case where external fields are applied which are coupled to the orientation of the grains, motion invariance cannot further be assumed and extensions are necessary (see Problem: Anisotropy in Sect. 5).

The Minkowski functional W_ν is homogeneous of order $d - \nu$, i.e. for a dilated domain λA one obtains

$$W_\nu(\lambda A) = \lambda^{d-\nu} W_\nu(A) \,. \tag{8}$$

This scaling property will be used below to study fractals (Sect. 3.4) and domain growth during phase separation (Sect. 3.2), for instance.

An important consequence of the additivity and the characterization theorem (7) is the possibility to calculate analytically certain integrals of Minkowski functionals. For instance, the **kinematic fundamental formula** [6,52]

$$\int_{\mathcal{G}} M_\nu(A \cap gB) dg = \sum_{\mu=0}^{\nu} \binom{\nu}{\mu} M_{\nu-\mu}(B) M_\mu(A) . \tag{9}$$

describe the factorization of the Minkowski functionals of the intersection $A \cap B$ of two grains A and B if one integrates over the motions $g = (\mathbf{r}, \Theta)$, i.e. translations $\mathbf{r}$ and rotations Θ of B. The integration $\int d\mathbf{g} = \int d\mathbf{r} \times \int d\Theta$ is the direct product of the integrations over all translations and orientations. Similar kinematic formula can be derived for oriented cuboids (see Eq. (86) in Sect. 5.2) and for configurations on lattices (Eq. (110), Sect. 5.4). Kinematic formulae are extremely useful tools to calculate mean values of Minkowski functionals for random distributions of grains (see Eqs. (61) and (74)).

In the case that B is a sphere $B_\epsilon(x)$ of radius ϵ centered at x and $A = K$ is a convex grain K the kinematic formula (9) for $\nu = d$ reproduces Steiner's formula for convex sets

$$V_\epsilon(K) = \int_{\mathbb{R}^d} \chi(K \cap B_\epsilon(x)) d^d x = \sum_{\nu=0}^{d} \binom{d}{\nu} W_\nu(K) \epsilon^\nu . \tag{10}$$

Interesting applications of Steiner's formula (10) for the excluded volume of a body concern an effective interaction potential for non-spherical molecules (introduced in Sect. 4.1) and the estimation of percolation thresholds (Sect. 4.2). Whereas in the Sects. 2 and 3 mainly the elementary properties of Minkowski functionals given by Eqs. (6) - (8) are used, physical applications of the kinematic formula are considered in the Sects. 4 and 5.

The finite sum for the parallel volume in Eq. (10) and the decomposition of the integral in Eq. (9) into products of Minkowski functionals is a direct consequence of the additivity relation (6). Additivity is the relevant condition in Hadwiger's theorem (7) which is the backbone of integral geometry. To summarize this short introduction an early attempt is made to answer the question why Minkowski functionals should be used in statistical physics:

(i) Curvatures are widespread in physics and integral geometry defines them even on edges of polyhedra (Sect. 2).
(ii) The additivity and scaling relations qualify them as order parameters for spatial structures (Sect. 3).
(iii) Statistical physics of fluids uses often the concept of parallel bodies which can be expressed by Minkowski functionals (Sect. 4).
(iv) Thermal averages require integrals over the group of motions so that the kinematic formula can be applied naturally (Sect. 5).

2.1   Curvature Energy of Membranes

Integrals over curvatures of a surface occur in many applications in physics: For instance, biological cells are bounded by membranes which consist of a bilayer of amphiphilic lipid molecules. From a physical point of view one can neglect not only attached proteins and other biological relevant details and consider solely the membrane as a flexible two-dimensional sheet. Cells and membranes show a remarkable variety of shapes which changes with temperature and the chemical nature of the surrounding fluid. The knowledge of the energy which is necessary to bend a membrane is essential for a physical study of such systems. Motivated by the elastic energy of a thin sheet, Helfrich [23] proposed the energy of a membrane

$$\mathcal{H} = \int_{\mathcal{M}} \sigma d\mathcal{S} + \int \frac{\kappa}{2}(H - H_0)^2 d\mathcal{S} + 2\pi\bar{\kappa} \int G d\mathcal{S} \quad = \quad \sum_{\nu=1}^{3} h_\nu W_\nu + \frac{\kappa}{2}\hat{H}^2 \quad (11)$$

given in terms of the mean and Gaussian curvature integrated over the membrane $\mathcal{M}$. Here, H_0 denotes a spontaneous mean curvature, σ the surface tension, and κ, $\bar{\kappa}$ bending rigidities determined by the molecular details of the membrane. Except of the integral over the squared mean curvature

$$\hat{H}^2 := \int H^2 d\mathcal{S} \,, \tag{12}$$

the energy (11) is additive and therefore given by the Minkowski functionals W_ν of the cell weighted by material constants h_ν. Although the Helfrich Hamiltonian (11) describes accurately the energy of membranes, one should notice that the curvature energy of a simple liquid interface is more difficult, i.e., contains additional 'non-additive' terms due to long range attrative interactions between the molecules [44].

Problem: Non-additivity of the Squared Mean Curvature $\hat{H}^2$ In many studies of microemulsion phases (Sect. 4.3) the Helfrich Hamiltonian (11) is used for random configurations of interfaces [17,18,30]. Using configurations on a lattice (Sect. 5.4) or a model of overlapping grains (Sect. 5.1) to generate such random interfaces (membranes) one has difficulties to define the integral of the squared mean curvatures due to singularities at the edges where two or three grains meet (Sect. 5). Unfortunately, the squared mean curvature $\hat{H}^2$ is not an additive functional, what causes several problems.

From a differential geometric point of view there is no relevant difference between the mean H and the squared mean curvature H^2. The essential difference occurs only within the context of integral geometry where one can define mean curvatures (but not H^2) even at singular edges by using Steiner's formula (10) or Crofton's definition of Minkowski functionals [43,52]. Of course, it is possible to define a mean curvature $\bar{H}(\mathbf{x}) = dW_2(K \cup K', \mathbf{x})$ locally at such an intersection point $\mathbf{x} \in \partial K \cap \partial K'$ and to integrate the square of the *local* mean curvature $\bar{H}(\mathbf{x})$ along the intersection line

$$\hat{H}^2 = \int\limits_{\partial K/K'} \left(H(\mathbf{x})\right)^2 dS + \int\limits_{\partial K'/K} \left(H(\mathbf{x})\right)^2 dS + \int\limits_{\partial K \cap \partial K'} \left(\bar{H}(\mathbf{x})\right)^2 dS . \quad (13)$$

But the properties of this functional are not studied yet and it is definitely not the only possible definition of $\hat{H}^2$ for surfaces with singular edges. Mean values of $\hat{H}^2$ have been calculated for the germ grain model of overlapping spheres so that at least thermal averages of the non-additive functional $\hat{H}^2$ can be used for applications in physics [35].

An interesting feature of the squared mean curvature $\hat{H}^2$ is revealed by studying thermal fluctuations of the membrane. Short-wavelength thermal undulations, i.e., capillary waves of wavelengths smaller than a given scale ξ yields a renormalization of the material constants h_ν, in particular, of the bending rigidity κ. Thus, on a scale ξ one observes an effective scale ξ-dependent bending energy $\mathcal{H}(\xi)$ obtained by replacing κ (h_ν, respectively) in the Helfrich Hamiltonian (11) with a renormalized rigidity $\kappa(\xi)$ ($h_\nu(\xi)$) to account for the reduction of the bare rigidity κ caused by the short-wavelength undulations. The scale-dependence of $\kappa(\xi)$ is known explicitly only to leading order in $k_B T/\kappa < 1$, and reads [37,46]

$$\kappa(\xi) = \kappa - \frac{3}{4\pi} k_B T \ln\left(\frac{\xi}{\delta}\right), \quad (14)$$

where the membrane thickness δ is used to provide a microscopic cutoff. The length ξ_κ at which $\kappa(\xi)$ has dropped to values of the order $k_B T$ is estimated from $\kappa(\xi_\kappa) = 0$. The persistence length ξ_κ defines the basic structural length scale in this approach. In the rigidity-dominated regime, $\xi < \xi_\kappa$, an isolated membrane is likely to be locally flat; for $\xi > \xi_\kappa$, so that $\kappa(\xi)$ is negligible compared with $k_B T$, thermal fluctuations dominate over the rigidity and the membrane is crumpled. Thus, for lengths beyond ξ_κ one can neglect the squared mean curvature H^2 in the Hamiltonian (11).

2.2 Capillary Condensation of Fluids in Porous Media

Another example where naturally integrals of curvature occur is the phase behavior of fluids in a porous medium. Foams, gels, and porous structures become increasingly important for technological applications due to their special material properties as spatially structured matter. The physical properties depend crucially on the morphology, i.e., on shape and connectivity of the pores [13]. For instance, knowledge of the dependence of percolation phenomena, in particular of the percolation threshold on the distribution of pores is necessary for many applications ranging from oil recovery to conductivities of modern materials (see Sect. 4.2 below).

An important phenomenon is *capillary condensation*, i.e., the reduction of the critical point and the shift of the equilibrium chemical potential and the equilibrium pressure towards lower values due to the interaction of the fluid with a substrate. In other words, a liquid starts to boil at a higher temperature if it is enclosed by a small box. This phenomenon is quite general and can be explained straightforwardly using a simple geometric configuration. Consider, for instance, two plates of distance D then the grand canonical potentials Ω of a homogeneous vapor and liquid phase are given by $\Omega_g = -pV + 2A\sigma_{sg}$ and $\Omega_l = -p^+V + 2A\sigma_{sl}$, respectively. Here, p is assumed to be the unaltered bulk vapor pressure whereas p^+ is the pressure of a metastable fluid phase stabilized by the substrate. A is the surface area of the substrate and σ denotes the surface tension between the solid substrate and the fluid phases. Thermodynamic equilibrium requires $\Omega_g = \Omega_l$ and one obtains the well-known Kelvin equation

$$p - p^+ = 2\frac{\sigma_{sg} - \sigma_{sl}}{D} \simeq 2\frac{\sigma_{lg}}{D} > 0 \tag{15}$$

for the coexistence of a fluid and a vapor phase between two planar walls of distance D. Thus, if the distance of the walls becomes small the pressure difference $p - p^+$ forces a vapor to condense while it remains gaseous in the bulk outside of the slit. But what happens in a real porous substrate?

Using a density functional theory (see the paper by H. Löwen in this volume and also Sects. 5.1 and 4.1) for fluids in a porous medium one can calculate the shift of the critical point and of the boiling temperature in terms of geometric measures of the porous substrate. The present analysis is based on a simple version of density functional theory for one-component fluids which consist of particles with a rotationally symmetric pair interaction potential $\phi(r)$. Within this approach the interaction potential $\phi(r) = w_s(r) + w(r)$ is split into a short-ranged repulsive part $w_s(r)$ and a long-ranged attractive part $w(r)$ [14]. The interaction between the fluid and the substrate is taken into account by a potential $V_S(\mathbf{r})$. The grand canonical density functional reads

$$\Omega\left[\rho(\mathbf{r})\right] = \int_V d^3r \left[f_h(\rho(\mathbf{r})) - \mu\rho(\mathbf{r}) + \rho(\mathbf{r})V_S(\mathbf{r})\right]$$
$$+ \frac{1}{2}\int_V d^3r \int_V d^3r' w(\|\mathbf{r} - \mathbf{r}'\|)\rho(\mathbf{r})\rho(\mathbf{r}') \tag{16}$$

where V is the volume of the sample, $\rho(\mathbf{r})$ the number density of the fluid particles at $\mathbf{r} = (x, y, z)$, $r = |\mathbf{r}|$, and $f_h(\rho)$ is the reference free energy of a system determined by the short-ranged contribution to the interaction potential $w_s(r)$. For these calculations we adopt the Carnahan-Starling expression $f_h(\rho) = k_BT\rho\left\{\ln(\rho\lambda^3) - 1 + \frac{4\eta - 3\eta^2}{(1-\eta)^2}\right\}$, where λ is the thermal de Broglie wavelength and $\eta = \frac{\pi}{6}\rho r_0^3$ the packing fraction (see [14] and the paper by H. Löwen in this volume).

Within this density functional approach the equilibrium density $\rho^{(eq)}(\mathbf{r})$ of the fluid inside the porous medium minimizes the functional $\Omega\left[\rho(\mathbf{r})\right]$ in Eq. (16) which yields the grand canonical potential $\Omega = \Omega\left[\rho^{(eq)}(\mathbf{r})\right]$. The equilibrium profile depends not only on the temperature T, the chemical potential μ, and the substrate potential $V_S(\mathbf{r})$ but also on the position $\mathbf{r}$ inside the pores. Nevertheless, one can show that the shift in the critical point and, accordingly, the difference in the equilibrium pressures of the fluid bulk phases is given by the simple expansion

$$(p - p^+)V = \sum_{\nu=1}^{3} h_\nu[\phi(r)]W_\nu + \frac{\kappa}{2} \int H^2 d\mathcal{S} + \dots \tag{17}$$

with coefficients h_ν depending on the system parameters and the interaction potential $\phi(x)$ of the fluid particles. For a slit of parallel flat walls of distance D one obtains $W_2 = W_3 = \hat{H}^2 = 0$ and recovers immediately the result given by Eq. (15). In other words, the relation (17) generalizes Kelvin's equation, which turns out to be the first term in a curvature expansion of the pressure difference $p - p^+$. The density functional approach gives also expressions for the coefficients $h_\nu[\phi(r)]$, i.e., for surface tensions σ_{sg} and bending rigidities in terms of the microscopic interaction potential $\phi(\mathbf{r})$. Assuming that the equilibrium fluid density $\rho^{(eq)}(\mathbf{r}) \equiv \rho$ is constant inside the pores one can derive an alternative expression to Eq. (17) based on the structure function $S(r)$ of the porous substrate $\mathcal{K}$. One finds, for instance, for the critical point shift

$$\delta T_c = \int_V d^3\mathbf{r} \frac{w(r)}{w(0)} e^{n v} (S(0) - S(r)) \tag{18}$$

and for Poisson distributed grains $S(r) = e^{-n v + n V(K \cap K_\mathbf{r})}$ (see the Boolean model in Sect. 5). Expanding the structure function in powers of the distance r [62]

$$S(r) = \frac{W_0(\mathcal{K})}{V} - \frac{3r}{4}\frac{W_1(\mathcal{K})}{V} - \frac{r^3}{32}\frac{W_3(\mathcal{K})}{V} + \frac{r^3}{32V} \int_{\partial\mathcal{K}} H^2\, dS + \mathcal{O}(r^5) \ . \tag{19}$$

one recovers an expression in terms of Minkowski functionals W_ν of the porous structure $\mathcal{K}$. Since the expansion (19) is only valid for sharp interfaces, the contribution proportional to the mean curvature H vanishes and the bending rigidities κ and $\bar{\kappa}$ are identical, which is not the case for smooth density profiles $\rho(\mathbf{r})$ across the interface. Inserting (19) in Eq. (18) one recovers a curvature expansion for the critical point shift $\delta T_c V$ analogous to the generalized Kelvin equation (17) with explicitly given coefficients

$$h_1[\phi(r)] = \pi \int_0^\infty dr\, r^3 w(r)\ , \qquad h_2[\phi(r)] = \frac{\pi}{24} \int_0^\infty dr\, r^5 w(r)\ . \tag{20}$$

Of course, the assumption of a homogeneous density of the fluid inside the pores is not valid near the critical point because the thickness of a fluid adsorption layer at the substrate wall is determined by the correlation length $\xi(T)$ which becomes large at T_c. Instead of the critical point shift $\delta T_c \sim D^{-1}$ as implied by

Eq. (18) for a fluid between walls of distance D, one obtains $\delta T_c \sim D^{-2}$ if density inhomogeneities of size ξ are taken into account. This might be done by applying the concept of parallel surfaces as indicated below. Although the expression (18) has to be improved it indicates a direction of future work, namely the prediction of thermodynamic properties of materials when the morphology is known. The structure function of a porous substrate, for instance, can be measured by scattering experiments independently from a calorimetric determination of the critical point.

Curvature expansions such as Eqs. (11), (17), and (18) are quite common in physics and very useful for practical purposes. Integral geometry and Minkowski functionals provide precisely the mathematical backbone and technical calculus for physical applications of curvature measures. In Sect. (5), for instance, densities of the geometric functionals W_ν are given for the Boolean model, so that explicit expressions for the phase behavior of fluids in porous media modeled by overlapping grains can be derived.

The complicated pore structure of an interconnected three-dimensional network of capillary channels of nonuniform sizes and shapes distinguishes a porous medium from any other solid or planar substrate. The connection of the two main features of fluids in porous media, namely morphology and interfacial effects such as surface energies and wettability, may help in future studies to understand the influence of the random geometric structure on phase behavior and transport properties, which are inherently determined by inhomogeneous spatial structures on all length scales of the porous material.

Problem: Parallel Surfaces and Non-additivity of Effective Properties

Starting from a microscopic density functional for inhomogeneous fluids in porous media the dependence of thermodynamical quantities on the geometry of the substrate has been determined and it was shown that the free energy, pressure, or the critical point of a fluid can be written approximatively as an additive functional of pore space. But so far wetting behavior of the fluid has been neglected although one knows that close to the substrate a fluid layer may form which influence the thermodynamic behavior. Consequently, one has to take into account inhomogeneouities of the fluid density which essentially depend on the distance to the substrate. Therefore, the integral geometric concept of parallel sets (see Steiner's theorem (10) and Sect. 3.1) may help to define an inhomogeneous fluid density to clarify how physical phenomena such as capillary condensation and two-phase flow depend on the wetting behavior of the fluid. Additionally, parallel sets can be used to define effective network models of porous structures in order to refine inversion percolation theories where well-defined morphological quantities such as pore volume distributions, throat sizes, and connectivities of pores are needed.

The parallel body A_ϵ of a structure A is defined as the set of all points with distances less than ϵ to A (see Fig. 2). Measuring the Minkowski functionals $M_\nu(A_\epsilon)$ one obtains detailed morphological information when ϵ is used as a diagnostic parameter. Changing ϵ corresponds to dilation ($\epsilon > 0$) and erosion

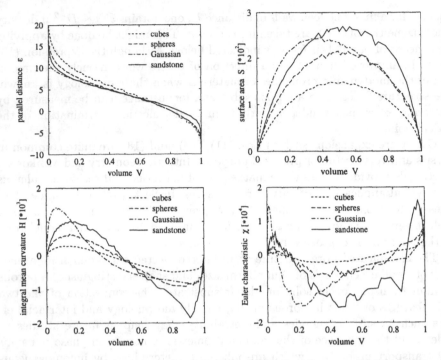

Fig. 3. Minkowski functions $M_\nu(\epsilon)$ of the parallel body (distance ϵ) of four different configurations: (A) Boolean model of overlapping oriented cubes (edge length $L = 16$) and (B) of spheres of radius $R = 8$; (C) a Gaussian random field and (D) of a sandstone - a real porous material. The three dimensional structures are digitized on a 128^3 lattice and lengths are given in units of the pixel size (for details see [2]). The Minkowski functions $M_\nu(\epsilon)$ as functions of the volume $V(\epsilon)$ of the parallel body are completely different for the four spatial structures and none of the models can reproduce the functions for the sandstone.

$(\epsilon > 0)$ of the spatial structure. For instance, in Fig. 3 the Minkowski functions $M_\nu(\epsilon)$ are shown for four different configurations: random overlapping but oriented cubes of edge length $L = 16$, overlapping spheres of radius $R = 8$, a Gaussian random field, and an image of a sandstone. Since experimental data are always digitized, a lattice of 128^3 pixels is used in order to digitize also the configurations of the models [38]. These functions may be used to characterize the spatial structure and to test the accuracy of model reconstructions of porous media (see paper by R. Hilfer in this volume). First results indicate that none of the models in Fig. 3 can reproduce the Minkowski functions of the sandstone [2]. Since not only the amplitudes but also the functional shape of the Minkowski functions $M_\nu(\epsilon)$ depend on many parameters such as the size of the grains and the porosity, a thorough analysis has to be done.

Another problem of applying Minkowski functionals is related to the measurement of the permeability $k = 10^{6.6} \left(\Phi^m V_p/S\right)^{2.1}$ or the NMR-relaxation time $T_1 = 10^{3.08} \left(V_p/S\right)^{0.9}$ in materials as function of porosity Φ and surface to volume ratio S/V_p of the pore space [56]. For a wide class of materials one finds for the elasticity $log E \approx b(V_p - V_c)$ with constants b and V_c. There are no explanations yet for the experimentally found relations which are obviously not additive functions. Hopefully it is possible to re-express these material properties (or their logarithms $\log k$, $\log T_1$) by taking into account all of the additive Minkowski functionals and to predict the conductivities and elasticity properties of multiple phases in porous materials more accurately. If it is possible, for instance, to relate the Minkowski functions shown in Fig. 3 to the physical properties of the sandstone, it would be sufficient for the prediction of experimental results to measure the geometry of the porous sample.

2.3   Spectral Density of the Laplace Operator

The short distance expansion (19) of the structure functions $S(r)$ corresponds to an asymptotic expansion in Fourier space

$$\tilde{S}(q) = \frac{S}{V}\frac{2\pi}{q^4} + \frac{\pi}{V}\frac{1}{q^6}\int_{\partial\mathcal{K}} dS\left(3H^2 - K\right) + \mathcal{O}(q^{-8}) \ . \tag{21}$$

Often one assumes that structure functions or 'propagators' in field theory are solutions of the Laplace equation

$$\Delta f(\mathbf{r}) = \gamma f(\mathbf{r}) \quad , \quad \mathbf{r} \in D \tag{22}$$

for a function $f(\mathbf{r})$ defined on a domain D. For instance, in quantum mechanics the Laplace operator is used for the kinetic energy of a classical particle. In condensed matter it describes so diverse physical phenomena as diffusion in porous medium D, the oscillation of a membrane (D), electro-magnetic fields and waves in cavities D, and even the scattering of particles.

In his seminal paper 'Can one hear the shape of a drum?' M. Kac asked if it is possible to determine the shape of a membrane D knowing the frequencies, i.e. the eigenvalues γ_n of the differential equation (22) with $f(\mathbf{r}) = 0$ for $\mathbf{r} \in \partial D$ [28]. He showed that the sum

$$Z := \sum_n e^{t\gamma_n} \sim \frac{M_0(D)}{4\pi t} - \frac{M_1(D)}{4}\sqrt{\frac{\pi}{t}} + M_2(D)\frac{\pi}{6} + \mathcal{O}(t) \tag{23}$$

does indeed contain information on the shape of the membrane but only the so-called Minkowski functionals $M_\nu(D)$. In particular, for a two-dimensional membrane they are the area $F = M_0$ of the membrane, the boundary length $U = 2\pi M_1$, and the Euler characteristic $\chi = \pi M_2$ as a measure of the connectivity of D. Since the Laplace equation (22) is an adequate description of many physical problems one may try to express physical quantities such as conductivities in porous structure D in terms of Minkowski functionals (Sect. 2.2).

Problem: Non-additivity of Edge Contributions The expansion (23) is only valid for smooth boundaries ∂D. But models of overlapping grains which are used widely in physics (see Fig. 1 and the Boolean model in Sect. 5) generate singular edges and corners where the boundaries of two or three grains, respectively, intersect. Additional terms in the expansion have then to be taken into account when overlapping grain models with singular edges are used or even lattice configurations as introduced in Sect. 5.4. These terms cannot be expressed in terms of Minkowski functionals and a careful geometric analysis is required. Similar problems occur for the asymptotic expansion of structure functions which are introduced in Sect. 2.2, because the expression (19) does not take into account the contribution due to singular edges of the grain boundaries in germ-grain models.

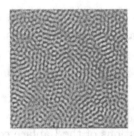

Fig. 4. Inhomogeneous density profiles of iodide occur spontaneously in a *chemical reaction (CIMA)* (dark regions correspond to high concentrations, [38]). One can observe in reaction-diffusions systems not only regular Turing patterns such as hexagonal cells (left) and parallel stripes (middle) but also irregular, turbulent structures (right pattern). The reversible transitions between these patterns can be quantitatively described by *morphological order parameters* which reduce the details of the spatial structures to relevant quantities without neglecting essential differences.

3   Morphology: Characterization of Spatial Structures

Spatial statistics, in particular, morphology and stereology have a wide field of applications in medicine, epidomology, and biology because of the enormous amount of spatial information in patterns occurring in biological systems. But the amount of spatial information is growing fast also in physics and material science. Scanning microscopy techniques, digital image recording, and computer simulations provide the scientist with spatial information which are usually not analyzed in a systematic way beyond standard methods such as correlation functions and Fourier transformations. The techniques to analyze spatial information developed in spatial statistics may be very useful for physicists in order to get more out of their data.

Most of the patterns considered in statistical physics contain an enormous amount of spatial information. Therefore, the physics of such spatial systems faces the problem of reducing the details to a finite number of relevant quantities, so-called order parameters. Minkowski functionals seems to be promising measures to work as such *morphological order parameters*. They can, for instance, distinguish quantitatively the turbulent and regular dissipative structures (Turing patterns) in chemical reaction-diffusion systems shown in Fig. 4 (Sect. 3.1). The functional form of the morphological order parameter change the symmetry if the pattern type is changed, so that pattern transitions can be classified analogously to thermodynamic phase transitions.

Another example is the morphological analysis of holes in thin polymer films which allow insight into the dynamical mechanisms which leads to the rupture of the film and finally to the visible stochastic point process (see Sect. 3.2 and Fig. 6). Also the different dynamical mechanisms of the phase separation during spinodal decomposition can be determined by measuring the time dependent morphological measures of the homogeneous domains (Sect. 3.2). An ambitious aim of the time dependent study is the formulation of *dynamic equations* for the morphological measures as relevant order parameters. Spinodal decomposition kinetics and the rupture of thin films are just first examples out of a huge number of dynamical pattern formation.

Since the Minkowski functionals in these cases depend on parameters such as a density threshold ρ (Sect. 3.1), the time t (Sect. 3.2), the radius of a sphere r (Sect. 3.3), or the scaling length L (Sect. 3.4) we use the notion of Minkowski *functions* in contrast to a *functional* which assigns a pure number to a spatial configuration. Nevertheless, the term functional is used whenever the mathematical properties of these measures are important and not the dependence on a parameter.

3.1 Minkowski Functions as Order Parameters

In 1952 A. Turing predicted the existence of inhomogeneous spatial patterns in chemical reactions when diffusion of the species plays a role [25]. These patterns consist of regular and stationary spatially structured concentration profiles of the reactants. He showed that the homogeneous solution of reaction-diffusion equations may become unstable and hexagonal or stripe patterns emerge. It was only in 1991 when a group in Bordeaux followed by another in Austin could report the first experimental realization of such Turing patterns. Moreover, also a turbulent, irregular, and time-dependent pattern was found which is shown in Fig. 4(c). A pattern converts reversible into another depending on system parameters, such as the temperature or the concentrations of the species. For example, defects occur in the hexagonal structure when the parameters are changed in such a way that the turbulent pattern becomes stable. Because of the proliferation of defects when the system is turning into the turbulent pattern, it is hard to tell whether the intermediate state of the pattern is hexagonal or turbulent already. The typical length scale or the correlation function do not change drastically. Naturally, the question arises how one can describe irregular patterns in order to characterize the patterns in a unique way and, in particular, the transitions between them. There is a need to find measures which are capable to describe the morphology and topology of the patterns and which are sensitive to the pattern transitions.

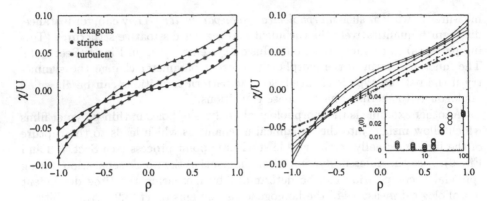

Fig. 5. The Euler characteristic per boundary length $P_\chi(\rho) = \chi/U$ (averaged local curvature) is shown as function of the black/white threshold parameter ρ for the same patterns as in Fig. 4. The thin full lines are best fits to the experimental data using cubic polynomials. $P_\chi(\rho)$ is negative (positive) if many disconnected black (white) components dominate the image. The asymmetry of the function for the hexagonal pattern (triangles) is due to the asymmetry of white dots isolated in a black connected structure shown in Fig. 4 whereas the stripe (circles) and turbulent patterns (stars) are symmetric in black and white. The polynomials $P_\chi(\rho)$ on the right indicates a transition from hexagonal point patterns to turbulent structures as a function of the concentration $[H_2SO_4]_0^B$ (mM) at constant $[ClO_2^-]_0^A = 20$ mM. The inset shows the value $P_\chi(\rho = 0)$ which can be used as an order parameter for the symmetry breaking of the polynomial. A vanishing Euler characteristic indicates a highly connected structure with equal amount of black and white components. The transition at a concentration of $17mM$ seems to be continuous with decreasing cubic and quadratic terms in the polynomial for hexagonal patterns which remain zero for turbulent patterns.

In order to study the spatial concentration profile of iodide in greater detail the concept of level contours can be introduced. The grey value at each pixel is reset to either white or black depending on whether the original value is larger or lower than a threshold value ρ, respectively. The qualitative features of the images varies drastically when the threshold parameter ρ is changed. For high thresholds ρ the regions of maximum concentration are studied, i.e. one obtains information concerning the shape of the peaks in the profile. For low thresholds the deep valleys of the concentration profile are examined and for intermediate ρ one obtains more or less the same visual impression as from the gray-scale pattern. Thus, the spatial information depends strongly on the threshold. For each threshold ρ one can calculate the area F of the white structure, the total length U of the boundaries, and the integral of the curvature along the boundaries, i.e. the Euler characteristic χ as functions of ρ [38]. Surprisingly, it turns out that the morphological quantities $F(\rho)$, $U(\rho)$, and $\chi(\rho)$ are given by cubic polynomials. That is, these functionals F, U, and χ may be described by a few parameters for different patterns since the functional form is universal in character.

The Euler characteristic per boundary length $P_\chi(\rho)$ is shown in Fig. 5 for a hexagonal pattern (triangles), a lamellar stripe structure (circles), and a turbulent state (stars). The lines are the best fits with a cubic polynomial to the data (symbols), which can hardly be distinguished from the experimental data. The accuracy is remarkable and could not be achieved with polynomials of lower orders nor be improved by using higher orders. Since the functional form of the measures F, U, and χ is always a cubic polynomial, the dependence of the morphology on the experimental conditions, i.e. on control parameters such as the concentration of malonic acid is reflected only in a finite number of coefficients. The most striking result of this integral geometric method for pattern analysis is the dependence of the coefficients of the polynomials on control parameters shown in the inset of Fig. 5. One can observe a symmetry breaking of the polynomials when the type of the pattern changes. Therefore, it is possible to describe the pattern transitions quantitatively using morphological measures and it may be possible to classify them in a way similar to thermodynamic phase transitions (for details see [38,40]). Concluding we emphasize that Minkowski functions, in particular the Euler characteristic, describe quantitatively irregular spatial patterns and their transitions in a morphological way and can be used as order parameters.

Problem: Parallel Surfaces of Non-convex Patterns

Related to the two main results one may focus on two questions: Why are the morphological measures $F(\rho)$, $U(\rho)$, and $\chi(\rho)$ polynomials in the threshold ρ? In many models for statistical geometries such as the Boolean model or Gaussian random fields, similar polynomial behaviors of Minkowski functionals do occur (Sect. 5). But what is an adequate statistical model for pattern formation in reaction-diffusion system, i.e., what is the connection of the morphological measures to the dynamical equations in such systems? The second problem is related to the observed symmetry breaking in the mean curvature, i.e., in the Euler characteristic of the patterns? Is it possible to formulate a mesoscopic theory of pattern-transitions analogous to the Landau theory for thermodynamic phase transitions? Is it possible to define Minkowski functions as order parameters in an effective theory.

The universal form of the morphological order parameter, i.e., the cubic polynomial, seems to be related to the additivity of the Minkowski functionals. The finite sum in the expression (10) for the parallel volume of konvex grains (Steiner's formula) and the decomposition of the kinematic integral (9) into products of Minkowski functionals is a direct consequence of the additivity relation (6). In Sect. 5 it turns out that for the Boolean model additive functionals are essentially polynomials where the coefficients depend on shape, orientation and correlations of the grains. Thus, the physical and morphological properties of configurations enter only into a finite number of relevant quantitites.

A fundamental problem of characterizing irregular non-convex patterns may be addressed when looking at Figs. 4 and 7 where the main feature is obviously the complex shape of intertwined black and white regions. For simplicity, one may consider first a single convex shape K and calculate the d-dimensional volume $V(\epsilon) = V(K_\epsilon)$ of its parallel body K_ϵ. Using Steiner's formula (10) one knows how this function looks like: it is a polynomial in ϵ where the coefficients are given by the Minkowski functionals $M_\nu(K)$ of the grain K. Thus, if one knows the Minkowski functionals M_ν of a convex structure, one knows the 'volume function' $V(\epsilon)$ and vice versa: measuring the parallel volume of a convex shape one knows its Minkowski functionals. Obviously, that is not the case for irregular, non-convex shapes A as shown in Figs. 1, 2 and 4, because the volume function $V(\epsilon) = V(A_\epsilon)$ contains more information about the spatial structure than just the Minkowski functionals $M_\nu(A)$ (Fig. 3). For instance, the parallel body A_ϵ includes all points of distance less than ϵ to A which depends on 'non-local' properties of A such as narrow throats or bottle necks where distant parts of A come close together. Thus, the parallel surface ∂A_ϵ contains spatial information about the embedding of the surface ∂A in space in addition to the Minkowski functionals $M_\nu(A)$ of the grain A itself (analogous to a generalized contact distribution function). In Fig. 3, for instance, the Minkowksi functions $M_\nu(\epsilon) = M_\nu(A_\epsilon)$ are shown for three stochastic models and for a sandstone. The differences in these geoemtric measures are remarkeable since the porosity $V(A)$ is the same for all configurations. Thus, parallel surfaces may be used to describe the morphology of irregular, non-convex patterns going beyond the additive Minkowski functionals. Measuring the parallel volume or, in general, the Minkowski functions of a non-convex shape one may characterize the spatial structure and be able to extended Steiner's formula to non-convex shapes. But nothing is gained without knowledge of the functional form of these functions for relevant configurational shapes. If it is possible to find a suitable 'basis set' of Minkowski functions for special non-convex structures (e.g. periodic minimal surfaces or the models shown in Fig. 2) one may decompose the Minkowski functions of an arbitrary spatial structure in terms of the basic structures. Of course, the linear superposition of modes like a Fourier analysis of periodic functions is not reasonable and feasible for Minkowski functions. But one might find at least signatures of the morphology of a spatial structure in the functional form of its Minkowski functions. A polynomial, for instance, would be a hint for a convex shape.

The number of possible applications of such a morphological analysis using parallel sets and Minkowski functions would be enormous. For instance, mesophases in systems of block-polymers can be described by periodic minimal surfaces. But in general one does not observe in an experiment a pure regular structure but a mixture of many different phases. Therefore, experimentalists face the problem of characterizing irregular mixtures of phases and of quantifying the content of a specific periodic minimal surface in a measured spatial structure. The knowledge of Minkowski functions is also important if one wants to calculate volume integrals $\int d\mathbf{x} F(u(\mathbf{x}))$ of functions $F(u)$ which depends only on the normal distance u to a non-convex surface. For instance, the density of a fluid in a porous medium is mainly a function of distance to the substrate wall and one needs such integrals over the pore volume for the description of transport and thermodynamic properties of the fluid (Sect. 2.2). In Sect. (3.3) parallel bodies are used to define Minkowski functions of point patterns which can then be used as alternative method to correlation functions to characterize the stochastic properties of point processes.

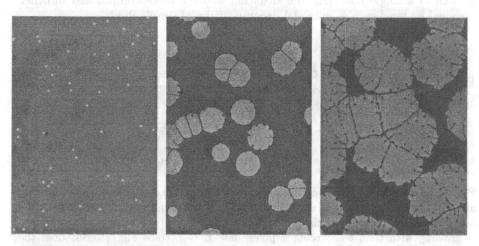

Fig. 6. Growing holes (white) in a thin (106 nm) polystyrene film on OTS-silizium ($T = 130^{\circ}C$) leads to a complex net-like structure of drops (dark; see the paper by K. Jacobs et al. in this volume). A morphological analysis reveals that the locations of the holes (white dots in the left figure) are distributed independently without any correlations and that the *dynamics of the morphology* does not change during the growth even if the holes are touching and merging (see Fig. 8).

3.2 Minkowski Functions as Dynamical Quantities

The rupture of a thin fluid film on a substrate, such as a water film on a freshly cleaned window, proceeds by the formation of holes whose radius grows in time until they finally merge which leads to a drying of the substrate (see the paper by K. Jacobs et al. in this volume and [24,27]. An example of such growing and finally merging circular holes in a thin polymer film is shown in Fig. 6. The study of this dynamical behavior by measuring the radius of the holes is limited since finally the holes coalesce. But measuring the Minkowski functionals of the undisturbed film yields the temporal behavior of the morphology of the holes which can now be recorded from the very beginning to the final state of an almost dry substrate. The time dependent Minkowski functions are shown in Fig. 8 and compared with a theoretical model (solid line) where one assumes that the centers of the holes follow a Poisson process (see D. Stoyan in this volume), that they remain circular, and that the only time dependence can be described by the radius $R(t)$. The latter is well studied for single holes yielding $R(t) \sim t$ in good agreement with the measured Minkowski functions shown in Fig. 8. The dynamical equations for the Minkowski functions of growing holes in thin liquid films turn out to be quite simple. Because the holes remain spherical discs even when they merge and because the dynamics does not change after merging the dependence of the functionals on the size, i.e. on the radius R of the holes remains unaltered and the dynamics of $R(t)$ is known already from the growth of a single hole. But the situation is much more complicated in other physical systems such as spinodal decomposition of simple fluids.

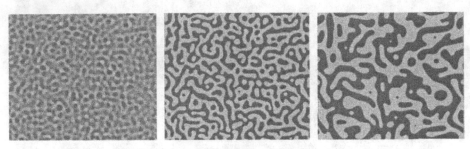

Fig. 7. *Spinodal decomposition:* a homogeneous fluid phase decomposes spontaneously into coexisting vapor (white) and liquid (dark) phases if the system is cooled rapidly below the boiling point. The characteristic length scale L of the homogeneous domains is growing with time t according a power law $L \sim t^\alpha$ (see Fig. 8). But does the morphology scale with this length L or does the shape of the domains change with time? Is it possible to formulate dynamic equations for morphological measures analogous to the dynamics of structure functions.

Phase separation kinetics is probably the most common way of producing irregular spatial patterns on a mesoscopic scale [7,20]. A fluid system above the critical temperature is usually homogeneous. But after a sudden quench below the critical point into the two-phase coexistence region the fluid separates into the coexisting liquid and vapor phase. A typical example of patterns that emerge during spinodal decomposition in a two-dimensional liquid system is shown in Fig. 7. In order to describe the morphology of the homogeneous phases [40,41], in particular the scaling behavior, Minkowski functions provide means to define the characteristic length scale $L(t)$ of the homogeneous domains, to study the time evolution of the morphology, and to define the cross-over from the early stage decomposition (figure on the left) to the late stage domain growth (figure on the right). For instance, the magnitudes of the morphological measures $U(t)$ and $\chi(t)$ shown in Fig. 8 increase during the early stage due to the formation of homogeneous domains, i.e. of sharp boundaries separating the phases. After domains have been formed, the boundary length $U(t)$ as well as the connectivity $\chi(t)$ decrease due to the growth of the domains. Thus, their maximum values $\bar{U}$ and $\bar{\chi}$ mark a transition time $\bar{t}$, i.e. the end of the phase decomposition and the beginning of domain growth. The details of a morphological study of spinodal decomposition can be found in [59]. In Fig. 8, for instance, the time dependence of $U^{-1}(t)$ and $\chi^{-1/2}(t)$, i.e., of the scaling length $L(t)$ (see Eq. (8)) is shown as function of time confirming the scaling relation $L(t) \sim t^\alpha$ with a scaling exponent α depending on the viscosity τ. One can use the scaling relation (8) of the Minkowski functionals to test the scaling behavior of the morphology during spinodal decomposition of fluid phases. It turns out that the morphology indeed does not scale in certain circumstances, for instance, at early times due to shape relaxation of the domains or even at late times in binary fluids where the symmetry of the fluid phases causes a breakdown of scale invariance [59,63].

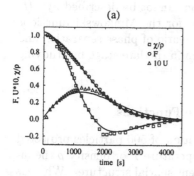
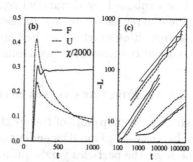

Fig. 8. (a) Time evolution of the morphological measures M_ν of holes in thin liquid polymer films (see Fig. 6). The solid lines are given by the dynamical evolution of the radius $R(t) \sim t$. (b) Time evolution of the morphological measures F, U, and χ during spinodal decomposition of a fluid at the mean density $\rho = 3$ and the viscosity $\tau = 0.6$ (see Fig. 7, [41]). Two different regimes of the time evolution show up in the functional form of the Minkowski functions: for early times one observes an increase of the measures due to the formation of homogeneous domains whereas in the late stage the measures decrease due to domain growth. (c) The boundary length $w_1/U(t)$ (dashed line), and the connectivity $w_2/\chi(t)^{-1/2}$ (thick solid line) exhibit the scaling relation (8) with $L \sim t^\alpha$ and the exponents $\alpha = 2/3$ ($\tau = 0.53$), $\alpha = 1/2$ ($\tau = 0.6$), and $\alpha = 1/3$ ($\tau = 1.5$) indicated by thin solid lines. The coefficients w_i are chosen to separate the curves.

Problem: Dynamical Equations for Minkowski Functions The linear dependence $R(t) = v \cdot t$ on time t of the radius R of a hole in a fluid film (see Fig. 6) can easily explained by the dynamical equation

$$\frac{\partial F(t)}{\partial t} = vU(t) \tag{24}$$

i.e., by the assumption that the rate of increase in the area F is strictly proportional to the perimeter U of the merging holes. Of course, it remains to be proven that the underlying microscopic dynamics can indeed be described by this macroscopic rule, but the important point is that Equation (24) is expressed solely in terms of Minkowski functionals.

For the spinodal decomposition dynamics the physics is much more complicated and requires the knowledge of the time dependent density correlation functions. Dynamical equations for such structure functions have been derived and tested experimentally in detail [7,20]. But is it possible to formulate dynamical equations for the Minkowski functionals of the homogeneous domains during spinodal decomposition analogous to the simple example (24) describing the growth of holes?

In the late stage the time dependence of the Minkowski functionals is given by a single scaling function $L(t)$ and by the scaling relation (8). But the details of the morphology in the beginning of the time evolution cannot be described by $L(t)$ and should be explained by dynamical equations for the Minkowski functions $M_\nu(t)$. Maybe differences between a nucleation scenario of phase separation and spinodal decomposition kinetics can be found even in the late stage by studying the morphology of the homogeneous domains.

3.3   Minkowski Functions versus Structure Functions

Figure 9 shows three snapshots of the spatial position of 500 particles projected onto a two dimensional plane. Depending on temperature T or density ρ the interaction between the particles yields quite different spatial structures. Whereas at low densities the particles are so distant that they merely influence the locations of each other yielding a vapor of particles, at high density the mutual repulsion leads to a regular order of the positions, i.e., to a crystalline phase. If the interaction is attractive a middle phase at intermediate densities can be stable resembling the properties of a liquid. Such phase transitions in Gibbsian ensembles are well studied and the papers by H.-O. Georgii and H. Löwen in this volume present analytic and numerical results. The difference in the position patterns of the vapor and the solid phase is obvious, but a thorough analysis is required in order to distinguish the vapor and liquid structures.

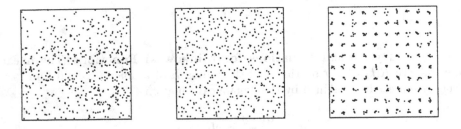

Fig. 9. Three-dimensional positions of 500 atoms in a *Lennard-Jones fluid* at different densities $\rho = 0.05$ (vapor), $\rho = 0.7$ (liquid), $\rho = 1.0$ (solid) are projected on a two-dimensional plane. Whereas the solid phase can clearly be distinguished from the fluid phases, a thorough morphological analysis is necessary in order to find differences between the vapor and the fluid phase. If the two-point structure function is essentially similar (Fig. 11) Minkowski functionals may help.

A very important method of point process statistics, i.e. statistical structure analysis of point patterns, are *structure functions* (point-point correlation functions) or, in general, second order characteristics. But in many experiments the number of observable points is too small in order to draw conclusions about the distribution of the points because of large variability. For instance, the number of galaxies for which three-dimensional distances are measured are yet a few thousand and in many catalogue even only a few hundred. The situation is similar for the observable number of holes in thin films (Fig. 6) or the number of particles in simulations of fluids in confined geometries (Fig. 9). Therefore, spatial statistics (see D. Stoyan in this volume and [61]) provides alternative methods and tools such as the nearest-neighbor distance distribution function $D(r)$ and the spherical contact distribution function $H_s(r)$ (or void probability). In this context also morphological measures such as the Minkowski functionals are helpful.

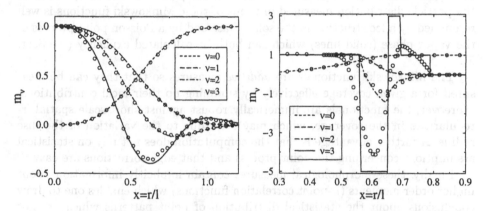

Fig. 10. (a) Minkowski functions $m_\nu(r) = W_\nu(\mathcal{A})/W_\nu(B_r)/\rho$ ($\nu = 1, 2, 3$) and $m_0 = 1 - W_\nu(\mathcal{A})$ in the vapor phase ($\rho = 0.01$, see Fig. 9), where l denotes the mean distance of the particles and $\mathcal{A} = \cup_i B_r(\mathbf{x}_i)$ the union of spheres of radius r centered at the position of the particles. The structure of the particle distribution measured in terms of these morphological functions (circles) is well resembled by a Poisson distribution of points (solid lines). In contrast to structure functions ($S(r) \approx 1$ in the vapor phase) the measures $m_\nu(r)$ also include higher correlations. Thus, they provide additional information about the statistical distribution of the particles beyond the second order moments. (b) Minkowski functions $m_\nu(r)$ in the solid phase (circles, $\rho = 1.1$) are close to the morphological measures of a perfect fcc structure (solid lines). The differences, in particular, for m_2 and m_3 are due to thermal fluctuations in an ensemble of 500 particles.

In order to apply the Minkowski functionals to point processes in $\mathbb{R}^d$ one may decorate each point at $\mathbf{x}$ with a d-dimensional sphere $B_r(\mathbf{x})$ of radius r centered at $\mathbf{x}$. The scale-dependent morphological features of the coverage $\mathcal{A}_r = \cup_i B_r(\mathbf{x}_i)$, i.e. of the parallel body A_r of the point pattern $A = \cup_i \mathbf{x}_i$ can then be explored by varying the radius of the spheres. Depending on the size of the sphere the pattern $\mathcal{A}_r$ exhibit quite different topological and geometric properties. For instance, the spheres may be disconnected (isolated) for small radius, whereas for large radius the grains overlap yielding a connected structure. The morphology of the emerging pattern can be characterized by the Minkowski functionals, i.e., by the covered volume W_0 (or area, respectively), the surface area $3W_1$ (or boundary length), the integral mean curvature $3W_2$, and the Euler characteristic W_3/ω_3, i.e., connectivity of the penetrating grains as functions of the radius r. Figure 10 shows these Minkowski functions $M_\nu(r) = M_\nu(\cup_i B_r(\mathbf{x}_i))$ for the vapor and solid configurations given in Fig. 9 (Note that $M_\nu(r)$ is closely related to the spherical contact distribution function). Obviously, the surface area and curvatures vanishes for large radii whereas the normalized covered volume approaches unity. In both phases one finds radii of the spheres where the Euler characteristic, for instance, is negative indicating a highly connected bicontinuous pattern. But the structure is quantitatively much more pronounced in the solid phase than in the vapor indicating a regular distribution of the particles. The morphology of the particle distribution measured in terms of these Minkowski functions is well resembled by a fcc structure in the solid phase and by a Poisson point process in the vapor phase (solid lines) which can both be calculated explicitly (see Sect. 5).

The Minkowski functionals are *additive* measures so that they can be measured for a given coverage effectively by summing up their local contributions. Moreover, the functionals are numerically robust against short-scale spatial irregularities in the coverage, which may arise due to the variation of the disc radius. A further advantage is that the computation does not rely on statistical assumptions concerning the point process and that edge corrections are easy.

Finally, these morphological measures contain implicitly information about higher order moments (n-point correlation functions) which enables one to draw conclusions about the statistical distribution of point patterns which are not accessible by two-point structure functions. For instance, the vapor and liquid phase shown in Fig. 9 show almost no difference in the structure function while the Minkowski functions reveal significant differences hidden in the higher correlations of the point distribution. A further disadvantage of structure functions is the lack of almost any information about the geometric shape (morphology) of a point pattern. The occurrence of one-dimensional filaments or two-dimensional sheets of points as obviously relevant geometric structures seen in the distribution of galaxies, for instance, is difficult to be revealed by the knowledge of the correlation functions (see the statistical analysis of large scale structures in the universe by M. Kerscher in this volume). It is even possible to construct completely different point distributions which nevertheless exhibit the same two-point structure function [3]. This example demonstrates that alternative statistical methods may be important in order to gain relevant spatial information in physical systems and that morphological measures may serve as a complementary method of statistical structure analysis in physics.

Morphological measures are important tools particularly for the *comparison of experiment and theory*. They reveal not only relevant difference between simulations of non-linear dynamical equations and experimental data of chemical reaction-diffusion systems (Sect. 3.1, [38]). Also the measures of the hole distribution in thin films are not consistent with the concept of spinodal decom-

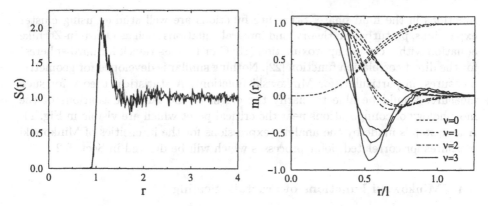

Fig. 11. Structure function $S(r)$ and Minkowski functions $m_\nu(r) = M_\nu(\cup_{i=1}^N B_r(\mathbf{x}_i))/(NM_\nu(B_r)$ for three different configurations of the Lennard-Jones fluid at the critical density $\rho_c = 0.01$ and temperature $T_c = 1.0$ (Fig. 9). Higher order correlations are visible in the fluctuations of the morphological measures which may be relevant for the understanding of critical behavior of fluids. The radius r is normalized by the mean distance l of the particles.

position but with a nucleation scenario (see Sect. 3.2 and K. Jacobs et al. in this volume). Cosmological models of the early universe can be discriminated because of the observed morphology of the present galaxy distribution (see M. Kerscher in this volume, [36]). These examples demonstrate the importance of spatial information and the reduction to relevant measures if one studies spatial structures of physical systems.

Finally, let us discuss the fluid structures shown in Fig. 9. Whereas the solid and vapor phases are well understood (see Fig. 10) the statistical physics of the fluid phase is still a challenging problem. In particular, the existence of a critical point where the difference of a vapor and liquid phase vanishes is a fascinating phenomenon. Figure 11 shows the structure function $S(r)$ and the Minkowski functions $m_\nu(r) = M_\nu(\cup_{i=1}^N B_r(\mathbf{x}_i))/(NM_\nu(B_r)$ for three different configurations of a Lennard-Jones fluid at the critical density $\rho_c = 0.01$ and temperature $T_c = 1.0$. Because of large scattering of the data one can observe no differences in the structure function $S(r)$ in contrast to the Minkowski functions which are statistically robust due to the additivity relation (6). The origin of the visible differences between the configurations at the critical point is yet not understood and requires a detailed study. Obviously, higher order correlations are visible which may be important to understand the critical behavior of a fluid. Because of the self-averaging of the Minkowski functionals even in small samples, structure differences of single configurations can be studied near the critical point where such spatial fluctuations become large.

Problem: Minkowski Functions of Fluid Phases and Critical Points In contrast to the solid and the vapor phase where the Minkowski functions can be calculated analytically (see Fig. 10), no expressions are known for the Minkowski functions in the fluid phase. Structure functions are well studied using cluster expansions, perturbation theory, and integral equations such as Ornstein-Zernike equation with a closure approximation (HNC or Percus-Yevick for hard spheres) for the direct correlation function [22]. Nothing similar is developed for geometric measures, in particular, for Minkowski functions. A statistical theory for such measures should be able to clarify the origin of the large fluctuations of the morphology of configurations near the critical point which are visible in Fig. 11. A first step is made by the analytic expressions for the intensities of Minkowski measures for correlated point processes which will be derived in Sect. 5.2.

3.4   Minkowski Functions of Fractals: Scaling

The Minkowski functionals $M_\nu(A)$ of a spatial configuration A are 'dimensional' measures which scale according to Eq. (8), i.e., they depend on the length scale L of structures in the pattern A. For instance, spatial domains of homogeneous fluid phases evolving during spinodal decomposition (see Fig. 7) exhibit an enormous amount of information which is normally reduced to a single time-dependent characteristic length $L(t)$ of a typical homogeneous domain. If the inhomogeneous pattern consists of homogeneous domains A with sharp interfaces the domain growth is due to a rearrangement the domains without changing the area F of the fluid phase. Because the geometric measures $W_\nu(A)$ are homogeneous functions of order $d - \nu$ (see (8)) one can assume for their densities the scaling behavior

$$f \sim 1 \quad , \quad u \sim L^{-1} \quad , \quad \chi \sim L^{-2} \tag{25}$$

with a scaling length L. This scaling is indeed confirmed by the time dependence shown in Fig. 8. One can test the scaling assumption further by changing system parameters such as surface tension and viscosity which should not influence the ratio U^2/χ, for instance.

But in the last 20 years a large variety of spatial structures has been detected in nature which exhibit a quite different scaling behavior of the volume density than the one given by Eq. (25). It is well known that fractals [19,32,33] can be characterized by a non-integer dimension which describes the increase of the 'content' of the spatial structures with the size of an observation window. Usually under 'content' the d-dimensional volume of the fractal is meant. But integral geometry provides us with a complete family of so-called 'intrinsic volumes' or Minkowski functionals in d-dimensional space which can be used to define a complete family of fractal dimensions based on the scaling property (8) of the Minkowski functionals.

Three different fractals generated as percolating clusters on a lattice Λ are shown in Fig. 13 for three different lattice models. Instead of occupying each lattice site of Λ independently one can introduce correlations by occupying the neighbored sites on a rectangular of edge lengths λ_1 and λ_2 simultaneously, in particular, of a square $\lambda_1 = \lambda_2$ or a stick of length λ_1 and $\lambda_2 = 1$. Examples of such constitutional shapes are shown in Fig. 12. It is necessary to introduce an upper cut-off L for structures by imposing periodic boundary conditions,

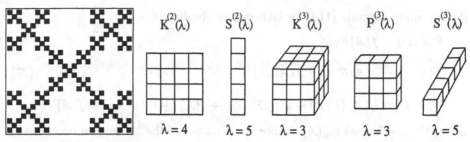

$$K^{(2)}(\lambda) \qquad S^{(2)}(\lambda) \qquad K^{(3)}(\lambda) \qquad P^{(3)}(\lambda) \qquad S^{(3)}(\lambda)$$

$$\lambda = 4 \qquad \lambda = 5 \qquad \lambda = 3 \qquad \lambda = 3 \qquad \lambda = 5$$

Fig. 12. A *deterministic* fractal and the constitutional grains of the lattice percolation models which are used to generate the *random* fractals shown in Fig. 13. In two dimensions two types are considered in order to study the dependence of percolation thresholds (Sect. 4.2) and of the fractal scaling behavior (Sect. 3.4) on shape and orientation: squares $K^{(2)}$ and sticks $S^{(2)}$ of edge length λ. On a three-dimensional lattice three shapes are possible: cubes $K^{(3)}$ (isotropic), plaquettes $P^{(3)}$ (oblate), and sticks $S^{(3)}$ (prolate) of size λ.

i.e. the spatial structure within a box Ω of edge length L is repeated in all space directions. Thus, each lattice configurations $\mathcal{A}$ is periodic and scaling of geometric measures of a fractal is expected only for lengths smaller than L. One finds that the number of occupied sites, i.e., the area f of the black structures shown in Fig. 13 scales according to

$$f(L) \sim L^{d_f - 2} \tag{26}$$

with the fractal dimension $d_f = 91/48$ in two dimensions instead of $d = 2$ for the scaling of homogeneous domains given by Eq. (25). The fractal dimension is universal and does not depend on the details of the percolation model, i.e., on the local correlations given by $\lambda_\nu > 1$ (see Fig. 12). The same relation (26) holds also for the total perimeter $u(L) \sim L^{d_f - 2}$ within a window of size L which can be found in many papers on fractals and percolation (see references in [33]). But this scaling relation for the perimeter is at least only half of the answer. In the following integral geometry is used to derive the scaling behavior of the Minkowski functionals $M_\nu(L)$. Without loss of generality we focus on lattice configurations in two dimensions. Since boundaries of fractals are typically huge and of the same size as the volume itself, it is better to be careful about boundary corrections defining geometric quantities and their densities. For this purpose an observation window D, i.e., a square of size l_D is introduced, so that one measures the covered area $F = V_0(\mathcal{A} \cap D)$, the boundary length $U = 4V_1(\mathcal{A} \cap D)$, and the Euler characteristic $X = V_2(\mathcal{A} \cap D)$ only for the 'visible' configuration $\mathcal{A} \cap D$ of a fractal $\mathcal{A}$. Because the observation window D is placed at random one is interested in the mean values of these measures $\bar{M}_\nu(\mathcal{A}, D) = \int_\Lambda M_\nu(\mathcal{A} \cap D)\, dD / \int_\Lambda dD$ defined by averaging over all positions of D on the lattice Λ, i.e., by moving a given point of D (e.g. the midpoint) on each lattice site of Λ. Using the kinematic formula (110) for lattices one obtains

$$\bar{F}(\mathcal{A}, D) = f(\mathcal{A})F(D)$$

$$\bar{U}(\mathcal{A}, D) = u(\mathcal{A})F(D) + f(\mathcal{A})U(D) \tag{27}$$

$$\bar{X}(\mathcal{A}, D) = \chi(\mathcal{A})\left(F(D) + \tfrac{1}{2}U(D) + 1\right) + u(\mathcal{A})\left(\tfrac{1}{8}U(D) + \tfrac{1}{2}\right) + f(\mathcal{A})$$

where the coefficients $m_\nu(\mathcal{A})$ denote the densities of the Minkowski functionals, i.e. the measures per unit volume of the lattice configuration. Integral geometry guarantees that these densities do not depend on the chosen observation window D but are intrinsic measures of the configuration $\mathcal{A}$. The contributions proportional to $M_\mu(D)$ in Eq. (27) describe effects due to the finite size and shape of the observation window D.

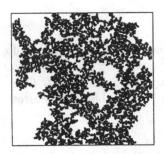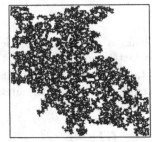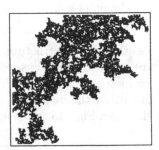

Fig. 13. Three percolating clusters at the critical volume fraction v_c with grains of size λ (see Fig. 12, [58]): (A) squares ($\lambda = 5$, $v_c = 0.615$), (B) sticks ($\lambda = 5$, $v_c = 0.422$), and (C) mixture of squares ($\lambda_1 = 2$, $\lambda_2 = 5$).

In contrast to non-fractal lattice configurations (considered below in Sect. 5.4) fractal configurations yield densities $m_\nu(\mathcal{A}, L)$ which depend on the size L of the largest structure. For instance, one obtains for the deterministic fractal shown in Fig. 12 (left) the mean values $\bar{F}(\mathcal{A}, D) = L^{y_0-2}l_D^2$, $\bar{U}(\mathcal{A}, D) = 4L^{y_0-2}(l_D^2 + l_D) - 4L^{y_1-2}l_D^2$, and $\bar{X}(\mathcal{A}, D) = L^{y_0-2}(l_D^2 + 2l_D + 2) - 2L^{y_1-2}(l_D^2 + 3l_D + 2) + 2(l_D + 1)^2$ of the Minkowski functionals and therefore the densities

$$f(\mathcal{A}, L) = L^{y_0-2}$$
$$u(\mathcal{A}, L) = 4L^{y_0-2} - 4L^{y_1-2} \tag{28}$$
$$\chi(\mathcal{A}, L) = -2L^{y_1-2} + 2$$

with the non-trivial exponents $y_0 = \log(5)/\log(3)$ and $y_1 = \log(2)/\log(3)$. Generally, one obtains the scaling relations

$$f(\mathcal{A}) = f^{(0)}L^{y_0-2}$$

$$u(\mathcal{A}) = u^{(0)}L^{y_0-2} + u^{(1)}L^{y_1-2} \tag{29}$$

$$\chi(\mathcal{A}) = \chi^{(0)}L^{y_0-2} + \chi^{(1)}L^{y_1-2} + \chi^{(2)}L^{y_2-2}$$

with coefficients $v_\nu^{(\mu)}$, $\mu = 0, \ldots, \nu$, and scaling exponents y_ν, $\nu = 0, \ldots, d$ (particularly $y_d = 0$). Thus, one finds d scaling dimensions in a d-dimensional space and not only the one given by the volume, $y_0 = d_f$. For random fractals such as the percolating clusters shown in Fig. 13 the scaling exponents are found to be simply related by $y_n = y_0 - n$ and $y_d = 0$. We emphasize that the scaling behavior of the boundary length and the Euler characteristic is not just a correction to the leading scaling given by Eq. (26) but an intrinsic property of theses morphological quantities. Assuming self similarity on all scales, i.e., a perfect scaling behavior for the volume M_0 we still get deviations of scaling for the other measures M_ν due to their geometric definition and the importance of boundaries for fractal geometries.

Problem: Universality of Scaling Amplitudes and Dimensions Although the scaling exponents y_n are found to be related to y_0 for the percolating clusters, the configurations shown in Fig. 13 show obvious differences in their morphology. Naturally one may ask if the *amplitudes* $u^{(\nu)}$ and $\chi^{(\nu)}$ of the scaling behavior given by Eq. (29) are universal? Or do the amplitudes characterize the morphology of fractals beyond their fractal dimension $y_0 = d_f$. Since the fractal dimensions y_n of a deterministic fractal are non-trivial due to different scaling behaviors of the occupied sites on the boundary of the observed fractal, one may ask under which circumstances models for random fractals can be defined with different, i.e. non-trivial scaling exponents y_n?

4   Integral Geometry: Statistical Physics of Fluids

Whereas in the previous Sects. 2 and 3 only elementary properties of Minkowski functionals have been used, we consider now applications of the 'completeness' of the additive functionals, the notion of a parallel body, and of the kinematic formulae. In Sect. 4.1 the kinematic formula (9) is used to describe the thermodynamic phase behavior of real fluids. Section 4.2 uses the excluded volume of a particle and Steiner's formula (10) to estimate the occurrence of an infinite connected cluster of particles. This percolation phenomenon is important for the transport of fluids in porous media, for instance, and depends on the morphology of the pores. The *excluded volume* of a pair of grains K and K' is defined as the volume around a given particle K which would be excluded for another particle K' if they were hard impenetrable objects. In terms of the Euler characteristic χ the excluded volume can be measured by the relative positions where an overlap occurs

$$V_{\text{ex}}(K, K') = \int_{\mathcal{G}} \chi(K \cap gK') dg . \tag{30}$$

The excluded volume $V_{\text{ex}}^{(d)}(K) = V_{\text{ex}}(K, K)$ can be calculated for any shape K in d dimensions by using the kinematic formula (9). One obtains in terms of the

Minkowski functionals $M_\nu(K)$ the explicit expressions

$$V_{ex}^{(2)}(K) = 2M_0(K) + 2\pi M_1(K)^2 \text{ and } V_{ex}^{(3)}(K) = 2M_0(K) + 8\pi M_1(K)M_2(K) . \tag{31}$$

Section 4.3 focuses on the characterization theorem (7) of additive measures. Thermodynamic potentials are extensive quantities which can usually be decomposed in volume, surface, and curvature contributions, i.e. in terms of Minkowski functionals. Assuming additivity of thermodynamic potentials on a mesoscopic scale one can define naturally a general Hamiltonian for mesophases and study the corresponding thermodynamic properties in terms of morphological measures.

We use the notion of Minkowski functionals again instead of functions as in the previous section where the dependence on parameters was relevant, because the mathematical properties of these morphological measures of spatial configurations shall be emphasized.

4.1 Kinematic Formula: Cluster Expansion and Density Functional Theory

One of the main goals of thermodynamics is the calculation of the equation of state. The pressure p of a fluid is given by the temperature T and the density ρ, for instance, $p = k_B T \rho$ for an ideal gas where molecular interactions between the fluid particles are neglected. This assumption is justified at low densities or high temperatures but not for real fluids like water at room temperature and not, in particular, in the vicinity of a phase transition such as condensation (vapor-liquid coexistence) or freezing. Statistical physics of condensed matter developed several methods in order to obtain equations of state for real fluids and to describe phase transitions. One of them are cluster expansions in terms of virial coefficients and another one are density functional theories yielding an adequate description of inhomogeneous fluids.

Virial Coefficients The equation of state, i.e., the dependence of the pressure p on the density ρ and temperature T of a real fluid can be expanded in powers of the density

$$p = k_B T \left(1 + B(T)\rho + C(T)\rho^2 + \mathcal{O}(\rho^3) \right) , \tag{32}$$

where the functions $B(T)$ and $C(T)$ are called the second and third virial coefficient, respectively. The ideal gas is given by $B(T) = C(T) = 0$ but in general the virial coefficients depend on the molecular interactions between the particles. In particular, for hard particles K one obtains for the second virial coefficient

$$B = \frac{1}{2} \int_{\mathcal{G}} \chi(K \cap gK)dg = V(K) + \frac{1}{4\pi}S(K)H(K) . \tag{33}$$

where the overlap of two particles K and gK are averaged over all relative positions and orientations $g \in \mathcal{G}$ of the group of motion $\mathcal{G}$. Thus, the second virial coefficient B is half of the excluded volume $V_{ex}(K)$ of K and using the kinematic formula (9) one obtains an explicit expression (31) in terms of Minkowski functionals of an arbitrarily shaped hard particle K. Of course, a realistic modeling of real fluids and a liquid-vapor phase transition requires not only a temperature independent hard repulsions of the molecules but also an attractive force. The attractive interaction $U(r)$ between spherical symmetric atoms is often well described by the Lennard-Jones potential. But for structured molecules this rotational invariant potential does not take into account the shape of the molecules. Therefore, T. Kihara [29] proposed that the shortest distance s between the molecules is the relevant parameter in the interaction potential $U(s)$ instead of the distance r of the centers of mass. The parallel bodies $K_{s/2}$ and $K'_{s/2}$ of two interacting molecules touches at the half of the minimal distance so that the second virial coefficient can given by the integral

$$B = 4 \int\limits_0^\infty ds \left(1 - e^{-\beta U(s)}\right) S_{||}(K_{s/2}) \tag{34}$$

where the parallel surface $S_{||}(K_{s/2})$ is given by the Minkowski functionals $M_\nu(K)$ and particularly in $d = 3$ by

$$S_{||}(K_{s/2}) = S(K)s + H(K)\frac{s}{2} + \frac{\pi}{2}s^3 . \tag{35}$$

Assuming a single point as the shape of a molecule one recovers with $V = S = H = 0$ and $\chi = 1$ the standard expression $B = 2\pi \int_0^\infty dr \left(1 - e^{-\beta U(r)}\right)$ for the second virial coefficient. This result may motivate to look for statistical theories for simple fluids by applying concepts of integral geometry, in particular, the kinematic formula (9). One prominent example is the density functional theory recently developed by Rosenfeld [51], which is the most elaborated and reliable functional for hard spheres.

Rosenfeld's Fundamental Measure Density Functional The main idea of density functional theory is to express the grand canonical free energy $\Omega[\rho(\mathbf{r})]$ as a functional of the averaged inhomogeneous density $\rho(\mathbf{r})$ of the particles. The equilibrium one-particle density minimizes $\Omega[\rho(\mathbf{r})]$ at a given chemical potential μ and one obtains the thermodynamic potential $\Omega(\mu)$. Density functional theory provides a unified picture of the solid and fluid phase as described by H. Löwen in this volume.

One of the most valuable and accurate functionals for hard spheres B_R of radius R is the 'fundamental measure theory' proposed by Rosenfeld. The main idea is to decompose the local excess free energy at low densities

$$\beta f_{ex}[\rho(\mathbf{r})] \sim -\tfrac{1}{2} \int d\mathbf{r}' \rho(\mathbf{r})\rho(\mathbf{r}') f(\mathbf{r} - \mathbf{r}') = \qquad -\left(\bar{V}(\mathbf{r})\bar{\chi}(\mathbf{r}) + \frac{\bar{S}(\mathbf{r})\bar{H}(\mathbf{r})}{4\pi} \right.$$
$$\left. + \frac{\bar{S}(\mathbf{r})\bar{H}(\mathbf{r})}{4\pi} \right) \tag{36}$$

in terms of the averaged Minkowski functionals (compare Eq. (2))

$$\bar{V}(\mathbf{r}) = \int\limits_{B_R(\mathbf{r})} d^3r'\rho(\mathbf{r}') \qquad , \bar{S}(\mathbf{r}) = \int\limits_{\partial B_R(\mathbf{r})} d^2r'\rho(\mathbf{r}')$$

$$\bar{H}(\mathbf{r}) = \int\limits_{\partial B_R(\mathbf{r})} d^2r' H(\mathbf{r}')\rho(\mathbf{r}') , \bar{\chi}(\mathbf{r}) = \tfrac{1}{4\pi} \int\limits_{\partial B_R(\mathbf{r})} d^2r'\rho(\mathbf{r}')G(\mathbf{r}') \qquad (37)$$

$$\bar{\mathbf{S}}(\mathbf{r}) = \int\limits_{\partial B_R(\mathbf{r})} d^2r'\rho(\mathbf{r}')\mathbf{n}(\mathbf{r}) \quad , \bar{\mathbf{H}}(\mathbf{r}) = \int\limits_{\partial B_R(\mathbf{r})} d^2r'\rho(\mathbf{r}')H(\mathbf{r}')\mathbf{n}(\mathbf{r})$$

with the normal vector $\mathbf{n}(\mathbf{r})$, the mean $H(\mathbf{r})$ and Gaussian $G(\mathbf{r})$ curvature at the position $\mathbf{r} \in \partial B_R(\mathbf{r})$ on the boundary of the sphere. This definition is possible because the Meyer f-function for spheres centered at $\mathbf{r}$ and $\mathbf{r}'$ is given by the Euler characteristic of the overlap $f(\mathbf{r} - \mathbf{r}') = \chi(B_R(\mathbf{r}) \cap B_R(\mathbf{r}'))$ so that the kinematic formula (9) can be applied. Notice: one immediately recovers for an homogeneous density $\rho(\mathbf{r}) \equiv \rho$ with $B = -\beta f_{ex}[\rho]/\rho^2$ the expression (33) for the second virial coefficient. The proposed density functional for the local excess free energy then reads [51]

$$\beta f_{ex}[\rho(\mathbf{r})] = -\bar{\chi}\ln(1 - \bar{V}) + \frac{\bar{S}\bar{H} - \bar{\mathbf{S}} \cdot \bar{\mathbf{H}}}{4\pi(1 - \bar{V})} + \frac{\bar{V}^3 - 3\bar{\mathbf{S}}^2\bar{S}}{24\pi(1 - \bar{V})^2} , \qquad (38)$$

which gives excellent results for thermodynamical quantities and for structures in the fluid phase. As a direct consequence of the geometric approach (38) to define a density functional one finds that the crossover to lower dimensional systems is straightforward and works very well.

Problem: Cluster Integrals and Non-spherical Soft Particles It is not known yet, if kinematic formulae or geometric measures can be used to evaluate not only the second virial coefficient but also cluster integrals of higher order in a cluster expansion (for instance, the third virial coefficient $C(T)$ in Eq. (32)). Is it possible to evaluate a whole subset of cluster integrals so that one might apply kinematic formulae on integral equations and closure relations [22]? Is it possible to give upper or lower bounds for cluster integrals and therefore for the equation of state, Eq. (32)?

The Rosenfeld density functional has only been applied for hard spheres and hard oriented cubes. Naturally the question arise if it is possible to use integral geometry to develop a general 'fundamental measure' functional for non-spherical and soft particles? It would be helpful to reformulate the existing papers on Rosenfeld's functional using integral geometry in order to emphasize the geometric features of the theory and to make future developments visible.

4.2   Excluded Volumes: Percolation Thresholds

The characterization and realistic modeling of random disordered materials as diverse as soils, sedimentary rocks, wood, bone, paper, polymer composites, catalysts, coatings, ceramics, foams, gels and concretes has been a major problem for physicists, material scientists, earth scientists and engineers for many years. Nevertheless, the prediction of mechanical and optical properties of the material, as well as the prediction of transport and phase behavior of fluids in porous structures from measures of the morphology and topology is still an unsolved problem. In Fig. 1 a sketch of a porous medium is shown where pores (black regions) of different size and shape are distributed in a solid material (white region). For practical purposes one wants to know whether the pores percolate or not, i.e. if water can flow inside the pores through the system from the upper edge to the bottom. Generally, one observes a threshold in the volume density of the pores above which the system supports water flow, whereas for densities below this critical density the pores do not percolate. A main aim of the integral geometric approach to porous media is the description of macroscopic transport properties such as the diffusion constant of the material in terms of the morphology of the pores. Although we focus on percolation in this section, we finally look for a general effective theory of porous media in terms of the morphology of its spatial structure [13,49].

Since 1957, the year of introduction of 'percolation processes' by Broadbent and Hammersley [8], percolation models became important for the understanding of many physical properties such as the fluid transport through a porous medium. Most of the effort did focus on critical exponents of the percolation transition, which exhibit a universal behavior and can therefore be described by the simplest model with a percolation threshold. But in designing composite materials it is more important to understand the non-universal behavior of transport quantities such as electrical and thermal conductivity, diffusion constants or elastic moduli. These non-universal features include the location of the critical threshold and also the dependence of physical quantities on the spatial details of the component phases away from the critical region. In particular, the prediction of the percolation threshold as a function of volume fraction, dispersity, shape, and orientation of the component phases remains a key problem in studying random multi-phase structures (for references see [16]). Also the dependence on interactions of the constituents of random materials and on the stochastic properties (correlations) of the spatial model are important for many practical applications. For spatial structures such as gels or cement-based materials which evolve in time the knowledge of the non-universal quantities even may provide insight into kinetic processes.

Already at the very beginning of percolation studies various criteria for the onset of percolation have been developed. In 1970 Scher and Zallen [53] proposed that the critical fractional occupied volume v_c (i.e., area in two dimensions) is almost constant in many percolation models and in 1985 Balberg [4] presented numerical limits for two and three-dimensional continuum percolation: $0.551 < v_c < 0.675$ for $d = 2$ and $0.084 < v_c < 0.295$ for $d = 3$. These limits can be rewritten in terms of the particle density in the so-called 'excluded volume' B_c defined by Eq. (30), i.e., by $v_c = 1 - e^{-B_c 2^{-d}}$ for spheres in d dimensions, yielding $3.2 < B_c < 4.5$ for $d = 2$ and $0.7 < B_c < 2.8$ for $d = 3$. In a series of papers [5] argued that the excluded volume, i.e., the average critical number B_c of bonds per particle is an approximate dimensional invariant and therefore, may be used as an estimate for the percolation threshold.

The prediction of the percolation threshold in composite media remained an unsolved problem in statistical physics. For instance, in the case of continuum percolation models, such as the random distribution of penetrable objects K, exact values of the percolation threshold are not known. Certainly, the threshold depends on size and shape of the pores K and on their statistical distribution. Therefore, we have to study the relevance of the morphology of the stochastic spatial structure for the percolation behavior. A widely used criterion for the percolation of randomly placed penetrable objects K is based on the mean number of neighbors $B = \rho V_{ex}(K)$ which is given by the density ρ of the objects and by the excluded volume V_{ex}, i.e., the volume around a given particle K which is excluded for another particle in the case of hard impenetrable objects. If the system percolates, each particle has at least one neighbor, therefore

$$B_c = \rho_c V_{ex} = 1 \tag{39}$$

is a reasonable percolation criterion. Thus, an estimate of the percolation threshold ρ_c is given by the excluded volume (30) and therefore by well-known geometric quantities of the distributed objects. Let us consider, for instance, sticks of length L in two dimensions with $M_0 = 0$ and $M_1 = L/\pi$. Using the criterion (39) and the expression (31) one finds the estimate $\rho_c = A/L^2$ with $A = \pi/2$. Computer simulations shows indeed over a wide range of lengths L a percolation threshold $\rho_c \sim 5.7/L^2$ [48], i.e., the same functional dependence but a somewhat larger coefficient A.

Although for a number of systems the excluded volume of a particle provides a first insight of the threshold dependence on size, shape and orientation of the grains in the ensemble, it is far from being satisfactory. Unfortunately, for arbitrary non-spherical particles the criterion in not very accurate and even for discs and spheres one wishes a better estimate. One finds numerically $0.5 < B_c < 3$ for a large number of shapes and $B_c = 2.8$ for spheres [4], so that a better estimate than that given by Eq. (39) is necessary. In particular, for continuous percolation a more accurate but nevertheless explicit expression of threshold estimates would be an important tool for various practical applications. In a more recent paper [1] proposed a more elaborate criteria which is more accurate but needs a tremendous effort to calculate it. They postulate that percolation occurs when the mean distance l between two connected neighbors and the average distance L between grains which have at least two grains connected to them obey $L = 2l$. Another approach applied topological arguments based on the observation that the Euler characteristic χ vanishes near the threshold [34]. It was argued that the zero $\chi(\tilde{v}_c) = 0$ may provide an interesting estimation of the percolation threshold v_c based on the topological connectivity of the configurations. Various percolation models were tested numerically in order to confirm the assumption that the analytical available zero $\tilde{v}_c$ of the Euler characteristic is a better estimate than any other proposed criteria. Of course, analytic calculations based, for instance, on the cluster expansion of the pair-connectedness function can in principle provide more accurate values, but they have to be performed for each model separately. The advantage of a heuristic criterion should be the explicit availability of the formula which does not need any further evaluations for specific models.

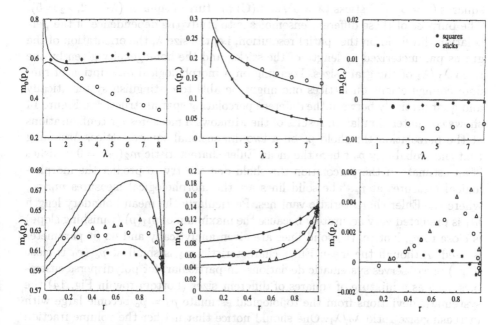

Fig. 14. (a) Minkowski functionals for squares (stars) and sticks (circles) of size λ at p_c [58]. Whereas $v_c \approx v_c^\infty$ is nearly constant for squares one finds the scaling behavior $v_c \approx \rho_c^\infty/\lambda$ for sticks. Numerical errors are smaller than the symbol sizes. The solid line is the prediction using the zero of the Euler characteristic ρ_0 and the mean values $v_\nu(\rho)$ of the morphological measures given by Eq. (118). (b) Minkowski functionals at criticality for a mixture of squares of sizes $\lambda_1 = 2$, $\lambda_2 = 4$, 6, 8 (stars, circles, triangles), where $r = p_1/(p_1 + p_2)$ denotes the density fraction at p_c. In contrast to the monodisperse systems shown in (a) the Euler characteristic at p_c deviates considerably from zero so that the prediction of the critical volume fraction $m_0(p_c)$ fails.

In order to study percolation thresholds in detail one can define a whole family of lattice models by introducing parameters for size, shape, orientation, and dispersity of the constituents. Let us consider three different types of ensembles on a two and three-dimensional lattice (see Fig. 12), namely random distributions of

(A) squares $K^{(2)}(\lambda)$ and cubes $K^{(3)}(\lambda)$ of edge length $\lambda \cdot a$, $\lambda = 1, 2, \ldots$.
(B) sticks $S^{(d)}(\lambda) = \bigcup_{i=1}^{\lambda} K_i^{(d)}(1)$, i.e., prolate unions of neighbored cells on the d-dimensional lattice where one edge length is chosen to be λ, $\lambda = 1, 2, \ldots$.
    Additionally, on the three-dimensional lattice we consider also oblate grains, i.e., plaquettes $P^{(3)}(\lambda)$ where two edge lengths are chosen λ.

(C) mixture of squares (cubes) $K^{(d)}(\lambda_1)$ and $K^{(d)}(\lambda_2)$ on the d-dimensional lattice, i.e. two grains with different edge lengths λ_1 and λ_2.

The details of the lattice models are described in Sect. 5.4. Figure 13 shows three percolating clusters at the critical threshold p_c with grains of size λ: namely (A) squares ($\lambda = 5$), (B) sticks ($\lambda = 5$), and (C) mixture of squares ($\lambda_1 = 2$, $\lambda_2 = 5$). The purpose of these different ensembles is to study the dependence of the percolation threshold on the spatial resolution, i.e, the size λ, the orientation of the grains parameterized by length of the sticks, and the homogeneity given by the ratio λ_1/λ_2 of the grain sizes. Focusing on a morphological description of random configurations on lattices one might be able to distinguish systematically and quantitatively between the different percolating spatial structures. Figure 14 shows computer simulation results of the Minkowski measures for configurations at the percolation threshold p_c on a two-dimensional square lattice. Assuming that the probability p_0 where the mean Euler characteristic $m_2(p_0) = 0$ vanishes is an estimate of the percolation threshold one can try to predict the morphological measures at p_c. The solid lines are the morphological measures $m_\nu(p_0)$ where the Euler characteristic vanishes. Particularly the mean boundary length m_1 is predicted very accurately because the maximum of $m_1(p)$ (vanishing slope) is close to p_c. Notice: the solid lines are given analytically and no fit parameter is used. Although the mean Euler characteristic m_2 is small at p_c (p_0 is close to p_c) one observes systematic deviations. In particular, for polydisperse ensembles such as a mixture of squares of different sizes (bottom row in Fig. 14) the systematic deviations from the topological estimate $p_0 = p_c$ become large with increasing size ratio λ_2/λ_1. One should notice that neither the volume fraction $v_c = m_0(p_c)$, nor the excluded volume argument $B_c = const$ provide a better description of the morphology or a better estimate of the percolation threshold. The excluded volume B_c is neither for an ensemble of sticks nor for a mixture of squares constant and varies considerably with λ, λ_2/λ_1, and r. Therefore, a more accurate estimate of percolation thresholds is still needed.

Problem: Accurate Estimation of Percolation Thresholds Three heuristic criteria to estimate the percolation thresholds were presented: the volume fraction [53], the excluded volume argument [4], and the zero of the Euler characteristic [34]. All three of them are based on additive measures, i.e, on the Minkowski functionals introduced in Sect. (3). Does another additive measure exist which is more accurate, i.e., which is almost invariant at p_c and does barely depend on other system parameters such as size, shape, and orientation of the grains? Is there a heuristic argument to determine the morphology near percolation thresholds? A detailed morphological analysis of spatial configurations at percolation thresholds as shown in Fig. 13 may help to find an estimate based on additive measures. A final goal would be a proof of useful ('sharp') upper and lower bounds on the percolation threshold based on Minkowski functionals.

4.3   Additivity and Completeness:
Curvature Model for Complex Fluids

In Sect. 2 is was shown that not only the curvature energy (11) of biological cells can be written as an additive functional but also the the pressure difference (17) of coexisting phases of fluids in porous media and the partition sum (23) of the eigenvalues of the Laplacian. Of course, the additivity of these quantities holds only approximatively, i.e., only for the leading terms of an expansion in curvature, distance, or time, respectively. But these examples may motivate to have a closer look on additivity in statistical physics and to examine which physical properties are related to it. In physics many quantities are known to be additive functions of subsystems. For instance, thermodynamic potentials such as free energy or entropy are extensive quantities if long range interactions can be ignored. For the union of two disjoint subsystems A and B one finds immediately the free energy $F(A \cup B) = F(A) + F(B)$. But also probabilities are additive functions, so that one may speculate if it is possible to relate the signed probability measures used in quantum mechanics (Wigner functions) to the additive Minkowski functionals. Here, only the thermodynamic consequences of an additive free energy of fluid systems are discussed.

Complex Fluids: Microemulsions and Colloidal Suspensions Amphiphilic surfactants added to oil and water tend to assemble spontaneously at the interface of the immiscible solvents and are capable to form polymorphic aggregates by separating oil-rich from water-rich domains. Apparently, the complex phase diagrams observed in these mixtures originate from the variety of sizes, shapes and topologies displayed by the spatial structure, i.e. by the interfacial patterns. Globular and bicontinuous microemulsions are particular examples among the structured liquid phases occurring at low surfactant concentration [31,45].

The attempt to work out the phase diagram of such mixtures by starting at the microscopic level with realistic molecular forces and covering the full range of compositions is still a problem for the future. However, considerable progress has already been made with idealized models constructed to explain salient features in the phase behavior of microemulsions in particular [12,17,18]. From a geometric point of view one should tie the thermodynamics of such composite media with the statistical morphology of their supra-molecular structures.

Therefore, let us consider an ensemble of random interfaces ∂A whose Hamiltonian

$$\mathcal{H} = \sum_{\nu=0}^{d} \varepsilon_\nu M_\nu(A) \tag{40}$$

is taken to be a linear combination of the geometric measures volume $V = M_0$, area $S = 8M_1$, integral mean curvature $H = 2\pi^2 M_2$, and Euler characteristic $\chi = \frac{4\pi}{3} M_3$ of the spatial domain A. The criteria underlying the choice of the Hamiltonian (40) are natural; in particular, the requirement of additivity leads to extensive thermal averages $\langle M_\nu \rangle \propto V$ for the model, and this secures a close linking of morphology with thermodynamics which will be considered below.

The bending energy of a membrane, arising from splay deformations and usually described by a mean curvature-squared term in the Helfrich Hamiltonian (11) is missing in the Hamiltonian (40), formally because this term is not additive; The resulting problems are already discussed in Sect. 2.1. Without going into the details [30,39,42], we note that this random surface model defined by the Hamiltonian (40) reproduces quite a few salient features observed in mixtures of oil, water and surfactants, namely a coexistence regime of oil-rich and water-rich phases with an isotropic bicontinuous middle phase, originating from a tricritical point; an ordered phase, known as "plumbers nightmare"; a peak in the liquid structure factor at a nonzero wave-vector; and a drastic reduction of surface tensions in the three-phase coexistence regime, caused by incipient critical endpoints. For details we refer to the mentioned literature and focus next on the general thermodynamic properties of a physical system if the Hamiltonian is given by the morphological expression (40). Since the statistical averages $< M_\nu(\mathcal{A}) >$ of these functionals qualify as extensive thermodynamic variables which may be interpreted as morphological order parameters to characterize the spatial structure of the configurations on a macroscopic scale it is important to relate thermodynamical quantities such as free energy, specific heat, and pressure to the morphology of the spatial structure. This will be done in the following Sect. 5 assuming that $\mathcal{A}$ is given by the union of convex grains.

Problem: Renormalization Group Theory for Minkowski Functionals

The completeness theorem (7) guarantees that the Hamiltonian (40) includes all additive geometrical invariants, whose thermal averages are extensive. However, the thermodynamic requirement of extensive internal energy does not imply that the Hamiltonian must be manifestly additive. The currently popular statistical models [17,18,30] for an ensemble of surfactants films ("membrane models") employ the Helfrich Hamiltonian (11) which is of the form (40) but contains an additional bending energy contribution $\mathcal{H}_{bend} = \frac{\kappa}{2}\hat{H}^2$ given by Eq. (12) where H is the mean curvature of the surfactants film S and κ is a (bare) bending rigidity. In the studies of microemulsion phases employing the Helfrich Hamiltonian the Euler characteristic is ignored and attention is focused on a scale ξ-dependent bending energy $\mathcal{H}_{bend}(\xi)$ obtained by replacing κ with a renormalized rigidity $\kappa(\xi)$ to account for the reduction of the bare rigidity κ caused by short-wavelength thermal film undulations (see Eq. (14)). The morphological model (40) incorporates the energetics of an ensemble of randomly folded, multiply connected and self-avoiding films, and thus deals with lengths beyond ξ_κ where $\mathcal{H}_{bend}(\xi)$ may indeed be omitted. Thus, one avoids the difficulty to define appropriately the integral $\hat{H}^2$ of the squared mean curvature for configurations on lattices (Sect. (Sect. 5.4) or for germ grain models (Sect. 5.1) if one renormalizes the bending rigidity κ.

In the morphological model (40) one assumes that the scale-dependence of the parameters ε_α, viewed as renormalized effective couplings, can be approximated by their "naive" scaling behavior, $\varepsilon_\alpha(\xi) \sim \xi^{\alpha-3}$ (compare Eq. (8)), since a non-perturbative renormalization theory as required for length scales exceeding the persistence length is still lacking. Although the additivity relation (6) and the completeness theorem (7) provide a promising frame for studying the Hamiltonian (40), a renormalization theory which includes the Minkowski functionals, in particular, the Euler characteristic, is still not yet developed [30]. In the following Sect. 5.4 integral geometry is defined on lattices which may open the possibility to renormalize the Minkowski functional and not only the bending rigidity κ [37].

5 Boolean Model: Thermal Averages

Homogeneous spatial domains of phases on a mesoscopic scale are a characteristic feature of many composite media such as porous materials shown in Fig. 1 or complex fluids discussed in the previous Sect. 4.3. The spatial structure of such composite media may be described by randomly distributed, overlapping spheres, i.e., by configurations of a germ grain model (Fig. 1). In contrast to the Boolean model used in stochastic geometry it is necessary for physical applications to introduce interactions between the grains in order to describe particle correlations. In contrast to microscopic pair interactions, the statistical model for a two-component system on a mesoscopic scale should rest on the morphology of the homogeneous domains and should take into account geometric properties such as the volume, shape, and connectivity of the domains, for instance, the white area shown in Fig. 1. For instance, colloidal hard particles undergo a phase separation (aggregation) induced by fluid wetting layers which can be described by overlapping spheres so that the interaction of the colloids are given by the bulk free energy of the film proportional to its volume (see contribution 'Fun with hard spheres' by H. Löwen in this volume). The Widom-Rowlinson model is an important example of such an interacting model of stochastic geometries used mainly in the statistical physics of fluids where the interaction is given by the volume of overlapping spheres [65]. A general extension of this model rest on the Minkowski functionals of the overlapped region in space, i.e., on surface tension and curvature energies additionally to the volume of the spheres [35,39,43]. A density functional theory for the Widom-Rowlinson model and for its extension in terms of Minkowski functionals is introduced in Sect. 5.1 where explicit expressions are required for the intensities of Minkowski measures for inhomogeneous distributions of grains. These density dependent mean values are derived in Sect. 5.2 for various special cases such as oriented cylinders and inhomogeneous distributed spheres.

Since the Minkowski functionals are extensive quantities and not well defined for infinite systems we use the notion of Minkowski *measure* in contrast to *functional*. Nevertheless, the physical notions of 'mean value' or 'variance' of a functional is still used instead of the appropriate mathematical terms 'intensity' or 'second order moment measure'. The term functional is also used whenever the mathematical properties of these measures are important.

5.1   Widom-Rowlinson Model and Density Functional Theory

The thermodynamics and bulk properties of such composite media depend often on the morphology of its constituents, i.e. on the spatial structure of the homogeneous domains. Therefore, a statistical theory should include morphological descriptors to characterize size, shape and connectivity of the aggregating mesophases. In the context of integral geometry one focus on the morphological aspects of two component media by employing the Minkowski functionals as suitable descriptors of spatial patterns. Such a morphological thermodynamics may be outlined as follows: Each configuration of component (I) is assumed to be the union of mutually penetrable convex bodies ('grains')

$$K^N = \bigcup_{i=1}^{N} g_i K \ .$$
(41)

embedded in the host component (II). The form of the grains is otherwise arbitrary; they may be balls, flat discs, thin sticks, etc. Typical configuration are shown in Fig. 1. Let $\mathcal{G}$ denote the group of motions (translations and rotations) in the Euclidean space $\mathbb{R}^d$. The location and orientation of the grains are specified by the action of $g_i \in \mathcal{G}$ on a tripod fixed at the centroid of each grain K, $K_i = g_i K$. The Boltzmann weights are specified by the Hamiltonian (40),

$$\mathcal{H}(K^N) = \sum_{\nu=0}^{d} h_\nu \left(M_\nu \left(\bigcup_{i=1}^{N} K_i \right) - N M_\nu(K) \right)$$
(42)

which is a linear combination of Minkowski functionals on the configuration space of the grains. We emphasize that the Hamiltonian (Eq. (42)) constitutes the most general model for composite media assuming additivity of the energy of the homogeneous, mesoscopic components. The configurational partition function is taken to be

$$Z_N(T,V) = \frac{1}{N! \Lambda^{Nd}} \int \exp \left\{ -\beta \mathcal{H} \left(\bigcup_{i=1}^{N} g_i K \right) \right\} \prod_{j=1}^{N} dg_j \ .$$
(43)

The integral denotes averages over the motions of the grains with dg being the invariant Haar measure on the group $\mathcal{G}$. The length Λ is a scale of resolution for the translational degrees of freedom of the grains which are restricted to a cube of volume V. The grand canonical partition sum

$$\Xi(T,\mu,V) = \sum_{N=0}^{\infty} e^{\beta \mu N} Z_N(T,V)$$
(44)

determines the grand canonical potential $\Omega(T,\mu,V) = -k_B T \log \Xi(T,\mu,V)$ as function of the chemical potential μ. The thermodynamics of such composite materials are then given in terms of additive, morphological measures of its constituents. Depending on the relative strength of the energies related to the volume, surface area, mean curvature, and Euler characteristic of the domains one finds qualitative different phase diagrams and spatial structures. Monte-Carlo simulations of the model and also applications to colloidal systems are work in progress.

Morphological Thermodynamics Let us first recall some basic facts from thermodynamics. Given the partition function $Z_N(T, D)$ of the spatial configurations $A \cap D$ in a domain D (volume $V = |D|$) at fixed particle number N and temperature temperature T, the free energy is given by the logarithm

$$F(T, N, D) = -k_B T \log Z_N(T, D) \ . \tag{45}$$

The thermodynamic potentials are related by Legendre transformations so that the grand canonical potential $\Omega(T, \mu, D) = F - \mu N = -k_B T \log \Xi(T, \mu, D)$ is given by the grand canonical partition sum (44). The internal energy is the expectation value of the energy given by the Hamiltonian (42), i.e.,

$$U(N, T, D) = \left. \frac{\partial \beta F}{\partial \beta} \right|_{NV} = <\mathcal{H}>_{TNV} = \sum_{\nu=0}^{d} h_\nu \bar{M}_\nu(N, T, V) \tag{46}$$

with the mean value $\bar{M}_\nu(N, T, V) =< M_\nu(A) >_{TNV}$ of the Minkowski functionals in the Gibbsian ensemble where averages at constant particle number N are denoted by $< \cdot >_{NTV}$. Naturally one can assume that the averaged Minkowski functionals are additive functionals of the domain D so that the thermodynamic potential

$$U(N, T, D) = \sum_{\mu=0}^{d} u_\mu(T, N) M_\mu(D) \ , \quad u_\mu(T, N) = \sum_{\nu=0}^{d-\mu} \binom{\mu + \nu}{\mu} \tilde{h}_{\mu+\nu}^{(\mu)} m_\nu(T, N) \tag{47}$$

is an additive quantity and can be written as a sum of volume M_0, surface M_1, and curvature terms of the domain D, where the related bulk internal energy $u_0(T, N)$, surface tension $u_1(T, N)$, and bending energies $u_2(T, N)$ and $u_3(T, N)$ depend explicitly on the intensities $m_\nu(T, N)$ of the morphological measures. The tilde $\tilde{h}^{(\mu)}$ denotes that the interaction parameters h_ν depend one the dimension $d - \mu$ of the boundary due to additional interactions between particles and boundary. Of course, we neglected that the interaction of the grains with the wall influences the equilibrium configurations of the homogeneous domains A and finally the coefficients m_ν, too. In principle, the geometric quantities m_ν have to be determined as function of temperature T and particle number N for an inhomogeneous distribution of grains. This will be done for the Boolean model in the following sections. Notice: all Minkowski measures m_ν and coefficients h_ν are needed for the bulk energy u_0 whereas the topological energy contribution u_d depend solely on the topological parameter h_d and the intensity $m_0(T, N)$ of the volume.

An explicit expression for the free energy $F = U - TS$ in terms of the geometric measures is more complicated due to the contribution of the entropy S. One obtains in the thermodynamic limit

$$f(T,\rho) = \lim_{N,V\to\infty} \frac{1}{V} F(T,N,V) = k_B T (\rho \log \rho - \rho) - k_B T \int_0^\rho d\rho' \log z(T,\rho') \tag{48}$$

with $\rho = N/V$ and the fugacity $z(T,\rho) = \rho Z_{N+1}/Z_N$. Then, the chemical potential

$$\mu(T,\rho) = \frac{\partial F}{\partial N}\bigg|_T = k_B T \log \rho - k_B T \log z(T,\rho) \tag{49}$$

and the pressure

$$p = -\frac{\partial F}{\partial V}\bigg|_T = \rho \frac{\partial f(T,\rho)}{\rho} - f(T,\rho) = k_B T \rho - k_B T \int_0^\rho d\rho' \frac{\rho'}{z(T,\rho')} \frac{\partial z(T,\rho')}{\partial \rho'} \tag{50}$$

can be expressed in terms of the fugacity $z(T,\rho)$. The second derivatives of the free energy, namely the specific heat

$$C_N = T \frac{\partial S}{\partial T}\bigg|_{NV} = -T \frac{\partial^2 F(T,N,V)}{\partial T^2}\bigg|_{NV} = \frac{<(\Delta\mathcal{H})^2>_k}{k_B T^2}$$
$$= \frac{V}{k_B T^2} \sum_{\nu,\mu=0}^d h_\nu h_\mu m_{\nu\mu}^{(k)}(\rho) \tag{51}$$

is given by the second order moment of the Hamiltonian (42), i.e., by the the second order moments

$$m_{\nu\mu}^{(k)}(N,T,V) = \frac{< M_\nu(\mathcal{A})M_\mu(\mathcal{A}) >_{NTV} - \bar{M}_\nu(N,T,V)\bar{M}_\mu(N,T,V)}{V} \tag{52}$$

of the Minkowski measures. Here, the index (k) refers to a canonical ensemble where the average is performed at constant particle number N and temperature T. Using the thermodynamic relation

$$C_\mu = T \frac{\partial S}{\partial T}\bigg|_{\mu V} = C_N + T \left(\frac{\partial \mu(T,N,V)}{\partial T}\bigg|_{NV} \right)^2 \bigg/ \frac{\partial \mu(T,N,V)}{\partial N}\bigg|_{TV} \tag{53}$$

between the specific heats at constant chemical potential C_μ and constant particle number C_N one obtains

$$C_\mu - C_N = \frac{N k_B z(T,\rho)}{z(T,\rho) - \rho \frac{\partial z(T,\rho)}{\partial \rho}} \left(\beta\mu + \beta \frac{\partial \log z(T,\rho)}{\partial \beta} \right)^2 . \tag{54}$$

Because the specific heat

$$C_\mu = -T \frac{\partial^2 \Omega}{\partial T^2}\bigg|_V = \frac{\mu(T,\rho)^2}{k_B T^2} < (\Delta N)^2 > + \frac{<(\Delta\mathcal{H})^2>}{k_B T^2} - \frac{2\mu(T,\rho)}{k_B T^2} < \Delta N \Delta \mathcal{H} > \tag{55}$$

is related to the second order moments

$$< (\Delta N)^2 >= -\left.\frac{\partial^2 \beta\Omega}{\partial\beta\mu^2}\right|_V = k_B T \left.\frac{\partial N}{\partial\mu}\right|_{TV} = \frac{Nz(T,\rho)}{z(T,\rho) - \rho\frac{\partial z(T,\rho)}{\partial\rho}}, \qquad (56)$$

$$< (\Delta\mathcal{H})^2 >_{gk}=< (\Delta N)^2 > \left(\frac{\partial\log z(T,\rho)}{\partial\beta}\right)^2 + < (\Delta\mathcal{H})^2 >_k, \qquad (57)$$

one obtains the relation

$$m_{\nu\mu}^{(gk)}(\rho) - m_{\nu\mu}^{(k)}(\rho) = \rho\frac{\partial m_\nu(\rho)}{\partial\rho}\frac{\partial m_\mu(\rho)}{\partial\rho} + \mathcal{O}(\beta^2) \qquad (58)$$

for the grand-canonical and canonical second order moments of the Minkowski measures. Thus, we derived a relation between the moments of morphological measures based only on thermodynamic arguments. Of course, it remains to calculate $m_{\nu\mu}(N,V)$ and $\bar{M}_\nu(N,V)$ in a canonical ensemble which will be done in the Sect. 5.3 for the Boolean model of overlapping grains.

Mean-Field Theory It was shown that one needs only to calculate the fugacity

$$z(T,\rho) = \rho\frac{Z_{N+1}}{Z_N} =< e^{\beta\mathcal{H}(\cup_{i=1}^N K_i \cap K)} >_{TN} \qquad (59)$$

as function of temperature and particle density in order to obtain the thermodynamical quantities. In particular, this important function is the generating functional for the cumulants of the geometric measures in a Gibbsian ensemble. Expanding in powers of the inverse temperature $\beta = 1/(k_B T)$ and using the kinematic formulae (9) for the average over the position and orientation of identical grains K in a homogeneous and isotropic ensemble one obtains the explicit expression

$$z(T,\rho) = 1 + \sum_{\nu=0}^d \beta h_\nu \sum_{\mu=0}^\nu \binom{\nu}{\mu} m_\mu(K)m_{\nu-\mu}(\rho) + \mathcal{O}(\beta^2). \qquad (60)$$

The intensities $m_\nu(\rho)$ do not depend on the temperature and can be calculated within the Boolean model using again the kinematic formula (9). Following the method in [34] one obtains for the mean values of the Minkowski measures (cf. Eq. (5)) per unit volume [61]

$$m_\nu(\rho) = \left.\frac{\partial^\nu}{\partial t^\nu}\left(1 - e^{-\rho\sum_{\alpha=0}^d m_\alpha(K)t^\alpha/\alpha!}\right)\right|_{t=0}$$

$$= \delta_{\nu 0} - e^{-\rho m_0(K)}\sum_{\mu=1}^\nu \frac{(-\rho)^{\bar{\mu}}}{\mu!}\bar{m}_\nu^\mu(K), \qquad (61)$$

in particular, $m_0(\rho) = 1 - e^{-\rho m_0(K)}$, $m_1(\rho) = \rho m_1(K)e^{-\rho m_0(K)}$, and $m_2(\rho) = (\rho m_2(K) - m_1(K)^2\rho^2)e^{-\rho m_0(K)}$. Notice: the coefficients

$$\bar{m}_\nu^\mu(K) \geq \bar{m}_\nu^\mu(B^{(d)}) \qquad (62)$$

for non-spherical convex grains K are always larger than the coefficients for a d-dimensional sphere $B^{(d)}$. This relation is directly related to the well-known isoperimetric inequalities

$$\left(\frac{M_0(K)}{M_0(B^{(d)})}\right)^{d-\nu} \leq \left(\frac{M_\nu(K)}{M_\nu(B^{(d)})}\right)^d , \quad \frac{M_\nu(K)}{M_\nu(B^{(d)})} \leq \frac{M_{d-1}(K)}{M_{d-1}(B^{(d)})} \tag{63}$$

for the Minkowski functionals of arbitrary convex grains K in d dimensions. Such inequalities causes a liquid drop to relax into a spherical shape which minimizes its free energy because of its minimal surface area. It would be interesting to find analogous 'laws of minimal shape' by applying the generalized inequalities (62) on the fugacity (60), i.e., on the free energy of mesoscopic phases.

Using the mean values (61) of the morphological measures for configurations $\mathcal{A}$ and the relation

$$\sum_{\mu=0}^{\nu} \binom{\nu}{\mu} m_\mu m_{\nu-\mu}(\rho) = m_\nu - \frac{\partial m_\nu(\rho)}{\partial\rho} \tag{64}$$

one obtains the pressure

$$p(T,\rho) = k_B T\rho + \sum_{\nu=0}^{d} h_\nu \left(\rho\frac{\partial m_\nu(\rho)}{\partial\rho} - m_\nu(\rho)\right) + \mathcal{O}(\beta) \tag{65}$$

as function of density and temperature. The details of the phase behavior given by the equation of state (65) can be found in [30,35,38,42] and [9]. Notice: the critical points of the phase diagram are given by the equations

$$\frac{\partial p}{\partial\rho} = 0 , \quad \frac{\partial^2 p}{\partial\rho^2} = 0 \quad \Rightarrow \quad \rho\frac{\partial z(T,\rho)}{\partial\rho} = z(T,\rho) , \quad \frac{\partial^2 z(T,\rho)}{\partial\rho^2} = 0 . \tag{66}$$

Thus, critical temperatures can be expressed within the mean field approximation (60) in terms of the morphological measures

$$k_B T_c = -\rho_c \sum_{\nu=0}^{d} h_\nu m_\nu''(\rho_c) , \tag{67}$$

where the critical density is determined by the equation $\sum_{\nu=0}^{d} h_\nu(\rho_c m_\nu'''(\rho_c) + m_\nu''(\rho_c)) = 0$. For overlapping grains in two and three dimension these equations yield two critical points if the coefficient h_d of the Euler characteristic is large. Such an energy contribution may induce an additional continuous phase transition and a triple point. Thus, the topology of the phase diagram changes qualitatively if the Hamiltonian contains terms proportional to the Euler characteristic χ of the configuration.

The main aim of this section was to relate thermodynamical quantities to the morphology of the spatial configuration. Of course, Eq. (67) is only valid within the high temperature expansion (60) and second order moments of the

Minkowski measures are required already for the first correction term proportional to β^2 in Eq. (60). Therefore, it remains to calculate variances $m_{\nu\mu}(N, T, V)$
and not only mean values $m_\nu(N, T, V)$ given by Eq. (61). This will be done in
Sect. 5.3 for the Boolean model and more conveniently for lattice models in Sect.
5.4. Monte-Carlo simulation indicates that the mean values $m_\nu(N, T, V) \sim \beta$ depend linearly on the inverse temperature for a large range of values for β so that
the second order moments should be capable to describe the thermodynamic
behavior even for low temperatures close to the critical point [9].

    Notice: physical quantities such as the variances (56) and the compressibil-
ity $\kappa_T \sim \kappa^{(0)}(T - T_c)^{-\gamma}$ diverges at the critical point with a universal exponents γ ($\gamma = 1$ for the mean field approximation). It is not known yet how the
non-universal amplitudes $\kappa^{(0)}$ depend on the morphology of typical spatial configurations at the critical point. A similar question was already asked in Sect.
3.4 where the scaling behavior of fractals were studied and non-trivial scaling
properties of the Minkowski functionals were found.

Density Functional Theory So far only phases with a homogeneous density ρ are studied. But Monte-Carlo simulations indicate that at high densities
the morphological model defined by the Hamiltonian (42) exhibit not only a
solid phase of regular arranged grains with a periodic density $\rho(\mathbf{r})$ but also an
interesting inhomogeneous 'glassy' phase similar to a dense packing of styrene
spheres[9]. Expanding the partition sum (43) of the Hamiltonian (42) in powers
of β one can define a grand canonical density functional by

$$\Omega\left[\rho(\mathbf{r})\right] = \Omega_{ref}\left[\rho(\mathbf{r})\right] + \sum_{\nu=0}^{d} h_\nu \int d^3r \left[m_\nu\left[\rho(\mathbf{r})\right] - \rho(\mathbf{r})m_\nu(K)\right] , \qquad (68)$$

where $\Omega_{ref}[\rho(\mathbf{r})]$ denotes the reference free energy of a system determined by
pair interactions additional to the morphological Hamiltonian (42). The reference
density functional for an ideal gas in an external potential $V(\mathbf{r})$, for instance,
reads $\Omega_{ref}[\rho(\mathbf{r})] = \int d^3r \left[\rho(\mathbf{r})\ln(\rho(\mathbf{r})\Lambda) - \rho(\mathbf{r}) + \mu\rho(\mathbf{r}) + V(\mathbf{r})\rho(\mathbf{r})\right]$ (chemical
potential μ, thermal de Broglie wavelength Λ). The inhomogeneous intensities
$m_\nu[\rho(\mathbf{r})]$ of the Minkowski measures are given below for an ensemble of independent overlapping spheres K (Eq. (99)). Instead of perturbing around the ideal
gas equation of state one may use hard-core particles as reference system so
that $\Omega_{ref}[\rho(\mathbf{r})]$ is given by Eq. (16). (compare Sect. 2.2, Rosenfeld's functional
in Sect. 4.1, and the paper by H. Löwen in this volume). Then the intensities
$m_\nu[\rho(\mathbf{r}), c(\mathbf{r}; \mathbf{r}')]$ of the Minkowski functionals depend not only on the mean density $\rho(\mathbf{r})$ of the fluid particles at $\mathbf{r}$ but also on the correlation functions $c(\mathbf{r}, \mathbf{r}')$ of
the system due to the two-particles interaction potential $\Phi(\mathbf{r})$. The equilibrium
density profile $\rho^{(eq)}(\mathbf{r})$ minimizes the density functional in Eq. (68) and yields
the grand canonical potential $\Omega(\mu)$ as function of the chemical potential μ. The
equilibrium profile depends not only on the temperature T and a possible external potential $V(\mathbf{r})$ but also on the morphological interaction parameters h_i.
Of course, the minimization of the density functional (68) is not feasible yet

and remains to be performed approximatively in future works. But first one has to derive explicit expressions for the inhomogeneous intensities $m_\nu \left[\rho(\mathbf{r}, c(\mathbf{r}; \mathbf{r}'))\right]$ of the Minkowski measures. This will be done in the next section for several special examples: homogeneous but correlated grain distributions ($\rho[\mathbf{r}] \equiv \rho$) and inhomogeneous but Poisson distributed grains ($c(\mathbf{r}; \mathbf{r}') = 0$).

Concluding we emphasize that integral geometry (kinematic formula) combined with the Boolean model enables us to find a general expression for the grand canonical free energy of complex fluids in terms of the density dependent volume, surface area, and curvatures of the spatial configurations. Applications for colloids and composite media are straightforward and we expect to find qualitative rich phase diagrams and spatial structures. But many problems occur if one tries to evaluate the partition sum (43) determined by the morphological Hamiltonian (42).

Problem: Evaluating Partition Sums

Besides mean field approximations and density functional theories as described above, computer simulations are one of the most important tools in statistical physics for an understanding of the configurational partition sum (43). Unfortunately, the computational cost to evaluate the energy (42) of a configuration is enormous so that efficient algorithms have to be invented in order to make Monte-Carlo simulations feasible. In two dimensions this has been done by Brodatzki [9] but computational cost is still large. The implementation of perfect simulation algorithm for the Hamiltonian (42) may help to reach equilibrium configurations effectively (see the paper by E. Thönnes in this volume). In usual Monte-Carlo Simulations the initial state converges with a characteristic time scale towards a configuration sampled from the equilibrium distribution. Unfortunately, whether or not such an equilibrium configuration is reached cannot be decided in most of the simulations. But the technique of perfect simulation allows the sampling of a configuration from the equilibrium distribution. Crucial for the practicality of such an algorithms is a way of determining complete coalescence of extreme states in an efficient manner. Because interactions in many models in spatial statistics and statistical physics allow a partial order of spatial configurations and a monotonic transition rule one can apply perfect simulation on standard statistical systems such as the Ising model [26] or the continuum Widom-Rowlinson model [65]. The perfect simulation technique might help to overcome the computational cost when simulating the interacting germ grain model [9], i.e. the generalized Widom-Rowlinson model defined by the morphological Hamiltonian (42).

A second problem concerns the critical points (Eq. (67)) obtained within a high temperature approximation of the partition sum (43). H.-0. Georgii showed in his contribution to this volume on phase transitions and percolation in Gibbsian particle models that the existence of critical points can be proven for a class of models including the Ising model and the the continuum Widom-Rowlinson model. Naturally, the question arises if it is possible to proof the existence of the second critical point of the generalized Widom-Rowlinson model (42).

A third problem is related to the renormalization of the energy parameters h_ν already discussed in Sect. 4.3. Because of the universal form of the mean values (61) of the Minkowski functionals for the Boolean model it might be possible to capture the temperature dependence of the intensity measures $< M_\nu(T, K) > \sim m_\nu(\rho)$ by the expressions given in Eq. (61) but with temperature dependent shapes $m_\nu(T, K)$ of single grains K instead of the constants $m_\nu(K)$. Is it possible to use relation (64), for instance, to find such temperature dependent effective mean values $m_\nu(\rho, T)$? Or can one use mean-field renormalization techniques to renormalize the shape of single grains yielding temperature dependent mean values. A renormalization group theory for Minkowski measures can be used to study the critical scaling behavior of the configurational morphology. A similar question has already been discussed in the context of the critical behavior of Lennard-Jones fluids in Sect. 3.3.

5.2 Correlated, Inhomogeneous and Anisotropic Distributions

In the previous Sect. 5.1 thermal averages of morphological measures are used to calculate thermodynamical quantities such as the pressure and free energy of a physical system. But the calculus was restricted to a high temperature expansion where only the intensity measures of the Boolean model are required. Adding additional interaction potentials between the particles such as the Lennard-Jones potential discussed in Sect. 4.1 (see Fig. 9) one needs to derive expressions analogous to Eq. (61) but for correlated grains. Also the density functional theory (68) requires intensity measures $m_\nu[\rho(\mathbf{r})]$ of the Boolean model for inhomogeneous and anisotropic distributed grains and not only for a constant density ρ as in Eq. (61). The main purpose of this section is to illustrate the capability of the calculus. Integral geometry provide not only an interesting concept for the description of spatial structures, but also powerful techniques to calculate Minkowski functionals, for instance. Although this section contains a lot of technical details, it hopefully also illustrates that the mathematical results might be helpful for the applications in statistical physics mentioned in the previous sections.

Mean Values for Correlated Grains Repeating the additivity relation (6) of Minkowski functionals one obtains the inclusion-exclusion formula

$$M_\nu(\bigcup_{i=1}^{N} K_i) = \sum_i M_\nu(K_i) - \sum_{i<j} M_\nu(K_i \cap K_j)$$

$$+ \ldots + (-1)^{N+1} M_\nu(K_1 \cap \ldots \cap K_N) . \tag{69}$$

Generally the statistics of a homogeneous point process in a domain Ω with volume $|\Omega|$ is specified by a sequence of density correlation functions $\rho^{(n)}(\mathbf{r}_1, \ldots, \mathbf{r}_n)$ with the homogeneous density $\rho \equiv \rho^{(1)}(\mathbf{r}_1)$. The intensity of the ν-th Minkowski measure for the augmented coverage per unit volume are then obtained from the additivity relation (6) in the form

$$\bar{M}_\nu \left[\left\{ \rho^{(n)} \right\} \right] = \sum_{n=1}^{\infty} \frac{(-1)^{n+1}}{n! |\Omega|} \int_\Omega d\Gamma^{(n)} \, M_\nu [\bigcap_{i=1}^{n} B_R(\mathbf{r}_i)] \, \rho^{(n)}(\Gamma^{(n)}) \qquad (70)$$

where we introduced, for convenience, the variable $\Gamma^{(n)} = (\mathbf{r}_1, \ldots, \mathbf{r}_n)$ with the integration measure $\int_\Omega d\Gamma^{(n)} = \prod_{i=1}^{n} \int_\Omega d\mathbf{r}_i$. Obviously, the Minkowski measures embody information from every order n of the correlation functions $\rho^{(n)}(\mathbf{r}_1, \ldots, \mathbf{r}_n)$. If the density correlation functions $\rho^{(n)}(\mathbf{r}_1, \ldots, \mathbf{r}_n)$ were independent of position, the integrals in Eq. (70) could be performed using the fundamental kinematic formula (9). But this is not the case for most physical applications.

An alternative and sometimes more convenient expression for the densities $\bar{M}_\nu$ can be obtained in terms of so-called *connected* correlation functions, i.e., the centered cumulants $k^{(n)}(\Gamma^{(n)})$ with $k^{(1)}(\mathbf{r}_1) = \rho$, $k^{(2)}(\mathbf{r}_1, \mathbf{r}_2) = \rho^{(2)}(\mathbf{r}_1, \mathbf{r}_2) - \rho^{(1)}(\mathbf{r}_1)\rho^{(1)}(\mathbf{r}_2)$, and, in general,

$$\rho^{(n)}(\Gamma^{(n)}) = \sum_{\{\mathcal{P}\}} \prod_{i=1}^{|\mathcal{P}|} k^{(m_i)}(\Gamma^{(m_i)}) \qquad (71)$$

Thus, the correlation functions of order n is given by a sum over all possible partitions $\mathcal{P}$ of $\Gamma^{(n)} = (\mathbf{r}_1, \ldots, \mathbf{r}_n)$ into $|\mathcal{P}|$ parts of m_i elements. Each vector $\mathbf{r}_j \in \Gamma^{(n)}$ occurs exactly once as an argument of a cumulant function $k^{(m_i)}$ on the right side, i.e., $\sum_{i=1}^{|\mathcal{P}|} m_i = n$.

Using additivity (6) and kinematic formula (9) of the Minkowski functionals one can follow the derivation in [34] so that one immediately obtains the expression for the intensities [35]

$$\bar{M}_\nu \left[\left\{ k^{(n)} \right\} \right] = \frac{\partial^\nu}{\partial t^\nu} \left(1 - e^{M(\rho, t)} \right) \Big|_{t=0} \qquad (72)$$

due to the factorization of the integral in Eq. (70). The averaged Minkowski polynomial

$$M(\rho, t) = \sum_{n=1}^{\infty} \frac{(-\rho)^n}{n!} \left(\prod_{i=1}^{n} \int_{\mathfrak{g}} d\mathbf{g}_1 \right) \sum_{\nu=0}^{d} \frac{t^\nu}{\nu!} M_\nu(K_{\mathbf{g}_1} \ldots \cap K_{\mathbf{g}_n}) k^{(n)}(\mathbf{r}_1, \ldots, \mathbf{r}_n) \qquad (73)$$

dependent now on the correlation functions $k^{(n)}(\mathbf{r}_1, \ldots, \mathbf{r}_n)$ in contrast to (61). Here, $\mathbf{g} = (\mathbf{r}, \Theta)$ denotes not only the location $\mathbf{r}$ of the grain K but also the orientations Θ. The integration $\int d\mathbf{g} = \int d\mathbf{r} \times \int d\Theta$ is the direct product of the integrations over all locations and orientations. In particular, one obtains (compare the analogous result (61) for Poisson distributed grains)

$$\begin{aligned}
m_0(\rho, \{k^{(n)}\}) &= 1 - e^{-\rho \bar{m}_0} \\
m_1(\rho, \{k^{(n)}\}) &= \rho \bar{m}_1 e^{-\rho \bar{m}_0} \\
m_2(\rho, \{k^{(n)}\}) &= (\rho \bar{m}_2 - \bar{m}_1^2 \rho^2) e^{-\rho \bar{m}_0} \\
m_3(\rho, \{k^{(n)}\}) &= (\rho \bar{m}_3 - 3\bar{m}_1 \bar{m}_2 \rho^2 + \bar{m}_1^3 \rho^3) e^{-\rho \bar{m}_0} .
\end{aligned} \qquad (74)$$

with the averaged volume $\bar{m}_0[k^{(n)}] = \bar{V}(\rho, \{k^{(n)}\})/\rho$ and the averaged Minkowski functionals

$$\bar{m}_\nu[k^{(n)}] = M_\nu(K) - \sum_{n=2}^\infty \frac{(-1)^n}{n!\rho} \prod_{i=2}^n \int_K dg_i \, M_\nu(K_0 \bigcap_{i=2}^n K_{g_i}) \, k^{(n)}(0, r_2, \ldots, r_n)$$

(75)

for single grains K. Poisson distributed grains have vanishing cumulants $k^{(n)}(\Gamma^{(n)}) = 0$ and one recovers $\bar{m}_\nu[k^{(n)}] = M_\nu(K)$. The void probability function of a domain K is simply $P_0(\rho, K) = \exp\{-\bar{V}\}$ and the probability to find exactly N points in the domain K is given by

$$P_N(\rho, K) = -\frac{\rho^N}{N!} \left(\frac{\partial}{\partial\rho}\right)^N P_0(\rho, K) \ .$$

(76)

Notice: we have used analogous expressions to Eq. (75) already for the formulation of Rosenfeld's fundamental measure functional (36) introduced in Sect. 4.1, where the Minkowski functionals are averaged over an inhomogeneous density $\rho(\mathbf{r})$.

The universal, i.e., polynomial form of the mean values (74) is related to the additivity of the Minkowski functionals due to the decomposition into *local* terms. Thus, the dependence on shape, orientation and correlations of the grains (see also Eqs. (89) and (94)) enters only into a finite number of relevant coefficients. We have used this property of additive functionals already in Sect. 3.1 to define morphological order parameters where the reduction of spatial information to a finite number of measures is essential.

An important approximation for many applications in physics is the assumption that the cumulants $k^{(n)}(\Gamma^{(n)})$ for $n \geq 3$ are small compared to the second centered two-point correlation function $\xi(\mathbf{r}) = k^{(2)}(0, \mathbf{r})/\rho^2$. Within this Gaussian approximation one obtains

$$\bar{m}_0[\xi(\mathbf{r})] = V(K) - \frac{\rho}{2} \int_{\mathcal{G}} dg V(K \cap gK) \xi(\mathbf{r})$$

$$\bar{m}_1[\xi(\mathbf{r})] = M_1(K) - \rho \int_{\mathcal{G}} dg M_1(K \cap gK, \partial K) \xi(\mathbf{r}) \ ,$$

$$\bar{m}_2[\xi(\mathbf{r})] = M_2(K) - \rho \int_{\mathcal{G}} dg (M_2(K \cap gK, \partial K)$$

$$+ \frac{1}{2} M_2(K \cap gK, \partial K \cap g\partial K)) \xi(\mathbf{r}) \ .$$

(77)

If the correlation function decays algebraicly $\xi(r) = k_2(\mathbf{r}_1, \mathbf{r}_2) = \left(\frac{r_0}{r}\right)^\gamma$, $r = |\mathbf{r}_1 - \mathbf{r}_2|$, with an scaling exponent γ and a correlation length r_0 one finds for the

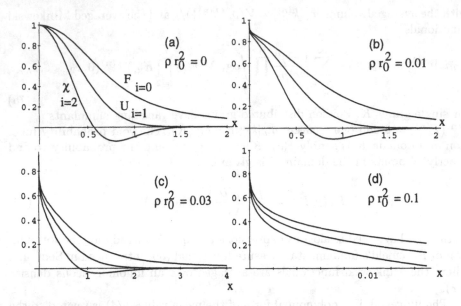

Fig. 15. Intensities $m_\nu(x)/m_\nu(x=0)$ of the area F ($\nu=0$), perimeter U ($\nu=1$), and Euler characteristic χ ($\nu=2$) for correlated discs of radius R in two dimensions as function of the density $x = \pi R^2 \rho$ [35]. The correlation function of the 'Poisson-Gauss process' is given by $\xi(r) = (r_0/r)^\gamma$ with $\gamma = 1.8$ - a commonly used expression for the distribution of galaxies in the universe (compare Sect. 3.1.1 and Fig. 7 in the paper by M. Kerscher in this volume).

intensities of the Minkowski measures (74) for homogeneously distributed discs $B_R^{(2)}$ of radius R in two dimensions

$$m_0(\rho) = 1 - \exp\left\{-\rho\pi R^2 + \rho^2 \pi R^2 r_0^2 \left(\frac{2R}{r_0}\right)^{2-\gamma}\left(\frac{\pi}{2-\gamma} - g(\gamma) - f(\gamma)\right)\right\}$$

$$m_1(\rho) = (1 - m_0(\rho))\rho R\left(1 - \rho r_0^2 \left(\frac{2R}{r_0}\right)^{2-\gamma}(\frac{\pi}{2-\gamma} - g(\gamma))\right)$$

$$m_2(\rho) = \frac{m_1(\rho)}{\pi R} - \frac{m_1(\rho)^2}{1 - m_0(\rho)} - (1 - m_0(\rho))\frac{\rho^2 r_0^2}{\pi}\left(\frac{2R}{r_0}\right)^{2-\gamma} g(\gamma)$$

$$(78)$$

with the functions

$$f(\gamma) = 2 \int\limits_0^1 dy\, y^{2-\gamma} \sqrt{1-y^2} = \frac{\Gamma(3/2)\Gamma(3/2-\gamma/2)}{\Gamma(3-\gamma/2)} \ ,$$

$$g(\gamma) = 2 \int\limits_0^1 dy\, y^{1-\gamma} \arcsin(y) \ .$$

The results are shown in Fig. 15 for $\gamma = 1.8$ and several correlation lengths $r_0^2 \rho$. In order to archive a positive covered volume $m_0 = 1 - P_0 > 0$ the density is constraint by the inequality $(2x)^{2-\gamma}(\frac{\pi}{2-\gamma} - g(\gamma) - f(\gamma)) < (\rho r_0^2)^{\gamma/2}$. For larger densities the Gaussian approximation fails so that the curves in Fig. 15 are plotted only for the density regime with $m_0(\rho) > 0$.

The main result of this Sect. is Eq. (74) which can immediately be used for the density functional theory (68) if the correlation function of the reference system is known. For hard spheres, for instance, one may use the Percus-Yevick approximation to obtain analytic expressions for $\xi(\mathbf{r})$ (see the paper by H. Löwen in this volume, [22]) and calculate $m_\nu[\rho, \xi(r)]$ within the Gaussian approximation. Further physical applications of the intensity measures (74) for correlated grains are obvious if one remembers the problems considered in the previous sections. The mean Euler characteristic (74) can be used, for instance, to estimate percolation thresholds for correlated processes (see Sect. 4.2). Using dynamical equations for the structure function $\xi(r,t)$ one immediately obtains the dynamics of the Minkowski functionals by applying the expressions (77). Of course, higher order correlations which are neglected in Eq. (77) should be taken into account in a complete dynamical theory for geometric measures (see Sect. 3.2). One can use the intensity measures (74) to analyze point processes as described in Sect. 3.3. For the configurations of the Lennard-Jones fluid shown in Fig. 9, for instance, or for an ensemble of hard particles (Fig. 1 on the right) one knows the two-point structure function very well theoretically by integral equations (see [22]) and also experimentally (scattering experiments with neutrons or X-rays). The Gaussian approximation (77) can be used to quantify the importance of higher order correlations in point patterns if the Minkowski functions can be measured directly for particle configurations as described in Sect. 3.3. It is not known yet if the Gaussian approximation (77) describes the Minkowski functions for fluids accurately or if an effective theory for the intensity measures for correlated processes such as an ensemble of hard spheres have to be developed (compare the problem considered in Sect. 3.3. For many applications in statistical physics it might be helpful to find explicit expressions for so-called hierarchical Poisson processes where the cumulants can be written as the product of two-point functions

$$k^{(n)}(\mathbf{x}_1, \ldots, \mathbf{x}_n) = c_n \sum_{r_j} \prod_{j=1}^{n-1} k^{(2)}(r_j) \ . \tag{79}$$

The sum runs over all n^{n-2} tree-graphs with n-1 lines $r_j = |\mathbf{x}_l - \mathbf{x}_m|$ so that all points $\mathbf{x}_j$ are connected. The dimensionless constant c_n depends on the model used, for instance, $n^{n-2}c_n = 1$, $n^{n-2}c_n = (n-1)!$, or $n^{n-2}c_n = (2n-3)!!$ [15].

Inhomogeneous and Anisotropic Distributions of Grains In many physical applications spatial structures are neither isotropic nor homogeneous. Consider, for example, magnetic particles in an external electro-magnetic field, or the influence of gravitation on the density of a fluid. In these cases the configurational averages have to be weighted with position dependent intensities which breaks the rotational and translational symmetry of the system. But the kinematic formulae (9) requires an integration over a motion-invariant measure so that they cannot be used for applications in inhomogeneous and anisotropic systems.

A first step towards a solution of this problem is the formulation of translational kinematic formulae introduced by Weil in his contribution 'Mixed measures and inhomogeneous Boolean models' to this volume. Nevertheless, the mixed measures which occur in these formulae are difficult to calculate for arbitrary shapes, so that special ensembles may be interesting to look at. In this section the configurations of the Boolean model are restricted first to an ensemble of oriented, overlapping cylinders with arbitrary basis, i.e., to shapes which are much more restricted in orientation then the configurations of the random coverage introduced in the beginning of this section (see Eq. (41). A special case are rectangular polyhedra (quader, cubes) which are oriented parallel to the coordination axes of the Euclidean space. Such statistical models have been used widely, for instance, as oriented percolation models [1], or models for colloidal particles [11].

Oriented Cylinders In a first step oriented cylinders $Z_i = A_i \oplus S_i(x)$ are considered where $A_i \in H$ denote arbitrary shaped, closed, and convex sets in a hyper-plane $H = \mathbb{R}^{d-1}$, $S_i(x)$ is a line segment in the orthogonal complement $H^\perp = \mathbb{R}$ centered at $x \in \mathbb{R}$, and $\oplus$ denotes the union of all points $\mathbf{z} = \mathbf{a} + \mathbf{s}$, which can be written as a sum of a point $a \in A_i$ and $s \in S_i$ contained in A_i and S_i, respectively. Using the relation $(A \oplus S) \cap (A' \oplus S') = (A \cap A') \oplus (S \cap S')$ one finds the expression [21]

$$\binom{d}{\mu} V_\mu(A \oplus S) = \sum_{\nu=0}^{\mu} \binom{p}{\nu}\binom{d-p}{\mu-\nu} V_\nu'(A) V_{\mu-\nu}'(S) , \tag{80}$$

for the Minkowski functionals $V_\mu = W_\mu/\omega_\mu$ of cylinders where the prime denotes that V_ν' is the Minkowski functional in the p-dimensional subspace H or in the complement $H^\perp$, respectively. The configurations of the Boolean model are given by unions of oriented cylinders Z_i $Z_N = \bigcup_{i=1}^{N} g_i A_i \oplus S_i(x_i)$, where g_i denotes translations and rotations in the hyper-plane $H = \mathbb{R}^{d-1}$. The intensity measures $v_\mu^{(d-1)}(\rho)$, i.e., the mean values of the Minkowski functionals per $d-1$-dimensional unit volume in H are given by Eq. (61). Using the one-dimensional kinematic formulae

$$\int_{\mathbb{R}} V_0'(S' \cap S(x)) dx = V_0'(S') V_0'(S)$$

$$\int_{\mathbb{R}} V_1'(S' \cap S(x)) dx = V_1'(S') V_0'(S) + V_1'(S) V_0'(S') , \qquad (81)$$

the additivity relation (6), and the decomposition rule (80) one finds for the intensity measure $(V_\mu = W_\mu / \omega_\mu, \; \rho = N/|\Omega|)$

$$v_\mu^{(cyl)}(\rho) = \prod_{i=1}^{N} \left(\frac{1}{|\Omega|} \int_{\mathcal{G}} dg_i \int_{\mathbb{R}} dx_i \right) V_\mu(\mathbb{Z}) , \quad |\Omega| = \int_{\mathcal{G}} dg \int_{\mathbb{R}} dx \qquad (82)$$

the relation [35]

$$v_\mu^{(cyl)}(\rho) = \frac{d-\mu}{d} v_\mu^{(d-1)}(a\rho) + \frac{\mu}{d} \frac{\partial}{\partial a} v_{\mu-1}^{(d-1)}(a\rho) \qquad (83)$$

with the mean length $a = \langle V_0'(S_i) \rangle$ of the line segments in $H^\perp$ and the intensities $v_\mu^{(d-1)}(\rho)$ for configurations $\cup_i A_i \subset H$ in the $d-1$-dimensional subspace H which are given by Eq. (61). Thus, mean values of Minkowski functionals for randomly distributed but oriented d-dimensional cylinders are determined by the mean values $v_\mu^{(d-1)}$ of the $(d-1)$-dimensional basis sets A_i.

Oriented d-dimensional Rectangles In a second step one can additionally restrict the orientation of the cylinders in the hyper-plane H. Let us consider d-dimensional oriented rectangles $Q^{(d)}$ of edge lengths q_i $(i = 1, \ldots, d)$, for instance, where the edges are parallel respectively to the axes of the coordination system in $\mathbb{R}^d$. Using the intensity measures $V_0^{(1)} = 1 - e^{-\rho q_1}$ and $V_1^{(1)} = \rho e^{-\rho q_1}$ for sticks of length q_1 in one dimension, one can apply relation (83) recursively yielding the explicit expression [35]

$$\bar{V}_\nu^{(d)}(\rho) = S_\nu \left(q_1^{-1}, \ldots, q_d^{-1} \right) \sum_{n=1}^{\nu} \left[\sum_{i=1}^{n} \frac{i^\nu (-1)^{i+1}}{(n-i)! i!} \right] \alpha^n e^{-\alpha} \qquad (84)$$

for the intensities (mean values) of the Minkowski measures for oriented rectangles $Q^{(d)}$ with $\alpha = \rho \prod_{i=1}^{d} q_i$ and the ν-th elementary symmetric function

$$S_\nu \left(q_1^{-1}, \ldots, q_d^{-1} \right) = \frac{1}{\binom{d}{\nu}} \sum_{i(1) < i(2) < \ldots < i(\nu)} \frac{1}{q_{i(1)} q_{i(2)} \cdots q_{i(\nu)}} \qquad (85)$$

of the inverse edge lengths q_i. In particular, one obtains $dS_1 = \sum_{i=1}^{d} \frac{1}{q_i}$, $\binom{d}{2} S_2 = \sum_{i<j} \frac{1}{q_i q_j}$, $\binom{d}{3} S_3 = \sum_{i<j<k} \frac{1}{q_i q_j q_k}$, and therefore, $\bar{V}_0 = 1 - e^{-\alpha}$, $\bar{V}_1 = S_1 \alpha e^{-\alpha}$, $\bar{V}_2 = S_2(\alpha - \alpha^2) e^{-\alpha}$, and $\bar{V}_3 = S_3(\alpha - 3\alpha^2 + \alpha^3) e^{-\alpha}$. These expressions differ from the solution (61) where the average was performed over arbitrary orientations of the rectangles. Notice: the intensities (84) do not depend on the dimension of space $\mathbb{R}^d$ in contrast to the orientational-averaged results (61).

Oriented Averaged Cuboids Of course, one may rotate the oriented rectangles by $90°$ motions so that the edges parallel to the coordinate axes permutate yielding $S_\nu(q_1^{-1}, .., q_d^{-1}) = q^{-\nu}$ with an averaged edge length $q = \frac{1}{d}\sum_i q_i$. Notice: in general $S_\nu(q_i^{-1}) \neq q^{-\nu}$ so that the Minkowski measures of oriented rectangles differ by an amplitude $q^\nu S_\nu(q_i^{-1})$ from an averaged ensemble where $90°$ rotations are performed. For this special case of oriented d-dimensional rectangles $Q^{d)}$ with permutated edges it is even possible to derive analogously to Eq. (9) the kinematic formula [35]

$$\int_{\mathcal{G}} V_\mu(\Omega \cap g\Omega')dg = \sum_{\nu=0}^{\mu} \binom{\mu}{\nu} V_\nu(\Omega) V_{\mu-\nu}(\Omega') \qquad (86)$$

for unions Ω, Ω' of rectangles. Here, $\mathcal{G}$ denotes the restricted group of motion, i.e., translations and $90°$-rotations (point group of a d-dimensional cube). Following the same method as for Eq. (61) one obtains for the mean values of the Minkowski measures per unit volume with $m = \sum_{i=0}^{d} q^{d-i} t^i/i!$ the recursion relation [34]

$$v_\nu(\rho) = \frac{\partial^\nu}{\partial t^\nu} e^{-\rho m}|_{t=0} = \frac{\partial^{\nu-1}}{\partial t^{\nu-1}} \frac{\rho}{q}(-m + \frac{t^d}{d!}) e^{-\rho m}|_{t=0} = \frac{\rho}{q} \frac{\partial v_{\nu-1}(\rho)}{\partial \rho} \qquad (87)$$

instead of Eq. (83). Solving this reduced recursion relation one immediately recovers the expression (84) with $S_\nu = q^{-\nu}$.

Orientational Distribution $P(\Theta)$ In the ensembles considered above the grains are strictly oriented parallel to perpendicular axes. In most physical situations this extreme case is not realized but only a partial orientation, i.e., the orientational degrees of freedom Θ of a grain K are weighted with a probability $P(\Theta)$. The orientation of a magnetic particle, for instance, depends gradually on the intensity of an external magnetic field. Since the intensities $m_\nu[\rho, P(\Theta)]$ do not only depend on the density ρ of the grains but also on the distribution of the orientations Θ we refer to them as *intensity functionals* of Minkowski measures.

In order to derive intensities of Minkowski measures for unions of oriented grains we extend the notion of Minkowski functionals to *local* functionals. Instead of using integrals over the total surface ∂A of a grain A (see Eq. 1) we define *local Minkowski functionals* [64]

$$W_\nu(A, B) = \frac{1}{\nu\binom{d}{\nu}} \int_{(\partial A) \cap B} S_{\nu-1} dS \ , \qquad W_0(A, B) = \int_{A \cap B} \chi(A \cap \mathbf{x}) d\mathbf{x} \qquad (88)$$

by integrating only over a part $\partial A \cap B$ of the surface ∂A. Then, one obtains for an ensemble of grains with density ρ the mean values
$m_\nu(\rho) = \delta_{\nu 0} - e^{-\rho m_0(K)} \sum_{\mu=1}^{\nu} \frac{(-\rho)^\mu}{\mu!} \bar{m}_\nu^\mu[P(\Theta)]$ (compare Eq. (61)) with
$\bar{m}_\nu^1[P(\Theta)] = M_\nu(K)$ and

$$\bar{m}_\nu^\mu[P(\Theta)] = \prod_{i=1}^{\mu} \int_{\mathcal{G}} dg_i \, P(\Theta_i) \, M_\nu(\bigcap_{i=1}^{\mu} g_i K_i, \bigcap_{i=1}^{\mu} g_i \partial K_i) \qquad (89)$$

where the integration over the motions $g_i = (\mathbf{r}_i, \Theta_i)$ of the grain K_i is weighted with respect to the orientation Θ_i [35]. Notice: the mean volume $m_0(\rho) = 1 - e^{-\rho V}$ and surface area $m_1(\rho) = m_1 \rho e^{-\rho V}$ do not depend on the probability $P(\Theta)$, i.e., on the orientational distribution. W. Weil shows in his paper in this volume that for the volume and surface area the kinematic formula (9) remains the same even for oriented grains where the integration is performed only over the translational part. This is not the case for the integral mean curvature H and the Euler characteristic so that the intensities $m_2(\rho; P(\Theta))$ and $m_3(\rho; P(\Theta))$ depend sensitively on the orientational distribution $P(\Theta)$. Notice: due to the kinematic formula (9) for $P(\Theta) = 1$ the coefficients $\bar{m}_\nu^\mu[P(\Theta)]$ factorize into products of Minkowksi functionals $M_\alpha(K)$ of single grains K so that one recovers the solution (61).

Let us consider independently distributed rectangles in two dimensions, for instance. Similar results can be obtained in d dimensions but with an inconvenient increase of notation. Then, expression (89) for the mean value of the Euler characteristic reduces to $m_2[\rho; P(\phi)] = e^{-\rho ab}\left(\pi\rho - \frac{\rho^2}{2}\bar{m}_2^2[P(\phi)]\right)$ with

$$\bar{m}_2^2[P(\phi)] = \int\int_0^{2\pi} \frac{d\phi d\phi'}{4\pi} P(\phi)P(\phi')\left[(a^2+b^2)\sin|\phi-\phi'| + 2ab\cos|\phi-\phi'|\right] \quad (90)$$

where a and b denotes the length of the edges [35]. One recovers the solutions (61), (84), and (87) for oriented rectangles by using the probabilities $P(\phi) = 1$, $P(\phi) = 2\pi\delta(\phi)$, and $P(\phi) = \pi/2(\delta(\phi) + \delta(\phi - \pi/2) + \delta(\phi - \pi) + \delta(\phi - 3\pi/2))$, respectively. The advantage of Eq. (90) is the explicit dependence on the distribution $P[\phi]$ so that it can be used within a density functional theory with orientational degrees of freedom of the particles (see Sects. 4.1 and 5.1).

Let us consider independent distributed particles in a constant electric or magnetic field $\mathbf{B} = B\mathbf{e}_z$, where $\mathbf{e}_z$ denotes the normal direction of the field. Then the Boltzmann factor, i.e., the probability of an orientation θ between the dipole $\mathbf{d}$ of the particle and $\mathbf{e}_z$ is given by $P(\theta) \sim e^{-\beta B \cos(\theta)}$ with $\cos(\theta) = \mathbf{d}\mathbf{e}_z$. Using the approximation

$$P(\theta) = \frac{B}{\sinh(B)}\begin{cases} e^{-B+2\theta B/\pi} & 0 \le \theta < \pi \\ e^{3B - 2\theta B/\pi} & \pi \le \theta < 2\pi \end{cases}, \quad (91)$$

for the Boltzmann factor $P(\theta)$ one obtains ($x = (2B/\pi)^2$)

$$w_2(\rho; B) = e^{-\rho ab}\left(\pi\rho - \frac{\rho^2}{2}\frac{B}{\sinh B}\frac{1}{1+x}\left[\frac{B}{\sinh B}(a^2+b^2+2ab\cosh B) + (a^2+b^2)\cosh B(1 + \frac{2x}{1+x}) + 2ab(1 + \frac{2x}{1+x}(1 + \sqrt{x}\sinh B))\right]\right) \quad (92)$$

In Fig. 16 the intensity measure $w_2(\rho; B)$ is shown as function of the magnetic field B. The Euler characteristic of an ensemble of rectangles (edge length a and b), i.e., the difference of components and holes in two dimensions, change even its sign when the magnetic field is increased. Oriented rectangles ($|B| \gg 1$, compare Eq. (84)) are less connected than random oriented grains (compare Eq.(61)). The difference can induce a phase transition in an ensemble of magnetic rod like particles what is well-known in the theory of liquid crystals.

Inhomogeneous Density $\rho(\mathbf{r})$ of Grains In order to apply the density functional (68) on inhomogeneous fluids one has to find first explicit expressions for the intensity measures $m_\nu[\rho(\mathbf{r})]$ on the density $\rho(\mathbf{r})$ of the grains. In the previous subsections oriented grains were studied, here, isotropic but spatially inhomogeneous distribution of grains are considered. Since the intensities depend on the density function $\rho(\mathbf{r})$ of the grains $m_\nu[\rho(\mathbf{r})]$ are *intensity functionals* of Minkowski measures. In contrast to the result (89) for a homogeneous but anisotropic distribution even the mean volume and surface area do depend on the local density. One obtains for the volume intensity

$$m_0[\rho(\mathbf{r})] = 1 - e^{-\bar{V}[\rho(\mathbf{r})]} \tag{93}$$

with the averaged volume $\bar{V}[\rho(\mathbf{r})] = \int_{\mathcal{g}} \chi(\mathbf{r} \cap gK)\rho(\mathbf{r})dg$ already used in Eq. (37) for the formulation of the Rosenfeld density functional in Sect. 4.1. Accordingly, the intensities of the Minkowski measures read ($\mathbf{g} = (\mathbf{r}, \Theta)$)

$$m_\nu[\rho(\mathbf{r})] = e^{-\bar{V}[\rho(\mathbf{r})]} \sum_{\mu=1}^{\nu} \frac{(-1)^{\mu+1}}{\mu!} \prod_{i=1}^{\mu} \int_{\mathcal{g}} dg_i \rho(\mathbf{r}_i) M_\nu(\bigcap_{i=1}^{\mu} g_i K_i, \mathbf{r} \bigcap_{i=1}^{\mu} g_i \partial K_i) \tag{94}$$

(see the paper by W. Weil in this volume and [35]). These expressions are similar in structure than the mean values (61) for a Poisson distribution of grains but with weighted integrals of local Minkowski functionals of μ grains. The local density functions $\rho(\mathbf{r}_i)$ make an explicit calculation via a kinematic formula impossible but for special shapes one can at least find explicit expressions for the densities $m_\nu[\rho(\mathbf{r})]$ which then can be evaluated numerically. Let us discuss a few examples illustrating the dependence of the Minkowski measures on the distribution of the grains, particularly of the Euler characteristic which is more sensitive to inhomogeneous features than the covered volume or the surface area. For sticks of length L in one dimension one obtains $w_0[\rho(\mathbf{r})] = 1 - e^{-L[\rho(\mathbf{r})]}$ and $2L\,w_1[\rho(\mathbf{r})] = L[\rho(\mathbf{r})]e^{-L[\rho(\mathbf{r})]}$ with the local length $L[\rho(\mathbf{r})] = \int_{-L/2}^{L/2} \rho(x-y)dy$. A homogeneous gravitational field $V = mgx$, $x \in [-D, D]$, yields the local density

$$\rho(\mathbf{x}) = \cosh^{-1}(\beta mgD) \cdot \begin{cases} e^{\beta mgD} & x_1 \leq -D \\ e^{-\beta mgx_1} & -D < x_1 < D \\ e^{-\beta mgD} & x_1 \geq D \end{cases} \tag{95}$$

for grains with mass m in the center and in thermal equilibrium at temperature $k_B T = \beta^{-1}$. The local length is then given by

$$
L(x) =
\begin{cases}
e^{\beta mgD} & x \leq -D - \frac{L}{2} \\
e^{-\beta mgx}\frac{1-e^{\beta mg(x+D+L/2)}}{1-e^{\beta mgL}} + e^{\beta mgD}(\frac{L}{2} - x - D) & -\frac{L}{2} < x + D < +\frac{L}{2} \\
c \cdot e^{-\beta mgx} & -D + \frac{L}{2} < x < D - \frac{L}{2} \\
c \cdot e^{-\beta mgx}\frac{1-e^{\beta mg(x-D-L/2)}}{1-e^{-\beta mgL}} + e^{-\beta mgD}(\frac{L}{2} + x - D) & -\frac{L}{2} < x - D < \frac{L}{2} \\
e^{-\beta mgD} & x \geq D + \frac{L}{2}
\end{cases}
$$

$$(96)$$

with $c = \frac{2\sinh(\beta mgL/2)}{\beta mgL}$ (Fig. 16). The transition from a homogeneous phase (ρ^+)

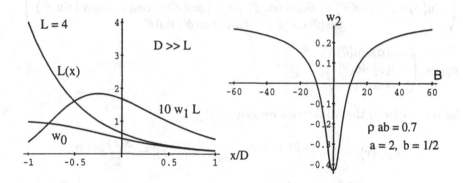

Fig. 16. Intensity measures w_0 and w_1 for one-dimensional sticks in a gravitational field $(L << D)$ and the mean Euler characteristic $w_2(B)$ of rectangles (edge lengths a and b) in a magnetic field B [35].

to another with lower density (ρ^-) is accompanied by an increase of the Euler characteristic w_1 (number of components in one dimension) in the interfacial zone $[-D, D]$, whereas the covered volume w_0 is decreasing monotonously.

For discs of radius R in two dimensions one obtains the averaged area and boundary length (compare Eq. (37))

$$
\bar{V}[\rho(\mathbf{r})] = \int_0^R \int_0^{2\pi} d\phi dx \, x\rho(\mathbf{r} + x\mathbf{e}_\phi) \qquad \bar{U}[\rho(\mathbf{r})] = R \int_0^{2\pi} d\phi \, \rho(\mathbf{r} + R\mathbf{e}_\phi) \quad (97)
$$

with the normal direction $\mathbf{e}_\phi = (\cos(\phi), \sin(\phi))$ and the intensity of the Euler characteristic

$$
w_2[\rho(\mathbf{r})] = e^{-\bar{V}(\mathbf{r})}\left(\frac{\bar{U}(\mathbf{r})}{2R} - \frac{R^2}{2} \int_0^{2\pi}\int_{-\pi}^{\pi} d\phi d\phi' \rho(\mathbf{r} + R\mathbf{e}_\phi)\rho(\mathbf{r} + R\mathbf{e}_{\phi+\phi'})\frac{\phi'}{2}\sin(\phi') \right).
$$

$$(98)$$

The factor $|\phi'|/2$ denotes the curvature at $\mathbf{r}$ where two discs centered at $R\mathbf{e}_\phi$ and $R\mathbf{e}_{\phi+\phi'}$ have a common boundary point. The factor $\sin|\phi'|$ is the Jacobian due to the coordinate transition to ϕ. Accordingly, for spheres of radius R in three dimensions one obtains the averaged volume and surface area (compare Eq. (37))

$$\bar{V}[\rho(\mathbf{r})] = \int_0^R \int_0^{2\pi} \int_0^\pi d\phi d\theta dx\, x^2 \sin(\theta)\rho(\mathbf{r}+x\mathbf{e}_{\phi\theta})\,,$$

$$\bar{S}[\rho(\mathbf{r})] = R^2 \int_0^{2\pi} \int_0^\pi d\phi d\theta \sin(\theta)\rho(\mathbf{r}+R\mathbf{e}_{\phi\theta})\,,$$

with the spherical surface element $d\Omega = \sin(\theta)d\phi d\theta$ and the normal directions

$$\mathbf{e}'_{\phi'\theta'} =$$
$$\begin{pmatrix} \cos(\phi)\sin(\theta)\cos(\theta') + \cos(\phi)\cos(\theta)\cos(\phi')\sin(\theta') - \sin(\phi)\sin(\phi')\sin(\theta') \\ \sin(\phi)\sin(\theta)\cos(\theta') + \sin(\phi)\cos(\theta)\cos(\phi')\sin(\theta') - \cos(\phi)\sin(\phi')\sin(\theta') \\ \cos(\theta)\cos(\theta') - \sin(\theta)\cos(\phi')\sin(\theta') \end{pmatrix}$$

$$\mathbf{e}_{\phi\theta} = \begin{pmatrix} \cos(\phi)\sin(\theta) \\ \sin(\phi)\sin(\theta) \\ \cos(\theta) \end{pmatrix}$$

the intensities of the Minkowski measures [35]

$$w_0[\rho(\mathbf{r})] = 1 - e^{-\bar{V}(\mathbf{r})} \quad , \quad w_1[\rho(\mathbf{r})] = \frac{\bar{S}(\mathbf{r})}{3}e^{-\bar{V}(\mathbf{r})}$$

$$w_2[\rho(\mathbf{r})] = e^{-\bar{V}(\mathbf{r})}\left(\frac{\bar{S}(\mathbf{r})}{3R} - \frac{R^4}{2}\iint d\Omega d\Omega'\rho(\mathbf{r}+R\mathbf{e}_{\phi\theta})\rho(\mathbf{r}+R\mathbf{e}'_{\phi'\theta'})\frac{\theta'}{6}\sin(\theta')\right)$$

$$w_3[\rho(\mathbf{r})] = e^{-\bar{V}(\mathbf{r})}\left(\frac{\bar{S}(\mathbf{r})}{3R^2} - \frac{R^4}{2}\iint d\Omega d\Omega'\rho(\mathbf{r}+R\mathbf{e}_{\phi\theta})\rho(\mathbf{r}+R\mathbf{e}'_{\phi'\theta'})\frac{4}{3R}\sin^2(\theta'/2)\right.$$

$$+\frac{R^6}{6}\iiint d\Omega d\Omega'd\Omega''\rho(\mathbf{r}+R\mathbf{e}_{\phi\theta})\rho(\mathbf{r}+R\mathbf{e}'_{\phi'\theta'})\rho(\mathbf{r}+R\mathbf{e}'_{\phi'+\phi''\theta''})$$

$$\left.\frac{|\Delta(\phi'',\theta',\theta'')|}{3}\sin(\theta')\sin(\theta'')\sin(\phi'')\right)\,.$$

$$(99)$$

The factor $|\Delta|/3$ denotes the Gaussian curvature at $\mathbf{r}$ where the boundary of three spheres centered at $R\mathbf{e}_{\phi\theta}$, $R\mathbf{e}'_{\phi'\theta'}$, and $R\mathbf{e}'_{\phi'+\phi''\theta''}$ meet. It is given by L'Huiliers formula

$$\tan^2\frac{\Delta}{4} = \tan\frac{\alpha_1+\alpha_2+\alpha_3}{4}\tan\frac{\alpha_1+\alpha_2-\alpha_3}{4}$$
$$\tan\frac{\alpha_1-\alpha_2+\alpha_3}{4}\tan\frac{-\alpha_1+\alpha_2+\alpha_3}{4} \quad (100)$$

in terms of the angles $\alpha_1 = 2\arcsin(|\mathbf{r}|/2R)$, $\alpha_2 = 2\arcsin(|\mathbf{y}|/2R)$, $\alpha_3 = 2\arcsin(|\mathbf{r}-\mathbf{y}|/2R)$ of a spherical triangle. Taking into account the relation $\Delta(\Phi,\theta',\theta'') + \Delta(\Phi,\pi-\theta',\pi-\theta'') = 2\Delta(\Phi,\pi/2,\pi/2) = 2\Phi$ one can perform the integration in Eq. (99) for homogeneous density $\rho(\mathbf{r}) \equiv \rho$ and one recovers the mean value (61). The factor $\sin(\theta')\sin(\theta'')\sin|\phi''|$ is a Jacobian due to the change to spherical coordinates ϕ, θ for the motion of a sphere.

The evaluation of the integrals in Eq. (99) is only necessary if the characteristic length σ of the inhomogeneous density is comparable to the size of the grains, i.e., of the radius R of the discs and spheres. In the limit $R \ll \sigma$ the dependence of the Minkowski functionals on the inhomogeneity is well described *locally* by $W_\nu[\rho, \mathbf{r}] = W_\nu(\rho(\mathbf{r}))$, i.e., by using the local density $\rho(\mathbf{r})$ instead of a global parameter ρ.

Possible applications of the main results (98) and (99) (or Eq. (94) in general) of this section are numerous. One may use them to analyze inhomogeneous distributions of galaxies in a cluster (see M. Kerscher in this volume), the estimation of percolation threshold in inhomogeneous porous rocks (Sect. 4.2), and the formulation of a density functional theory for the morphological model of complex fluids introduced in Sect. 5.1. Several variational problems, for instance, the minimization of surface energies (Minkowski functionals) with respect of the density of grains or the determination of the shape of vesicles or drops in an inhomogeneous environment are now possible.

Problem: Effective Approximations The above formulae are often quite useless since the cumulants are not known or difficult to calculate. So for many important physical models effective expressions for the mean values of the Minkowski measures are not known. For instance, it would be very helpful for many applications to have at least approximative expressions for the mean values for distributions of hard spheres or discs where the cumulant expression in the Gaussian approximation (77) completely fails.

5.3   Second-Order Moments of Minkowski Measures

In Sect. 5.1 a high temperature expansion for the partition sum of the generalized Widom-Rowlinson model were derived. Thermodynamical quantities such as specific heats or susceptibilities are related to second order moments, i.e., variances of geometric measures. Thus, starting from the Hamiltonian (42) it is necessary to calculate not only the mean values but also the second order moments

$$m_{\nu\mu}(\rho) = \lim_{N\to\infty} \frac{< M_\nu(\mathcal{A})M_\mu(\mathcal{A}) > - < M_\nu(\mathcal{A}) >< M_\mu(\mathcal{A}) >}{|\Omega|} \tag{101}$$

of the Minkowski measures $M_\nu(\mathcal{A})$. For the Boolean model of independent distributed grains K these second order moments are given by the integrals [35]

$$m_{\nu\mu}(\rho) = \sum_{\lambda_1=1}^{\nu} \sum_{\lambda_1=1}^{\mu} \int_{\mathbb{R}^d} d\mathbf{x} \, M_{\nu\mu}^{\lambda_1\lambda_2}(N; 0, \mathbf{x}) \tag{102}$$

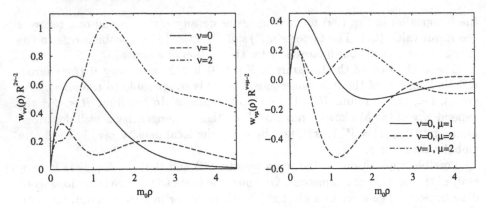

Fig. 17. Grand-canonical second order measures $w_{\nu\mu}(\rho)$ for the area F ($\nu = 0$), perimeter U ($\nu = 1$), and Euler characteristic χ ($\nu = 2$) of Poisson distributed discs of radius R ($m_0 = \pi R^2$) in two dimensions as function of the density $\pi R^2 \rho$. Notice: the functional form of the variances becomes more complex with increasing index ν. The boundary length w_{11} exhibit two fluctuation maxima and the Euler characteristic shows a single (large) maxima but two additional shoulders.

with the structure functions

$$
M_{\nu_1\nu_2}^{\lambda_1\lambda_2}(N;\mathbf{x}_1,\mathbf{x}_2) = -(-1)^{\lambda_1+\lambda_2}\binom{N}{\lambda_1}\binom{N}{\lambda_2}\frac{M_{\nu_1\nu_2;0}^{\lambda_1\lambda_2}(\mathbf{0},\infty)}{|\Omega|^{\lambda_1+\lambda_2}}\left(1-\frac{V}{|\Omega|}\right)^{2N-\lambda_1-\lambda_2}
$$

$$
+ (-1)^{\lambda_1+\lambda_2}\sum_{l=0}^{\lambda_2}\binom{N}{\lambda_1}\binom{\lambda_1}{l}\binom{N-\lambda_1}{\lambda_2-l}
$$

$$
\frac{M_{\nu_1\nu_2;l}^{\lambda_1\lambda_2}(\mathbf{x}_1,\mathbf{x}_2)}{|\Omega|^{\lambda_1+\lambda_2-l}}\left(1-\frac{V(\mathbf{x}_1,\mathbf{x}_2)}{|\Omega|}\right)^{N-\lambda_1-\lambda_2+l} \tag{103}
$$

and ($\lambda_i > 0$, $Y_j^{(1)} := r_j\partial\hat{K}_{\mathbf{x}_1}/r_j\hat{K}_{\mathbf{x}_2}$, $Y_j^{(2)} := r_j\partial\hat{K}_{\mathbf{x}_2}/r_j\hat{K}_{\mathbf{x}_1}$, $Y_j^{(12)} := r_j\partial\hat{K}_{\mathbf{x}_1} \cap r_j\partial\hat{K}_{\mathbf{x}_2}$)

$$
M_{\nu_1\nu_2;l}^{\lambda_1\lambda_2}(\mathbf{x}_1,\mathbf{x}_2) = \prod_{i=1}^{\lambda_1+\lambda_2-l}\int_{\mathcal{R}}dr_i \prod_{j=\lambda_1-l+1}^{\lambda_1}\int_{Y_j^{(12)}}\frac{d\mathbf{y}_j}{\sin\Phi_j}\prod_{j'=1}^{\lambda_1-l}\int_{Y_{j'}^{(1)}}d\mathbf{y}_{j'}\prod_{j''=\lambda_1+1}^{\lambda_1+\lambda_2-l}\int_{Y_{j''}^{(2)}}d\mathbf{y}_{j''}
$$

$$
\times\Delta_{11}(\mathbf{n}_{\mathbf{y}_1}^1\mathbf{n}_{\mathbf{y}_2}^1)\Delta_{21}(\mathbf{n}_{\mathbf{y}_1}^1\mathbf{n}_{\mathbf{y}_2}^1\mathbf{n}_{\mathbf{y}_3}^1)\cdots\Delta_{(\lambda_1-1)1}(\mathbf{n}_{\mathbf{y}_1}^1\cdots\mathbf{n}_{\mathbf{y}_{\lambda_1}}^1)
$$

$$
\times\Delta_{11}(\mathbf{n}_{\mathbf{y}_{\lambda_1-l+1}}^2\mathbf{n}_{\mathbf{y}_{\lambda_1-l+2}}^2)\cdots\Delta_{(\lambda_2-1)1}(\mathbf{n}_{\mathbf{y}_{\lambda_1-l+1}}^2\cdots\mathbf{n}_{\mathbf{y}_{\lambda_2}}^2)
$$

$$
\times M_{\nu_1}(\bigcap_{j=1}^{\lambda_1}r_jK_{\mathbf{y}_j},\mathbf{x}_1)M_{\nu_2}(\bigcap_{j=\lambda_1-l+1}^{\lambda_1+\lambda_2-l}r_jK_{\mathbf{y}_j},\mathbf{x}_2) \tag{104}
$$

where Φ_j is the angle between the normals $\mathbf{n}_j^1$ and $\mathbf{n}_j^2$ of $\hat{K}_{\mathbf{x}_1}$ and $\hat{K}_{\mathbf{x}_2}$ at $\mathbf{y}_j$. The Jacobians Δ_{ij} are given explicitly by Santaló [52], particularly $\Delta_{11} = |\sin\Phi|$. For a distribution of Poisson distributed sticks of length r in one spatial dimension the expressions (102)-(104) reduce to

$$m_{00}(\rho) = \tfrac{2}{\rho}\left(e^{-\rho r} - (1+\rho r)e^{-2\rho r}\right)$$
$$m_{10}(\rho) = -\left(e^{-\rho r} - (1+2\rho r)e^{-2\rho r}\right)/2 \qquad (105)$$
$$m_{11}(\rho) = \rho\left(e^{-\rho r} - 2\rho r e^{-2\rho r}\right)/4$$

which are shown in Fig. 18. For discs of radius R and density ρ in two dimensions one obtains with $\bar{V}(\bar{x}_1, \bar{x}_2) = \pi R^2 + 2R^2 \arcsin(s) + 2R^2 s\sqrt{1-s^2}$, $V = \pi R^2$, $s = (|\mathbf{x}_1 - \mathbf{x}_2|)/(2R)$, and $\sin(\Phi/2) = s$ $(0 \le \Phi \le \pi)$ the Minkowski functions $(\nu = 1, 2)$

$$M_{\nu 0}^{10} = R^{2-\nu}\pi^{1-\nu}\rho e^{-2\rho V} - R^{2-\nu}\frac{\pi+\Phi}{2\pi^\nu}\rho e^{-\rho\bar{V}} \ ,$$

$$M_{20}^{20} = \frac{\rho^2 R^2}{4\pi^2}e^{-\rho\bar{V}}\left(4\pi\Phi + 4 + 4\cos\Phi - 2(\pi-\Phi)\sin\Phi\right) - R^2\rho^2 e^{-2\rho V}$$

$$M_{\nu\mu}^{11} = R^{2-\nu-\mu}\left(R^2\rho^2\frac{(\pi+\Phi)^2}{4\pi^{\nu+\mu}}e^{-\rho\bar{V}} + \frac{\rho}{2\pi^{\nu+\mu}\sin\Phi}e^{-\rho\bar{V}} - \pi^{2-\nu-\mu}R^2\rho^2 e^{-2\rho V}\right)$$

$$M_{2\mu}^{21} = -R^{4-\mu}\rho^3\frac{\pi+\Phi}{2\pi^{2+\mu}}e^{-\rho\bar{V}}\left(\pi\Phi + 1 + \cos\Phi - \frac{\pi-\Phi}{2}\sin\Phi\right)$$
$$- \frac{R^{2-\mu}}{2\pi^{2+\mu}\sin\Phi}\rho^2 e^{-\rho\bar{V}}\left(2\pi - \sin\Phi - (\pi-\Phi)\cos\Phi\right) + \pi^{1-\mu}R^{4-\mu}\rho^3 e^{-2\rho V}$$

$$M_{22}^{22} = \left(\frac{R\rho}{\pi}\right)^4 e^{-\rho\bar{V}}\left(\pi\Phi + 1 + \cos\Phi - \frac{\pi-\Phi}{2}\sin\Phi\right)^2 + \rho^2\left(\frac{\pi-\Phi}{2\pi^2}\right)^2 e^{-\rho\bar{V}}$$
$$+ \rho^3\frac{R^2}{2\pi^4\sin\Phi}e^{-\rho\bar{V}}\left(2\pi - \sin\Phi - (\pi-\Phi)\cos\Phi\right)^2 - (R\rho)^4 e^{-2\rho V}$$
$$+ \tfrac{1}{2}\rho^2 R^2\left(\tfrac{1}{\pi} - \tfrac{4}{\pi^3}\right)e^{-\rho V}\delta(\mathbf{x}_1 - \mathbf{x}_2) \ .$$

$$(106)$$

In Fig. 17 the exact grand-canonical second order moments of the Minkowski measures are shown for discs in two dimensions. Notice: the variances for the Euler characteristic w_{22} are large. It is well known that the difference $\delta C = C_p -$

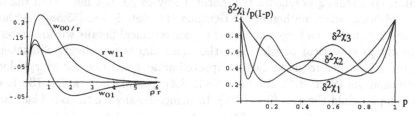

Fig. 18. (a) Second order measures $w_{\nu\mu}(\rho)$ of length L $(\nu = 0)$ and Euler characteristic χ $(\nu = 1)$ for Poisson distributed sticks of length r in one dimension as function of the density ρr (see Eq. (105)). (b) Second order measures $\delta^2\chi_d/(p(1-p))$ of the Euler characteristic χ on a d-dimensional cubic lattice Λ (see Eq. (119)). Notice: the variances as functions of the probability p that a lattice site is occupied becomes more and more complex with increasing index ν. In contrast to the results (106) for random distributions in space $\mathbb{R}^d$ (Fig. 17), the second order moments for lattice configurations are more convenient to calculate.

$C_V > 0$ between the specific heat at constant pressure C_p and at constant volume C_v is given by $\delta C = T \left(\frac{\partial p}{\partial T}\right)_V \left(\frac{\partial V}{\partial T}\right)_p$. Accordingly, one finds for the difference $\delta m_{\nu_1\nu_2}(\rho) = m_{\nu_1\nu_2}^{(gk)} - m_{\nu_1\nu_2}^{(k)}$ of the moments at constant grain number (canonical ensemble, Bernoulli distribution) and at constant density but fluctuating number (grand canonical ensemble, Poisson distribution) the expression

$$\delta m_{\nu_1\nu_2}(\rho) = \sum_{\mu_1=1}^{\nu_1} \sum_{\mu_2=1}^{\nu_2} \delta M_{\nu_1\nu_2}^{\mu_1\mu_2} = \rho \frac{\partial m_{\nu_1}(\rho)}{\partial \rho} \frac{\partial m_{\nu_2}(\rho)}{\partial \rho} , \qquad (107)$$

given by the derivatives of the densities $m_\nu(\rho)$ (see Eq. (58) in Sect. 5.1).

Problem: Dependence on Shape, Orientation, and Interaction of the Grains Although it is possible to derive the explicit expression (103) for second order moments of Minkowski functionals, the actual evaluation is quite difficult for non-spherical shapes. In order to study the dependence on shape and orientation of the grains similar general results as for the mean values (see Sect. 5.2) or at least reliable approximations would be useful for many applications in statistical physics. Recent Monte-Carlo simulations of hard spheres decorated with larger penetrable spherical shells (compare the statistical model for colloidal particles introduced in Sect. 4.3; see Fig. 1) showed that second order moments of the Minkowski measures depend essentially on the interactions between the grains [10].

5.4   Lattices: Models for Complex Fluids and Percolation

Since Lenz and Ising introduced the so-called Ising model in the early 1920s, lattice models became the backbone of statistical physics (see the papers by H.-0. Georgii and by E. Thönnes in this volume). Therefore, introducing integral geometry for lattice configurations, i.e., for spatial sets generated by unions of regular, space-filling polyhedra is a natural way to go. Let me sketch the main ideas without being mathematical rigorous (for details see [35,38]). Although this section contains a large amount of rather technical details without referring explicitly to statistical physics. But the main aim is to provide a mathematical basis in order to make lattice approximations feasible for morphological image analysis (Sect. 3.1), fractals (Sect. 3.4), percolation models (Sect. 4.2), and complex fluid theories (Sect. 4.3). In principle, any continuous black/white pattern in the Euclidean space $\mathbb{R}^d$ can be discretized by applying a mesh of finite lattice spacing a and erasing or filling a lattice cells completely. This procedure is done automatically by a digital recording of a continuous shape anyway so that every picture of a realistic structure is actually a lattice configuration. Such discretized shapes can be translated and rotated according finite lattice spacings and angles. Integral geometry guarantees that curvature measures are well-defined on lattices irrespectively of edges and corners of lattice configurations. Although realistic models for non-universal properties should rely on a continuum description of materials, lattice models exhibit all possibilities for a systematic study of the dependence of physical quantities on dispersity, shape, and orientation of the constituents (see Fig. 12). Lattice models allow a systematic study of orientational effects though isotropy is not attained. The advantage of lattice configurations in contrast to continuum models is the straightforward implementation and calculation of configurations and clusters yielding efficient and fast algorithms for the accurate determination of physical quantities such as Minkowski functionals as order parameters, size and shape of connected fractal clusters, percolation thresholds, or phase diagrams of fluids.

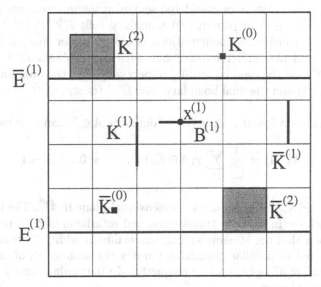

Fig. 19. Lattice configurations on a 2-dimensional lattice $\Lambda^{(2)}$ ($L = 6a$): ν-dimensional cubes (plaquettes) $K^{(2)}$, edges $K^{(1)}$, corners $K^{(0)}$, ν-dimensional lattice planes $E^{(\nu)}$ and the dual lattice configurations K^*, E^*, respectively.

In the following, we try to adopt the notions and definitions of integral geometry for lattice configurations and lattice groups of motions (finite translations and 90^0-rotations, point group of a hypercube). In particular, Minkowski functionals are defined and fundamental kinematic formulae are derived. Let us consider a d-dimensional hyper-cubic lattice $\Lambda^{(d)} \equiv a\mathbb{Z}^d$, where a denotes the lattice constant. At each point $\mathbf{x}_i^{(d)} \in \mathbb{Z}^d$ one can fix a d-dimensional cube $K_i^{(d)}$ of edge length a with $\mathbf{x}_i^{(d)}$ as midpoint, i.e., the Wigner-Seitz or unit cell of the lattice, where $i = 1, \ldots L^d$ numerates the lattice sites inside a hyper-cubic window of size L (see Fig. 19). Let $E^{(\nu)} \subset \Lambda^{(d)}$ denote a ν-dimensional planar sublattice, i.e. a ν-dimensional plane cutting $\Lambda^{(d)}$ with lattice points $\mathbf{x}_j^{(\nu)} \in E^{(\nu)}$, $j = 1, \ldots L^\nu$. Analogous, we denote by $K_j^{(\nu)}(E^{(\nu)})$ a ν-dimensional unit cells of $E^{(\nu)}$ located at $\mathbf{x}_j^{(\nu)}$. Notice: the lattice sites $\mathbf{x}_j^{(d)} = K_j^{(0)} \in \mathbb{Z}^d$ can be considered as the corners $\bar{E}^{(0)}$ of a hyper-cubic unit cell $\bar{K}_j^{(d)}$ of the dual lattice $\bar{\Lambda}$ with sites $\bar{x}_j^{(d)}$ (Fig. 19). One can define ν-dimensional dual lattice planes $\bar{E}^{(\nu)}$ and analogously unit cells $\bar{K}_j^{(\nu)} \subset \bar{E}^{(\nu)}$. Of course, one finds that $\bar{x}_i^{(d)} = \bar{K}_i^{(0)}$ is a corner $E^{(0)}$ of a unit cell $K_j^{(d)}$.

Let us now define the spatial configurations which constitutes the stage for our model. $\mathcal{R}$ denotes the class of all subsets A of the Euclidean space $\mathbb{R}^d$, which can be represented in terms of a finite union of unit cells $K^{(\nu)}$ or boundary cells $B^{(\mu)}$ of any dimension on a d-dimensional hyper-cubic lattice; also, $\emptyset \in \mathcal{R}$. The intersection of two cells $B^{(\mu)} = K^{(\nu)} \cap K^{(\nu')} \in \mathcal{R}$ is a boundary cell with $\mu < \nu, \nu'$, i.e., belongs to the class $\mathcal{R}$ of considered sets, also the intersections of boundary cells $B^{(\nu)} \cap B^{(\nu')} \in \mathcal{R}$ or of unit and boundary cells $K^{(\nu)} \cap B^{(\nu')} \in \mathcal{R}$. Thus, the class $\mathcal{R}$ of considered configurations is closed under mutual intersections and unions, and also under intersections with hyper-planes $E^{(\nu)}$ and $\bar{E}^{(\nu)}$ of the lattice Λ^d and the dual lattice $\bar{\Lambda}^d$, respectively. But the dual unit cells $\bar{K}^{(\nu)}$ (except of $\bar{K}^{(0)}$) and the dual boundary cells $\bar{B}^{(\nu)}$ (except of $\bar{B}^{(0)}$) do not belong to $\mathcal{R}$.

The Minkowski functionals $V_\nu(A)$ of domains $A \in \mathcal{R}$ can now be defined by

$$V_\nu(A) = \frac{1}{\binom{d}{\nu}} \sum_{E_\nu} \chi(A \cap E_\nu) \quad , \quad \nu = 0, \ldots, d-1 \qquad (108)$$

and $V_d(A) = \chi(A)$. Here, E_ν is a ν-dimensional plane in $\mathbb{R}^d$. The integral runs over all positions (induced by translations and rotations) of E_ν, From definition (108) it is clear that the Minkowski functionals inherit additivity from χ and that they are related to familiar quantities, namely the number V_0 of occupied unit cells, the number dV_1 of boundary plaquettes. In particular, for a j-dimensional cube $K^{(j)}$ one obtains

$$V_\nu(K^{(j)}) = \frac{\nu! j!}{d!(\nu+j-d)!} \quad , \quad d-j \leq \nu \leq d \qquad (109)$$

and $V_\nu(K^{(j)}) = 0$ for $0 \leq \nu < d-j$.

Analogous to the kinematic formula (9) for the continuous motion of grains one finds for the integral over the lattice group of motion the kinematic formula [35]

$$\mathfrak{J}_\mu(K, K') = \int V_\mu(K \cap K') dK' = \sum_{\nu=0}^{\mu} \sum_{\kappa=0}^{\nu} \binom{\mu}{\nu} \binom{\nu}{\kappa} V_\nu(K) V_{\mu-\kappa}(K') \quad , \qquad (110)$$

in particular, for the integrated volume $\mathfrak{J}_0(K, K') = V_0(K) V_0(K')$ and surface area $\mathfrak{J}_1(K, K') = V_0(K) V_1(K') + V_1(K) V_0(K') + V_1(K) V_1(K')$. The kinematic formula (110) can be used to obtain immediately the excluded volume $V_{ex}^{(d)}(K) =$

$\mathfrak{I}_d(K,K)$ in d dimensions for convex grains K

$$V_{\text{ex}}^{(2)}(K) = 2V_0(K) + 2V_1(K)^2 + 4V_1(K) + 1$$

$$V_{\text{ex}}^{(3)}(K) = 2V_0(K) + 6V_1(K) + 6V_2(K) + 6V_1(K)V_2(K) + 6V_2^2(K) + 1 .$$
$$\tag{111}$$

Let us now consider random distributions of convex grains $K \in \mathcal{R}$ with density ρ on the lattice. In Fig. 13 three clusters of random distributed grains on the lattice are shown, namely configurations of squares of edge length $\lambda = 5$ (see Fig. 12), sticks of length $\lambda = 5$, and mixtures of squares of different size ($\lambda = 2$ and $\lambda = 5$). The probability that a grain K is not touched by any other is then given by the void probability

$$P_{\text{void}}(K) = (1-p)^{V_{\text{ex}}(K)} , \tag{112}$$

where the probability $p = 1 - e^{-\rho}$, $q = 1 - p$ is defined that a specific orientation and location of a grain is occupied at least once. Applying Eq. (110) the void probability is given by the Minkowski functionals $V_\nu(K)$ of the grains K. Also the probabilities

$$\begin{aligned}
P(A) &= q^{V_0+2V_1}p \\
P(B) &= q^{V_0+V_1}(1 - q^{V_1+1} - 2q^{V_1}p) \\
P(C) &= q^{V_0+2V_1-V_2}p^2 \\
P(D) &= q^{V_0}(1 + q^{2V_1+1} + q^{2V_1}p \\
     &\quad - q^{2V_1-1}p^2 - 2q^{V_1}) \\
P(E) &= q^{V_0+2V_1+1} \\
P(F) &= 1 - P(E) - 4P(A) - 4P(B) - 2P(C) - 4P(D)
\end{aligned} \tag{113}$$

of the local coverages X of a unit cell $\bar{K}^{(2)}$ shown in Fig. 20 are expressible in terms of the morphological measures $V_\nu(K)$ of the single grains K. The graphs indicate if one (A), two (B,C) or three (D) quarters of a unit cell $\bar{K}^{(2)}$ are covered. The first term is the void probability of the empty lattice sites at the corners of the cells (white sites in Fig. 20). Such probabilities are needed for lattice approximations and real-space renormalization group theory for the generalized Widom-Rowlinson model introduced in Sect. 5.1). The mean value of the Euler characteristic, for instance, can now be expressed by the sum $\bar{\chi}(\rho) = v_2(\rho) = P(A) - B(C) - P(D)$.

Using Morse's theorem of counting critical points one can define the Euler characteristic χ_d in d dimensions recursively by

$$\chi_d(A) = \sum_{\bar{E}^{(d-1)}} \chi_{d-1}(A \cap \bar{E}^{(d-1)}) - \sum_{E^{(d-1)}} \chi_{d-1}(A \cap E^{(d-1)}) . \tag{114}$$

Repeating Eq. (114) and using the definition (108) one obtains the decomposition of the morphological measures in terms of edge and corner contributions

$$V_\alpha(A) = \sum_{\nu=d-\alpha}^{d} \frac{(-1)^{d-\alpha+\nu}\alpha!\nu!}{d!(\alpha+\nu-d)!} \sum_{\mathbf{x}^{(\nu)}} \chi(A \cap \mathbf{x}^{(\nu)}) , \tag{115}$$

where the sum runs of all midpoints $\mathbf{x}^{(\nu)}$ of the ν-dimensional boundary cell $B^{(\nu)}$. Applying the kinematic formula (110) the differential equation [35]

$$\frac{\partial v_\mu(\rho)}{\rho} = V_\mu - \sum_{\nu=0}^{\mu}\sum_{\kappa=0}^{\nu}\binom{\mu}{\nu}\binom{\nu}{\kappa}v_\nu(\rho)V_{\mu-\kappa}. \tag{116}$$

for the mean values $v_\mu(\rho)$ of the Minkowski functionals can be derived. With the initial condition $v(\rho = 0) = 0$ the differential equation (116) can be solved readily for any dimension d

$$v_\nu(\rho) = \delta_{0\nu} - \sum_{\mu=0}^{\nu}(-1)^\mu\binom{\nu}{\mu}e^{-\rho\sum_{\mu=0}^{\nu}\binom{\nu}{\mu}V_\mu(K)}. \tag{117}$$

Thus, the intensities of the Minkowski measures for random distributed grains K on a hypercubic lattice are given by

$$\begin{aligned}
v_0(\rho) &= 1 - e^{-\rho V_0} \\
v_1(\rho) &= e^{-\rho V_0}\left(1 - e^{-\rho V_1}\right) \\
v_2(\rho) &= e^{-\rho V_0}\left(-1 + 2e^{-\rho V_1} - e^{-2\rho V_1 - \rho V_2}\right) \\
v_3(\rho) &= e^{-\rho V_0}\left(1 - 3e^{-\rho V_1} + 3e^{-\rho(2V_1+V_2)} - e^{-\rho(3V_1+3V_2+V_3)}\right)
\end{aligned} \tag{118}$$

where V_ν denotes the functionals for the single grains K.

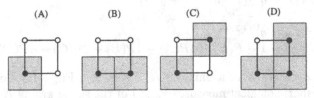

Fig. 20. Local coverages in a unit cell $\bar{K}^{(2)}$ of a two-dimensional lattice. The graphs indicate if one (A), two (B,C) or three (D) quarters of a cell $\bar{K}^{(2)}$ are covered. The complete empty (E) and occupied (F) cell, i.e., the configurations without a boundary within the cell are not depicted.

Morse's theorem does not only help to calculate probabilities of local coverages and mean values, but also second order moments of the Minkowski functionals. One obtains, for instance, on a d-dimensional cubic lattice for unit cubes K_i of edge length $\lambda = 1$ the variance of the Euler characteristic [35]

$$\delta\chi_d^2 = \overline{\chi^{(d)}(\mathcal{A})\chi^{(d)}(\mathcal{A})} - \overline{\chi^{(d)}(\mathcal{A})}\,\overline{\chi^{(d)}(\mathcal{A})} = \sum_{i=0}^{d}\binom{d}{i}f_i^2(q^{2^{d-i}})\,g_{d-i}(q^{2^{d-i}}) \tag{119}$$

with the functions $f_i(q) = \sum_{\nu=0}^{i}\binom{i}{\nu}(-2)^\nu q^{2^\nu}$

and $g_i(q) = (-1)^{i+1} + \sum_{\nu=0}^{i}\binom{i}{\nu}(-2)^\nu q^{-2^{-\nu}}$. In Fig. (18) the second order moments of the Euler characteristic

$$\delta\chi_1^2 = q(1-q)(1-3q+3q^2)$$
$$\delta\chi_2^2 = q(1-q)(1-6q+14q^2+q^3-23q^4+5q^5+9q^6)$$
$$\delta\chi_3^2 = q(1-q)(1-9q+33q^2+3q^3-117q^4+39q^5+99q^6$$
$$-3q^7+61q^8-59q^9-107q^{10}+7q^{11}+7q^{12}+19q^{13}+27q^{14})$$

(120)

are shown for the dimensions $d = 1, 2, 3$, respectively.

The mathematical definition of lattice configurations, kinematic formula (110) and excluded volumes (111), as well as the explicit expression of mean values (118) of Minkowski measures, second order moments (120), and even probabilities of local spatial configurations (113) are only first steps towards physical applications. Statistical models for microemulsions [30, Sect. 4.3], percolation [58, Sect. 4.2], and fluids in porous media [2, Sect. 2.2] can now be studied within a lattice approximation.

Problem: Lattice Approximation The problems mentioned already in the Sects. 5.2 and 5.3 occur analogously when lattice configurations are considered. It is necessary to calculate not only mean values and second order moments for independent and homogeneously distributed grains but also correlated, inhomogeneous, and anisotropic distributions. Additionally a renormalization group theory for lattice models including the Minkowski functionals is required for the study of complex fluids. As shown earlier, straightforward real-space methods such as Migdal-Kadanof bond moving schemes do not work due to the inherent geometric features of the Minkowski functionals which should be preserved under the renormalization group [37,35]. But finite-cell approximations might be possible since local probabilities (113) are known for grains of arbitrary size.

Integral Geometry	Statistical Physics	Problems
convex grains K unions of grains $A = \cup_i K_i$ (convex ring) **group of motions** (translations, rotations)	spatial configurations (Figs. 1 and 2, Sects. 4.2 and 5.1) partition sum (Sec. 5.1)	hyperbolic surfaces inhomogeneity
Minkowski functionals M_ν intrinsic volumes $V_\nu = M_{d-\nu}$	spectral density of Laplace operator (scattering, diffusion, see Sec. 2.3)	edge contributions
	curvature energy of membranes (2.1)	squared curvature $\hat{H}^2$
volume $V \sim M_0$	capillary condensation of fluids (2.2)	non-convex parallel body
surface area $S \sim M_1$	order parameters for patterns (3.1)	effective theory
mean curvature $H \sim M_2 \sim \int dS \left(\frac{1}{R_1} + \frac{1}{R_2} \right)$	generalized contact distribution	dynamical equations
	n-point correlation functions (3.3) (galaxy distribution, holes in thin films)	non-convex parallel body
Euler characteristic $\chi \sim M_d \sim \int \frac{dS}{R_1 R_2}$	percolation criteria $\chi(\rho_c) \approx 0$ (4.2)	accuracy

Integral Geometry	Statistical Physics	Problems
additivity $M(K \cup K') = M(K)$ $+M(K') - M(K \cap K')$	robustness (critical fluid, Sec. 3.3) thermodynamic potentials (entropy) signed probability measures QM: projections, Wigner functions	critical fluctuations
completeness of additive functionals $M(K) = \sum\limits_{\nu=0}^{d} c_\nu M_\nu(K)$	general Hamiltonian for fluids (4.3,5.1) (colloids, complex fluids)	squared curvature $\hat{H}^2$ renormalization
scaling behavior $M_\nu(\lambda K) = \lambda^{d-\nu} M_\nu(K)$	fractals (3.4), critical phenomena (3.3) spinodal decomposition (3.2)	universality dynamical equations
kinematic formulae $\int\limits_{\mathcal{G}} M_\nu(A \cap gB)dg =$ $\sum\limits_{\mu=0}^{\nu} \binom{\nu}{\mu} M_\mu(A) M_{\nu-\mu}(B)$	thermal averages (Sec. 5) second order moments (5.3) Rosenfeld density functional (4.1)	correlation (hard particle) non-isotropic distributions non-spherical particles
excluded volume $V_{ex} = \int \chi(K \cap K')dK'$ Steiner's formula: $V(K_\epsilon) = V(K) + S(K)$ $+\epsilon^2 H + \frac{4\pi}{3}\epsilon^3$	percolation threshold $\rho_c \approx V_{ex}^{-1}$ (4.2) second virial coefficient (4.1)	polydispersity cluster expansion
Crofton's formula	stereology, porous media (2.2)	
curved spaces	cosmology	(3+1)-Minkowski space
lattice geometry (Sec. 5.4)	models for complex fluids + percolation	squared curvature $\hat{H}^2$

References

1. Alon, U., I. Balberg, A. Drory (1991): 'New, heuristic, percolation criterion for continuum systems', *Phys. Rev. Lett.* **66**, pp. 2879–2882; *Phys. Rev.* **A 42**, pp. 4634–4638
2. Arns, C.H., M.A. Knackstedt, K. Mecke (2000):'Euler-Poincaré characteristics of classes of disordered media', preprint
3. Baddeley, A.J., B.W. Silverman (1984): 'A Cautionary Example on the Use of Second-Order Methods for Analyzing Point Patterns', *Biometrics* **40**, pp. 1089–1093
4. Balberg, I. (1985): ' 'Universal' percolation-threshold limits in the continuum', *Phys. Rev.* B **31**, pp. 4053–4055
5. Balberg, I., N. Binenbaum (1987): 'Scher and Zallen criterion: Applicability to composite systems', *Phys. Rev.* B **35**, pp. 8749–8752; I. Balberg (1987), *Phil. Mag.* B. **56**, pp. 991–1003
6. Blaschke, W. (1936): *Integralgeometrie*, erstes Heft, (Bernd G. Teubner, Leipzig, Berlin)
7. Bray, A.J. (1994): 'Theory of phase-ordering kinetics', *Adv. Phys.* **43**, pp. 357–459
8. Broadbent, S.R., J.M. Hammersley (1957): 'Percolation Processes', *Proc. Cambridge Philos. Soc.* **53**, pp. 629–645
9. Brodatzki, U. (1997): Monte-Carlo Simulationen stochastischer Geometrien. Diplomarbeit, Bergische Universität Wuppertal
10. Brodatzki, U., K. Mecke (2000): 'Monte-Carlo Simulation of the Boolean Model', preprint 2000
11. Cuesta, J.A. (1996): 'Fluid Mixtures of Parallel Hard Cubes', *Phys. Rev. Lett.* **76**, pp. 3742–3745
    J.A. Cuesta, Y. Martinez-Raton (1997): 'Fundamental measure theory for mixtures of parallel hard cubes', *J. Chem. Phys.* **107**, pp. 6379–6389
12. Dawson, K.A. (1992): p. 265 in: *Structure and Dynamics of Strongly Interacting Colloids and Supra- molecular Aggregates in Solution*, ed. by S.-H. Chen, J.S. Huang, P. Tartaglia (Kluwer, Dordrecht)
13. Dullien, F.A.L. (1992): *Porous Media, Fluid Transport and Pore Structure* (Academic Press, San Diego)
14. Evans R. (1992): *Density Functionals in the Theory of Nonuniform Fluids*, p. 85 in: *Fundamentals of Inhomogeneous Fluids*, ed. by D. Henderson (Dekker, New York); also R. Evans (1979): *Adv. Phys.* **28**, pp. 143–200
15. Fry, J.N. (1988): 'Voids and Scaling in Cosmology', pp.1336-1339 in: *The Lawrence Workshop on Cosmological Topology 1988*, ed. by A.L. Melott; Publ. Astr. Soc. Pacific **100**, pp. 1336–1339
16. Garboczi, E.J., K.A. Snyder, J.F. Douglas, M.F. Thorpe (1995): 'Geometrical percolation threshold of overlapping ellipsoids', *Phys. Rev.* E **52**, pp. 819–828
17. Gelbart, W., Ben-Shaul, A., Roux, D. (Eds.) (1994): *Micelles, Membranes, Microemulsions, and Monolayers* (Springer-Verlag, New York)
18. Gompper, G., M. Schick (1994): 'Self-Assembling Amphiphilic Systems', in: *Phase Transitions and Critical Phenomena*, ed. by C. Domb, J. Lebowitz, Vol. 16 (Academic Press, London)
19. Gouyet, J.-F. (1996): *Physics and Fractal Structures* (Springer-Verlag, New York)
20. Gunton, J.D., M. Miguel, P.S. Sahni (1983): 'The dynamics of first-order phase transitions', pp. 269–466 in: *Phase Transition and Critical Phenomena*, vol. 8, ed. by C. Domb, J.L. Lebowitz (Academic Press, New York)

21. Hadwiger, H. (1957): *Vorlesungen über Inhalt, Oberfläche und Isoperimetrie* (Springer-Verlag, Heidelberg)
22. Hansen J.P., I.R. McDonald (1986): *Theory of simple liquids* (Academic Press, London)
23. Helfrich, W. (1973): 'Elastic Properties of Lipid Bilayers: Theory and Possible Experiments', *Zeitschrift für Naturforschung* **28c**, pp. 693–703
24. Herminghaus, St., K. Jacobs, K. Mecke, J. Bischof, A. Frey, M. Ibn-Elhaj, St. Schlagowski (1998): 'Spinodal Dewetting in Liquid Crystal and Liquid Metal Films', *Science* **282**, pp. 916–919
25. Hilborn, R.C. (1994): *Chaos and nonlinear dynamics: an introduction for scientists and engineers* (Oxford Univ. Press, New York)
26. Ising, E. (1925): 'Beitrag zur Theorie des Ferromagnetismus', *Zeitschrift für Physik* **31**, pp. 253–258
27. Jacobs, K., S. Herminghaus, K. Mecke (1998): 'Thin Liquid Polymer Films Rupture via Defects', *Langmuir* **14**, pp. 965–969
28. Kac, M. (1966): 'Can one hear the shape of a drum?', *Amer. Math. Monthly* **73**, pp. 1–24
29. Kihara, T. (1953): 'Virial Coefficients and Models of Molecules in Gases', *Rev. Mod. Phys.* **25**, pp. 831–843
30. Likos, C.N., K. Mecke, H. Wagner (1995): 'Statistical morphology of random interfaces in microemulsions', *J. Chem. Phys.* **102**, pp. 9350–9361
31. Lipowsky, R., D. Richter, K. Kremer (1992): *The Structure and Conformation of Amphiphilic Membranes* (Springer-Verlag, Berlin)
32. Mandelbrot, B.B. (1982): *The Fractal Geometry of Nature* (Freeman, San Francisco)
33. Meakin, P. (1998): *Fractals, scaling and growth far from equilibrium* (Cambridge University Press, Cambridge)
34. Mecke, K., H. Wagner (1991): 'Euler characteristic and related measures for random geometric sets', *J. Stat. Phys.* **64**, pp. 843–850
35. Mecke, K. (1994): *Integralgeometrie in der Statistischen Physik*, Reihe Physik Bd. 25 (Verlag Harri Deutsch, Frankfurt)
36. Mecke, K., Th. Buchert, H. Wagner (1994): 'Robust morphological measures for large-scale structure in the universe', *Astron. Astrophys.* **288**, pp. 697–704
37. Mecke, K. (1995): 'Bending rigidity of fluctuating membranes', *Z. Phys.* **B 97**, pp. 379–387
38. Mecke, K. (1996a): 'Morphological Characterization of Patterns in Reaction-Diffusion Systems', *Phys. Rev. E* **53**, pp. 4794–4800
39. Mecke, K. (1996b): 'Morphological Model for Complex Fluids', *J. Phys.: Condensed Matter* **8**, pp. 9663–9667
40. Mecke, K. (1997): 'Morphology of Spatial Patterns: Porous Media, Spinodal Decomposition, and Dissipative Structures', *Acta Physica Polonica B* **28**, pp. 1747–1782
41. Mecke, K., V. Sofonea (1997): 'Morphology of Spinodal Decomposition', *Phys. Rev. E* **56**, pp. R3761–R3764
42. Mecke, K. (1998a): 'Morphological Thermodynamics of Composite Media', *Fluid Phase Equilibria* **150/151**, pp. 591–598
43. Mecke, K. (1998b): 'Integral Geometry and Statistical Physics: Porous Media, Spinodal Decomposition, and Dissipative Structures', *International Journal of Modern Physics B* **12**, pp. 861–899
44. Mecke K., S. Dietrich (1999): 'Effective Hamiltonian for liquid-vapor interfaces', *Phys. Rev. E* **59**, pp. 6766–6784

45. Meunier, J., D. Langevin, N. Boccara (1987): *Physics of Amphiphilic Systems* , Springer Proceedings in Physics Vol. 21 (Springer-Verlag, Berlin)

46. Nelson, D., T. Piran, S. Weinberg (1989): *Statistical Mechanics of Membranes and Surfaces* (World Scientific, Singapore)

47. Peebles, P.J.E. (1980): *The Large-scale Structure of the Universe* (Princeton University Press)

48. Pike, G.E., C.H. Seager (1974): 'Percolation and conductivity: A computer study', *Phys. Rev. B* **10**, pp. 1421–1434

49. Reiss, H. (1992): 'Statistical geometry in the study of fluids and porous media', *J. Phys. Chem.* **96**, pp. 4736–4747

50. Rosenfeld, A., A.C. Kak (1976): *Digital Picture Processing* (Academic Press, New York)

51. Rosenfeld, Y. (1995): 'Free energy model for the inhomogeneous hard-body fluid: application of the Gauss-Bonnet theorem', *Mol. Phys.* **86**, pp. 637–647

52. Santaló, L. A. (1976): *Integral Geometry and Geometric Probability* (Addison - Wesley, Reading)

53. Scher, H., R. Zallen (1970): 'Critical density in percolation processes', *J. Chem. Phys.* **53**, pp. 3759–3761

54. Schneider, R. (1993): *Convex bodies: The Brunn-Minkowski theory* (Cambridge University Press, Cambridge)

55. Schoen, M. (1993): *Computer Simulation of Condensed Phases in Complex Geometries* (Springer-Verlag, Heidelberg

56. Sen, P.N., C. Straley, W.E. Kenyon (1990): 'Surface-to-volume ratio, charge density, nuclear magnetic relaxation, and permeability in clay-bearing sandstones', *Geophysics* **55**, pp. 61–69

57. Serra, J. (1982): *Image Analysis and Mathematical Morphology* (Academic Press, New York)

58. Seyfried, A., K. Mecke (2000): 'Topological Percolation Criterion', preprint

59. Sofonea, V., K. Mecke (1999): 'Morphological Characterization of Spinodal Decomposition Kinetic', *European Physical Journal* **8**, pp. 99–112

60. Stauffer D., A. Aharony (1992): *Introduction to Percolation Theory* (Taylor & Francis, London)

61. Stoyan, D., W.S. Kendall, J. Mecke (1995): *Stochastic Geometry and its Applications* (John Wiley & Sons, Chichester)

62. Teubner, M. (1990): 'Scattering from two-phase random media', *J. Chem. Phys.* **92**, pp. 4501–4507

63. Wagner, A.J., J.M. Yeomans (1998): 'Breakdown of Scale Invariance in the Coarsening of Phase-Separating Binary Fluids', *Phys. Rev. Lett.* **80**, pp. 1429–1432

64. Weil, W. (1983): *Stereology, A survey for Geometers*, in: *Convexity and its Applications*, ed. by P.M. Gruber, J.M. Wills (Birkhäuser)

65. Widom, B., J.S. Rowlinson (1970): 'New model for the study of liquid-vapor phase transitions', *J. Chem. Phys.* **52**, pp. 1670–1684

Considerations About the Estimation of the Size Distribution in Wicksell's Corpuscle Problem

Joachim Ohser[1] and Konrad Sandau[2]

[1] Institute of Industrial Mathematics, Erwin-Schrödinger-Straße
   D-67663 Kaiserslautern, Germany
[2] Faculty of Mathematics and Science, University of Applied Sciences, Schöfferstr. 3
   D-64295 Darmstadt, Germany

Abstract. Wicksell's corpuscle problem deals with the estimation of the size distribution of a population of particles, all having the same shape, using a lower dimensional sampling probe. This problem was originary formulated for particle systems occurring in life sciences but its solution is of actual and increasing interest in materials science. From a mathematical point of view, Wicksell's problem is an inverse problem where the interesting size distribution is the unknown part of a Volterra equation. The problem is often regarded ill-posed, because the structure of the integrand implies unstable numerical solutions. The accuracy of the numerical solutions is considered here using the condition number, which allows to compare different numerical methods with different (equidistant) class sizes and which indicates, as one result, that a finite section thickness of the probe reduces the numerical problems. Furthermore, the relative error of estimation is computed which can be split into two parts. One part consists of the relative discretization error that increases for increasing class size, and the second part is related to the relative statistical error which increases with decreasing class size. For both parts, upper bounds can be given and the sum of them indicates an optimal class width depending on some specific constants.

1   Introduction

Wicksell's corpuscle problem is one of the classical problems in stereology [1,29, 30]. A set of spheres having an unknown distribution of diameters is hit by a section, see Figs. 1a) to 1d). The unknown distribution has to be estimated using the observable distribution of the diameters of the section profiles. A large number of papers have been published since the first paper of Wicksell in 1925. Different types of solutions and different kinds of generalizations are studied in those papers. The reviews of Exner [6] and Cruz-Orive [5] can give a first impression, reflecting also the historical development and the amount of publications. The various methods are summarized in the book [28], where also more recent developments are considered. One kind of generalization deals with the thickness of the probes and with the dimension of sampling. We will distinguish here sections of a given thickness (thin sections) and ideal sections (planar sections). The cases of sampling are shown in Table 1.

Table 1. Survey of sectioning and sampling used for stereological estimation of particle size distribution. Using planar sampling design, a microstructure is observed in a planar window while linear sampling design uses test segments (commonly a system of parallel equidistant segments).

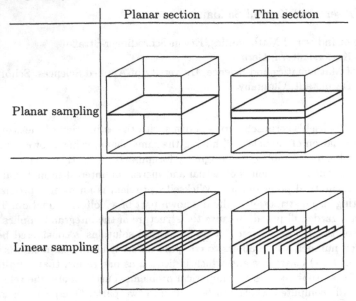

Another kind of generalization concerns the shape of the objects (or particles). In materials science it is convenient to use particle models, more often than in life sciences. The following models are in use:

- Spheres, spheroids, ellipsoids,
- cubes, regular polyhedra, prisms,
- lamellae, cylinders, and
- irregular convex polyhedra (with certain distribution assumptions).

A survey of stereological methods for systems of non-spherical particles is given in [21].

Assuming that only one of the mentioned shapes appears in a particle population but with varying size, it is possible to completely transfer the principles of the solution, see [16] and [22], which are shown below for spheres. It should be mentioned that the section profiles of all other shapes are more informative than the sections of a sphere. In particular, if the particles are cubic in shape then the section profiles form convex polygons, and from the size and shape of any section polygon with more than three vertices the edge length of the corresponding cube can be determined, see [19] and [22].

Stereological methods always depend on the assumptions of randomness. In materials analysis the spatial particle system is assumed to be homogeneous and isotropic. The principle behind this assumption of randomness is called the model-based approach. In this case information about the spatial structure can be obtained from an arbitrarily chosen planar or thin section. We use this model-based approach here instead of the design-based approach. The latter requires a deterministic aggregate of non-overlapping particles sampled by means of a planar section randomized in position and direction, see [11] and [12].

In the last 15 years other stereological methods, working without assuming a particular shape were often preferred in applications. A recently published book about stereology does not even mention Wicksell's classical problem [9]. However, if stronger assumptions on the particle system are justified, they can help to get comparable results with less effort. Methods, which estimate the size distribution of particles without shape assumption need considerably more effort in measuring than methods assuming spherical shape of the particles. Even probing and sampling can be simplified if stronger assumptions are fulfilled.

There is a considerable number of mathematically orientated papers dealing with Wicksell's corpuscle problem. This becomes understandable, looking to the interesting mathematical part of the problem in more detail. Considering the probability of appearing profiles in the section one yields an integral equation, determining the wanted distribution of the particle size, but only in an implicit form. We call this the direct equation. Wicksell's problem is often called an inverse problem because inversion of the equation leads to an explicit integral representation of the diameter distribution of the spheres. However, the kernel of the integration has now a singularity, and because of the resulting numerical problems it is called ill-posed or improperly posed. Jakeman and Anderssen [10] characterize this property in the following way: *Small perturbations of the data can correspond to arbitrarily large perturbations of the solution.*

The solution given here, follows in a first step the proposal of Saltykov [25], who solved the implicit equation numerically. To do this, the range of the size must be partitioned into classes. The size of the classes plays an important role. The smaller the class size, the better the approximation of the wanted distribution by a step function. The larger the class size, the more each class is occupied with data, thereby improving the statistical estimation of the class probability. Finally, the larger the class size, the better also the numerical stability. Obviously, the advantages go into different directions and thus we have to seek an optimum size.

In this paper we want to show some relations, illuminating in which way this optimum is affected. Discretization and regularization of the integral changes the direct equation to a matrix equation i.e. a system of linear equations. The relative deviation of the estimation of the outcome has an upper bound, depending on the product of relative deviation of the input and the so-called condition number. The condition number is a specific number for the matrix of a system of linear equations indicating the behavior of the system under small deviations of the input vector (i.e. the right-hand side of the linear equation). A smaller condition number indicates an outcome with smaller deviation. We use the condition number to evaluate different methods and different class sizes. The results we show in §4 are in contradiction to the numerical supposition that a regularization with a quadrature rule of higher order should be preferred [15]. This is demon-

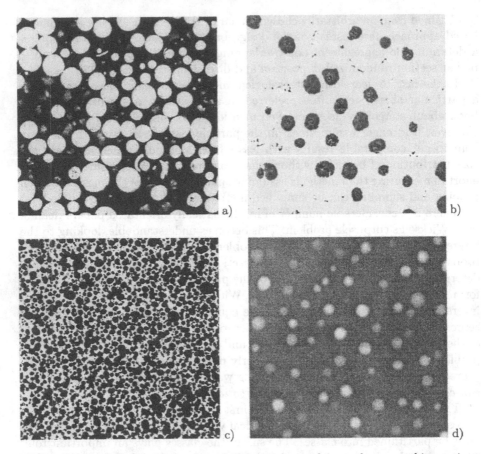

Fig. 1. Systems of spherically shaped objects observed in a planar or thin section. a) A metal powder embedded in a resin matrix. The section profiles of the spherical particles are observed in a planar section. b) Cast iron with spherolitic graphite, planar section. c) Pores in gas concrete, planar section. d) Particles in the nickel-base superalloy Nimonic PE16, orthogonal projection of a thin section.

strated here for rectangular and trapezoidal quadrature rule. The results using the condition number also quantify the reduction of instability with respect to growing section thickness.

An extension of this classical method is offered by the EM algorithm, see [14], which is first applied in stereology by Silverman et al. [27]. This algorithm is presented here as an alternative method for the solution of the discretized stereological equation. In the EM algorithm, an approximation is calculated not only with respect to numerical but also with respect to the statistical aspects. Practical experience shows that the EM algorithm works even under weak assumptions [28, p. 359]. The results based on the condition number can be transferred in a similar way to the EM algorithm. In conclusion, these results suggest that the recommended method is in particular well suited for practical use.

2   The Stereological Equation

In the following we confine ourselves to the classical case where the particles
form a system of spheres having random diameters. Other shape assumptions
are touched on only briefly. Figures 1a) to 1d) show typical examples of spherically shaped particles (or pores) occurring in materials science. The Cu-35Sn
powder shown in Fig. 1a) has been embedded in an opaque resin in order to estimate the sphere diameter distribution by means of stereological methods. Cast
iron with spherolitic graphite, see Fig. 1b), is the playground for the application
of stereology in materials science. The graphite form occurs by inoculation with
metals such as manganese and cerium. The spherolites consist of a number of
crystallites which grow outward from a common center. The microstructure of
gas concrete, see Fig. 1c), can be described by a system of overlapping spheres.
The spherically shaped pores were infiltrated with black resin. Since the spherical pores touch each other, it is a delicate problem of image analysis to isolate
the section circles. Anyway, if a representative sample of circle diameters is
available, the sphere diameter distribution can be estimated by means of stereological methods. Fig. 1d) shows a dark-field transmission electron micrographs
of coherent γ'-particles in the nickel-base superalloy Nimonic PE16 (undeformed
state). The image can be considered as an orthogonal projection of a thin section
through a system of opaque spheres lying in a matrix which is transparent with
respect to the applied radiation. For a sphere with its center lying in the slice,
the diameter is equal to the diameter of its projected circle. In the other case,
when the sphere center is outside the slice, a section circle is observed. In the
projection image the two cases cannot be distinguished.

   We are interested in the stereological determination of the following two
spatial characteristics:

N_V – the mean number of spheres per unit volume (density of spheres),
F_V – the distribution function of the diameters of the spheres, $F_V(u) :=$
   $I\!P$(diameter $\leq u$).

   It is possible to estimate these two quantities using data sampled in images
of planar or thin sections, and in both cases planar or linear sampling designs
can be applied. As a first step, one usually estimates the following quantities:

λ – the density of the section profiles,
F – the size distribution function of the section profiles, $F(s) := I\!P$(profile size $\leq$
   s).

   The second step consists of the solution of a stereological equation which can
be given in the form

$$\lambda\left[1 - F(s)\right] = N_V \int\limits_{s}^{\infty} p(u, s)\, \mathrm{d}F_V(u), \quad s \geq 0. \tag{1}$$

This integral equation is a Volterra equation of the first kind. Here $F_V(u)$ is
the unknown distribution while the left-hand side $\lambda\left[1 - F(s)\right]$ is known from

measurement in a planar or thin section. The function of two variables $p(u, v)$ is called the *kernel* of the integral equation.

The analytical expression for the kernel $p(u, s)$ of (1) depends on the applied sampling design, see Table 2. The kernel also carries information about the shape of the objects, which are in our case spheres. Kernels for non-spherical particles are given e.g. in [4,8] and [22]. The function $F(s)$ in (1) stands for the distribution function $F_A(s)$ of the circle diameters and the distribution function $F_L(s)$ of the chord lengths, respectively. In the case of planar sampling design, the density λ is the mean number of circles per unit area, $\lambda = N_A$, and for linear sampling design the density λ is the mean number of chords per unit length.

Table 2. The kernel $p(u, s)$ of the stereological integral equations for $0 \leq s \leq u$ corresponding to the four sampling methods applied in stereology, see Table 1, in the case of spherically shaped particles. The parameter t is the thickness of the slice.

	Planar section	Thin section
Planar sampling	$\sqrt{u^2 - s^2}$	$\sqrt{u^2 - s^2} + t$
Linear sampling	$\frac{\pi}{4}(u^2 - s^2)$	$\frac{\pi}{4}(u^2 - s^2) + t\sqrt{u^2 - s^2}$

The problem of estimation of F_V is said to be an *inverse problem*, and when using this term one is tempted to ask for the *direct problem*. Consider for example planar sampling design in a planar section. Let $\lambda = N_A$ be the density of the section circles (the mean number of section circles per unit area). Using

$$N_A[1 - F_A(s)] = N_V \int_s^\infty \sqrt{u^2 - s^2}\, dF_V(u), \quad s \geq 0, \qquad (2)$$

the circle diameter distribution function F_A can be straightforwardly computed from F_V. This is the direct problem. In the given case an analytical solution of the inverse problem is known:

$$N_V[1 - F_V(u)] = N_A \frac{2}{\pi} \int_u^\infty \frac{dF_A(s)}{\sqrt{s^2 - u^2}}, \quad u \geq 0, \qquad (3)$$

i.e. we obtain analytically F_V from F_A. Since the formulation of one problem involves the other one, the computation of F_A from F_V and the computation of F_V from F_A are inverse to each other. Frequently, the solutions of such inverse problems do not depend continuously on the data. Furthermore, given any distribution function F_A then the formal solution F_V of the inverse problem is not necessarily a distribution function. Mathematical problems having these undesirable properties are referred to as *ill-posed problems*. Indeed, the stereo-logical estimation of sphere diameter distribution is often criticized because of the ill-posedness of the corresponding stereological integral equation.

Applying a kernel to a function, as on the right-hand side of (1), is generally a smoothing operation (convolution). Therefore, numerical solution (deconvolution) which requires inverting the smoothing operator will be sensitive to small errors (noise) of the experimentally determined left-hand side, and thus integral equations such as Fredholm equations of the first kind are often ill-conditioned. However, the Volterra equation (1) tends not to be ill-conditioned since the lower limit of the integral yields a sharp step that destroys the smoothing properties of the kernel function. Hence, stereological estimation of sphere diameter distribution is a 'good-natured' ill-posed problem. As we will see, the condition numbers of the operators which correspond to the kernels in Table 2 are relatively small.

3   Numerical Solutions of the Stereological Equation

There is a close correspondence between linear integral equations which specify linear relations among distribution functions and linear equations which specify analogous relations among vectors of relative frequencies. Thus, for the purpose of numerical solution, the stereological integral equation (1) is usually transformed into a matrix equation

$$y = P\vartheta \tag{4}$$

where ϑ is the vector of frequencies of the sphere diameters and y is the vector of frequencies of the profile sizes. Comparing with (1), we see that the kernel of the Volterra integral equation corresponds to a matrix $P = (p_{ki})$ that is upper triangular, with zero entries below the diagonal. This matrix equation is straightforwardly soluble by backward substitution.

To facilitate reconstruction, we introduce a discretization of the size s, i.e. the estimation of the distribution function $F(s)$ is carried out in histogram form. For simplification we consider a discretization where the bins (classes) are equally spaced. Given a constant class width Δ such that $y_k = \lambda[F(k\Delta) - F((k-1)\Delta)]$, $k = 1, \ldots, n$, from (1) one can easily derive the formula

$$y_k = N_V \int_s^\infty [p(u, (k-1)\Delta) - p(u, k\Delta)] \, dF_V(u), \quad k = 1, \ldots, n. \tag{5}$$

The next step in the solution of (1) is a discretization of $F_V(u)$. A possible discretization is based on the Nystrom method which requires the choice of some approximate quadrature rule of the kind $\int f(x) \, dx = \sum w_i f(x_i)$, where the w_i are the weights of the quadrature rule. This method involves $O(n^3)$ operations, and so from the numerical point of view the most efficient methods tend to use high-order quadrature rules like Simpson's rule or Gaussian quadrature to keep n as small as possible. Mase [15] suggests using the Gauss-Chebychev quadrature rule. For smooth, nonsingular problems, Press et al. [23, p. 792] concluded that nothing is better than Gaussian quadrature. However, as pointed out by Nippe and Ohser [19], for many stereological problems concerned with non-spherical particles, analytical expressions for the corresponding kernel are not available, and the coefficients p_{ki} of P have to be computed numerically, see also [20]. For this reason, a geometric or probabilistic interpretation of the p_{ki} may be helpful. This is in fact possible only for the rectangular quadrature rule. Since in practical application experimental noise dominates, the accuracy of approximation is not as important as in numerical mathematics, and thus low-order quadrature rules are preferred to solve stereological integral equations. Furthermore, as shown in the following, low-order quadrature rules involve some kind of regularization. Two simple examples of low-order methods are given in the following, the repeated rectangular quadrature rule and the repeated trapezoidal quadrature rule.

3.1   Repeated Rectangular Quadrature Rule

We will focus attention on one particular interval $[(i-1)\Delta, i\Delta)$. The sphere diameter is assumed to be a discrete random variable, i.e. the quadrature rule described here is based on the assumption that $F_V(u)$ is constant within each bin,

$$F_V^{app}(u) = a_i, \qquad (i-1)\Delta \le u < i\Delta, \ i = 1,\ldots,n, \tag{6}$$

with $a_0 = 0$, $a_{i-1} < a_i$ and $a_n = 1$. Let ϑ_i be the mean number of spheres per unit volume with diameters equal to $i\Delta$, $\vartheta_i = N_V[a_i - a_{i-1}]$. Then (1) can be rewritten as a linear equation system

$$y_k = N_V \sum_{i=1}^{n} p_{ki}[a_i - a_{i-1}] = \sum_{i=1}^{n} p_{ki}\vartheta_i, \quad k = 1,\ldots,n, \tag{7}$$

whose matrix analog is as (4). The coefficients p_{ki} are given by

$$p_{ki} = p(i\Delta, (k-1)\Delta) - p(i\Delta, k\Delta), \quad i \ge k,$$

and $p_{ki} = 0$ otherwise. Note that for $p(u,s) = \sqrt{u^2 - s^2}$ the well-known Scheil-Saltykov-Schwartz method is obtained, see e.g. [26], while $p(u,s) = (\pi/4)(u^2 - s^2)$ yields the Bockstiegel method [3].

3.2   Repeated Trapezoidal Quadrature Rule

Now we turn to linear interpolation of $F_V(u)$ in the interval $[(i-1)\Delta, i\Delta)$,

$$F_V^{app}(u) = a_i + b_i[u - (i-1)\Delta],$$

i.e. the sphere diameter distribution is assumed to be uniform within the single histogram class. Furthermore, we suppose that the distribution function $F_V(u)$ is continuous, even at the points $i\Delta$, i.e. $a_i + b_i\Delta = a_{i+1}$. Hence $\vartheta_i = N_V[a_{i+1} - a_i]$ and

$$N_V b_i = N_V \frac{a_{i+1} - a_i}{\Delta} = \frac{\vartheta_i}{\Delta}.$$

In this particular case the integral equation will be solved with a repeated trapezoidal quadrature rule:

$$y_k = N_V \sum_{i=1}^{n} b_i \int_{(i-1)\Delta}^{i\Delta} [p(u, (k-1)\Delta) - p(u, k\Delta)] \, du$$

$$= \sum_{i=1}^{n} p_{ki} \vartheta_i, \quad k = 1, \ldots, n.$$

We obtain an equation system analogous to (7), but now the coefficients p_{ki} are given by

$$p_{ki} = \frac{1}{\Delta} \int_{(i-1)\Delta}^{i\Delta} [p(u, (k-1)\Delta) - p(u, k\Delta)] \, du, \quad i \geq k,$$

and $p_{ki} = 0$ otherwise. With $q(u, s) = \int_0^u p(u', s) du'$ this formula can be rewritten as

$$p_{ki} = \frac{1}{\Delta} \left[q\left(i\Delta, (k-1)\Delta\right) - q\left(i\Delta, k\Delta\right) - q\left((i-1)\Delta, (k-1)\Delta\right) + q\left((i-1)\Delta, k\Delta\right) \right]$$

for $i \geq k$.

Notice that the repeated trapezoidal quadrature rule was first suggested for the special case of planar sampling in a planar section by Blödner et al. [2]. For this particular case the function $q(u, s)$ is given by

$$q(u, s) = \frac{1}{2} \left(u\sqrt{u^2 - s^2} - s^2 \operatorname{arccosh} \frac{u}{s} \right), \quad 0 \leq s \leq u,$$

and $q(u, s) = 0$ otherwise.

3.3   Application of the EM Algorithm

The stereological estimation of the vector ϑ of relative frequencies of sphere diameters entails both statistical and numerical difficulties. A natural statistical approach is by maximum likelihood, conveniently implemented using the EM algorithm. The EM algorithm was first applied in stereology by Silverman et al. [27]. In the EM approach, y is said to be the vector of *incomplete data* obtained by planar or linear sampling, while ϑ represents the parameter vector to be estimated. The EM algorithm is an iterative procedure which increases the log-likelihood of a current estimate ϑ^λ of ϑ.

Each iteration step consists of two substeps, an E-substep and an M-substep, where E stands for 'expectation' and M for 'maximization'. Let c_{ki} be the number of events occurring in bin i which contribute to the count in bin k; thus (c_{ki}) is called the matrix of complete data. The E-substep yields the expected value of the c_{ki}, given the incomplete data y, under the current estimate ϑ^ν of the parameter vector ϑ. The M-substep yields a maximum likelihood estimate of the parameter vector ϑ using the estimated complete data c from the previous E-substep.

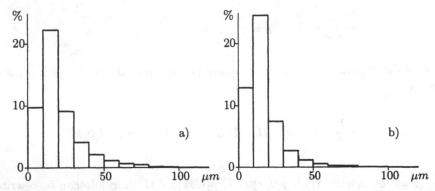

Fig. 2. Stereological estimation of the sphere diameter distribution for the cast iron with spherolitic graphite shown in Fig. 1b): a) Histogram of the sizes of section profiles observed in a planar section, b) histogram of the particle size. The sample size was 6995 and the total area of the sampling windows was $A(W) = 18.207\,\text{mm}^2$. (The data have been obtained from 144 sampling windows.) This yields $N_A = 384\,\text{mm}^{-2}$, and the particle density N_V is obtained from $N_V = \sum_i \vartheta_i = 17\,545\,\text{mm}^{-3}$. The coefficients of the discretized stereological equation system have been obtained from the repeated rectangular quadrature rule, and the equation system has been solved by means of the EM-algorithm (32 EM-steps).

Define q_i and r_k to be $q_i = \sum_{k=1}^{i} p_{ki}$ and $r_k = \sum_{i=k}^{n} p_{ki}\,\vartheta_i^\nu$, respectively. Then the E-substep is $c_{ki} = \vartheta_i^\nu\, y_k\, p_{ki}/r_k$ for each i and k, and the M-substep is given by $\vartheta_i^{\nu+1} = \frac{1}{q_i}\sum_{k=1}^{i} c_{ki}$, $i = 1, \ldots, n$. Combining these two substeps we obtain an EM-step given by the updating formula

$$\vartheta_i^{\nu+1} = \frac{\vartheta_i^\nu}{q_i} \sum_{k=1}^{i} \frac{p_{ki}}{r_k}\, y_k, \quad i = 1, \ldots, n. \tag{8}$$

This formula yields a sequence $\{\vartheta^\nu, \nu = 0, 1, \ldots\}$ of solutions for the vector ϑ. The convergence of the algorithm is guaranteed in theory, but the solution depends on the chosen initial value ϑ_i^0. We propose using $\vartheta_i^0 = y_i$ for $i = 1, \ldots, n$, see [20]. The choice of non-negative initial values ensures that each of the ϑ^ν is automatically non-negative. Furthermore, when ν is not too big the EM algorithm includes some kind of regularization.

An example of application of the EM algorithm is shown in Fig. 2.

We remark that the M-substep can be theoretically justified as a maximization step when the sphere diameters are independent and the sphere centers form a Poisson point field. In practice these assumptions are frequently unrealistic.

However, the EM algorithm has been found by long experience to be reasonably adequate for practical purposes (cf. also the remarks in [28], p. 359).

4   Accuracy of Estimation and Choise of Bin Size of Stereological Estimation

The question most frequently asked of a statistican by the experimental researcher is: *How large a sample size do I need?* Vice versa, given the sample size, one may ask the question: *What is the optimal bin size?* In fact, the answer can be given if the estimation error is known. Unfortunately, there is no simple way to derive an expression for the error in the stereological estimation. A practical method for the error estimation is to divide the sample into two subsamples of equal size, say, and treat the difference between the two estimates as a conservative estimate of the error in the result obtained for the total sample size. However, this method cannot be used to compare properties of stereological estimation methods in general.

4.1   The Condition Number of the Operator P

In numerical mathematics it is common to describe the behavior of a linear equation system like (4) by the condition number of the matrix P. We attempt to give a statistical interpretation: Let $\widehat{y}$ be an estimator of the vector y. The difference $\widehat{y} - y$ is the experimental measurement noise, and $\mathbb{E}|\widehat{y} - y|/|y|$ is the relative statistical error of $\widehat{y}$. The estimator $\widehat{\vartheta}$ of ϑ may be obtained via $\widehat{\vartheta} = P^{-1}\widehat{y}$ where P^{-1} is the inverse matrix of P. A well-known estimate of the relative statistical error $\mathbb{E}|\widehat{\vartheta} - \vartheta|/|\vartheta|$ of $\widehat{\vartheta}$ is given by

$$\frac{\mathbb{E}|\widehat{\vartheta} - \vartheta|}{|\vartheta|} \leq \text{cond}(P)\,\frac{\mathbb{E}|\widehat{y} - y|}{|y|}. \tag{9}$$

The condition number is defined as $\text{cond}(P) = |P| \cdot |P^{-1}|$ where the matrix norm corresponds to the Euclidean vector norm. It is an essential quantity giving an indication of the accuracy of numerical solutions of stereological problems; it describes the accuracy of numerical solutions independently of assumptions made for the sphere diameter distribution, see [7,13,18].

The Tables 3 and 4 present numerical values of the condition number of the matrix P based on the spectral norm $|P| = \max\limits_{x \neq 0} \frac{|Px|}{|x|}$. To compare the statistical accuracy of several numerical methods, the data space is divided into n bins of constant width Δ. There are significant differences in the condition numbers:

1. With decreasing class width Δ (increasing number n of classes) the statistical accuracy decreases.
2. The trapezoidal quadrature rule has no clear advantages over the rectangular one: the approximation is surely improved but the statistical error of the estimates increases rapidly. Hence, for a small sample size the repeated

rectangular quadrature rule is preferable to higher-order quadrature rules. Obviously, one cannot improve approximation and regularization simultaneously.

3. With increasing slice thickness t the statistical accuracy increases. Formally, the slice thickness t can be considered as a parameter of regularization. However, in the orthogonal projection of a 'thick' slice, overlapping cannot be avoided and overlapping destroys the information available from the slice.

4. In general, the condition numbers corresponding to linear sampling design are greater than those corresponding to planar sampling design. Thus, stereological estimation based on linear sampling design is more unstable than that based on planar sampling design (cf. also the discussion in [24], p. 212).

Table 3. The condition number of the matrix P for the repeated rectangular quadrature rule, depending on the class number n and a normalized slice thickness $t_0 = t/d_V$ where d_V is the mean sphere diameter. The left table gives numerical values of the condition number for planar sampling design, and the right one presents those for linear sampling design.

n	$t_0 = 0$	$t_0 = \frac{1}{4}$	$t_0 = \frac{1}{2}$	$t_0 = 1$	n	$t_0 = 0$	$t_0 = \frac{1}{4}$	$t_0 = \frac{1}{2}$	$t_0 = 1$
4	3.10	2.04	1.69	1.41	4	10.82	6.40	5.19	4.30
8	5.68	2.53	1.91	1.50	8	38.30	14.53	10.77	8.42
12	8.28	2.77	2.00	1.53	12	82.69	23.05	16.42	12.55
16	10.88	2.92	2.05	1.55	16	143.98	31.72	22.10	16.69
20	13.48	3.01	2.09	1.56	20	222.19	40.47	27.79	20.83
24	16.08	3.08	2.11	1.57	24	317.30	49.25	33.49	24.97
28	18.68	3.14	2.13	1.58	28	429.31	58.06	39.19	29.12
32	21.29	3.18	2.14	1.58	32	558.25	66.88	44.90	33.26

Table 4. The condition number of the matrix P for the repeated trapezoidal quadrature rule, depending on the class number n and a normalized slice thickness $t_0 = t/d_V$ where d_V is the mean sphere diameter. The left table gives numerical values of the condition number for planar sampling design, and the right one presents those for linear sampling design.

n	$t_0 = 0$	$t_0 = \frac{1}{4}$	$t_0 = \frac{1}{2}$	$t_0 = 1$	n	$t_0 = 0$	$t_0 = \frac{1}{4}$	$t_0 = \frac{1}{2}$	$t_0 = 1$
4	5.49	2.39	1.81	1.44	4	33.67	13.53	10.11	7.97
8	11.22	2.86	2.01	1.52	8	156.38	32.35	22.55	17.07
12	16.92	3.05	2.08	1.55	12	374.37	51.40	34.97	26.13
16	22.60	3.16	2.12	1.57	16	691.13	70.55	47.39	35.16
20	28.28	3.22	2.14	1.58	20	1 108.98	89.76	59.82	44.19
24	33.96	3.27	2.16	1.58	24	1 629.63	108.99	72.24	53.21
28	39.63	3.30	2.17	1.59	28	2 254.70	128.23	84.66	62.24
32	45.30	3.32	2.18	1.59	32	2 985.27	147.49	97.08	71.26

In comparison with other ill-conditioned equations which occur in practice, the condition numbers in Table 3 and Table 4 are rather small and thus the numerical methods for solving the stereological integral equations presented above can be considered as relatively stable. If experimental measurement noise does not dominate then one can expect that the variances of stereological estimators of sphere diameter distribution function are also small.

Finally, we remark that the consideration of weighted sphere diameter distributions can provide smaller condition numbers than the classical (number-weighted) sphere diameter distribution considered in this paper, see [18].

4.2   The Optimal Discretization Parameter

Consider now the discretization error of stereological estimation. The discretization error can be expressed in terms of the $\mathcal{L}^2$-norm of the difference between the probability densities of the distribution functions $F_V^{app}(u)$ and $F_V(u)$, respectively, where $F_V^{app}(u)$ is a discretization of the sphere diameter distribution. The statistical error and the discretization error – considered as functions of the bin size Δ – are schematically shown in Fig. 3. The errors behave in an opposite way. The relative total error of statistical estimation is the sum of the relative statistical error and the relative discretization error. It depends on the sample size and the unknown sphere diameter distribution.

Without loss of generality, let $[0, d]$ be the interval covered by the n bins, i.e. $d = n\Delta$, and let w be the size of the sampling window (the area of a planar window or the length of a test segment). Then an upper bound of the relative total error can be given by the expression

$$\frac{d}{\sqrt{3}} c \Delta + \left(1 + \frac{d}{\sqrt{3}} c \Delta\right) \operatorname{cond}(P) \sqrt{\frac{2d}{\pi \lambda w \Delta}},$$

see Appendix, where c is a constant, $c > 0$. The graphs of the relative errors shown in Fig. 3 are obtained for $\lambda w = 1000$, $d = 1$, and $c = 8$.

The choice of the optimal discretization parameter Δ_{opt} would minimize the total error, but even in optimal circumstances one always gets a loss of accuracy. Unfortunately, the optimal discretization parameter cannot be obtained from only the sample data. In conclusion, the results give some hints for designing the statistics and it turns out that the 'simple ways are good ways'. That means that the bin size should not be chosen to be too small and a low-order quadrature rule can be taken. The condition numbers confirm that planar sampling design gives more stable results than linear sampling design, and a positive slice thickness reduces the estimation error as well.

The computation of the upper bound of the error in the stereological estimation presented in the Appendix is based on the particular case of trapezoidal quadrature rule. These considerations can be transferred to other quadrature rules. Finally, we remark that the EM algorithm can be understood as a statistically motivated iterative method for solving the linear equation (4). The

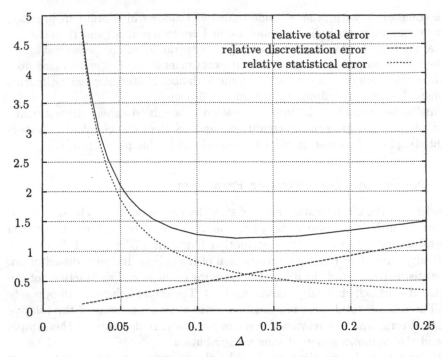

Fig. 3. Errors of stereological estimation of particle size distribution based on the upper bounds computed in the Appendix. If the bin size Δ becomes too small, the relative statistical error increases. Vice versa, if Δ is too large, the relative discretization error increases. There is an 'optimal' discretization parameter $\Delta_{opt} \approx 0.125$ (8 bins) which can only be computed explicitly in special cases since it depends on unavailable information on the relative measurement error as well as the distribution function $F_V(u)$ to be estimated.

application of the EM algorithm does not change any properties of the operator P. Thus, in principle, the results of this section can also be extended to maximum likelihood estimators.

References

1. Bach, G. (1963): 'Über die Bestimmung von charakteristischen Größen einer Kugelverteilung aus der Verteilung der Schnittkreise', *Z. wiss. Mikrosk.* **65**, p. 285
2. Blödner, R., P. Mühlig, W. Nagel (1984): 'The comparison by simulation of solutions of Wicksell's corpuscle problem', *J. Microsc.* **135**, pp. 61–64
3. Bockstiegel, G. (1966): 'Eine einfache Formel zur Berechnung räumlicher Größenverteilungen aus durch Linearanalyse erhaltenen Daten', *Z. Metallkunde* **57**, pp. 647–656
4. Cruz-Orive, L.-M. (1976): 'Particle size-shape distributions: The general spheroid problem, I. Mathematical model', *J. Microsc.* **107**, pp. 235–253

5. Cruz-Orive, L.-M. (1983): 'Distribution-free estimation of sphere size distributions from slabs showing overprojection and truncation, with a review of previous methods', *J. Microsc.* **131** pp. 265–290

6. Exner, H.E. (1972): 'Analysis of grain and particle size distributions in metallic materials', *Int. Metall. Rev.* **17**, p. 25

7. Gerlach, W., J. Ohser (1986): 'On the accuracy of numerical solutions such as the Wicksell corpuscle problem', *Biom. J.* **28**, pp. 881–887

8. Gille, W. (1988): 'The chord length distribution density of parallelepipeds with their limiting cases', *Experimentelle Technik in der Physik* **36**, pp. 197–208

9. Howard, C.V., M.G. Reed (1998): *Unbiased Stereology, Three-dimensional Measurement in Microscopy* (Bios Scientific Publisher)

10. Jakeman, A.J., R.S. Anderssen (1975): 'Abel type integral equations in stereology. I. General discussion', *J. Microsc.* **105**, pp. 121–133

11. Jensen, E.B.V. (1984): 'A design-based proof of Wicksell's integral equation', *J. Microsc.* **136**, pp. 345–348

12. Jensen, E.B.V. (1998): *Local Stereology* (World Scientific, Singapore, New Jersey, Hong Kong)

13. Kanatani, K.I., O. Ishikawa (1985): 'Error analysis for the stereological estimation of sphere size distribution: Abel type integral equation', *J. Comput. Physics* **57**, pp. 229–250

14. Little, R.J.A., D.B. Rubin (1987): *Statistical Analysis with Missing Data* (J. Wiley & Sons, New York)

15. Mase, S. (1995): 'Stereological estimation of particle size distributions', *Adv. Appl. Prob.* **27**, pp. 350–366

16. Mehnert, K., J. Ohser, P. Klimanek (1998): 'Testing stereological methods for the estimation of grain size distributions by means of a computer-simulated polycrystalline sample', *Mat. Sci. Eng.* **A246**, pp. 207–212

17. Nippe, M. (1998): Stereologie für Systeme homothetischer Partikel. Ph.D. thesis, TU Bergakademie Freiberg

18. Nagel, W., J. Ohser (1988): 'On the stereological estimation of weighted sphere diameter distributions', *Acta Stereol.* **7**, pp. 17–31

19. Nippe, M., J. Ohser (1999): 'The stereological unfolding problem for systems of homothetic particles', *Pattern Recogn.* **32**, pp. 1649–1655

20. Ohser, J., F. Mücklich (1995): 'Stereology for some classes of polyhedrons', *Adv. Appl. Prob.* **27**, pp. 384–396

21. Ohser, J., F. Mücklich (2000): *Statistical Analysis of Materials Structures* (J. Wiley & Sons, Chichester)

22. Ohser, J., M. Nippe (1997): 'Stereology of cubic particles: various estimators for the size distribution', *J. Microsc.* **187**, pp. 22–30

23. Press, W.H., S.A. Teukolsky, W.T. Vetterling, B.P. Flannery (1994): *Numerical Recipes in C*, 2nd ed. (Cambridge University Press)

24. Ripley, B.D. (1981): *Spatial Statistics* (J. Wiley & Sons, Chichester)

25. Saltykov, S.A. (1967): 'The determination of the size distribution of particles in an opaque material from a measurement of the size distribution of their sections'. In: *Proceedings of the Second International Congress for Stereology*, ed. by H. Elias

26. Saltykov, S.A. (1974): *Stereometrische Metallographie* (Deutscher Verlag für Grundstoffindustrie, Leipzig)

27. Silverman, B.W., M.C. Jones, D.W. Nychka, J.D. Wilson (1990): 'A smoothed EM approach to indirect estimation problems, with particular reference to stereology and emission tomography', *J. R. Statist. Soc.* **B52**, pp. 271–324

28. Stoyan, D., W.S. Kendall, J. Mecke, (1995): *Stochastic Geometry and its Applications*, 2nd ed. (J. Wiley & Sons, Chichester)
29. Weibel, E.R. (1980): *Stereological Methods*, Vol. 1: *Practical Methods for Biological Morphology*, Vol. 2: *Theoretical Foundations* (Academic Press, London)
30. Wicksell, S.D. (1925): 'The corpuscle problem I', *Biometrica* **17**, pp. 84–89

Appendix

Let the probability density of F_V be f_V. We assume $f_V = 0$ outside the interval $[u_0, d]$, and f_V is Lipschitz with the Lipschitz constant $c > 0$. The interval $[0, d]$ is divided into n equally spaced bins of width Δ, so that $n \Delta = d$. Let b be a piecewise constant approximation of f_V,

$$b(u) = \sum_{i=1}^{n} b_i \, \mathbf{1}_{[(i-1)\Delta, i\Delta)}(u), \qquad u \geq 0,$$

where $\mathbf{1}_A(\cdot)$ is the indicator function of the set A, and b_i are suitable constants as described in Section 3 for F_V^{app}. Section 3 also shows the relation between the relative frequencies (class probabilities) ϑ_i with respect to the distribution function F_V, the relative frequencies y_i with respect to F and furthermore the relation $\vartheta_i = N_V \Delta b_i$, $i = 1, \ldots, n$.

A sample of sizes of section profiles gives an estimate $\widehat{y}$ of y which gives an estimate $\widehat{\vartheta}$ of ϑ where $\widehat{\vartheta} = P^{-1}\widehat{y}$, and $\widehat{\vartheta}$ gives an estimate $\widehat{b}(u)$ of the function $b(u)$.

We are interested to find upper bounds on the relative error. These bounds illustrate the interaction between the statistical error and the discretization error using the $\mathcal{L}^2$-norm $\| \cdot \|$. There is an correspondence between the piecewise constant function b, and the values b_i that form a vector $\underline{b}$ which can be considered with respect to the Euclidean vector norm $|\cdot|$. These cases shall be distinguished here. Clearly, for a function b as defined above, the relationship $\|b\| = \sqrt{\Delta}\,|\underline{b}|$ holds.

We get the following chain of inequalities

$$\frac{\mathbb{E}\|f_V - \widehat{b}\|}{\|f_V\|} \leq \frac{\|f_V - b\|}{\|f_V\|} + \frac{\mathbb{E}\|b - \widehat{b}\|}{\|f_V\|} = \frac{\|f_V - b\|}{\|f_V\|} + \frac{\|b\|}{\|f_V\|}\frac{\mathbb{E}\|b - \widehat{b}\|}{\|b\|}$$

$$= \frac{\|f_V - b\|}{\|f_V\|} + \frac{\|b\|}{\|f_V\|}\frac{\mathbb{E}|\underline{b} - \widehat{\underline{b}}|}{|\underline{b}|}$$

$$\leq \frac{\|f_V - b\|}{\|f_V\|} + \frac{\|b\|}{\|f_V\|}\frac{\mathbb{E}|\vartheta - \widehat{\vartheta}|}{|\vartheta|}$$

$$\leq \frac{\|f_V - b\|}{\|f_V\|} + \left(1 + \frac{\|f_V - b\|}{\|f_V\|}\right)\frac{\mathbb{E}|\vartheta - \widehat{\vartheta}|}{|\vartheta|}.$$

Now the two parts are considered separately.

For the term $I\!\!E|\vartheta - \widehat{\vartheta}|/|\vartheta|$ an upper bound is obtained from (9). To find a global bound of $I\!\!E|y - \widehat{y}|/|y|$, we give bounds of numerator and denominator separately.

A lower bound of the denominator can be obtained in the following way:

$$|y| = \sqrt{\sum_{i=1}^{n} y_i^2} \geq \sqrt{\sum_{i=1}^{n} \left(\frac{\lambda}{n}\right)^2} = \frac{\lambda}{\sqrt{n}}$$

where λ is the density of the section profiles, $\lambda = \sum_{i=1}^{n} y_i$.

An upper bound of the numerator can be get using the extreme case of only one class ($n = 1$). Then it is

$$I\!\!E\,|y - \widehat{y}| \leq I\!\!E\,\sqrt{(\widehat{\lambda} - \lambda)^2} = I\!\!E\,|\widehat{\lambda} - \lambda|.$$

Using Stirling's formula one gets the approximation

$$I\!\!E\,|\widehat{\lambda} - \lambda| \approx \sqrt{\frac{2}{\pi}}\,\frac{\sqrt{w\,\lambda}}{w} = \sqrt{\frac{2\lambda}{\pi\,w}},$$

see Nippe (1998, p. 45). Here, w is the size of the sampling window W (for planar sampling design the area of W, and for linear sampling design the length of W).

From the above relationships we get

$$\frac{I\!\!E\,|y - \widehat{y}|}{|y|} \lesssim \frac{\sqrt{n}}{\lambda}\sqrt{\frac{2\lambda}{\pi\,w}} = \sqrt{\frac{2n}{\pi\,\lambda\,w}}.$$

The second term $\|f_V - b\|/\|f_V\|$ is considered using the Lipschitz condition and the elementary property that there is a $\xi_i \in ((i-1)\Delta, i\Delta)$ with $f_V(\xi_i) = b_i$. Then

$$\|f_V - b\|^2 = \int_0^d [f_V(u) - b(u)]^2 \, du$$

$$= \sum_{i=1}^{n} \int_{(i-1)\Delta}^{i\Delta} [f_V(u) - f_V(\xi_i)]^2 \, du \leq \sum_{i=1}^{n} \int_{(i-1)\Delta}^{i\Delta} c^2[u - \xi_i]^2 \, du$$

$$= \sum_{i=1}^{n} \frac{c^2}{3} \left[(i\Delta - \xi_i)^3 - ((i-1)\Delta - \xi_i)^3\right] \leq \frac{c^2}{3}\Delta^3 n = \frac{c^2}{3} d\,\Delta^2$$

and this results yields

$$\frac{\|f_V - b\|}{\|f_V\|} \leq \sqrt{\frac{d}{3}}\,\frac{c}{\|f_V\|}\,\Delta$$

which is a bound, depending on $\|f_V\|$. For an independent bound we are looking for a lower bound of $\|f_V\|$ which is obviously larger than the norm of the uniform distribution f_V^u on the closed interval $[0, d]$ because the Lipschitz condition must be satisfied. So we have for all probability densities f_V

$$\|f_V\| \geq \|f_V^u\| = \frac{1}{\sqrt{d}}.$$

Bringing all these results together we obtain

$$\frac{\mathbb{E}\,\|f_V - b\|}{\|f_V\|} \leq \frac{\|f_V - b\|}{\|f_V\|} + \left(1 + \frac{\|f_V - b\|}{\|f_V\|}\right) \frac{\mathbb{E}\,|\vartheta - \widehat{\vartheta}|}{|\vartheta|}$$

$$\leq \sqrt{\frac{d}{3}}\,\frac{c}{\|f_V\|}\Delta + \left(1 + \sqrt{\frac{d}{3}}\,\frac{c}{\|f_V\|}\Delta\right) \mathrm{cond}(P)\,\frac{\mathbb{E}\,|y - \widehat{y}|}{|y|}$$

$$\lesssim \frac{d}{\sqrt{3}}\,c\,\Delta + \left(1 + \frac{d}{\sqrt{3}}\,c\,\Delta\right) \mathrm{cond}(P)\,\sqrt{\frac{2n}{\pi\,\lambda\,w}}$$

$$= \frac{d}{\sqrt{3}}\,c\,\Delta + \left(1 + \frac{d}{\sqrt{3}}\,c\,\Delta\right) \mathrm{cond}(P)\,\sqrt{\frac{2d}{\pi\,\lambda\,w\,\Delta}}\,.$$

Local Porosity Theory and Stochastic Reconstruction for Porous Media

Rudolf Hilfer

ICA-1, Universität Stuttgart, 70569 Stuttgart, Germany
Institut für Physik, Universität Mainz, 55099 Mainz, Germany

Abstract. The paper reviews recent developments in local porosity theory, and discusses its application to the analysis of stochastic reconstruction models for sedimentary rocks. Special emphasis is placed on the geometric observables in local porosity theory and their relation with the Hadwiger theorem from stochastic geometry. In addition recent results for the exact calculation of effective physical transport properties are given for a Fontainebleau sandstone. The calculations pertain to potential type problems such as electrical conduction, dielectric relaxation, diffusion or Darcy flow. The exact results are compared to the approximate parameterfree predictions from local porosity, and are found to be in good agreement.

1 Introduction

An important subclass of heterogeneous and disordered systems are porous materials which can be loosely defined as mixtures of solids and fluids [1,20,30,55]. Despite a long history of scientific study the theory of porous media or, more generally, heterogeneous mixtures (including solid-solid and fluid-fluid mixtures) continues to be of central interest for many areas of fundamental and applied research ranging from geophysics [26], hydrology [7,43], petrophysics [36] and civil engineering [19,21] to the materials science of composites [17].

My primary objective in this article is to review briefly the application of local porosity theory, introduced in [27,28,30], to the geometric characterization of porous or heterogeneous media. A functional theorem of Hadwiger [23, p.39] emphasizes the impr   nce of four set-theoretic functionals for the geometric characterization porous media (see also the paper by Mecke in this volume). In contrast herewith local porosity theory has emphasized geometric observables, that are not covered by Hadwigers theorem [25,29,31]. Other theories have stressed the importance of correlation functions [60,63] or contact distributions [38,46,61] for characterization purposes. Recently advances in computer and imaging technology have made threedimensional microtomographic images more readily available. Exact microscopic solutions are thereby becoming possible and have recently been calculated [11,66,68]. Moreover, the availability of threedimensional microstructures allows to test approximate theories and geometric models and to distinguish them quantitatively.

Distinguishing porous microstructures in a quantitative fashion is important for reliable predictions and it requires apt geometric observables. Examples of important geometric observables are porosity and specific internal surface area

[6,20]. It is clear however, that porosity and specific internal surface area alone are not sufficient to distinguish the infinite variety of porous microstructures.

Geometrical models for porous media may be roughly subdivided into the classical capillary tube and slit models [6], grain models [61], network models [15,22], percolation models [16,54], fractal models [34,53], stochastic reconstruction models [1,49] and diagenetic models [4,51]. Little attention is usually paid to match the geometric characteristics of a model geometry to those of the experimental sample, as witnessed by the undiminished popularity of capillary tube models. Usually the matching of geometric observables is limited to the porosity alone. Recently the idea of stochastic reconstruction models has found renewed interest [1,50,70]. In stochastic reconstruction models one tries to match not only the porosity but also other geometric quantities such as specific internal surface, correlation functions, or linear and spherical contact distributions. Similar ideas have been proposed in spatial statistics [61]. As the number of matched quantities increases one expects that also the model approximates better the given sample. My secondary objective in this review will be to compare simple stochastic reconstruction models and physically inspired diagenesis models with the experimental microstructure obtained from computer tomography [11].

2   Problems in the Theory of Porous Media

2.1   Physical Problems

Many physical problems in porous and heterogeneous media can be formulated mathematically as a set of partial differential equations

$$\boldsymbol{F_P}(r,t,\boldsymbol{u},\partial\boldsymbol{u}/\partial t,\dots,\boldsymbol{\nabla}\cdot\boldsymbol{u},\boldsymbol{\nabla}\times\boldsymbol{u},\dots)=0, \qquad r\in\mathbb{P}\subset\mathbb{R}^3, t\in\mathbb{R} \qquad (1a)$$

$$\boldsymbol{F_M}(r,t,\boldsymbol{u},\partial\boldsymbol{u}/\partial t,\dots,\boldsymbol{\nabla}\cdot\boldsymbol{u},\boldsymbol{\nabla}\times\boldsymbol{u},\dots)=0, \qquad r\in\mathbb{M}\subset\mathbb{R}^3, t\in\mathbb{R} \qquad (1b)$$

for a vector of unknown fields $\boldsymbol{u}(r,t)$ as function of position and time coordinates. Here the two-component porous sample $\mathbb{S}=\mathbb{P}\cup\mathbb{M}$ is defined as the union of two closed subsets $\mathbb{P}\subset\mathbb{R}^3$ and $\mathbb{M}\subset\mathbb{R}^3$ where $\mathbb{P}$ denotes the pore space (or component 1 in a heterogeneous medium) and $\mathbb{M}$ denotes the matrix space (or component 2). In (1) the vector functionals $\boldsymbol{F_P}$ and $\boldsymbol{F_M}$ may depend on the vector $\boldsymbol{u}$ of unknowns and its derivatives as well as on position r and time t. A simple example for (1) is the time independent potential problem

$$\boldsymbol{\nabla}\cdot\boldsymbol{j}(r)=0, \qquad r\in\mathbb{S} \qquad (2)$$

$$\boldsymbol{j}(r)+C(r)\boldsymbol{\nabla}u(r)=0, \qquad r\in\mathbb{S} \qquad (3)$$

for a scalar field $u(r)$. The coefficients

$$C(r)=C_\mathbb{P}\chi_{_\mathbb{P}}(r)+C_\mathbb{M}\chi_{_\mathbb{M}}(r) \qquad (4)$$

contain the material constants $C_\mathbb{P}\neq C_\mathbb{M}$. Here the characteristic (or indicator) function $\chi_{_\mathbb{G}}(r)$ of a set $\mathbb{G}$ is defined as

$$\chi_{_\mathbb{G}}(r)=\begin{cases}1 & \text{for } r\in\mathbb{G}\\ 0 & \text{for } r\notin\mathbb{G}.\end{cases} \qquad (5)$$

Hence $C(r)$ is not differentiable at the internal boundary $\partial_P = \partial_M$, and this requires to specify boundary conditions

$$\lim_{s \to r} n \cdot j(r + s) = \lim_{s \to r} n \cdot j(r - s), \qquad r \in \partial \mathbb{P} \qquad (6)$$

$$\lim_{s \to r} n \times \nabla u(r + s) = \lim_{s \to r} n \times \nabla u(r - s), \qquad r \in \partial \mathbb{P} \qquad (7)$$

at the internal boundary. In addition, boundary conditions on the sample boundary $\partial \mathbb{S}$ need to be given to complete the formulation of the problem. Initial conditions may also be required. Several concrete applications can be subsumed under this formulation depending upon the physical interpretation of the field u and the current j. An overview for possible interpretations of u and j is given in Table 2.1. It contains hydrodynamical flow, electrical conduction, heat conduction and diffusion as well as cross effects such as thermoelectric or electrokinetic phenomena.

Table 1. Overview of possible interpretations for the field u and the current j produced by its gradient according to (3).

$j \setminus u$	pressure	el. potential	temperature	concentration
volume	Darcy's law	electroosmosis	thermal osmosis	chemical osmosis
el. charge	streaming pot.	Ohm's law	Seebeck effect	sedim. electricity
heat	thermal filtration	Peltier effect	Fourier's law	Dufour effect
particles	ultrafiltration	electrophoresis	Soret effect	Fick's law

The physical problems in the theory of porous media may be divided into two categories: direct problems and inverse problems. In direct problems one is given partial information about the pore space configuration $\mathbb{P}$. The problem is to deduce information about the solution $u(r, t)$ of the boundary and/or initial value problem that can be compared to experiment. In inverse problems one is given partial information about the solutions $u(r, t)$. Typically this information comes from various experiments or observations of physical processes. The problem is to deduce information about the pore space configuration $\mathbb{P}$ from these data.

Inverse problems are those of greatest practical interest. All attempts to visualize the internal interface or fluid content of nontransparent heterogeneous media lead to inverse problems. Examples occur in computer tomography. Inverse problems are often ill-posed due to lack of data [39,52]. Reliable solution of inverse problems requires a predictive theory for the direct problem.

2.2 Geometrical Problems

The geometrical problems arise because in practice the pore space configuration $\chi_p(r)$ is usually not known in detail. The direct problem, i.e. the solution of a physical boundary value problem, requires detailed knowledge of the internal boundary, and hence of $\chi_p(r)$.

While it is becoming feasible to digitize samples of several mm^3 with a resolution of a few μm this is not possible for larger samples. For this reason the true pore space $\mathbb{P}$ is often replaced by a geometric model $\widetilde{\mathbb{P}}$. One then solves the problem for the model geometry and hopes that its solution $\widetilde{u}$ obeys $\widetilde{u} \approx u$ in some sense. Such an approach requires quantitative methods for the comparison of $\mathbb{P}$ and the model $\widetilde{\mathbb{P}}$. This raises the problem of finding generally applicable quantitative geometric characterization methods that allow to evaluate the accuracy of geometric models for porous microstructues. The problem of quantitative geometric characterization arises also when one asks which geometrical characteristics of the microscctructure $\mathbb{P}$ have the greatest influence on the properties of the solution u of a given boundary value problem.

Some authors introduce more than one geometrical model for one and the same microstructure when calculating different physical properties (e.g. diffusion and conduction). It should be clear that such models make it difficult to extract reliable physical or geometrical information.

3   Geometric Characterizations

3.1   General Considerations

A general geometric characterization of stochastic media should provide macroscopic geometric observables that allow to distinguish media with different microstructures quantitatively. In general, a stochastic medium is defined as a probability distribution on a space of geometries or configurations. Distributions and expectation values of geometric observables are candidates for a general geometric characterization.

A general geometric characterization should fulfill four criteria to be useful in applications. These four criteria were advanced in [30]. First, it must be well defined. This obvious requirement is sometimes violated. The so called "pore size distributions" measured in mercury porosimetry are not geometrical observables in the sense that they cannot be determined from knowledge of the geometry alone. Instead they are capillary pressure curves whose calculation involves physical quantities such as surface tension, viscosity or flooding history [30]. Second, the geometric characterization should be directly accessible in experiments. The experiments should be independent of the quantities to be predicted. Thirdly, the numerical implementation should not require excessive amounts of data. This means that the amount of data should be mangeable by contemporary data processing technology. Finally, a useful geometric characterization should be helpful in the exact or approximate theoretical calculations.

For simplicity only two-component media will be considered throughout this paper, but most concepts can be generalized to media with an arbitrary finite number of components.

3.2   Geometric Observables

Well defined geometric observables are the basis for the geometric characterization of porous media. A perennial problem in all applications is to identify those macroscopic geometric observables that are relevant for distinguishing between classes of microstructures. One is interested in those properties of the microstructure that influence the macroscopic physical behaviour. In general this depends on the details of the physical problem, but some general properties of the microstructure such as volume fraction or porosity are known to be relevant in many situations. Hadwigers theorem [23] is an example of a mathematical result that helps to identify an important class of such general geometric properties of porous media. It will be seen later, however, that there exist important geometric properties that are not members of this class.

A two component porous (or heterogenous) sample $S \subset \mathbb{R}^d$ consists of two closed subsets $\mathbb{P} \subset \mathbb{R}^d$ and $M \subset \mathbb{R}^d$ called pore space $\mathbb{P}$ and matrix M such that $S = \mathbb{P} \cup M$. Its internal boundary is denoted as $\partial \mathbb{P} = \partial M = \mathbb{P} \cap M$. The boundary ∂G of a set is defined as the difference between the closure and the interior of G where the closure is the intersection of all closed sets containing G and the interior is the union of all open sets contained in G. A geometric observable f is a mapping (functional) that assigns to each admissible $\mathbb{P}$ a real number $f(\mathbb{P}) = f(\mathbb{P} \cap S)$ that can be calculated from $\mathbb{P}$ without solving a physical boundary value problem. A functional whose evaluation requires the solution of a physical boundary value problem will be called a physical observable.

Before discussing examples for geometric observables it is necessary to specify the admissible geometries $\mathbb{P}$. The set $\mathcal{R}$ of admissible $\mathbb{P}$ is defined as the set of all finite unions of compact convex sets [23,44,57,58,61] (see also the papers by M. Kerscher and K. Mecke in this volume). Because $\mathcal{R}$ is closed under unions and intersections it is called the convex ring. The choice of $\mathcal{R}$ is convenient for applications because digitized porous media can be considered as elements from $\mathcal{R}$ and because continuous observables defined for convex compact sets can be continued to all of $\mathcal{R}$. The set of all compact and convex subsets of $\mathbb{R}^d$ is denoted as $\mathcal{K}$. For subsequent discussions the Minkowski addition of two sets $A, B \subset \mathbb{R}^d$ is defined as

$$A + B = \{x + y : x \in A, y \in B\}. \tag{8}$$

Multiplication of A with a scalar is defined by $aA = \{ax : x \in A\}$ for $a \in \mathbb{R}$.

Examples of geometric observables are the volume of $\mathbb{P}$ or the surface area of its boundary $\partial \mathbb{P}$. Let

$$V_d(\mathbb{K}) = \int_{\mathbb{R}^d} \chi_{\mathbb{P}}(r) d^d r \tag{9}$$

denote the d-dimensional Lebesgue volume of the compact convex set $\mathbb{K}$. The volume is hence a functional $V_d : \mathcal{K} \to \mathbb{R}$ on $\mathcal{K}$. An example of a compact convex set is the unit ball $\mathbb{B}^d = \{x \in \mathbb{R}^d : |x| \leq 1\} = \mathbb{B}^d(0,1)$ centered at the origin 0 whose volume is

$$\kappa_d = V_d(\mathbb{B}^d) = \frac{\pi^{d/2}}{\Gamma(1 + (d/2))}. \tag{10}$$

Other functionals on $\mathcal{K}$ can be constructed from the volume by virtue of the following fact. For every compact convex $\mathbb{K} \in \mathcal{K}$ and every $\varepsilon \geq 0$ there are numbers $V_j(\mathbb{K}), j = 0, \ldots, d$ depending only on $\mathbb{K}$ such that

$$V_d(\mathbb{K} + \varepsilon \mathbb{B}^d) = \sum_{j=0}^{d} V_j(\mathbb{K}) \varepsilon^{d-j} \kappa_{d-j} \tag{11}$$

is a polynomial in ε. This result is known as Steiners formula [23,61]. The numbers $V_j(\mathbb{K}), j = 0 \ldots, d$ define functionals on $\mathcal{K}$ similar to the volume $V_d(\mathbb{K})$. The quantities

$$W_i(\mathbb{K}) = \frac{\kappa_i V_{d-i}(\mathbb{K})}{\binom{d}{i}} \tag{12}$$

are called quermassintegrals [57]. From (11) one sees that

$$\lim_{\varepsilon \to 0} \frac{1}{\varepsilon}(V_d(\mathbb{K} + \varepsilon \mathbb{B}^d) - V_d(\mathbb{K})) = \kappa_1 V_{d-1}(\mathbb{K}), \tag{13}$$

and from (10) that $\kappa_1 = 2$. Hence $V_{d-1}(\mathbb{K})$ may be viewed as half the surface area. The functional $V_1(\mathbb{K})$ is related to the mean width $w(\mathbb{K})$ defined as the mean value of the distance between a pair of parallel support planes of $\mathbb{K}$. The relation is

$$V_1(\mathbb{K}) = \frac{d\kappa_d}{2\kappa_{d-1}} w(\mathbb{K}) \tag{14}$$

which reduces to $V_1(\mathbb{K}) = w(\mathbb{K})/2$ for $d = 3$. Finally the functional $V_0(\mathbb{K})$ is evaluated from (11) by dividing with ε^d and taking the limit $\varepsilon \to \infty$. It follows that $V_0(\mathbb{K}) = 1$ for all $\mathbb{K} \in \mathcal{K} \setminus \{\emptyset\}$. One extends V_0 to all of $\mathcal{K}$ by defining $V_0(\emptyset) = 0$. The geometric observable V_0 is called Euler characteristic.

The geometric observables V_i have several important properties. They are Euclidean invariant (i.e. invariant under rigid motions), additive and monotone. Let $T_d \cong (\mathbb{R}^d, +)$ denote the group of translations with vector addition as group operation and let $SO(d)$ be the matrix group of rotations in d dimensions [5]. The semidirect product $E_d = T_d \odot SO(d)$ is the Euclidean group of rigid motions in $\mathbb{R}^d$. It is defined as the set of pairs (a, A) with $a \in T_d$ and $A \in SO(d)$ and group operation

$$(a, A) \circ (b, B) = (a + Ab, AB). \tag{15}$$

An observable $f : \mathcal{K} \to \mathbb{R}$ is called euclidean invariant or invariant under rigid motions if

$$f(a + A\mathbb{K}) = f(\mathbb{K}) \tag{16}$$

holds for all $(a, A) \in E_d$ and all $\mathbb{K} \in \mathcal{K}$. Here $A\mathbb{K} = \{Ax : x \in \mathbb{K}\}$ denotes the rotation of $\mathbb{K}$ and $a + \mathbb{K} = \{a\} + \mathbb{K}$ its translation. A geometric observable f is called additive if

$$f(\emptyset) = 0 \tag{17a}$$

$$f(\mathbb{K}_1 \cup \mathbb{K}_2) + f(\mathbb{K}_1 \cap \mathbb{K}_2) = f(\mathbb{K}_1) + f(\mathbb{K}_2) \tag{17b}$$

holds for all $\mathbb{K}_1, \mathbb{K}_2 \in \mathcal{K}$ with $\mathbb{K}_1 \cup \mathbb{K}_2 \in \mathcal{K}$. Finally a functional is called monotone if for $\mathbb{K}_1, \mathbb{K}_2 \in \mathcal{K}$ with $\mathbb{K}_1 \subset \mathbb{K}_2$ follows $f(\mathbb{K}_1) \leq f(\mathbb{K}_2)$.

The special importance of the functionals $V_i(\mathbb{K})$ arises from the following theorem of Hadwiger [23]. A functional $f : \mathcal{X} \to \mathbb{R}$ is euclidean invariant, additive and monotone if and only if it is a linear combination

$$f = \sum_{i=0}^{d} c_i V_i \qquad (18)$$

with nonnegative constants $c_0, \ldots, c_d$. The condition of monotonicity can be replaced with continuity and the theorem remains valid [23]. If f is continuous on $\mathcal{X}$, additive and euclidean invariant it can be additively extended to the convex ring $\mathcal{R}$ [58]. The additive extension is unique and given by the inclusion-exclusion formula

$$f\left(\bigcup_{i=1}^{m}\mathbb{K}_1\right) = \sum_{\mathbb{I}\in\mathcal{P}(m)} (-1)^{|\mathbb{I}|-1} f\left(\bigcap_{i\in\mathbb{I}}\mathbb{K}_i\right) \qquad (19)$$

where $\mathcal{P}(m)$ denotes the family of nonempty subsets of $\{1, \ldots, m\}$ and $|\mathbb{I}|$ is the number of elements of $\mathbb{I} \in \mathcal{P}(m)$. In particular, the functionals V_i have a unique additive extension to the convex ring $\mathcal{R}$ [58], which is again be denoted by V_i.

For a threedimensional porous sample with $\mathbb{P} \in \mathcal{R}$ the extended functionals V_i lead to two frequently used geometric observables. The first is the porosity of a porous sample $\mathbb{S}$ defined as

$$\phi(\mathbb{P}\cap\mathbb{S}) = \phi_3(\mathbb{P}\cap\mathbb{S}) = \frac{V_3(\mathbb{P}\cap\mathbb{S})}{V_3(\mathbb{S})}, \qquad (20)$$

and the second its specific internal surface area which may be defined in view of (13) as

$$\phi_2(\mathbb{P}\cap\mathbb{S}) = \frac{2V_2(\mathbb{P}\cap\mathbb{S})}{V_3(\mathbb{S})}. \qquad (21)$$

The two remaining observables $\phi_1(\mathbb{P}) = V_1(\mathbb{P}\cap\mathbb{S})/V_3(\mathbb{S})$ and $\phi_0(\mathbb{P}) = V_0(\mathbb{P}\cap\mathbb{S})/V_3(\mathbb{S})$ have received less attention in the porous media literature. The Euler characteristic V_0 on $\mathcal{R}$ coincides with the identically named topological invariant. For $d = 2$ and $\mathbb{G} \in \mathcal{R}$ one has $V_0(\mathbb{G}) = c(\mathbb{G}) - c'(\mathbb{G})$ where $c(\mathbb{G})$ is the number of connectedness components of $\mathbb{G}$, and $c'(\mathbb{G})$ denotes the number of holes (i.e. bounded connectedness components of the complement).

3.3   Definition of Stochastic Porous Media

For theoretical purposes the pore space $\mathbb{P}$ is frequently viewed as a random set [30,61]. In practical applications the pore space is usually discretized because of measurement limitations and finite resolution. For the data discussed below the set $\mathbb{S} \subset \mathbb{R}^3$ is a rectangular parallelepiped whose sidelengths are M_1, M_2 and M_3 in units of the lattice constant a (resolution) of a simple cubic lattice. The position vectors $r_i = r_{i_1 \ldots i_d} = (a i_1, \ldots, a i_d)$ with integers $1 \le i_j \le M_j$ are used to label the lattice points, and r_i is a shorthand notation for $r_{i_1...i_d}$. Let V_i denote a cubic volume element (voxel) centered at the lattice site r_i. Then the discretized sample may be represented as $\mathbb{S} = \bigcup_{i=1}^N V_i$. The discretized pore space $\widetilde{\mathbb{P}}$ defined as

$$\widetilde{\mathbb{P}} = \bigcup_{\{i : \chi_{\mathbb{P}}(r_i) = 1\}} V_i. \tag{22}$$

is an approximation to the true pore space $\mathbb{P}$. For simplicity it will be assumed that the discretization does not introduce errors, i.e. that $\widetilde{\mathbb{P}} = \mathbb{P}$, and that each voxel is either fully pore or fully matrix. This assumption may be relaxed to allow voxel attributes such as internal surface or other quermassintegral densities. The discretization into voxels reflects the limitations arising from the experimental resolution of the porous structure. A discretized pore space for a bounded sample belongs to the convex ring $\mathcal{R}$ if the voxels are convex and compact. Hence, for a simple cubic discretization the pore space belongs to the convex ring. A configuration (or microstructure) Z of a 2-component medium may then be represented in the simplest case by a sequence

$$Z = (Z_1, \ldots, Z_N) = (\chi_{\mathbb{P}}(r_1), \ldots, \chi_{\mathbb{P}}(r_N)) \tag{23}$$

where r_i runs through the lattice points and $N = M_1 M_2 M_3$. This representation corresponds to the simplest discretization in which there are only two states for each voxel indicating whether it belongs to pore space or not. In general a voxel could be characterized by more states reflecting the microstructure within the region V_i. In the simplest case there is a one-to-one correspondence between $\mathbb{P}$ and Z given by (23). Geometric observables $f(\mathbb{P})$ then correspond to functions $f(Z) = f(z_1, \ldots, z_N)$.

As a convenient theoretical idealization it is frequently assumed that porous media are random realizations drawn from an underlying statistical ensemble. A discretized stochastic porous medium is defined through the discrete probability density

$$p(z_1, \ldots, z_N) = \text{Prob}\{(Z_1 = z_1) \wedge \ldots \wedge (Z_N = z_N)\} \tag{24}$$

where $z_i \in \{0, 1\}$ in the simplest case. It should be emphasized that the probability density p is mainly of theoretical interest. In practice it is usually not known. An infinitely extended medium or microstructure is called stationary or statistically homogeneous if p is invariant under spatial translations. It is called isotropic if p is invariant under rotations.

3.4   Moment Functions and Correlation Functions

A stochastic medium was defined through its probability distribution p. In practice p will be even less accessible than the microstructure $\mathbb{P} = Z$ itself. Partial information about p can be obtained by measuring or calculating expectation values of a geometric observable f. These are defined as

$$\langle f(z_1, \ldots, z_N) \rangle = \sum_{z_1=0}^{1} \cdots \sum_{z_N=0}^{1} f(z_1, \ldots, z_N) p(z_1, \ldots, z_N) \tag{25}$$

where the summations indicate a summation over all configurations. Consider for example the porosity $\phi(\mathbb{S})$ defined in (20). For a stochastic medium $\phi(\mathbb{S})$ becomes a random variable. Its expectation is

$$\langle \phi \rangle = \frac{\langle V_3(\mathbb{P}) \rangle}{V_3(\mathbb{S})} = \frac{1}{V_3(\mathbb{S})} \int_{\mathbb{S}} \left\langle \chi_{\mathbb{P}}(\boldsymbol{r}) \right\rangle d^3 r$$

$$= \frac{1}{V_3(\mathbb{S})} \sum_{i=1}^{N} \langle z_i \rangle V_3(\mathbb{V}_i) = \frac{1}{N} \sum_{i=1}^{N} \langle z_i \rangle$$

$$= \frac{1}{N} \sum_{i=1}^{N} \mathrm{Prob}\{z_i = 1\} = \frac{1}{N} \sum_{i=1}^{N} \mathrm{Prob}\{\boldsymbol{r}_i \in \mathbb{P}\} \qquad (26)$$

If the medium is statistically homogeneous then

$$\langle \phi \rangle = \mathrm{Prob}\{z_i = 1\} = \mathrm{Prob}\{\boldsymbol{r}_i \in \mathbb{P}\} = \left\langle \chi_{\mathbb{P}}(\boldsymbol{r}_i) \right\rangle \qquad (27)$$

independent of i. It happens frequently that one is given only a single sample, not an ensemble of samples. It is then necessary to invoke an ergodic hypothesis that allows to equate spatial averages with ensemble averages.

The porosity is the first member in a hierarchy of moment functions. The n-th order moment function is defined generally as

$$S_n(\boldsymbol{r}_1, \ldots, \boldsymbol{r}_n) = \left\langle \chi_{\mathbb{P}}(\boldsymbol{r}_1) \cdots \chi_{\mathbb{P}}(\boldsymbol{r}_n) \right\rangle \qquad (28)$$

for $n \leq N$. (If a voxel has other attributes besides being pore or matrix one may define also mixed moment functions $S_{i_1 \ldots i_n}(\boldsymbol{r}_1, \ldots, \boldsymbol{r}_n) = \langle \phi_{i_1}(\boldsymbol{r}_1) \cdots \phi_{i_n}(\boldsymbol{r}_n) \rangle$ where $\phi_i(\boldsymbol{r}_j) = V_i(\mathbb{P} \cap \mathbb{V}_j)/V_i(\mathbb{V}_j)$ for $i = 1, \ldots d$ are the quermassintegral densities for the voxel at site $\boldsymbol{r}_j$.) For stationary media $S_n(\boldsymbol{r}_1, \ldots \boldsymbol{r}_n) = g(\boldsymbol{r}_1 - \boldsymbol{r}_n, \ldots, \boldsymbol{r}_{n-1} - \boldsymbol{r}_n)$ where the function g depends only on $n-1$ variables. Another frequently used expectation value is the correlation function which is related to S_2. For a homogeneous medium it is defined as

$$G(\boldsymbol{r}_0, \boldsymbol{r}) = G(\boldsymbol{r} - \boldsymbol{r}_0) = \frac{\left\langle \chi_{\mathbb{P}}(\boldsymbol{r}_0) \chi_{\mathbb{P}}(\boldsymbol{r}) \right\rangle - \langle \phi \rangle^2}{\langle \phi \rangle (1 - \langle \phi \rangle)} = \frac{S_2(\boldsymbol{r} - \boldsymbol{r}_0) - (S_1(\boldsymbol{r}_0))^2}{S_1(\boldsymbol{r}_0)(1 - S_1(\boldsymbol{r}_0))} \qquad (29)$$

where $\boldsymbol{r}_0$ is an arbitrary reference point, and $\langle \phi \rangle = S_1(\boldsymbol{r}_0)$. If the medium is isotropic then $G(\boldsymbol{r}) = G(|\boldsymbol{r}|) = G(r)$. Note that G is normalized such that $G(0) = 1$ and $G(\infty) = 0$.

The hierarchy of moment functions S_n, similar to p, is mainly of theoretical interest. For a homogeneous medium S_n is a function of $n - 1$ variables. To specify S_n numerically becomes impractical as n increases. If only 100 points are required along each coordinate axis then giving S_n would require $10^{2d(n-1)}$ numbers. For $d = 3$ this implies that already at $n = 3$ it becomes economical to specify the microstructure $\mathbb{P}$ directly rather than incompletely through moment or correlation functions.

3.5   Contact Distributions

An interesting geometric characteristic introduced and discussed in the field of stochastic geometry are contact distributions [18,61, p. 206]. Certain special cases of contact distributions have appeared also in the porous media literature [20]. Let $\mathbb{G}$ be a compact test set containing the origin $\mathbf{0}$. Then the contact distribution is defined as the conditional probability

$$H_{\mathbb{G}}(r) = 1 - \text{Prob}\{\mathbf{0} \notin \mathbb{M} + (-r\mathbb{G})|\mathbf{0} \notin \mathbb{M}\} = 1 - \frac{\text{Prob}\{\mathbb{M} \cap r\mathbb{G} = \emptyset\}}{\phi} \quad (30)$$

If one defines the random variable $R = \inf\{s : \mathbb{M} \cap s\mathbb{G} \neq \emptyset\}$ then $H_{\mathbb{G}}(r) = \text{Prob}\{R \leq r|R > 0\}$ [61].

For the unit ball $\mathbb{G} = \mathbb{B}(0,1)$ in three dimensions $H_{\mathbb{B}}$ is called spherical contact distribution. The quantity $1 - H_{\mathbb{B}}(r)$ is then the distribution function of the random distance from a randomly chosen point in $\mathbb{P}$ to its nearest neighbour in $\mathbb{M}$. The probability density

$$p(r) = \frac{\mathrm{d}}{\mathrm{d}r}(1 - H_{\mathbb{B}}(r)) = -\frac{\mathrm{d}}{\mathrm{d}r}H_{\mathbb{B}}(r) \quad (31)$$

was discussed in [56] as a well defined alternative to the frequently used pore size distrubution from mercury porosimetry.

For an oriented unit interval $\mathbb{G} = \mathbb{B}^1(0,1;e)$ where e is the a unit vector one obtains the linear contact distribution. The linear contact distribution written as $L(re) = \phi(1 - H_{\mathbb{B}^1(0,1;e)}(r))$ is sometimes called lineal path function [70]. It is related to the chord length distribution $p_{cl}(x)$ defined as the probability that an interval in the intersection of $\mathbb{P}$ with a straight line containing $\mathbb{B}^1(0,1;e)$ has length smaller than x [30,61, p. 208].

3.6   Local Porosity Distributions

The idea of local porosity distributions is to measure geometric observables inside compact convex subsets $\mathbb{K} \subset \mathbb{S}$, and to collect the results into empirical histograms [27]. Let $\mathbb{K}(r,L)$ denote a cube of side length L centered at the lattice vector r. The set $\mathbb{K}(r,L)$ is called a measurement cell. A geometric observable f, when measured inside a measurement cell $\mathbb{K}(r,L)$, is denoted as $f(r,L)$ and called a local observable. An example are local Hadwiger functional densities $f = \sum_{i=1}^{d} c_i \psi_i$ with coefficients c_i as in Hadwigers theorem (18). Here the local quermassintegrals are defined using (12) as

$$\psi_i(\mathbb{P} \cap \mathbb{K}(r,L)) = \frac{W_i(\mathbb{P} \cap \mathbb{K}(r,L))}{V_d(\mathbb{K}(r,L))} \quad (32)$$

for $i = 1,\dots,d$. In the following mainly the special case $d = 3$ will be of interest. For $d = 3$ the local porosity is defined by setting $i = 0$,

$$\phi(r,L) = \psi_0(\mathbb{P} \cap \mathbb{K}(r,L)). \quad (33)$$

Local densities of surface area, mean curvature and Euler characteristic may be defined analogously. The local porosity distribution, defined as

$$\mu(\phi; r, L) = \langle \delta(\phi - \phi(r, L)) \rangle, \tag{34}$$

gives the probability density to find a local porosity $\phi(r, L)$ in the measurement cell $\mathbb{K}(r, L)$. Here $\delta(x)$ denotes the Dirac δ-distribution. The support of μ is the unit interval. For noncubic measurement cells $\mathbb{K}$ one defines analogously $\mu(\phi; \mathbb{K}) = \langle \delta(\phi - \phi(\mathbb{K})) \rangle$ where $\phi(\mathbb{K}) = \phi(\mathbb{P} \cap \mathbb{K})$ is the local observable in cell $\mathbb{K}$.

The concept of local porosity distributions (or more generally "local geometry distributions" [28,30]) was introduced in [27] and has been generalized in two directions [30]. Firstly by admitting more than one measurement cell, and secondly by admitting more than one geometric observable. The general n-cell distribution function is defined as [30]

$$\mu_{n;f_1,\ldots,f_m}(f_{11},\ldots,f_{1n};\ldots;f_{n1},\ldots,f_{nm};\mathbb{K}_1,\ldots,\mathbb{K}_n) =$$
$$\langle \delta(f_{11} - f_1(\mathbb{K}_1)) \ldots \delta(f_{1n} - f_1(\mathbb{K}_n)) \ldots \delta(f_{m1} - f_1(\mathbb{K}_1)) \ldots \delta(f_{mn} - f_m(\mathbb{K}_n)) \rangle \tag{35}$$

for n general measurement cells $\mathbb{K}_1,\ldots,\mathbb{K}_n$ and m observables $f_1,\ldots,f_m$. The n-cell distribution is the probability density to find the values f_{11} of the local observable f_1 in cell $\mathbb{K}_1$ and f_{12} in cell $\mathbb{K}_2$ and so on until f_{mn} of local observable f_m in $\mathbb{K}_n$. Definition (35) is a broad generalization of (34). This generalization is not purely academic, but was motivated by problems of fluid flow in porous media where not only ψ_0 but also ψ_1 becomes important [28]. Local quermassintegrals, defined in (32), and their linear combinations (Hadwiger functionals) furnish important examples for local observables in (35), and they have recently been measured [40].

The general n-cell distribution is very general indeed. It even contains p from (24) as the special case $m = 1, f_1 = \phi$ and $n = N$ with $\mathbb{K}_i = \mathbb{V}_i = \mathbb{K}(r_i, a)$. More precisely one has

$$\mu_{N;\phi}(\phi_1,\ldots,\phi_N;\mathbb{V}_1,\ldots,\mathbb{V}_N) = p(\phi_1,\ldots,\phi_N) \tag{36}$$

because in that case $\phi_i = z_i = 1$ if $\mathbb{V}_i \in \mathbb{P}$ and $\phi_i = z_i = 0$ for $\mathbb{V} \notin \mathbb{P}$. In this way it is seen that the very definition of a stochastic geometry is related to local porosity distributions (or more generally local geometry distributions). As a consequence the general n-cell distribution $\mu_{n;f_1,\ldots,f_m}$ is again mainly of theoretical interest, and usually unavailable for practical computations.

Expectation values with respect to p have generalizations to averages with respect to μ. Averaging with respect to μ will be denoted by an overline. In the special case $m = 1$, $f_1 = \phi$ and $\mathbb{K}_i = \mathbb{V}_i = \mathbb{K}(r_i, a)$ with $n < N$ one finds [30]

$$\overline{\phi(r_1, a) \cdots \phi(r_n, a)}$$

$$= \int_0^1 \cdots \int_0^1 \phi_1 \cdots \phi_n \mu_{n;\phi}(\phi_1, \ldots, \phi_n; \mathbb{V}_1, \ldots, \mathbb{V}_n) \mathrm{d}\phi_1 \cdots \mathrm{d}\phi_n$$

$$= \int_0^1 \cdots \int_0^1 \phi_1 \cdots \phi_n \mu_{N;\phi}(\phi_1, \ldots, \phi_N; \mathbb{V}_1, \ldots, \mathbb{V}_N) \mathrm{d}\phi_1 \cdots \mathrm{d}\phi_N$$

$$= \int_0^1 \cdots \int_0^1 \phi_1 \cdots \phi_n \langle \delta(\phi_1 - \phi(r_1, a)) \cdots \delta(\phi_N - \phi(r_N, a)) \rangle \, \mathrm{d}\phi_1 \cdots \mathrm{d}\phi_N$$

$$= \langle \phi(r_1, a) \cdots \phi(r_n, a) \rangle$$

$$= \left\langle \chi_{\mathbb{P}}(r_1) \ldots \chi_{\mathbb{P}}(r_n) \right\rangle$$

$$= S_n(r_1, \ldots, r_n) \tag{37}$$

thereby identifying the moment functions of order n as averages with respect to an n-cell distribution.

For practical applications the 1-cell local porosity distributions $\mu(r, L)$ and their analogues for other quermassintegrals are of greatest interest. For a homogeneous medium the local porosity distribution obeys

$$\mu(\phi; r, L) = \mu(\phi; 0, L) = \mu(\phi; L) \tag{38}$$

for all lattice vectors r, i.e. it is independent of the placement of the measurement cell. A disordered medium with substitutional disorder [71] may be viewed as a stochastic geometry obtained by placing random elements at the cells or sites of a fixed regular substitution lattice. For a substitutionally disordered medium the local porosity distribution $\mu(r, L)$ is a periodic function of r whose period is the lattice constant of the substitution lattice. For stereological issues in the measurement of μ from thin sections see [64].

Averages with respect to μ are denoted by an overline. For a homogeneous medium the average local porosity is found as

$$\overline{\phi}(r, L) = \int_0^1 \mu(\phi; r, L) \mathrm{d}\phi = \langle \phi \rangle = \overline{\phi} \tag{39}$$

independent of r and L. The variance of local porosities for a homogeneous medium defined in the first equality

$$\sigma^2(L) = \overline{(\phi(L) - \overline{\phi})^2} = \int_0^1 (\phi(L) - \overline{\phi})^2 \mu(\phi; L) d\phi$$

$$= \frac{1}{L^3} \langle \phi \rangle (1 - \langle \phi \rangle) \left(1 + \frac{2}{L^3} \sum_{\substack{r_i, r_j \in K(r_0, L) \\ i \neq j}} G(r_i - r_j) \right) \tag{40}$$

is related to the correlation function as given in the second equality [30]. The skewness of the local porosity distribution is defined as the average

$$\kappa_3(L) = \frac{\overline{(\phi(L) - \overline{\phi})^3}}{\sigma(L)^3} \tag{41}$$

The limits $L \to 0$ and $L \to \infty$ of small resp. large measurement cells are of special interest. In the first case one reaches the limiting resolution at $L = a$ and finds for a homogeneous medium [27,30]

$$\mu(\phi; a) = \overline{\phi}\delta(\phi - 1) - (1 - \overline{\phi})\delta(\phi). \tag{42}$$

The limit $L \to \infty$ is more intricate because it requires also the limit $\mathbb{S} \to \mathbb{R}^3$. For a homogeneous medium (40) shows $\sigma(L) \to 0$ for $L \to 0$ and this suggests

$$\mu(\phi, L \to \infty) = \delta(\phi - \overline{\phi}). \tag{43}$$

For macroscopically heterogeneous media, however, the limiting distribution may deviate from this result [30]. If (43) holds then in both limits the geometrical information contained in μ reduces to the single number $\overline{\phi} = \langle \phi \rangle$. If (42) and (43) hold there exists a special length scale L^* defined as

$$L^* = \min\{L : \mu(0; L) = \mu(1; L) = 0\} \tag{44}$$

at which the δ-components at $\phi = 0$ and $\phi = 1$ vanish. In the examples below the length L^* is a measure for the size of pores.

The ensemble picture underlying the definition of a stochastic medium is an idealization. In practice one is given only a single realization and has to resort to an ergodic hypothesis for obtaining an estimate of the local porosity distributions. In the examples below the local porosity distribution is estimated by

$$\widetilde{\mu}(\phi; L) = \frac{1}{m} \sum_r \delta(\phi - \phi(r, L)) \tag{45}$$

where m is the number of placements of the measurement cell $\mathbb{K}(r, L)$. Ideally the measurement cells should be far apart or at least nonoverlapping, but in practice this restriction cannot be observed because the samples are not large enough. In the results presented below $\mathbb{K}(r, L)$ is placed on all lattice sites which are at least a distance $L/2$ from the boundary of $\mathbb{S}$. This allows for

$$m = \prod_{i=1}^{3}(M_i - L + 1) \tag{46}$$

placements of $\mathbb{K}(r, L)$ in a sample with side lengths M_1, M_2, M_3. The use of $\widetilde{\mu}$ instead of μ can lead to deviations due to violations of the ergodic hypothesis or simply due to oversampling the central regions of $\mathbb{S}$ [10,11].

3.7   Local Percolation Probabilities

Transport and propagation in porous media are controlled by the connectivity of the pore space. Local percolation probabilities characterize the connectivity [27]. Their calculation requires a threedimensional pore space representation, and early results were restricted to samples reconstructed laboriously from sequential thin sectioning [32]

Consider the functional $\Lambda : \mathcal{K} \times \mathcal{K} \times \mathcal{R} \to \mathbb{Z}_2 = \{0, 1\}$ defined by

$$\Lambda(\mathbb{K}_0, \mathbb{K}_\infty; \mathbb{P} \cap \mathbb{S}) = \begin{cases} 1: & \text{if } \mathbb{K}_0 \rightsquigarrow \mathbb{K}_\infty \text{ in } \mathbb{P} \\ 0: & \text{otherwise} \end{cases} \tag{47}$$

where $\mathbb{K}_0 \subset \mathbb{R}^3, \mathbb{K}_\infty \subset \mathbb{R}^3$ are two compact convex sets with $\mathbb{K}_0 \cap (\mathbb{P} \cap \mathbb{S}) \neq \emptyset$ and $\mathbb{K}_\infty \cap (\mathbb{P} \cap \mathbb{S}) \neq \emptyset$, and "$\mathbb{K}_0 \rightsquigarrow \mathbb{K}_\infty$ in $\mathbb{P}$" means that there is a path connecting $\mathbb{K}_0$ and $\mathbb{K}_\infty$ that lies completely in $\mathbb{P}$. In the examples below the sets $\mathbb{K}_0$ and $\mathbb{K}_\infty$ correspond to opposite faces of the sample, but in general other choices are allowed. Analogous to Λ defined for the whole sample one defines for a measurement cell

$$\Lambda_\alpha(r, L) = \Lambda(\mathbb{K}_{0\alpha}, \mathbb{K}_{\infty\alpha}; \mathbb{P} \cap \mathbb{K}(r, L)) = \begin{cases} 1: & \text{if } \mathbb{K}_{0\alpha} \rightsquigarrow \mathbb{K}_{\infty\alpha} \text{ in } \mathbb{P} \\ 0: & \text{otherwise} \end{cases} \tag{48}$$

where $\alpha = x, y, z$ and $\mathbb{K}_{0x}, \mathbb{K}_{\infty x}$ denote those two faces of $\mathbb{K}(r, L)$ that are normal to the x direction. Similarly $\mathbb{K}_{0y}, \mathbb{K}_{\infty y}, \mathbb{K}_{0z} \mathbb{K}_{\infty z}$ denote the faces of $\mathbb{K}(r, L)$ normal to the y- and z-directions. Two additional percolation observables Λ_3 and Λ_c are introduced by

$$\Lambda_3(r, L) = \Lambda_x(r, L)\Lambda_y(r, L)\Lambda_z(r, L) \tag{49}$$

$$\Lambda_c(r, L) = \text{sgn}(\Lambda_x(r, L) + \Lambda_y(r, L) + \Lambda_z(r, L)). \tag{50}$$

Λ_3 indicates that the cell is percolating in all three directions while Λ_c indicates percolation in x- or y- z-direction. The local percolation probabilities are defined as

$$\lambda_\alpha(\phi; L) = \frac{\sum_r \Lambda_\alpha(r, L)\delta_{\phi, \phi(r, L)}}{\sum_r \delta_{\phi, \phi(r, L)}} \tag{51}$$

where

$$\delta_{\phi,\phi(r,L)} = \begin{cases} 1 : & \text{if } \phi = \phi(r, L) \\ 0 : & \text{otherwise.} \end{cases} \tag{52}$$

The local percolation probability $\lambda_\alpha(\phi; L)$ gives the fraction of measurement cells of sidelength L with local porosity ϕ that are percolating in the "α"-direction. The total fraction of cells percolating along the "α"-direction is then obtained by integration

$$p_\alpha(L) = \int_0^1 \mu(\phi; L)\lambda_\alpha(\phi; L)d\phi. \tag{53}$$

This geometric observable is a quantitative measure for the number of elements that have to be percolating if the pore space geometry is approximated by a substitutionally disordered lattice or network model. Note that neither Λ nor Λ_α are additive functionals, and hence local percolation probabilities have nothing to do with Hadwigers theorem.

It is interesting that there is a relation between the local percolation probabilities and the local Euler characteristic $V_0(\mathbb{P} \cap \mathbb{K}(r, l))$. The relation arises from the observation that the voxels $\mathbb{V}_i$ are closed, convex sets, and hence for any two voxels $\mathbb{V}_i, \mathbb{V}_j$ the Euler characteristic of their intersection

$$V_0(\mathbb{V}_i \cap \mathbb{V}_j) = \begin{cases} 1 : & \text{if } \mathbb{V}_i \cap \mathbb{V}_j \neq \emptyset \\ 0 : & \text{if } \mathbb{V}_i \cap \mathbb{V}_j = \emptyset \end{cases} \tag{54}$$

indicates whether two voxels are nearest neighbours. A measurement cell $\mathbb{K}(r, L)$ contains L^3 voxels. It is then possible to construct a $(L^3 + 2) \times (L^3 + 2)^2$-matrix B with matrix elements

$$(B)_{i\ (i,j)} = V_0(\mathbb{V}_i \cap \mathbb{V}_j) \tag{55}$$
$$(B)_{i\ (j,i)} = -V_0(\mathbb{V}_i \cap \mathbb{V}_j) \tag{56}$$

where $i, j \in \{0, 1, \ldots, L^3, \infty\}$ and the sets $\mathbb{V}_0 = \mathbb{K}_0$ and $\mathbb{V}_\infty = \mathbb{K}_\infty$ are two opposite faces of the measurement cell. The rows in the matrix B correspond to voxels while the columns correspond to voxel pairs. Define the matrix $A = BB^T$ where B^T is the transpose of B. The diagonal elements $(A)_{ii}$ give the number of voxels to which the voxel $\mathbb{V}_i$ is connected. A matrix element $(A)_{ij}$ differs from zero if and only if $\mathbb{V}_i$ and $\mathbb{V}_j$ are connected. Hence the matrix A reflects the local connectedness of the pore space around a single voxel. Sufficiently high powers of A provide information about the global connectedness of $\mathbb{P}$. One finds

$$\Lambda(\mathbb{K}_0, \mathbb{K}_\infty; \mathbb{P} \cap \mathbb{K}(r, L)) = \text{sgn}\left(|(A^m)_{0\infty}|\right) \tag{57}$$

where $(A^m)_{0\infty}$ is the matrix element in the upper right hand corner and m is arbitrary subject to the condition $m > L^3$. The set $\mathbb{P} \cap \mathbb{K}(r, L)$ can always be decomposed uniquely into pairwise disjoint connectedness components (clusters)

$\mathbb{B}_i$ whose number is given by the rank of B. Hence

$$V_0(\mathbb{P} \cap \mathbb{K}(r, L)) = \sum_{i=1}^{\text{rank}B} V_0(\mathbb{B}_i) \tag{58}$$

provides an indirect connection between the local Euler characteristic and the local percolation probabilities mediated by the matrix B. (For percolation systems it has been conjectured that the zero of the Euler characteristic as a function of the occupation probability is an approximation to the percolation threshold [45].)

4 Stochastic Reconstruction of Microstructures

4.1 Description of Experimental Sample

The experimental sample, denoted as $\mathbb{S}_{\text{EX}}$, is a threedimensional microtomographic image of Fontainebleau sandstone. This sandstone is a popular reference standard because of its chemical, crystallographic and microstructural simplicity [13,14]. Fontainebleau sandstone consists of monocrystalline quartz grains that have been eroded for long periods before being deposited in dunes along the sea shore during the Oligocene, roughly 30 million years ago. It is well sorted containing grains of around 200μm in diameter. The sand was cemented by silica crystallizing around the grains. Fontainebleau sandstone exhibits intergranular porosity ranging from 0.03 to roughly 0.3 [13].

Table 2. Overview of geometric properties of the four microstructures displayed in Figs. 1 through 4

Properties	$\mathbb{S}_{\text{EX}}$	$\mathbb{S}_{\text{DM}}$	$\mathbb{S}_{\text{GF}}$	$\mathbb{S}_{\text{SA}}$
M_1	300	255	256	256
M_2	300	255	256	256
M_3	299	255	256	256
$\phi(\mathbb{P} \cap \mathbb{S})$	0.1355	0.1356	0.1421	0.1354
$\phi_2(\mathbb{P} \cap \mathbb{S})$	10.4mm^{-1}	10.9mm^{-1}	16.7mm-1	11.06mm^{-1}
L^*	35	25	23	27
$1 - \lambda_c(0.1355, L^*)$	0.0045	0.0239	0.3368	0.3527

The computer assisted microtomography was carried out on a micro-plug drilled from a larger original core. The original core from which the micro-plug was taken had a measured porosity of 0.1484, permability of $1.3D$ and formation factor 22.1. The porosity $\phi(\mathbb{S}_{\text{EX}})$ of the microtomographic data set is only 0.1355(see Table 2). The difference between the porosity of the original core and that of the final data set is due to the heterogeneity of the sandstone and to the difference in sample size. The experimental sample is referred to as EX in the following. The pore space of the experimental sample is visualized in Fig. 1.

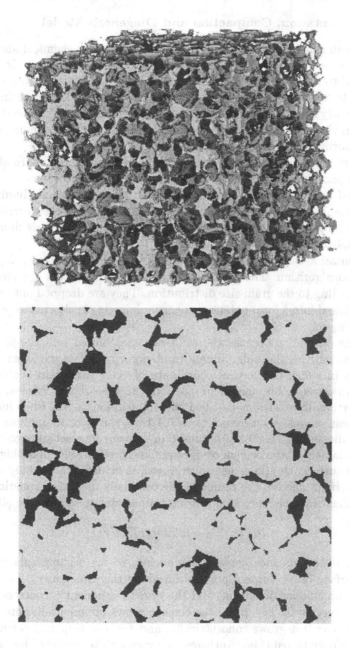

Fig. 1. Sample EX: Threedimensional pore space $\mathbb{P}_{EX}$ of Fontainebleau sandstone. The resolution of the image is $a = 7.5\mu$m, the sample dimensions are $M_1 = 300$, $M_2 = 300$, $M_3 = 299$. The pore space is indicated opaque, the matrix space is transparent. The lower image shows the front plane of the sample as a twodimensional thin section (pore space black, matrix grey).

4.2   Sedimentation, Compaction and Diagenesis Model

Fontainebleau sandstone is the result of complex physical, chemical and geological processes known as sedimentation, compaction and diagenesis. It is therefore natural to model these processes directly rather than trying to match general geometrical characteristics. This conclusion was also obtained from local porosity theory for the cementation index in Archie's law [27]. The diagenesis model abbreviated as DM in the following, attempts model the main geological sandstone-forming processes [4,48].

In a first step porosity, grain size distribution, a visual estimate of the degree of compaction, the amount of quartz cement and clay contents and texture are obtained by image analysis of backscattered electron/cathodo-luminescence images made from thin sections. The sandstone modeling is then carried out in three main steps: grain sedimentation, compaction and diagenesis described in detail in [4,48].

Sedimentation begins by measuring the grain size distribution using an erosion-dilation algorithm. Then spheres with random diameters are picked randomly according to the grain size distribution. They are dropped onto the grain bed and relaxed into a local potential energy minimum or, alternatively, into the global minimum.

Compaction occurs because the sand becomes buried into the subsurface. Compaction reduces the bulk volume (and porosity). It is modelled as a linear process in which the vertical coordinate of every sandgrain is shifted vertically downwards by an amount proportional to the original vertical position. The proportionality constant is called the compaction factor. Its value for the Fontainebleau sample is estimated to be 0.1 from thin section analysis.

In the diagenesis part only a subset of known diagenetical processes are simulated, namely quartz cement overgrowth and precipitation of authigenic clay on the free surface. Quartz cement overgrowth is modeled by radially enlarging each grain. If R_0 denotes the radius of the originally deposited spherical grain, its new radius along the direction r from grain center is taken to be [48,59]

$$R(r) = R_0 + \min(b\ell(r)^\gamma, \ell(r)) \tag{59}$$

where $\ell(r)$ is the distance between the surface of the original spherical grain and the surface of its Voronoi polyhedron along the direction r. The constant b controls the amount of cement, and the growth exponent γ controls the type of cement overgrowth. For $\gamma > 0$ the cement grows preferentially into the pore bodies, for $\gamma = 0$ it grows concentrically, and for $\gamma < 0$ quartz cement grows towards the pore throats [48]. Authigenic clay growth is simulated by precipitating clay voxels on the free mineral surface. The clay texture may be pore-lining or pore-filling or a combination of the two.

The parameters for modeling the Fontainebleau sandstone were 0.1 for the compaction factor, and $\gamma = -0.6$ and $b = 2.9157$ for the cementation parameters. The resulting model configuration of the sample DM is displayed in Fig. 2.

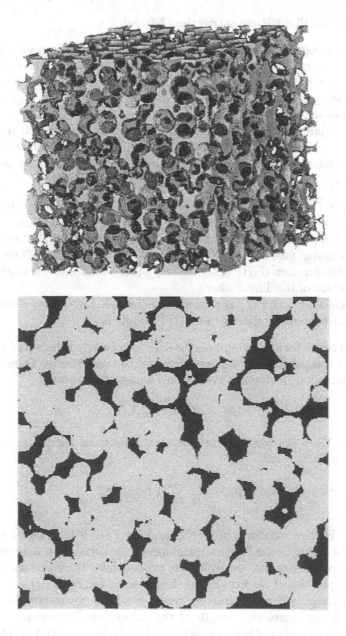

Fig. 2. Sample DM: Threedimensional pore space $\mathbb{P}_{DM}$ of the sedimentation and diagenesis model described in the text. The resolution is $a = 7.5\mu m$, the sample dimensions are $M_1 = 255$, $M_2 = 255$, $M_3 = 255$. The pore space is indicated opaque, the matrix space is transparent. The lower image shows the front plane of the sample as a twodimensional thin section (pore space black, matrix grey)

4.3   Gaussian Field Reconstruction Model

A stochastic reconstruction model attempts to approximate a given experimental sample by a randomly generated model structure that matches prescribed stochastic properties of the experimental sample. In this and the next section the stochastic property of interest is the correlation function $G_{\mathsf{EX}}(r)$ of the Fontainebleau sandstone.

The Gaussian field (GF) reconstruction model tries to match a reference correlation function by filtering Gaussian random variables [1,2,49,69]. Given the reference correlation function $G_{\mathsf{EX}}(r)$ and porosity $\phi(\mathsf{S}_{\mathsf{EX}})$ of the experimental sample the Gaussian field method proceeds in three main steps:

1. Initially a Gaussian field $X(r)$ is generated consisting of statistically independent Gaussian random variables $X \in \mathbb{R}$ at each lattice point r.
2. The field $X(r)$ is first passed through a linear filter which produces a correlated Gaussian field $Y(r)$ with zero mean and unit variance. The reference correlation function $G_{\mathsf{EX}}(r)$ and porosity $\phi(\mathsf{S}_{\mathsf{EX}})$ enter into the mathematical construction of this linear filter.
3. The correlated field $Y(r)$ is then passed through a nonlinear discretization filter which produces the reconstructed sample S_{GF}.

Step 2 is costly because it requires the solution of a very large set of non-linear equations. A computationally more efficient method uses Fourier Transformation [1]. The linear filter in step 2 is defined in Fourier space through

$$Y(k) = \alpha(G_Y(k))^{\frac{1}{2}}X(k), \tag{60}$$

where $M = M_1 = M_2 = M_3$ is the sidelength of a cubic sample, $\alpha = M^{\frac{d}{2}}$ is a normalisation factor, and

$$X(k) = \frac{1}{M^d}\sum_r X(r)e^{2\pi i k \cdot r} \tag{61}$$

denotes the Fourier transform of $X(r)$. Similarly $Y(k)$ is the Fourier transform of $Y(r)$, and $G_Y(k)$ is the Fourier transform of the correlation function $G_Y(r)$. $G_Y(r)$ has to be computed by an inverse process from the correlation function $G_{\mathsf{EX}}(r)$ and porosity of the experimental reference (details in [1]).

The Gaussian field reconstruction requires a large separation $\xi_{\mathsf{EX}} \ll N^{1/d}$ where ξ_{EX} is the correlation length of the experimental reference, and $N = M_1 M_2 M_3$ is the number of sites. ξ_{EX} is defined as the length such that $G_{\mathsf{EX}}(r) \approx 0$ for $r > \xi_{\mathsf{EX}}$. If the condition $\xi_{\mathsf{EX}} \ll N^{1/d}$ is violated then step 2 of the reconstruction fails in the sense that the correlated Gaussian field $Y(r)$ does not have zero mean and unit variance. In such a situation the filter $G_Y(k)$ will differ from the Fourier transform of the correlation function of the $Y(r)$. It is also difficult to calculate $G_Y(r)$ accurately near $r = 0$ [1]. This leads to a discrepancy at small r between $G_{\mathsf{GF}}(r)$ and $G_{\mathsf{EX}}(r)$. The problem can be overcome by choosing large M. However, in $d = 3$ very large M also demands prohibitively large memory. In

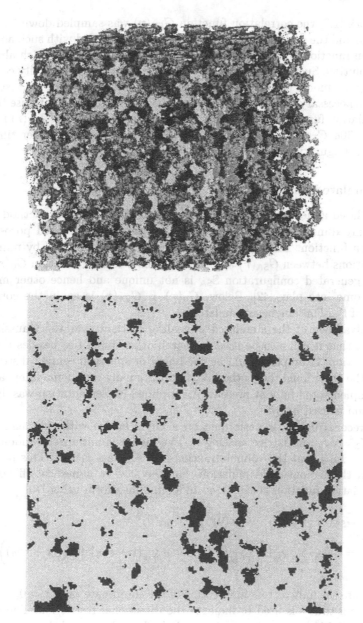

Fig. 3. Sample GF: Threedimensional pore space $\mathbb{P}_{GF}$ with $G_{GF}(r) \approx G_{EX}(r)$ constructed by filtering Gaussian random fields. The resolution is $a = 7.5\mu m$, the sample dimensions are $M_1 = 256$, $M_2 = 256$, $M_3 = 256$. The pore space is indicated opaque, the matrix space is transparent. The lower image shows the front plane of the sample as a twodimensional thin section (pore space black, matrix grey)

earlier work [1,2] the correlation function $G_{EX}(r)$ was sampled down to a lower resolution, and the reconstruction algorithm then proceeded with such a rescaled correlation function. This leads to a reconstructed sample S_{GF} which also has a lower resolution. Such reconstructions have lower average connectivity compared to the original model [9]. For a quantitative comparison with the microstructure of S_{EX} it is necessary to retain the original level of resolution, and to use the original correlation function $G_{EX}(r)$ without subsampling. Because $G_{EX}(r)$ is nearly 0 for $r > 30a$ $G_{EX}(r)$ was truncated at $r = 30a$ to save computer time. The resulting configuration S_{GF} with $M = 256$ is displayed in Fig. 3.

4.4   Simulated Annealing Reconstruction Model

The simulated annealing (SA) reconstruction model is a second method to generate a threedimensional random microstructure with prescribed porosity and correlation function. The method generates a configuration S_{SA} by minimizing the deviations between $G_{SA}(r)$ and a predefined reference function $G_0(r)$. Note that the generated configuration S_{SA} is not unique and hence other modeling aspects come into play [42]. Below, $G_0(r) = G_{EX}(r)$ is again the correlation function of the Fontainebleau sandstone.

An advantage of the simulated annealing method over the Gaussian field method is that it can also be used to match other quantities besides the correlation function. Examples would be the linear or spherical contact distributions [42]. On the other hand the method is computationally very demanding, and cannot be implemented fully at present. A simplified implementation was discussed in [70], and is used below.

The reconstruction is performed on a cubic lattice with side length $M = M_1 = M_2 = M_3$ and lattice spacing a. The lattice is initialized randomly with 0's and 1's such that the volume fraction of 0's equals $\phi(S_{EX})$. This porosity is preserved throughout the simulation. For the sake of numerical efficiency the autocorrelation function is evaluated in a simplified form using [70]

$$\widetilde{G}_{SA}(r)\left(\widetilde{G}_{SA}(0) - \widetilde{G}_{SA}(0)^2\right) + \widetilde{G}_{SA}(0)^2 =$$

$$= \frac{1}{3M^3} \sum_r \chi_M(r)\left(\chi_M(r + re_1) + \chi_M(r + re_2) + \chi_M(r + re_3)\right) \quad (62)$$

where e_i are the unit vectors in direction of the coordinate axes, $r = 0, \ldots, \frac{M}{2} - 1$, and where a tilde $\tilde{}$ is used to indicate the directional restriction. The sum $\sum_r$ runs over all M^3 lattice sites r with periodic boundary conditions, i.e. $r_i + r$ is evaluated modulo M.

A simulated annealing algorithm is used to minimize the "energy" function

$$E = \sum_r \left(\widetilde{G}_{SA}(r) - G_{EX}(r)\right)^2, \quad (63)$$

defined as the sum of the squared deviations of $\widetilde{G}_{SA}$ from the experimental correlation function G_{EX}. Each update starts with the exchange of two pixels, one

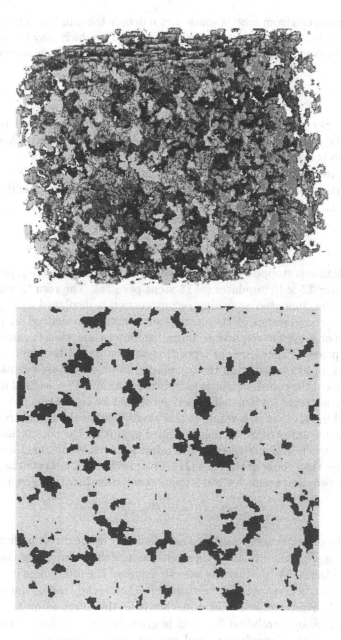

Fig. 4. Sample SA: Threedimensional pore space $\mathbb{P}_{SA}$ with $G_{SA}(r) = G_{EX}(r)$ constructed using a simulated annealing algorithm. The resolution is $a = 7.5\mu m$, the sample dimensions are $M_1 = 256$, $M_2 = 256$, $M_3 = 256$. The pore space is indicated opaque, the matrix space is transparent. The lower image shows the front plane of the sample as a twodimensional thin section (pore space black, matrix grey)

from pore space, one from matrix space. Let n denote the number of the proposed update step. Introducing an acceptance parameter T_n, which may be interpreted as an n-dependent temperature, the proposed configuration is accepted with probability

$$p = \min \left(1, \exp \left(-\frac{E_n - E_{n-1}}{T_n E_{n-1}} \right) \right). \tag{64}$$

Here the energy and the correlation function of the configuration is denoted as E_n and $\widetilde{G}_{SA,n}$, respectively. If the proposed move is rejected, then the old configuration is restored.

A configuration with correlation G_{EX} is found by lowering T. At low T the system approaches a configuration that minimizes the energy function. In the simulations T_n was lowered with n as

$$T_n = \exp \left(-\frac{n}{100000} \right). \tag{65}$$

The simulation was stopped when 20000 consecutive updates were rejected. This happened after 2.5×10^8 updates (≈ 15 steps per site). The resulting configuration $\mathbb{S}_{SA}$ for the simulated annealing reconstruction is displayed in Fig. 4.

A complete evaluation of the correlation function as defined in (29) for a threedimensional system requires so much computer time, that it cannot be carried out at present. Therefore the algorithm was simplified to increase its speed [70]. In the simplified algorithm the correlation function is only evaluated along the directions of the coordinate axes as indicated in (62). The original motivation was that for isotropic systems all directions should be equivalent [70]. However, it was found in [41] that as a result of this simplification the reconstructed sample may become anisotropic. In the simplified algorithm the correlation function of the reconstruction deviates from the reference correlation function in all directions other than those of the axes [41]. The problem is illustrated in Figs. 5(a) and 5(b) in two dimensions for a reference correlation function given as

$$G_0(r) = e^{-r/8} \cos r. \tag{66}$$

In Fig. 5a the correlation function was matched only in the direction of the x- and y-axis. In Fig. 5b the correlation function was matched also along the diagonal directions obtained by rotating the axes 45 degrees. The differences in isotropy of the two reconstructions are clearly visible. In the special case of the correlation function of the Fontainebleau sandstone, however, this effect seems to be smaller. The Fontainebleau correlation function is given in Fig. 7 below. Figure 6a and 6b show the result of twodimensional reconstructions along the axes only and along axes plus diagonal directions. The differences in isotropy seem to be less pronounced. Perhaps this is due to the fact that the Fontainebleau correlation function has no maxima and minima.

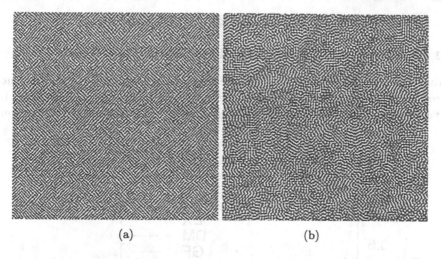

Fig. 5. Twodimensional stochastic reconstruction for the correlation function of $G_0(r) = e^{-r/8} \cos r$ (a) for the direction of the x- and y-coordinate axes only, and (b) for the directions of the coordinate axes plus diagonal directions.

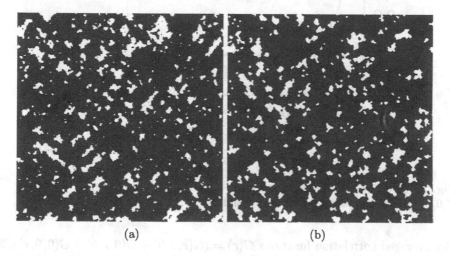

Fig. 6. A Twodimensional stochastic reconstruction for the correlation function $G_0(r) = G_{EX}(r)$ displayed as the solid line in Fig. 7a along the direction of the x- and y-coordinate axes only, and Fig. 7b along the directions of the coordinate axes plus diagonal directions.

5   Quantitative Comparison of Microstructures

5.1   Conventional Observables and Correlation Functions

Table 2 gives an overview of several geometric properties for the four microstructures discussed in the previous section. Samples GF and SA were constructed to have the same correlation function as sample EX. Figure 7 shows the direction-

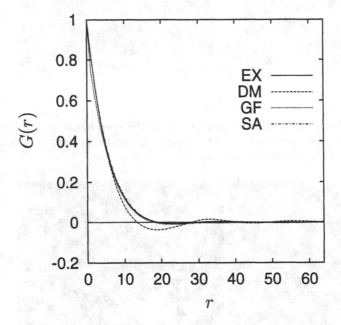

Fig. 7. Directionally averaged correlation functions $G(r) = (G(r,0,0) + G(0,r,0) + G(0,0,r))/3$ of the samples EX,DM,GF and SA

ally averaged correlation functions $G(r) = (G(r,0,0) + G(0,r,0) + G(0,0,r))/3$ of all four microstructures where the notation $G(r_1, r_2, r_3) = G(r)$ was used.

The Gaussian field reconstruction $G_{GF}(r)$ is not perfect and differs from $G_{EX}(r)$ for small r. The discrepancy at small r reflects the quality of the linear filter, and it is also responsible for the differences of the porosity and specific internal surface. Also, by construction, $G_{GF}(r)$ is not expected to equal $G_{EX}(r)$ for r larger than 30. Although the reconstruction method of sample $\mathbb{S}_{SA}$ is intrinsically anisotropic the correlation function of sample SA agrees also in the diagonal directions with that of sample EX. Sample $\mathbb{S}_{DM}$ while matching the porosity and grain size distribution was not constructed to match also the correlation function. As a consequence $G_{DM}(r)$ differs clearly from the rest. It reflects the grain structure of the model by becoming negative. $G_{DM}(r)$ is also anisotropic.

If two samples have the same correlation function they are expected to have also the same specific internal surface as calculated from

$$S = -4\langle\phi\rangle(1 - \langle\phi\rangle)\frac{dG(r)}{dr}\bigg|_{r=0}.\tag{67}$$

The specific internal surface area calculated from this formula is given in Table 2 for all four microstructures.

If one defines a decay length by the first zero of the correlation function then the decay length is roughly $18a$ for samples EX, GF and SA. For sample DM it is somewhat smaller mainly in the x- and y-direction. The correlation length, which will be of the order of the decay length, is thus relatively large compared to the system size. Combined with the fact that the percolation threshold for continuum systems is typically around 0.15 this might explain why models GF and SA are connected in spite of their low value of the porosity.

In summary, the samples S_{GF} and S_{SA} were constructed to be indistinguishable with respect to porosity and correlations from S_{EX}. Sample SA comes close to this goal. The imperfection of the reconstruction method for sample GF accounts for the deviations of its correlation function at small r from that of sample EX. Although the difference in porosity and specific surface is much bigger between samples SA and GF than between samples SA and EX sample SA is in fact more similar to GF than to EX in a way that can be quantified using local porosity analysis. Traditional characteristics such as porosity, specific surface and correlation functions are insufficient to distinguish different microstructures. Visual inspection of the pore space indicates that samples GF and SA have a similar structure which, however, differs from the structure of sample EX. Although sample DM resembles sample EX more closely with respect to surface roughness it differs visibly in the shape of the grains.

5.2   Local Porosity Analysis

The differences in visual appearance of the four microstructures can be quantified using the geometric observables μ and λ from local porosity theory. The local porosity distributions $\mu(\phi, 20)$ of the four samples at $L = 20a$ are displayed as the solid lines in Figs. 8a through 8d. The ordinates for these curves are plotted on the right vertical axis.

The figures show that the original sample exhibits stronger porosity fluctuations than the three model samples except for sample SA which comes close. Sample DM has the narrowest distribution which indicates that it is most homogeneous. Figures 8a–8d show also that the δ-function component at the origin, $\mu(0, 20)$, is largest for sample EX, and smallest for sample GF. For samples DM and SA the values of $\mu(0, 20)$ are intermediate and comparable. Plotting $\mu(0, L)$ as a function of L shows that this remains true for all L. These results indicate that the experimental sample EX is more strongly heterogeneous than the models, and that large regions of matrix space occur more frequently in sample EX. A similar conclusion may be drawn from the variance of local porosity

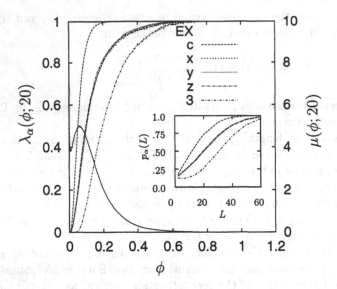

Figure 8a

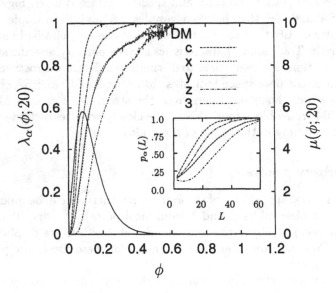

Figure 8b

Fig. 8. Local percolation probabilities $\lambda_\alpha(\phi, 20)$ (broken curves, values on left axis) and local porosity distribution $\mu(\phi, 20)$ (solid curve, values on right axis) at $L = 20$ for sample EX(Fig. 8a), sample DM(Fig. 8b), sample GF(Fig. 8c), and sample SA(Fig. 8d). The inset shows the function $p_\alpha(L)$. The line styles corresponding to $\alpha = c, x, y, z, 3$ are indicated in the legend.

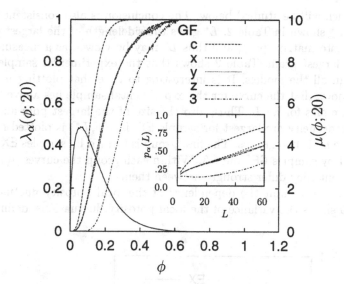

Figure 8c

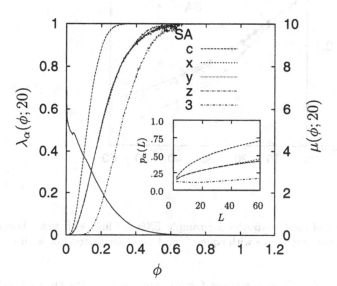

Figure 8d

fluctuations which will be studied below. The conclusion is also consistent with the results for L^* shown in Table 2. L^* gives the sidelength of the largest cube that can be fit into matrix space, and thus L^* may be viewed as a measure for the size of the largest grain. Table 2 shows that the experimental sample has a larger L^* than all the models. It is interesting to note that plotting $\mu(1, L)$ versus L also shows that the curve for the experimental sample lies above those for the other samples for all L. Thus, also the size of the largest pore and the pore space heterogeneity are largest for sample EX. If $\mu(\phi, L^*)$ is plotted for all four samples one finds two groups. The first group is formed by samples EX and DM, the second by samples GF and SA. Within each group the curves $\mu(\phi, L^*)$ nearly overlap, but they differ strongly between them.

Figures 9 and 10 exhibit the dependence of the local porosity fluctuations on L. Figure 9 shows the variance of the local porosity fluctuations, defined in

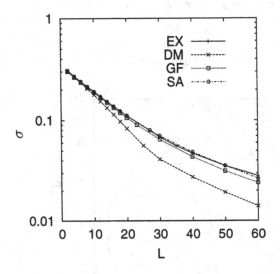

Fig. 9. Variance of local porosities for sample EX(solid line with tick), DM(dashed line with cross), GF(dotted line with square), and SA(dash-dotted line with circle).

(40) as function of L. The variances for all samples indicate an approach to a δ-distribution according to (43). Again sample DM is most homogeneous in the sense that its variance is smallest. The agreement between samples EX, GF and SA reflects the agreement of their correlation functions, and is expected by virtue of eq. (40). Figure 10 shows the skewness as a function of L calculated from (41). κ_3 characterizes the asymmetry of the distribution, and the difference between the most probable local porosity and its average. Again samples GF and SA behave similarly, but sample DM and sample EX differ from each other, and from the rest.

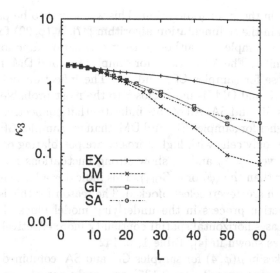

Fig. 10. Skewness of local porosities for sample EX(solid line with tick), DM(dashed line with cross), GF(dotted line with square), and SA(dash-dotted line with circle).

At $L = 4a$ the local porosity distributions $\mu(\phi, 4)$ show small spikes at equidistantly spaced porosities for samples EX and DM, but not for samples GF and SA. The spikes indicate that models EX and DM have a smoother surface than models GF and SA. For smooth surfaces and small measurement cell size porosities corresponding to an interface intersecting the measurement cell produce a finite probability for certain porosities because the discretized interface allows only certain volume fractions. In general whenever a certain porosity occurrs with finite probability this leads to spikes in μ.

5.3 Local Percolation Analysis

Visual inspection of Figs. 1 through 4 does not reveal the degree of connectivity of the various samples. A quantitative characterization of connectivity is provided by local percolation probabilities [10,27], and it is here that the samples differ most dramatically.

The samples EX, DM , GF and SA are globally connected in all three directions. This, however, does not imply that they have similar connectivity. The last line in Table 2 gives the fraction of blocking cells at the porosity 0.1355 and for L^*. It gives a first indication that the connectivity of samples DM and GF is, in fact, much poorer than that of the experimental sample EX.

Figures 8a through 8d give a more complete account of the situation by exhibiting $\lambda_\alpha(\phi, 20)$ for $\alpha = 3, c, x, y, z$ for all four samples. First one notes that sample DM is strongly anisotropic in its connectivity. It has a higher connectivity in the z-direction than in the x- or y-direction. This was found to be partly due to the coarse grid used in the sedimentation algorithm [47]. $\lambda_z(\phi, 20)$ for sample DM differs from that of sample EX although their correlation functions in the z-direction are very similar. The λ-functions for samples EX and DM rise much more rapidly than those for samples GF and SA. The inflection point of the λ-curves for samples EX and DM is much closer to the most probable porosity (peak) than in samples GF and SA. All of this indicates that connectivity in cells with low porosity is higher for samples EX and DM than for samples GF and SA. In samples GF and SA only cells with high porosity are percolating on average. In sample DM the curves λ_x, λ_y and λ_3 show strong fluctuations for $\lambda \approx 1$ at values of ϕ much larger than the $\langle\phi\rangle$ or $\phi(\mathbb{S}_{DM})$. This indicates a large number of high porosity cells which are nevertheless blocked. The reason for this is perhaps that the linear compaction process in the underlying model blocks horizontal pore throats and decreases horizontal spatial continuity more effectively than in the vertical direction, as shown in [4], Table 1, p. 142.

The absence of spikes in $\mu(\phi, 4)$ for samples GF and SA combined with the fact that cells with average porosity (≈ 0.135) are rarely percolating suggests that these samples have a random morphology similar to percolation.

The insets in Figs. 8a through 8d show the functions $p_\alpha(L) = \overline{\lambda_\alpha(\phi, L)}$ for $\alpha = 3, x, y, z, c$ for each sample calculated from (53). The curves for samples EX and DM are similar but differ from those for samples GF and SA. Figure 11 exhibits the curves $p_3(L)$ of all four samples in a single figure. The samples fall

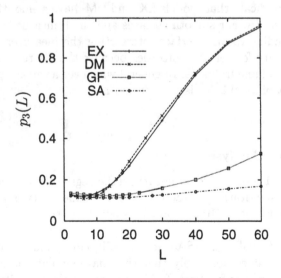

Fig. 11. $p_3(L)$ for sample EX(solid line with tick) DM(dashed line with cross) GF(dotted line with square), and SA(dash-dotted line with circle).

into two groups {EX,DM} and {GF,SA} that behave very differently. Figure 11 suggests that reconstruction methods [1,70] based on correlation functions do not reproduce the connectivity properties of porous media. As a consequence, one expects that also the physical transport properties will differ from the experimental sample, and it appears questionable whether a pure correlation function reconstruction can produce reliable models for the prediction of transport.

Preliminary results [42] indicate that these conclusions remain unaltered if the linear and/or spherical contact distribution are incorporated into the simulated annealing reconstruction. It was suggested in [70] that the linear contact distribution should improve the connectivity properties of the reconstruction, but the reconstructions performed by [42] seem not to confirm this expectation.

6 Physical Properties

6.1 Exact Results

One of the main goals in studying the microstructure of porous media is to identify geometric observables that correlate strongly with macroscopic physical transport properties. To achieve this it is not only necessary to evaluate the geometric observables. One also needs to calculate the effective transport properties exactly, in order to be able to correlate them with geometrical and structural properties. Exact solutions are now becoming available and this section reviews exact results obtained recently in cooperation with J. Widjajakusuma [10,65,67]. For the disordered potential problem, specified above in equations (2) through (7), the effective macroscopic transport parameter $\overline{C}$ is defined by

$$\langle \boldsymbol{j}(\boldsymbol{r}) \rangle = -\overline{C} \langle \boldsymbol{\nabla} u(\boldsymbol{r}) \rangle \tag{68}$$

where the brackets denote an ensemble average over the disorder defined in (25). The value of $\overline{C}$ can be computed numerically [33,66]. For the following results the material parameters were chosen as

$$C_{\mathrm{P}} = 1, \qquad C_{\mathrm{M}} = 0. \tag{69}$$

Thus in the usual language of transport problems the pore space is conducting while the matrix space is chosen as nonconducting. Equations (2) through (7) need to be supplemented with boundary conditions on the surface of $\mathbb{S}$. A fixed potential gradient was applied between two parallel faces of the cubic sample $\mathbb{S}$, and no-flow boundary condition were enforced on the four remaining faces of $\mathbb{S}$.

The macroscopic effective transport properties are known to show strong sample to sample fluctuations. Because calculation of $\overline{C}$ requires a disorder average the four microsctructures were subdivided into eight octants of size $128 \times 128 \times 128$. For each octant three values of $\overline{C}$ were obtained from the exact solution corresponding to application of the potential gradient in the x-, y- and z-direction. The values of C obtained from dividing the measured current by the applied potential gradient were then averaged. Table 3 collects the mean and the standard deviation from these exact calculations. The standard deviations in Table 3 show that the fluctuations in $\overline{C}$ are indeed rather strong. If the system is ergodic then one expects that $\overline{C}$ can also be calculated from the exact solution for the full sample. For sample EX the exact transport coefficient for the full sample is $\overline{C}_x = 0.02046$ in the x-direction, $\overline{C}_y = 0.02193$ in the y-direction, and $\overline{C}_z = 0.01850$ in the z-direction [65]. All of these are seen to fall within one standard deviation of $\overline{C}$. The numerical values have been confirmed independently by [47].

Finally it is interesting to observe that $\overline{C}$ seems to correlate strongly with $p_3(L)$ shown in Fig. 11. This result emphasizes the importance of non-Hadwiger functionals because by construction there is no relationship between $\overline{C}$ and porosity, specific surface and correlation functions.

Table 3. Average and standard deviation σ for effective macroscopic transport property $\overline{C}$ calculated from subsamples (octants) for $C_P = 1$ and $C_M = 0$.

	S_{EX}	S_{DM}	S_{GF}	S_{SA}
$\overline{C}$	0.01880	0.01959	0.00234	0.00119
σ	±0.00852	±0.00942	±0.00230	±0.00234

6.2   Mean Field Results

According to the general criteria discussed above in Section 3.1 a geometrical characterization of random media should be usable in approximate calculations of transport properties. In practice the full threedimensional microstructure is usually not available in detail, and only approximate calculations can be made that are based on partial geometric knowledge.

Local porosity theory [27,28] was developed as a generalized effective medium approximation for $\overline{C}$ that utilizes the partial geometric characterization contained in the quantities μ and λ. It is therefore useful to compare the predictions from local porosity theory with those from simpler mean field approximations. The latter will be the Clausius-Mossotti approximation with $\mathbb{P}$ as background phase

$$\overline{C}_c(\overline{\phi}) = C_P \left(1 - \frac{1 - \overline{\phi}}{(1 - C_M/C_P)^{-1} - \overline{\phi}/3} \right)$$

$$= C_P \left(\frac{3C_M + 2\overline{\phi}(C_P - C_M)}{3C_P - \overline{\phi}(C_P - C_M)} \right), \tag{70}$$

the Clausius-Mossotti approximation with M as background phase

$$\overline{C}_b(\overline{\phi}) = C_M \left(1 - \frac{\overline{\phi}}{(1 - C_P/C_M)^{-1} - (1 - \overline{\phi})/3} \right)$$

$$= C_M \left(\frac{2C_M + C_P + 2\overline{\phi}(C_P - C_M)}{2C_M + C_P - \overline{\phi}(C_P - C_M)} \right),$$

(71)

and the self-consistent effective medium approximation [35,37]

$$\overline{\phi}\frac{C_P - \overline{C}}{C_P + 2\overline{C}} + (1 - \overline{\phi})\frac{C_M - \overline{C}}{C_M + 2\overline{C}} = 0$$

(72)

which leads to a quadratic equation for $\overline{C}$. The subscripts b and c in (71) and (70) stand for "blocking" and "conducting". In all of these mean field approximations the porosity $\overline{\phi}$ is the only geometric observable representing the influence of the microstructure. Thus two microstructures having the same porosity $\overline{\phi}$ are predicted to have the same transport parameter $\overline{C}$. Conversely, measurement of $\overline{C}$ combined with the knowledge of C_M, C_P allows to deduce the porosity from such formulae.

If the microstructure is known to be homogeneous and isotropic with bulk porosity $\overline{\phi}$, and if $C_P > C_M$, then the rigorous bounds [8,24,62]

$$\overline{C}_b(\overline{\phi}) \leq \overline{C} \leq \overline{C}_c(\overline{\phi})$$

(73)

hold, where the upper and the lower bound are given by the Clausius-Mossotti formulae, eqs. (71) and (70). For $C_P < C_M$ the bounds are reversed.

The proposed selfconsistent approximations for the effective transport coefficient of local porosity theory reads [27]

$$\int\limits_0^1 \frac{\overline{C}_c(\phi) - \overline{C}}{\overline{C}_c(\phi) + 2\overline{C}}\lambda_3(\phi, L)\mu(\phi, L)d\phi + \int\limits_0^1 \frac{\overline{C}_b(\phi) - \overline{C}}{\overline{C}_b(\phi) + 2\overline{C}}(1 - \lambda_3(\phi, L))\mu(\phi, L)d\phi = 0$$

(74)

where $\overline{C}_b(\phi)$ and $\overline{C}_c(\phi)$ are given in eqs. (71) and (70). Note that (74) is still preliminary, and a generalization is in preparation. A final form requires generalization to tensorial percolation probabilities and transport parameters. Equation (74) is a generalization of the effective medium approximation. In fact, it reduces to eq. (72) in the limit $L \to 0$. In the limit $L \to \infty$ it also reduces to eq. (72) albeit with $\overline{\phi}$ in eq. (72) replaced with $\lambda_3(\overline{\phi})$. In both limits the basic assumptions underlying all effective medium approaches become invalid. For small L the local geometries become strongly correlated, and this is at variance with the basic assumption of weak or no correlations. For large L on the other hand the assumption that the local geometry is sufficiently simple becomes invalid [27]. Hence one expects that formula (74) will yield good results only for intermediate L. The question which L to choose has been discussed in the literature

[3,10,12,33,66]. For the results in Table 4 the so called percolation length L_p has been used which is defined through the condition

$$\frac{d^2 p_3}{dL^2}\bigg|_{L=L_p} = 0 \tag{75}$$

assuming that it is unique. The idea behind this definition is that at the inflection point the function $p_3(L)$ changes most rapidly from its trivial value $p_3(0) = \overline{\phi}$ at small L to its equally trivial value $p_3(\infty) = 1$ at large L (assuming that the pore space percolates). The length L_p is typically larger than the correlation length calculated from $G(r)$ [10,11].

The results obtained by the various mean field approximations are collected in Table 4 [65,67]. The exact result is obtained by averaging the three values for the full sample EX given in the previous section. The additional geometric information contained in μ and λ seems to give an improved estimate for the transport coefficient.

Table 4. Effective macroscopic transport property $\overline{C}$ calculated from Clausius-Mossotti approximations ($\overline{C}_c$, $\overline{C}_b$), effective medium theory $\overline{C}_{EMA}$ and local porosity theory $\overline{C}_{LPT}$ compared with the exact result $\overline{C}_{exact}$ (for $C_P = 1$ and $C_M = 0$).

$\overline{C}_c$	$\overline{C}_b$	$\overline{C}_{EMA}$	$\overline{C}_{LPT}$	$\overline{C}_{exact}$
0.094606	0.0	0.0	0.025115	0.020297

ACKNOWLEDGEMENT: Most results reviewed in this paper were obtained in cooperation with B. Biswal, C. Manwart, J. Widjajakusuma, P.E. Øren, S. Bakke, and J. Ohser. I am grateful to all of them, and to the Deutsche Forschungsgemeinschaft as well as Statoil A/S Norge for financial support.

References

1. Adler, P. (1992): *Porous Media* (Butterworth-Heinemann, Boston)
2. Adler, P., C. Jacquin, C., J. Quiblier (1990): 'Flow in simulated porous media', *Int.J.Multiphase Flow* **16**, p. 691
3. Andraud, C., A. Beghdadi, E. Haslund, R. Hilfer, J. Lafait, B. Virgin (1997): 'Local entropy characterization of correlated random microstructures', *Physica A* **235**, p. 307
4. Bakke, S., P. Øren (1997): '3-d pore-scale modeling of sandstones and flow simulations in pore networks', *SPE Journal* **2**, p. 136
5. Barut, A., R. Raczka (1986): *Theory of Group Representations and Applications* (World Scientific, Singapore)
6. Bear, J. (1972): *Dynamics of Fluids in Porous Media* (Elsevier Publ. Co., New York)
7. Bear, J., A. Verruijt (1987): *Modeling Groundwater Flow and Pollution* (Kluwer Academic Publishers, Dordrecht)

8. Bergman, D. (1982): 'Rigorous bounds for the complex dielectric constant of a two-component composite', *Ann. Phys.* **138**, p. 78

9. Biswal, B., R. Hilfer (1999): 'Microstructure analysis of reconstructed porous media', *Physica A* **266**, p. 307

10. Biswal, B., C. Manwart, R. Hilfer, R. (1998): 'Threedimensional local porosity analysis of porous media', *Physica A* **255**, p. 221

11. Biswal, B., C. Manwart, R. Hilfer, S. Bakke, P. Øren (1999): 'Quantitative analysis of experimental and synthetic microstructures for sedimentary rock', *Physica A* **273**, p. 452

12. Boger, F., J. Feder, R. Hilfer, R., T. Jøssang (1992): 'Microstructural sensitivity of local porosity distributions', *Physica A* **187**, p. 55

13. Bourbie, T., O. Coussy, B. Zinszner (1987): *Acoustics of Porous Media* (Editions Technip, Paris)

14. Bourbie, T., B. Zinszner (1995): 'Hydraulic and acoustic properties as a function of porosity in Fontainebleau snadstone', *J. Geophys. Res.* **90**, p. 11524

15. Bryant, S., D. Mellor, D., C. Cade (1993): 'Physically representative network models of transport in porous media', *AIChE Journal* **39**, p. 387

16. Chatzis, I., F. Dullien (1977): 'Modelling pore structure by 2-d and 3-d networks with applications to sandstones', *J. of Canadian Petroleum Technology*, p. 97

17. Crivelli-Visconti, I. (ed.) (1998): *ECCM-8 European Conference on Composite Materials* (Woodhead Publishing Ltd, Cambridge)

18. Delfiner, P. (1972): 'A generalization of the concept of size', *J. Microscopy* **95**, p. 203

19. Diebels, S., W. Ehlers (1996): 'On fundamental concepts of multiphase micropolar materials', *Technische Mechanik* **16**, p. 77

20. Dullien, F. (1992): *Porous Media - Fluid Transport and Pore Structure* (Academic Press, San Diego)

21. Ehlers, W. (1995): 'Grundlegende Konzepte in der Theorie poröser Medien', Technical report, Institut f. Mechanik, Universität Stuttgart, Germany

22. Fatt, I. (1956): 'The network model of porous media I. capillary pressure characteristics', *AIME Petroleum Transactions* **207**, p. 144

23. Hadwiger, H. (1955): *Altes und Neues über konvexe Körper* (Birkhäuser, Basel)

24. Hashin, Z., S. Shtrikman (1962): 'A variational approach to the theory of effective magnetic permeability of multiphase materials', *J. Appl. Phys.* **33**, p. 3125

25. Haslund, E., B. Hansen, R. Hilfer, B. Nøst (1994): 'Measurement of local porosities and dielectric dispersion for a water saturated porous medium', *J. Appl. Phys.* **76**, p. 5473

26. Hearst, J., P. Nelson (1985): *Well Logging for Physical Properties* (McGraw-Hill, New York)

27. Hilfer, R. (1991): 'Geometric and dielectric characterization of porous media', *Phys. Rev. B* **44**, p. 60

28. Hilfer, R. (1992): 'Local porosity theory for flow in porous media', *Phys. Rev. B* **45**, p. 7115

29. Hilfer, R. (1993): 'Local porosity theory for electrical and hydrodynamical transport through porous media', *Physica A* **194**, p. 406

30. Hilfer, R. (1996): 'Transport and relaxation phenomena in porous media', *Advances in Chemical Physics* **XCII**, p. 299

31. Hilfer, R., B. Nøst, E. Haslund, Th. Kautzsch, B. Virgin, B.D. Hansen (1994): 'Local porosity theory for the frequency dependent dielectric function of porous rocks and polymer blends', *Physica A* **207**, p. 19

240   Rudolf Hilfer

32. Hilfer, R., T. Rage, B. Virgin (1997): 'Local percolation probabilities for a natural sandstone', *Physica A* **241**, p. 105
33. Hilfer, R., J. Widjajakusuma, B. Biswal (1999): 'Macroscopic dielectric constant for microstructures of sedimentary rocks', *Granular Matter*, in print
34. Katz, A., A. Thompson (1986): 'Quantitative prediction of permeability in porous rock', *Phys. Rev. B* **34**, p. 8179
35. Kirkpatrick, S. (1973): 'Percolation and conduction', *Rev. Mod. Phys.* **45**, p. 574
36. Lake, L. (1989): *Enhanced Oil Recovery* (Prentice Hall, Englewood Cliffs)
37. Landauer, R. (1978): 'Electrical conductivity in inhomogeneous media', in: *Electrical Transport and Optical Properties of Inhomogeneous Materials*, ed. by J. Garland, D. Tanner (American Institute of Physics, New York), p. 2
38. Levitz, P., D. Tchoubar (1992): 'Disordered porous solids: From chord distributions to small angle scatterin', *J. Phys. I France* **2**, p. 771
39. Louis, A. (1989): *Inverse und schlecht gestellte Probleme* (Teubner, Stuttgart)
40. Manwart, C., R. Hilfer (1999a): to be published
41. Manwart, C., R. Hilfer (1999b): 'Reconstruction of random media using Monte Carlo methods', *Physical Review E* **59**, p. 5596
42. Manwart, C., S. Torquato, R. Hilfer (1999): 'Stochastic reconstruction of sandstones', preprint
43. Marsily, G. (1986): *Quantitative Hydrogeology – Groundwater Hydrology for Engineers* (Academic Press, San Diego)
44. Mecke, K. (1998): 'Integral geometry and statistical physics', *Int. J. Mod. Phys. B* **12**, p. 861
45. Mecke, K., H. Wagner (1991): 'Euler characteristic and related measures for random geometric sets', *J. Stat. Phys.* **64**, p. 843
46. Muche, L., D. Stoyan (1992): 'Contact and chord length distributions of the poisson voronoi tessellation', *Journal of applied probability* **29**, p. 467
47. Øren, P. (1999): private communication
48. Øren, P., S. Bakke, O. Arntzen (1998): 'Extending predictive capabilities to network models', *SPE Journal*, p. SPE 38880
49. Quiblier, J. (1984): 'A new three dimensional modeling technique for studying porous media', *J. Colloid Interface Sci.* **98**, p. 84
50. Roberts, A. (1997): 'Statistical reconstruction of three-dimensional porous media from two-dimensional images', *Phys.Rev.E* **56**, p. 3203
51. Roberts, J., L. Schwartz (1985): 'Grain consolidation and electrical conductivity in porous media', *Phys. Rev. B* **31**, p. 5990
52. Roy, D. (1991): *Methods of Inverse problems in Physics* (CRC Press, Boca Raton)
53. Roy, S., S. Tarafdar, S. (1997): 'Archies's law from a fractal model for porous rock', *Phys.Rev.B* **55**, p. 8038
54. Sahimi, M. (1993): 'Flow phenomena in rocks: From continuum models to fractals, percolation, cellular automata and simulated annealing', *Rev. Mod. Phys.* **65**, p. 1393
55. Sahimi, M. (1995): *Flow and Transport in Porous Media and Fractured Rock* (VCH Verlagsgesellschaft mbH, Weinheim)
56. Scheidegger, A. (1974): *The Physics of Flow Through Porous Media* (University of Toronto Press, Toronto)
57. Schneider, R. (1993): *Convex Bodies: The Brunn-Minkowski Theory* (Cambridge University Press, Cambridge)
58. Schneider, R., W. Weil (1992): *Integralgeometrie* (Teubner, Stuttgart)
59. Schwartz, L., S. Kimminau (1987): 'Analysis of electrical conduction in the grain consolidation model', *Geophysics* **52**, p. 1402

60. Stell, G. (1985) 'Mayer-Montroll equations (and some variants) through history for fun and profit', in: *The Wonderful World of Stochastics*, ed. by M. Shlesinger, G. Weiss (Elsevier, Amsterdam), p. 127

61. Stoyan, D., W. Kendall, J. Mecke (1987): *Stochastic Geometry and its Applications* (Akademie-Verlag / Wiley, Berlin / Chichester)

62. Torquato, S. (1991): 'Random heterogeneous media: Microstructure and improved bounds on effective properties', *Applied mechanics reviews* **44**, p. 37

63. Torquato, S., G. Stell (1982): 'Microstructure of Two Phase Random Media I: The *n*-Point Probability Functions', *J. Chem. Phys.* **77**, p. 2071

64. Virgin, B., E. Haslund, R. Hilfer (1996): 'Rescaling relations between two- and three dimensional local porosity distributions for natural and artificial porous media', *Physica A* **232**, p. 1

65. Widjajakusuma, J., B. Biswal, R. Hilfer (1999a): 'Predicting transport parameters of heterogeneous media', preprint

66. Widjajakusuma, J., B. Biswal, R. Hilfer (1999b): 'Quantitative prediction of effective material properties of heterogeneous media', *Comp. Mat. Sci.* **16**, p. 70

67. Widjajakusuma, J., R. Hilfer (2000): 'Local Porosity Theory for Sound Propagation in Porous Media'

68. Widjajakusuma, J., C. Manwart, B. Biswal, R. Hilfer (1999): 'Exact and approximate calculations for conductivity of sandstones', *Physica A* **16**, p. 70

69. Yao, J., P. Frykman, F. Kalaydjian, P. Thovert, P. Adler (1993): 'High-order moments of the phase function for real and reconstructed model porous media: A comparison', *J. of Colloid and Interface Science* **156**, p. 478

70. Yeong, C., S. Torquato (1998): 'Reconstructing random media', *Phys.Rev. E* **57**, p. 495

71. Ziman, J. (1982): *Models of Disorder* (Cambridge University Press, Cambridge)

Stochastic Models as Tools for the Analysis of Decomposition and Crystallisation Phenomena in Solids

Helmut Hermann

Institut für Festkörper- und Werkstofforschung Dresden,
Postfach 270016, D-01171 Dresden

Abstract. Random models are explained that are capable to describe decomposition and crystallisation phenomena in solids. The description is based on the physical nucleation-and-growth model and on methods developed within the framework of stochastic geometry. It is shown that the Boolean model easily reproduces the theory proposed by Kolmogorov, Johnson, Mehl, and Avrami for the kinetics of random second phase formation in solids. Additionally, the mathematical properties of the Boolean model are used to formulate an approach to the transformation kinetics of systems with interacting components. This approach is applied to the formation of nanocrystalline materials from the amorphous state. Finally, random tessellations are considered that are useful for the description of mosaic structures such as three-dimensional and planar polycrystals.

1 Introduction

Physical properties of many technologically important materials such as mechanical strength or superconducting and magnetic properties are essentially controlled by their microstructure. It is, therefore, advantageous to have available experimental methods which allow to generate solids with well-tailored microstructural parameters. Thermally activated decomposition and crystallisation of metastable solids such as supersaturated alloys or amorphous materials belong to the most important type of processes to attain this object.

Decomposition phenomena may show spinodal or binodal behaviour. If the starting material is unstable against small periodic fluctuations of the chemical concentration and if fluctuations with a certain characteristic length increase exponentially with time, the mechanism is called spinodal. Figure 1a illustrates this behaviour (see also Sect. 3.2 of the paper by K. Mecke in this volume). A detailed review of experimental and theoretical aspects of spinodal decomposition is given in [5]. The binodal mechanism is characterised by the existence of a phase boundary already at the very beginning of the decomposition process. Local fluctuations of chemical composition lead to the nucleation of small precipitates. If the size of a precipitate exceeds a certain critical value it grows according to a given growth law (see Fig. 1b) whereas it disappears otherwise.

In this paper, we consider binodal meachnisms, i.e. nucleation and growth of precipitates following [52], who summarised the current physical ideas which are important for the understanding of nucleation and growth of particles in a solid

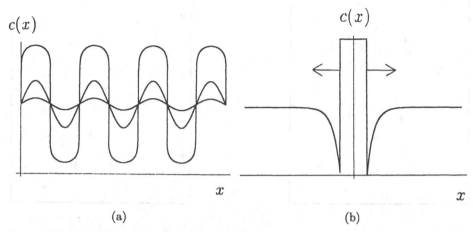

Fig. 1. Spinodal (a) and binodal (b) decomposition; $c(x)$ is the chemical composition; x is a spatial coordinate which is perpendicular to the fluctuation front in (a) and crosses the centre of a growing precipitate in (b).

matrix. The origin of nucleation of precipitates and the derivation of growth laws for existing particles are subject of atomistic considerations (see, e.g. [18]). Here, we assume the existence of a nucleation rate, $\mu(t)$, and a growth law, $r(t')$. The nucleation process starts at $t = 0$. The size of a particle created at τ, $0 \leq \tau \leq t$, and measured at time t is given by $r(t - \tau)$. The quantities $\mu(t)$ and $r(t')$ describe intrinsic properties of the material considered. That means the properties $\mu(t)$ and $r(t')$ are valid if the volume fraction, $\eta(t)$, of precipitates is small. With increasing volume fraction the microstructure may get significant influence on nucleation rate and growth law, and the corresponding effective properties will depend both on the intrinsic ones and on the properties of the microstructure.

The evolution of a microstructure can be observed by means of different experimental methods, and parameters such as volume fraction, mean size and size distribution of precipitates (or particles), correlation functions and specific interface area can be measured. It is, of course, not possible to give a complete description of the microstructure because only few structure parameters can be experimentally estimated. Therefore, it is helpful to develop models for the decomposition (and/or crystallisation) process. The experimental data available can be used to estimate the parameters of the model. Once the model is specified for the material under consideration, one can draw conclusions with respect to type and details of the decomposition mechanisms, and it may be possible to estimate structure parameters which are not directly accessible by experimental methods.

We consider models which are based on random point fields (point processes). The points give the nucleation sites where decomposition and/or crystallisation

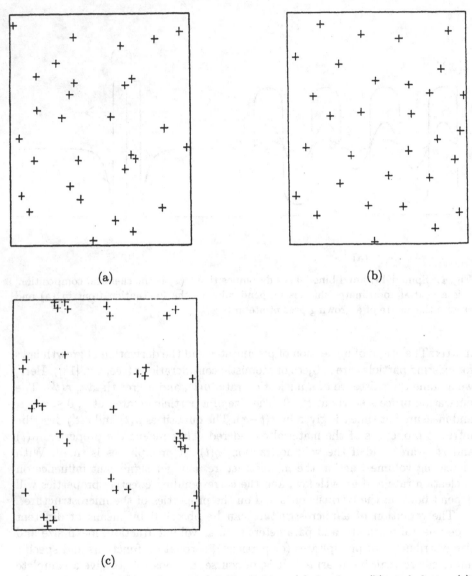

Fig. 2. Examples for random point fields: Poisson (a), hardcore (b), and cluster (c) point field.

may start. The number density, $\lambda(t)$, of points (or the intensity of the point field) at time t is given by the integral

$$\lambda(t) = \int_0^t \mu(\tau)d\tau. \tag{1}$$

There are different types of point processes. Among them, the homogeneous Poisson point field plays an exceptional role (see Stoyan, this volume, or [46]).

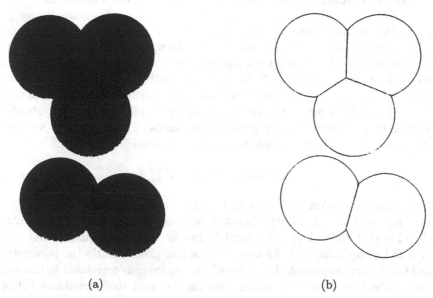

Fig. 3. Impinging of neighbouring particles during growth. Set-theoretical union of grains (a) and particles plus interface (b).

Its distribution is completely determined by the parameter λ, the mean number density of points. In this process there is no interaction between the points. If interactions have to be taken into account, e.g. by introducing pair potentials (see Stoyan, this volume), the situation becomes more difficult. Hardcore and cluster point fields are examples for such interactions (see Fig. 2).

Consider some nuclei created at neighbouring points of an arbitrary point field. They start to grow getting particles and after a certain time they will impinge. Then, their growth is finished at the common interface but continues at the free part of their surfaces (see Fig. 3). Constructing models for structures which develop according to this process one has to decide whether the interface between impinging particles should be considered in the model or not. If not, random germ-grain models are the right choice. If the interface is essential, one arrives at tessellations (or mosaic structures). The most powerful germ-grain model is probably the Boolean model, which is based on a Poisson process of germs. Among tessellations, the Poisson line and plane model, the Voronoi mosaic and the Johnson-Mehl tessellation are the most important ones.

In Sect. 2 the Boolean model is considered and used to re-derive the formulae for the well-known Kolmogorov-Johnson-Mehl-Avrami model. Despite the fact that the Boolean model describes random systems without any interaction, its properties are helpful to formulate an approach to the decomposition kinetics of materials with interacting components. This is demonstrated in Sect. 3. Finally, models for the generation of random tessellations are discussed in Sect. 4.

2   The Boolean Model and Crystallisation Kinetics

The kinetics of decomposition or crystallisation is often analysed by measuring the transformed volume fraction and its evolution during isothermal annealing. The standard model for the interpretation of the corresponding data was developed by Kolmogorov, Johnson, Mehl, and Avrami (KJMA model) in the thirties. Several authors, e.g. [6], re-derived the model and discussed it from actual points of view. The KJMA model has been successfully applied to the analysis of the transformation kinetics of very different materials. The continuous success is probably attributed to the simple form of the expression

$$\eta(t) = 1 - \exp(-kt^n) \tag{2}$$

for the volume fraction η transformed during the time interval $(0, t)$. When a set $(\eta_i, t_i), i = 1, 2, ..., k$, of experimental data is given one can plot the data as $\log \log(1 - \eta)^{-1}$ versus $\log(t)$. If a straight line is obtained the model may be assumed to be applicable and the ascent of the line gives directly the parameter n called the Avrami exponent. The value of the exponent is correlated to the mechanism of the transformation process and can be used to discriminate between diffent possible physical processes. Recent examples for experimental investigations and applications of the KJMA model are given in [4,10,11,12,31,36,49].

In recent papers the suppositions and the range of validity of the KJMA model have been discussed. Clearly, equation (2) is exact if neither during nucleation nor during growth of particles any interaction occurs, and if the growth rate, i.e. the velocity of the propagation of the transformation front, is constant (see [21,38,48,51]). If the transformation process is controlled by diffusion processes the growth rate of a given particle decreases with increasing size. In this case, which is quite important, the KJMA model can, however, be used as an approximation because the deviation of expression (2) for $\eta(t)$ from correct data obtained by computer simulations [50] is less than 1%. The deviations become more serious if the decrease of the growth rate gets stronger or if the condition of absence of any interaction between particles is violated, e.g. if the centres of particles follow a hardcore or a cluster process [50].

Recent attempts to apply the ideas of the KJMA model to more general theoretical descriptions of transformation phenomena concern non-isothermal and non-equilibrium processes [9,10,13,17,24,32,43,44,45]. There are also alternative models for decomposition and crystallisation phenomena, see, e.g., [22-26] and [39-41].

The main result of the KJMA model is the expression (2) for the volume fraction transformed. The volume fraction is a structure parameter reducing all information of the system to a single number. So it is not surprising that conclusions drawn from the interpretation of corresponding experimental data in terms of any model are not unambiguous. In case of the KJMA model the Avrami exponent n is calculated for a series of typical possible situations. Concerning the nucleation rate the following limiting cases exist: (i) All nuclei are preformed at $t = 0$ and start to grow at the same instant. (ii) The number of nuclei created

Fig. 4. Planar Boolean model with circular grains of uniform distribution of diameters. Such a structure corresponds to the physical situation of a partially crystallised thin film, randomly distributed nuclei and constant growth rate of crystallites.

at random is uniformly distributed in the time interval $(0, t)$, i.e. continuous nucleation.

Considering the growth of the particles one has to distinguish between growth in one (rods), two (platelets), and three (spheres *etc.*) spatial dimensions. The size of the particles (counted in one, two, or three spatial directions) increases proportional to time (interface controlled growth) or, if the growth is controlled by diffusion, proportional to $\sqrt{t}$. (Remember that the latter case is not treated exactly in the KJMA model but may be considered as a reasonable approximation.) In three-dimensional case, it is $3 \leq n \leq 4$ for interface controlled growth while $1.5 \leq n \leq 2.5$ is obtained for diffusion. The lower and upper limit of n corresponds to the nucleation rate according to (*i*) and (*ii*), respectively. The application of expression (2) to a certain set of experimental data yields a value for n which can be related to one or more possible mechanisms.

Now we consider the Boolean model. Its construction is explained in Stoyan (this volume). A detailed description and a summary of its properties is given in [46]. Applications to problems of materials science and solid state physics are discussed [19]. Figure 4 shows a two-dimensional realisation of the Boolean model.

The volume fraction, $\eta(t)$, occupied by the grains of the Boolean model is

$$\eta(t) = 1 - \exp[-\lambda(t)\bar{V}(t)]. \tag{3}$$

In terms of the Boolean model, $\lambda(t)$ is the intensity of the Poisson point field, i.e. the mean number of points created until t per unit volume, and $\bar{V}(t)$ is the mean volume of the grains. The parameter $\lambda(t)$ is related to the nucleation rate, $\mu(\tau)$, by (1). The mean volume of the grains is calculated in the following way: A nucleus created at time τ grows, and at time t the size, $\sigma(t,\tau)$, of this particle is given by

$$\sigma(t,\tau) = \begin{cases} 0 , t < \tau \\ r(t-\tau) , t \geq \tau \end{cases} \tag{4}$$

and its volume is

$$V(t,\tau) = \omega_d \sigma(t,\tau)^d. \tag{5}$$

(For spherical particles, $\sigma(t,\tau)$ is the radius, $d = 3$, and $\omega_3 = \frac{4}{3}\pi$.) The most important special cases for (4) concern interface controlled growth, i.e. $r \propto (t-\tau)$, and diffusion controlled growth, i.e. $r \propto \sqrt{t - \tau}$. Remember that for the latter case the KJMA model and also the Boolean model represents an approximation and expression (2) is not exact. In this case, an unphysical but formally permitted nucleus created in a region already crystallised may grow into noncrystallised regions and, consequently, contribute to the crystallised volume fraction. The reason for this is that at diffusion controlled growth the growth rate of an older crystallite is less than that of a younger one.

The volume of a single grain develops according to (5) and the mean volume, $\bar{V}(t)$, of all grains created in the time interval $(0,t)$ is given by the average of the volume $V(t,\tau)$ of all grains created in the time interval $(0,t)$ weighted by the fraction $\mu(\tau)d\tau/\lambda(t)$ of grains nucleated during the time increment $d\tau$. Therefore,

$$\lambda(t)\bar{V}(t) = \int_0^t \mu(\tau)V(t,\tau)d\tau. \tag{6}$$

Expressions (3) and (6) correspond to (2) developed for the JMAK model. In case of instantaneous creation of nuclei, i.e. $\mu(t) = \mu_0\delta(t)$, one obtains (2) from (3) and (6) with $n = 3$ and $n = 1.5$ for interface and diffusion controlled growth, respectively. ($\delta(t)$ is Dirac's delta function). Nucleation with constant rate $\mu = \mu_1$ yields $n = 4$ and $n = 2.5$ for the same growth rates.

3 Decomposition and Interaction

3.1 Microstructure and Nucleation Rate

The limitations of the KJMA model consist in the supposition that there is not any interaction between the structural components of the decomposing system. Additionally, the growth rate of the particles must obey the condition $r(t - \tau) \propto (t - \tau)^a, a \geq 1$.

In many cases, the processes controlling the decomposition of the material considered do not correspond to these suppositions. This is evident for amorphous alloys that transform into a nanocrystalline state during annealing. There, the growth rate of the crystallites tends rapidly to zero after nucleation. The final size distribution of the particles becomes quite narrow with a mean value of the order of 10nm, and the microstructure is more similar to a dense random packing of spheres than to a Boolean model realized with spherical grains. This points to the existence of non-negligible interactions between the particles during nucleation and growth.

The theoretical treatment of a system of interacting particles is always difficult and, in any case, it requires some approximations. The approximation made here is to describe the arrangement of the particles at a certain time by a Boolean model having the same volume fraction and the same specific surface area as a hardcore arrangement. The increment of transformed volume due to nucleation and growth is, however, not calculated via growth of the grains of the Boolean model but through the consideration of particles interacting with the existing microstructure. So, the increase of transformed volume can be described reasonably despite the details of the microstructure are considered only within the approximation explained. This approximation is reasonable at least at low and medium volume fraction, and in this range of validity it goes significantly beyond the KJMA model. It takes into account the interaction of particles formed in the amorphous matrix and considers also that the composition of the matrix changes during decomposition.

The amorphous alloy, i.e. the starting material, is supposed to be homogeneous at a certain length scale. Below a characteristic scale, fluctuations of the chemical composition appear. The fluctuations are modelled by a space-filling mixture of three components distributed at random. There is a random set A_s that characterises the geometrical distribution of regions where the local chemical composition is favourable for the spontaneous nucleation of crystallites whithout the necessity of diffusion. We call this *active chemical composition* where the (intrinsic) nucleation rate $\mu_s(t)$ applies. A_f describes those regions where the chemical composition deviates from that realised in A_s. There, nucleation is not possible and it is called *passive chemical composition*. Finally, A_n denotes the set of so-called *inhibitors*, i.e. of atoms which diffuse very slowly compared to the mean diffusity of the atoms forming the crystallites. Additionally, the inhibitors are not allowed to be incorporated in the atomic lattice of crystallites. The union of these components is denoted by A_0 and represents the amorphous sample.

$$A_0 = A_s \cup A_f \cup A_n. \tag{7}$$

When crystallisation goes on, an additional component, A_η, appears which describes the geometrical distribution of the crystallites. Then,

$$A = A_s \cup A_f \cup A_n \cup A_\eta \tag{8}$$

where $A(\eta = 0) = A_0$. In the model discussed here, the inhibitors ,A_n, are assumed to be point-like forming a Poisson process, whereas A_f and A_η are described by Boolean models with identical spherical grains (radii r_f and r_η, respectively). Figure 5 shows possible realisations of the structural components explained above. The number density of the grains and the volume fraction, V_V,

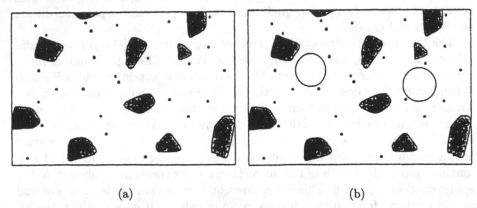

(a)   (b)

Fig. 5. Illustration of the structure of the amorphous (a) and the partially crystallised (b) state. Dark area: passive regions; points: inhibitors; circles: crystallites.

are the parameters of the models, i.e. $\lambda_f, f = V_V(A_f)$ and $\lambda_\eta, \eta = V_V(A_\eta)$ for A_f and A_η, respectively. Remembering that

$$\eta = 1 - \exp(-\lambda_\eta \frac{4\pi}{3} r_\eta^3) \tag{9}$$

is the volume fraction of the model A_η one can also use the parameter set (η, r_η). This choice is more appropriate because the correlation length of chemical fluctuations can (in principle) be measured by scattering experiments whereas it is rather difficult to get an estimate for the number density of the grains. The same argument holds for A_f, and the parameter set (f, r_f) is used instead of (f, λ_f). During crystallisation, the volume fraction occupied by the passive chemical composition increases outside the crystallised area, A_η, since the total volume of A_f is distributed over the decreasing volume available to the remaining amorphous matrix. Denoting the fraction of the total volume of the passive regions in the sample by δ, one obtains for the local volume fraction of A_f in the non-crystallised part of the sample, i.e. in the amorphous matrix, the expression

$$f(t) = \frac{\delta}{1 - \eta(t)}. \tag{10}$$

The quantity $(1 - \delta)$ gives the upper limit for the crystalline volume fraction $\eta(t)$ after long annealing time, t, and $\delta = f(0)$.

The nucleation rate that would be valid for a hypothetical homogeneous amorphous sample with active chemical composition is denoted by $\mu_s(t)$. It is called the intrinsic nucleation rate. This rate is modified by the presence of the components A_f, A_η, and A_n in the real material. It is well-known (see, e.g., [52]) that a nucleus created disappears if its radius is smaller than a critical value r_c. A nucleus can only exist and grow if its radius is greater than r_c and if it fulfils the following conditions:

(i) The nucleus can only be created outside the crystallised regions and its midpoint must have a minimum distance r_{min} to the boundary of the next-nearest crystallite. Clearly, this distance must be larger than or equal to the critical radius, r_c. If the growth of particles is controlled by diffusion then depletion zones around the particles appear where the chemical composition is not suitable for the creation of nuclei. This depletion zones may also contribute to r_{min}.

Let P be the probability that a random point thrown into A has the mentioned property and, therefore, may act as a nucleation centre for a new crystallite. Then, P is equal to the probability that the distance of the midpoint of the nucleus to the centre of the next-nearest crystallite is greater than $(r_{min} + r_\eta)$. Since the crystallised part of the alloy is approximated by the Boolean model A_η, the centres of the existing crystallites are supposed to form a Poisson point process characterised by parameter λ_η. Eliminating the number density λ_η by means of expression (3) and $\bar{V} = \frac{4}{3}\pi(r_{min} + r_\eta)^3$ one obtains

$$P = (1 - \eta)^a, \quad a = \left(1 + \frac{r_{min}}{r_\eta}\right)^3. \tag{11}$$

(ii) A nucleus matching condition (i) is subjected to further requirements. It can only be created in a region with active composition, and the smallest distance of its midpoint to the passive region must be at least r_c. The geometrical distribution of the passive component is approximated by a Boolean model with parameters f and r_f. That means, the probability, Q, that a nucleus created at random in the amorphous matrix has the property required can be calculated by means of the Poisson distribution in the same way as P. One obtains

$$Q = (1 - f)^b, \quad b = \left(1 + \frac{r_c}{r_f}\right)^3. \tag{12}$$

(iii) A new nucleus should not contain any of the inhibitors since this would cause strong distortions of the local atomic order, and the distortions would prevent the creation of a locally well-ordered crystalline structure. The probability, R, of finding a region G containing no inhibitor and having volume $V(G) = \frac{4}{3}\pi r_c^3$ of a nucleus with critical radius r_c is given by the emptiness formula for the Poisson point process as

$$R = \exp\left(-\lambda_n \frac{4}{3}\pi r_c^3\right). \tag{13}$$

The inhibitors are accumulated in the amorpous matrix where two different limiting cases of their arrangement are possible. Either they are all distributed randomly in the amorphous matrix according to a Poisson point process, or the inhibitors which were situtated previously in the volume occupied by a crystallite are deposited at the surface of this crystallite. In the first case the number density of inhibitors in the amorphous matrix is enhanced according to

$$\lambda_n(t) = \frac{\nu}{1 - \eta(t))} \tag{14}$$

where $\nu = \lambda_n(0)$. In the second case the number density of inhibitors in the amorphous matrix keeps constant and

$$\lambda_n(t) = \nu. \tag{15}$$

Expression (13) is valid for point-like inhibitors. There might be situations where the size of inhibitors should be taken into account, e.g. if the critical radius, r_c, of the nuclei is very small or if the inhibitors are not realised by single atoms but by clusters or aggregates of atoms. Then, the same arguments as used in items (i) and (ii) apply and r_c in expression (13) has to be replaced by $r_c + r_n$ where r_n is the radius of an inhibitor.

The conditions (i) to (iii) characterise the influence of the microstructure of the decomposing and crystallising alloy on the nucleation rate. In homogeneous regions with suitable (active) chemical composition the (intrinsic) nucleation rate is given by $\mu_s(t)$. This rate depends on specific properties of the alloy and on the temperature realised during the annealing process. The existence of different microstructural components leads to a modification of $\mu_s(t)$. Since the random events described by $\mu_s(t)$, P, Q, and R are statistically independent, the nucleation rate in a partially crystallised alloy with fluctuating chemical composition and randomly distributed inhibitors is given by

$$\mu(t) = P(t)\,Q(t)\,R(t)\mu_s(t). \tag{16}$$

Inserting the expressions (11) to (14) into (16) one obtains

$$\mu(t) = [1 - \eta(t)]^{a-b}\,[1 - \eta(t) - \delta]^b \exp\left(-\lambda_n \frac{4}{3}\pi r_c^3\right)\mu_s(t). \tag{17}$$

3.2   Growth Rate of Crystallites

There are two important intrinsic processes which may define the growth of a particle, the interface controlled growth $r \propto t$ and diffusion, $r \propto \sqrt{t}$. The latter relation is a solution of the diffusion equation derived by Zener (see, e.g., [1]) for a single particle in a matrix the chemical composition of which is kept constant at large distances from the particle. If many particles exist and grow in a matrix there is only a limited volume available for each particle. This case was considered by [36] who introduced time-dependent boundary conditions with respect to the chemical composition of the matrix. A similar model was developed by [35]. The result consists in growth rates that are reduced compared to the Zener's solution. Further aspects of size-dependent growth rates of particles were analysed, e.g., by Kelton [24] and Shneidman and Weinberg [42].

Here we use the expression

$$r(t) = r_b\{1 - (1 - \frac{r_c}{r_b}) \exp[-r_0(t)/r_b]\}. \tag{18}$$

for the effective growth of crystallites in the amorphous matrix. Parameter r_b denotes the maximum size of crystallites ($\lim_{t\to\infty} r(t) = r_b$) and $r_0(t)$ characterizes the intrinsic growth law. For $r_b \to \infty$, and zero critical radius, $r_c = 0$, equation (18) can be expanded and one obtains $r(t) = r_c + r_0(t)$. Expression (18) was derived in [21] and used in [20] for the interpretation of experimental data on nanocrystalline metallic alloys.

3.3   Transformed Volume Fraction

The calculation of the volume fraction of a Boolean model with spherical grains is straightforward. This is a consequence of the random distribution of the centres of the grains and the neglection of interaction effects. The model explained in Sect. 3.1, however, takes into account distance correlations between the particles already formed and the nucleus under consideration.

As pointed out in Sect. 3.1 the centre of a new nucleus is subjected to the condition that it should have at least the distance r_{min} to the region already crystallised, i.e. to the phase boundary. While the radius, r, of the new particle is less than its distance to the crystallised region it contributes its entire volume to the increment of the crystallised fraction of the sample. Otherwise, there will be an overlap volume, δV_{ov}, common to the considered particle and the crystallised region already existing. Of course, this overlap is fictitious because the growth of the particle stops if it impinges on another crystallite.

Clearly, the contribution of the considered crystallite to the increment of the crystallised volume of the sample is ($\frac{4\pi}{3}r^3 - \delta V_{ov}$). This concerns each crystallite the distance, ρ, of which to the phase boundary is between the minimum distance r_{min} and its radius r, $r_{min} < \rho < r$. Assuming the distance ρ to be uniformly distributed in the interval $r_{min} < \rho < r$ and neglecting the curvature of the intersectional area between the crystallite considered and the existing crystallised region, the mean overlap volume is given by

$$\delta V_{ov} = \frac{1}{r - r_{min}} \int_{r_{min}}^{r} \frac{\pi}{3}(r - \rho)^2(2r + \rho)d\rho \tag{19}$$

and

$$\delta V_{ov} \approx \frac{\pi}{4}r^3 \left(1 - \frac{5r_{min}}{3r}\right) \quad \text{for} \quad r_{min}/r << 1. \tag{20}$$

If the centre of the crystallite considered is situated in that part of the amorphous matrix where its distance to the crystallised region, A_η, of the sample is larger than r, it contributes the increment $\frac{4\pi}{3}r^3$ to A_η. If $r_{min} < \rho < r$, the increment to A_η is given by ($\frac{4\pi}{3}r^3 - \delta V_{ov}$). The part of the amorphous matrix where $\rho > r_{min}$ and $\rho > r$ applies is denoted by C_c and C_r, respectively. The probability of finding the centre of the considered crystallite within C_r is given by $V(C_r)/V(C_c)$, where $V(C)$ is the volume of C. With this probability, the considered crystallite contributes the amount $\frac{4\pi}{3}r^3$ to A_η. The probability that the centre of the crystallite has a distance between r_{min} and r to the amorphous-crystalline interface is $[1 - V(C_r)/V(C_c)]$ and a particle having this property contributes $(\frac{4\pi}{3}r^3 - \delta V_{ov})$. The mean contribution of a crystallite is therefore $\frac{4\pi}{3}r^3 f(r, r_{min}, r_\eta, \eta)$ with

$$f(r, r_{min}, r_\eta, \eta) = \frac{V(C_r)}{V(C_c)} + \left(1 - \frac{V(C_r)}{V(C_c)}\right) \frac{\frac{4\pi}{3}r^3 - \delta V_{ov}}{\frac{4\pi}{3}r^3}. \tag{21}$$

Expression (21) can be evaluated remembering that the crystallised region of the sample is approximated by a Boolean model with spherical grains of radius r_η. According to (3) the volume fraction, $(1 - \eta)$, of the amorphous part of the alloy is given by $\exp(-\lambda\frac{4\pi}{3}r_\eta^3)$. Replacing the spheres of this Boolean model by spherical grains with radius $(r_\eta + r)$ and $(r_\eta + r_{min})$ one obtains models decribing the complements of the regions C_r and C_c, respectively, and the corresponding volume fractions. Then,

$$\frac{V(C_r)}{V(C_c)} = \frac{\exp\left[-\frac{4\pi}{3}\lambda(r_\eta + r)^3\right]}{\exp\left[-\frac{4\pi}{3}\lambda(r_\eta + r_{min})^3\right]}, \tag{22}$$

and replacing λ by means of the expression for the volume fraction of the Boolean model with spherical grains of radius r_η, i.e. the model for the crystallised region, one obtaines

$$\frac{V(C_r)}{V(C_c)} = (1 - \eta)^\varepsilon, \quad \varepsilon = \left(1 + \frac{r}{r_\eta}\right)^3 - \left(1 + \frac{r_{min}}{r_\eta}\right)^3. \tag{23}$$

Inserting (23) into (22) and using (20) one obtains the expression

$$f(r, r_{min}, r_\eta, \eta) = \frac{13}{16} + \frac{3}{16}\left(1 - \frac{5r_c}{3r}\right)(1 - \eta)^\varepsilon + \frac{5r_c}{16r}, \quad r_{min}/r \ll 1 \tag{24}$$

for the fractional contribution of a new crystallite to the increment of the crystallised part of the considered alloy. A nucleus created at time τ and growing according to (18) contributes the amount $f(r, r_{min}, r_\eta, \eta)\frac{4\pi}{3}[r(t - \tau)]^3$ at time t, $t > \tau$, to the increment of the crystallised part of the alloy. Taking into account that at time τ the nucleation rate is given by expression (17), the integration over all crystallites in the amorphous matrix yields the value of the volume fraction of crystallites

$$\eta(t) = \int_0^t \mu(\tau)f(r, r_{min}, r_\eta, \eta)\frac{4\pi}{3}[r(t - \tau)]^3 \, d\tau. \tag{25}$$

3.4   Discussion and Application

The integral equation (25) is a Volterra type II equation. It is nonlinear but it can be solved numerically using standard methods. Figure 6 shows numerical

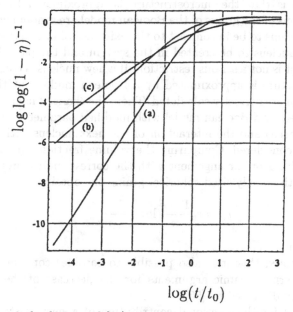

Fig. 6. Numerical solutions of (25) specified for the amorphous metallic alloy of the composition $Fe_{75.5}Si_{15.5}B_7Cu_1Nb_1$ and with parameters $r_f = 1.5nm$, $\delta = 0.25$, $r_\eta = 30nm$, and $r_c = 0$ (a), $r_c = 0.2nm$ (b), $r_c = 0.6nm$ (c).

solutions of (25). There, the value of the critical radius, r_c, of nuclei is varied. It is shown that the critical radius controls essentially the value for the Avrami exponent n that can be estimated in the range of low volume fraction.

For the derivation of expression (17) it was assumed that the particles have a spherical shape. This supposition was made because it makes the derivation quite simple and lucid. It should be noted that a more general derivation is possible if one uses the concept of contact distribution functions (see, e.g., [46]). The spherical contact distribution function is a measure for the distribution of distances, r, between a point chosen at random outside the set B of a Boolean model and the next nearest point of B (B is the set covered by the grains of the Boolean model). It can be defined for all types of grains suitable for the construction of Boolean models, e.g. also for random polyhedra.

On the other hand, expression (17) implies two approximations. One consists in the description of the crystallised regions of the alloy by means of a Boolean model with equally sized grains. At a certain stage of the evolution of the system there are, however, particles which have already reached their final diameter but other ones are still growing and have therefore smaller diameters. Thus, the particles obey a certain size distribution. It would be possible to incorporate the actual size distribution in the derivation of the expression for the nucleation but this would make the calculation more complicated. However, the essential effects are already taken into account in the present simple version.

Another point is that the microstructure is approximated by a Boolean model. The positions of the grains of the Boolean model are random without any interaction. This seems to be in contrast to the existence of interactions between the position of a nucleus to be created in the system and the particles already formed. However, it is not a serious restriction. If a new nucleus is considered the existing microstructure is approximated by a Boolean model, but the interactions between them are correctly calculated. Then, the changed microstructure is again approximated by a Boolean model with modified parameters (the volume fraction has increased) and the interaction of the next nucleus with the actual microstructure is considered. Comparing the volume fraction and the internal surface area of a hardcore arrangement with the corresponding properties of a Boolean model with equally sized spherical grains one obtains

$$r_\eta = r_b \frac{1-\eta}{\eta} \log \frac{1}{1-\eta} \tag{26}$$

for the grain radius.

It should be noted that it is also possible to consider complementary approaches using thermodynamic arguments for the decrease of the nucleation rate. This has been discussed by [55].

Expression (24) for the fractional contribution of a new crystallite to the increment of the crystallised volume fraction is an approximation which is valid for low and medium values of η. This is consistent with the approximation made for the description of the microstructure of the partially crystallised alloy discussed above. It is important that there are no restrictions to the type of the growth law of the crystallites. For the present purpose, (18) is suitable but other expressions may also be used.

Now we consider amorphous metallic alloys of the composition $Fe_{77.5-x-y}Si_{15.5}B_7Cu_xI_y$ (I = V, Mo, W, Ta, Nb, Hf, Nd), $x = 0$ to 1, $y = 0$ to 5, prepared by rapid quenching from the liquid state. The amorphous ribbons obtained are subjected to isothermal annealing whereby they transform into a nanocrystalline state. The nanocrystals have the composition Fe_3Si. They are embedded in a remaining amorphous matrix. The volume fraction of the crystallites ranges from 50% to 75% in the final stage and their diameter is between about 10nm and 30nm. Figure 7a shows the evolution of the crystallised volume fraction determined by X-ray scattering measurements (see [20] and references therein). The data do not follow straight lines as predicted by the KJMA model except at the early stage of the crystallisation process. The values for the Avrami exponent estimated there are about $n \approx 1$, which is outside the range of validity of the KJMA model ($n \geq 1.5$ for three-dimensional growth, see Sect. 2). In Fig. 7b the mean particle diameter is plotted vs. time for the alloy with I = Nb, $x = 1, y = 3$. The experimental data show the behaviour predicted by (18) with $r_b = 8$nm and $r_0(t) - r_c \propto \sqrt{t}$, i.e. diffusion controlled intrinsic growth. As discussed in Sect. 2 decrasing growth rates of the particles are not consistent with assumptions made in the KJMA model. The present theoretical approach is, however, applicable because decreasing growth rates of the particles are allowed. The low values of the parameter n estimated by formal application of

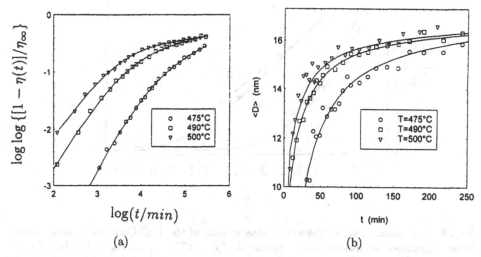

Fig. 7. Experimental data for the crystallisation of 1at%Cu3at%Nb alloys connected by a solid line for each annealing temperature; Avrami plot of the crystallised volume fraction (a) and mean diameter of crystallites (b).

(2) to the data at the early stage of the crystallisation process can already be understood if the existence of a critical radius of the nuclei is taken into account (see Fig. 6).

From X-ray scattering experiments, data for the mean size of the particles and the crystallised volume fraction are available, and one can estimate the number and, consequently, the nucleation rate of particles. Defining $\mu_a(\eta) = \mu(t)/[1 - \eta(t)]$ we compare now the experimental values for the 1at%Cu3at%Nb alloy with expression (17). Figure 8 shows the result for two samples crystallised at different annealing temperatures. For the theoretical curves the parameter values $\delta = 0.25, r_c/r_b = 0.1, r_b = r_f = 8$nm are chosen according to experimental data, and r_{min}/r_b is varied in order to obtain a quantitative agreement with the experimental curves. The values for r_{min}/r_b (0.8 and 0.3 at $T_A = 475°C$ and $565°C$, respectively) show that at high temperatures the minimum distance between a nucleus to be created and existing crystallites is smaller than at low annealing temperatures.

4   Random Tessellations

A completely crystallised solid may be considered as a random or partially ordered tessellation of the 3- (bulk materials) or 2-dimensional (thin films) space. The crystallites are described by the cells of the mosaic, and the grain boundaries between adjacent crystallites form the faces (or the edges in 2-dimensional case) of the mosaic. Both the atomic arrangement at the grain boundaries (see, e.g.,

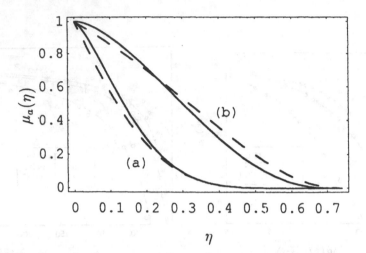

Fig. 8. Nucleation rate of the amorphous matrix of the 1at%Cu3at%Nb alloy. Solid lines: experimental; dashed line: theoretical. $T_A = 475°C$, $r_{min}/r_b = 0.8$ (a), $T_A = 565°C$, $r_{min}/r_b = 0.3$ (b).

[3,7,8,47]) and the geometric characteristics of the mosaic (see, e.g., [2,34,56]) are of great scientific and technological interest. Mathematical models that can be used as approximate descriptions of polycrystalline microstructures are the Poisson plane and line mosaics, the Dirichlet-Voronoi mosaic and the Johnson-Mehl tessellation model. Of course, these models cannot explain details of the microstructure of specific materials. Models for real materials require, in general, computer simulations where properties such as the relationship between the grain boundary energy and the crystallographic orientation of adjacent crystallites or the anisotropy of growth rates can be taken into account (see, e.g., [16]). The dynamics of mosaic structures can be simulated, e.g. by minimising the interface energy of the system considered. Frost [14] summarised physical models describing the evolution of the grain size distribution during grain growth and considered peculiarities of grain growth in thin films ([15,53]). In recent studies of recrystallisation processes ([29]) the concept of cellular automata (see [54]) has been used to simulate the evolution of mosaic structures.

Despite the rapid development of computer simulation techniques the mathematical models mentioned above are keeping their significance. One of the reasons for this is that there is a series of exact results that can be used either as a base for the construction of more realistic models or as special case for testing of simulation methods.

4.1   Poisson Line and Plane Tessellations

The Poisson line mosaic is generated by a random distribution of straight lines in the plane where the distribution of the lines is isotropic and stationary. The

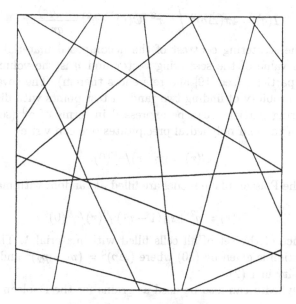

Fig. 9. Poisson line mosaic.

model is characterised by the mean number, ρ, of lines intersected by a test line line of unit length. Figure 9 illustrates the model. The Poisson plane model is produced by a random distribution of planes in the three-dimensional space. Then, parameter ρ is the number density of intersection points of planes on the test line. The mathematical foundations of the model are given in [30] and [37]. A compilation of the results obtained for characteristic quantities of the model can be found in [46]. This concerns mean values and second-order moments of parameters such as area, perimeter, edge number of cells and the probability that a polygon has n sides, and corresponding parameters for the Poisson plane model. Here we consider the distance probability function

$$\gamma^0(r) = \langle \frac{1}{4\pi} \int_\Omega \int_V s(\boldsymbol{x}+\boldsymbol{r})s(\boldsymbol{x})dV\,d\Omega \rangle \tag{27}$$

where

$$s(\boldsymbol{x}) = \begin{cases} 1 \,, \boldsymbol{x} \in B_0 \\ 0 \,, \text{otherwise} \end{cases} \tag{28}$$

is the shape function of a single Poisson polyhedron, B_0, and $\langle ... \rangle$ denotes the average over all polyhedra of the mosaic. The result is

$$\gamma^0(r) = \frac{6}{\pi\rho^3} \exp(-\rho r). \tag{29}$$

Expression (29) is quite important for applications. It can be used, e.g., for analyses of small-angle scattering data. The small-angle scattering intensity, $I(q)$, is given by

$$I(q) = 4\pi(\Delta p)^2 \int_0^\infty r^2 \left[C(r) - \eta^2\right] \frac{\sin(qr)}{qr} dr \qquad (30)$$

where Δp is the scattering contrast of the decomposed material considered, q is the absolute value of the scattering vector and η is the volume fraction of the scattering particles (see [19] and references therein). The covariance, $C(r)$, defined by the probility of finding two random test points with distance r both lying in scattering particles, can be expressed in terms of γ^0 (see [46]). For a dilute system of random polyhedral precipitates one can write

$$C(r) = \eta\gamma^0(r)/\gamma^0(0). \qquad (31)$$

If the cells of the Poisson plane model are filled at random with material of type 1 or 2 then

$$C_1(r) = \eta_1^2 + \eta_1(1 - \eta_1)\gamma^0(r)/\gamma^0(0) \qquad (32)$$

is the covariance of the set of all cells filled with material 1. The small-angle scattering function is given by (30) where $(\Delta p)^2 = (p_1 - p_2)^2$ and p_1 (p_2) is the scattering density in 1 (2).

Poisson polyhedra can also be used as grains for the Boolean model. Then, (see[46])

$$C(r) = 2\eta - 1 + (1 - \eta)^2 \exp\left[\lambda\gamma^0(r)\right]. \qquad (33)$$

4.2 Voronoi and Johnson-Mehl Tessellations

The process of nucleation and growth of particles is often accompanied by the impingement of growing particles (see Fig. 3). During this process the volume fraction, η, of the particles and the number of interfaces created by impingement of neighbouring particles increase. At $\eta = 1$ the microstructure consists of a spacefilling arrangement of particles separated by their interfaces. Such types of microstructures can be modelled by Voronoi and Johnson-Mehl tessellations. (The two-dimensional Voronoi tessellation is also called Dirichlet tessellation.) If all particles nucleate at the same time and if the growth rate is constant and the same for all particles one obtains a Voronoi mosaic. Note that the increase of the volume fraction during this process is described by the KJMA model and (2). In this case it is not necessary to consider details of the nucleation-and-growth process for the construction of the Voronoi mosaic. The mosaic can be generated immediately starting from the distribution of the nucleation sites. The cell belonging to an arbitrary nucleation site comprises all points of the space that are closer to this site than to any other one. There are no restrictions to the distribution of the nucleations sites. Figure 10 shows a two-dimensional example for a Voronoi mosaic where the distribution of the nucleation sites corresponds to a Poisson point field.

The Poisson-Voronoi mosaic was analysed by many authors (see, e.g., [27,28]). The main results are summarised in [46]. It should be noted that the cells of the Poisson-Voronoi mosaic can be used also as grains for the Boolean model. This is discussed in [19].

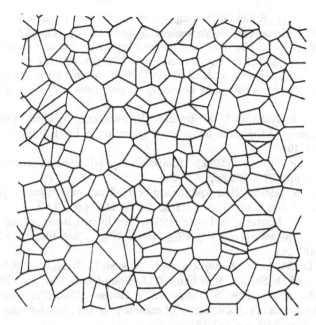

Fig. 10. Two-dimensional Poisson-Voronoi tessellation.

More general than the Voronoi mosaic is the Johnson-Mehl tessellation. It can be used to simulate arbitrary nucleation-and-growth processes. Exact results are, however, available only for special cases. Møller [33] studied random Johnson-Mehl tessellations generated by time-inhomogeneous Poisson processes and established results for first- and second-order moments for various characteristics. The most important results are summarised in [46].

Acknowledgement

Sincere thanks are due to D. Stoyan and K. Mecke for critical and helpful comments on an earlier version of this article.

References

1. Aaron, H. B., D. Fainstein, G. Kotler (1970): 'Diffusion-limited phase transformations: A comparison and critical evaluation of the mathematical approximations', J. Appl. Phys. **41**, pp. 4404-4410
2. Abbruzzese, G., P. Brozzo (Eds.) (1992): *Grain Growth in Polycrystalline Materials* (Trans Tech Publications, Zurich)
3. Alber, U., H. Müllejans, M. Rühle (1999): 'Bismuth segregation at copper grain boundaries', Acta mater. **47**, pp. 4047-4060
4. Bhargava, A., I. P. Jain (1994): 'Growth kinetics of the crystalline phase in $Se_{70}Te_{30}$ glass', J. Phys. D: Appl. Phys. **27**, pp. 830-833

262   Helmut Hermann

5. Binder, K. (1991): 'Spinodal Decomposition'. In: *Phase Transformations in Materials*, ed. by P. Haasen (VCH Verlagsges., Weinheim), pp. 405-472
6. Cahn, J. W. (1996): 'The time cone method for nucleation and growth kinetics on a finite domain', Mat. Res. Soc. Symp. Proc. **398**, pp. 425-437
7. Campbell, G. H., J. Belak, J. A. Moriarty (1999): 'Atomic structure of the Sigma-5 (319)/[001] symmstry tilt grain boundary in molybdenum', Acta mater. **47**, pp. 3977-3985
8. Chang, L.-S., E. Rabkin, B. B. Straumal, B. Baretzky, W. Gust (1999): 'Thermodynamic aspects of the grain boundary segregation in Cu(Bi) alloys', Acta mater. **47**, pp. 4041-4046
9. Chen, L. C., F. Spaepen (1991): 'Analysis of calorimetric measurement of grain growth', J. Appl. Phys. **69**, pp. 679-688
10. Clavaguera, N. (1993): 'Non-equilibrium crystallization, critical cooling rates and transformation diagrams', J. Non-Cryst. Solids **162**, pp. 40-50
11. Danzig, A., N. Mattern, S. Doyle (1995): 'An in-situ investigation of the Fe_3Si crystallization in amorphous $Fe_{73.5}Si_{15.5}B_7Cu_1Nb_3$', Nucl. Instr. Meth. Phys. Res. B **97**, pp. 465-567
12. Dimitrov, D., M. A. Ollazarizqueta, C. N. Afonso, N. Starbov (1996): 'Crystallization kinetics of Sb_xSe_{100-x} thin films', Thin Solid Films **280**, pp. 278-283
13. Dobreva, A., A. Stoyanov, S. Tzuparska, I. Gutzov (1996): 'Non-steady-state effects in the kinetics of crystallization of organic polymer glass-forming melts', Thermochimica Acta **280**, pp. 127-151
14. Frost, H. J. (1992): 'Stochastic models of grain growth', Materials Sci. Forum **94-96**, pp. 903-908
15. Frost, H. J., C. V. Thompson, D. T. Walton (1992): 'Abnormal grain growth in thin films due to anisotropy of free.surface energies', Materials Sci. Forum **94-96**, pp. 543-550
16. Gottstein, G., L. S. Shvindlerman (1999): *Grain Boundary Migration in Metals: Thermodynamics, Kinetics, Applications* (CRC Press, Boca Raton)
17. Graydon, J. W., S. J. Thorpe, D. W. Kirk (1994) Determination of the Avrami exponent for solid state transformations from nonisothermal differential scanning calorimetry. J. Non-Cryst. Solids **55**, pp. 14071-14073
18. Haasen, P. (Ed.) (1991): *Phase Transformations in Materials* (VCH Verlagsges., Weinheim)
19. Hermann, H. (1991): *Stochastic Models of Heterogeneous Materials* (Trans Tech Publications, Zurich)
20. Hermann, H., N. Mattern (1999): 'Nanocrystallization in amorphous metallic alloys - experiment and theory', Recent Res. Devel. Nanostructures **1**, pp. 97-113
21. Hermann, H., N. Mattern, S. Roth, P. Uebele (1997): 'Simulation of crystallization processes in amorphous iron-based alloys', Phys. Rev. B **56**, pp. 13888-13897
22. Jou, H. J., M. T. Lusk (1997): Comparison of Johnson-Mehl-Avrami-Kolmogorov kinetics with a phase field model for microstructural evolution driven by substructure energy. Phys. Rev. B **55**, pp. 8114-8121
23. Kelton, K. F. (1993): 'Numerical model for isothermal and non-isothermal crystallization of liquids and glasses', J. Non-Cryst. Solids **163**, pp. 283-296
24. Kelton, K. F. (1997): 'Analysis of crystallization kinetics', Mater. Sci. Eng. A **226**, pp. 142-150
25. Kelton, K. F., K. L. Narayan, L. E. Levine, T. C. Cull, C. S. Ray (1996): 'Computer modelling of nonisothermal crystallization,' J. Non-Cryst. Solids **204**, pp. 13-31
26. Kožíšek, Z. (1991): 'Influence of the form factor on homgeneous nucleation kinetics', Cryst. Res. Technol. **26**, pp. 11-17

27. Lorz, U. (1990): 'Cell-area distributions of planar sections of spatial Voronoi mosaics', Mater. Char. **3**, pp. 297-311

28. Lorz, U. (1995): 'Statistics for the spatial Poisson-Voronoi tessellation'. In: *Complex Stochastic Systems and Engineering*, ed. by D. M. Titterington (Clarendon Press, Oxford), pp. 141-153

29. Marx, V., F. R. Reher, G. Gottstein (1999): 'Simulation of primary recrystallization using a modified 3-dimensional cellular automaton', Acta mater. **47**, pp. 1219-1230

30. Matheron, G. (1975): *Random Sets and Integral Geometry* (J. Wiley & Sons, New York)

31. Mattern, N., A. Danzig, M. Müller, U. Kühn, S. Doyle (1996): 'High-temperature X-ray investigations of the crystallization of amorphous Fe-based alloys', Mater. Sci. Forum **225-227**, pp. 341-346

32. Michaelsen, C., M. Dahms (1996): 'On the determination of nucleation and growth kinetics by calorimetry', Thermochimica Acta **288**, pp. 9-27

33. Møller, J. (1992): 'Random Johnson-Mehl tessellations', Adv. Appl. Prob. **24**, pp. 814-844

34. Novikov, V. (1997): *Grain Growth and Control of Microstructure and Texture in Polycrystalline Materials* (CRC Press, Boca Raton)

35. Pradell, T., D. Crespo, N. Clavaguera, M. T. Clavaguera-Mora (1999): 'Nanocrystallisation in Finemet alloys with different Si/B ratio', J. Metastable Nanocryst. Mater. **1**, pp. 83-88

36. Roth, S. (1996): 'The growth of α-Fe(Si) particles in an amorphous $Fe_{73.5}Mo_3Cu_1Si_{13.5}B_9$ alloy', J. Magn. Magn. Mat. **160**, pp. 266-268

37. Serra, J. (1982): *Image Analysis and Mathematical Morphology* (Acad. Press, London)

38. Sessa, V., M. Fanfoni, M. Tomellini (1996): 'Validity of Avrami kinetics for random and nonrandom distribution of germs', Phys. Rev. B **54**, pp. 836-841

39. Shepilov, M. P. (1996): 'A model for calculation of isothermal kinetics of the nucleation-and-growth type phase separation in the course of one-step heat treatment', J. Non-Cryst- Solids **208**, pp. 64-80

40. Shepilov, M. P., D. S. Baik (1994): 'Computer simulation of crystallization kinetics for the model with simulataneous nucleation of randomly-oriented ellipsoidal crystals', J. Non-Cryst. Solids **171**, pp. 141-156

41. Shepilov, M. P., V. B. Bochkariov (1990): 'Computer simulation of the crystal volume distribution in a fully crystallized glass. Application to the estimation of the precipitate size distribution after the completion of the stage of diffusion-limited growth from supersaturated solid solution', J. Non-Cryst. Solids **125**, pp. 161-167

42. Shneidman, V. A., M. C. Weinberg (1993): 'The effect of transient nucleation and size-dependent growth rate on phase transformation kinetics', J. Non-Cryst. Solids **160**, pp. 89-98

43. Shneidman, V. A., M. C. Weinberg (1996): 'Crystallization of rapidly heated amorphous solids', J. Non-Cryst. Solids **194**, pp. 145-154

44. Starink, M. J., A. M. Zahra (1997): 'An analysis method for nucleation and growth controlled reactions at constant heating rate', Thermochimica Acta **292**, pp. 159-168

45. Stojanova, L., K. Russew (1997): 'Crystallization behaviour and viscous flow of $Zr_{50}Ni_{50-x}Cu_x$ metallic glasses under constant heating rate conditions', Mater. Sci. Eng. A **226**, pp. 483-486

46. Stoyan, D., W. S. Kendall, J. Mecke (1995): *Stochastic Geometry and its Applications*, 2nd edition (Wiley, Chichester)

264   Helmut Hermann

47. Sutton, A. P., R. W. Baluffi (1995): *Interfaces in Crystalline Materials* (Oxford Sci. Publishers)
48. Tomellini, M., M. Fanfoni (1997): 'Why phantom nuclei must be considered in the Johnson-Mehl-Avrami-Kolmogorov kinetics', Phys. Rev. B **55**, pp. 14071-14073
49. Tomic, P., M. Davidovic (1996): 'The study of relaxational properties of the metallic glass $Fe_{80}Si_{10}B_{10}$', J. Non-Cryst. Solids **204**, pp. 32-37
50. Uebele, P., H. Hermann (1996): 'Computer simulation of crystallization kinetics with non-Poisson distributed nuclei', Modelling Simul. Mater. Sci. Eng. **4**, pp. 203-214
51. VanSiclen, C. D. (1996): 'Random nucleation and growth kinetics', Phys. Rev. B **54**, pp. 11845-11848
52. Wagner, R., R. Kampmann (1991): 'Homogeneous Second Phase Precipitation'. In: *Phase Transformations in Materials*, ed. by P. Haasen (VCH Verlagsges., Weinheim), pp. 213-304
53. D. T. Walton, H. J. Frost, C. V. Thompson (1992): 'Modelling of grain growth in thin film strips', Materials Sci. Forum **94-96**, pp. 531-536
54. Wolfram, S. (1986): *Theory and Application of Cellular Automata* (World Scientific Publishing, Singapore)
55. Yavari, A. R., D. Negri (1999): 'Primary phase nucleation regime in amorphous alloys is non-steady-state', J. Metastable Nanocryst. Mater. **1**, pp. 63-68
56. Yoshinaga, H., T. Watanabe et al. (Eds.) (1996): *Grain Growth in Polycrystalline Materials* (Trans Tech Publications, Zurich)

Part III

Phase Transitions and Simulations
of Hard Particles

Phase Transition and Percolation in Gibbsian Particle Models

Hans-Otto Georgii

Mathematisches Institut der Universität München, Theresienstr. 39
D-80333 München, Germany

Abstract. We discuss the interrelation between phase transitions in interacting lattice or continuum models, and the existence of infinite clusters in suitable random-graph models. In particular, we describe a random-geometric approach to the phase transition in the continuum Ising model of two species of particles with soft or hard interspecies repulsion. We comment also on the related area-interaction process and on perfect simulation.

1 Gibbs Measures: General Principles

This section contains a brief introduction to the basic physical and stochastic ideas leading to the concept of Gibbs measures. The principal question is the following:

> Which kind of stochastic model is appropriate for the description of spatial random phenomena involving a very large number of components which are coupled together by an interaction depending on their relative position?

To find an answer we will start with a spatially discrete situation; later we will proceed to the continuous case. Consider the phenomenon of ferromagnetism. A piece of ferromagnetic material like iron or nickel can be imagined as consisting of many elementary magnets, the so-called spins, which are located at the sites of a crystal lattice and have a finite number of possible orientations (according to the symmetries of the crystal). The essential point is that these spins interact with each other in such a way that neighboring spins prefer to be aligned. This interaction is responsible for the phenomenon of spontaneous magnetization, meaning that at sufficiently low temperatures the system can choose between several distinct macrostates in which typically all spins have the same orientation.

How can one find a mathematical model for such a ferromagnet? The first fact to observe is that the number of spins is very large. So, probabilistic experience with the law of large numbers suggests to approximate the large finite system by an infinite system in order to get clear-cut phenomena. This means that we should assume that the underlying crystal lattice is infinite. The simplest case to think of is the d-dimensional hypercubic lattice $\mathbb{Z}^d$. (As the case $d = 1$ is rather trivial, we will always assume that $d \geq 2$.) On the other hand, to keep the model simple it is natural to assume that each spin has only finitely many possible orientations. In other words, the random spin ξ_i at lattice site i takes values in a finite state space S. The set of all possible spin configurations $\xi = (\xi_i)_{i \in \mathbb{Z}^d}$ is then the product space $\Omega = S^{\mathbb{Z}^d}$. This so-called *configuration space* is equipped with the Borel σ-algebra $\mathcal{F}$ for the natural product topology on Ω. Since the spins are random, we are interested in probability measures P on $(\Omega, \mathcal{F})$. Such probability spaces are known as *lattice systems*. For any $\xi \in \Omega$ and $\Lambda \subset \mathbb{Z}^d$ we write $\xi_\Lambda = (\xi_i)_{i \in \Lambda}$ for the part of the configuration that occurs in Λ. By abuse of notation, we use the same symbol ξ_Λ for the projection from Ω onto S^Λ.

Which kind of probability measure on $(\Omega, \mathcal{F})$ can serve as a model of a ferromagnet? As we have seen above, the essential feature of a ferromagnet is the interaction between the spins. We are thus interested in probability measures P on Ω for which the spin variables ξ_i, $i \in \mathbb{Z}^d$, are *dependent*. A natural way of describing dependencies is to prescribe certain conditional probabilities. This idea, which is familiar from Markov chains, turns out to be suitable also here. Since our parameter set $\mathbb{Z}^d$ admits no natural linear order, the conditional probabilities can, of course, not lead from a past to a future. Rather we *prescribe the behavior of a finite set of spins when all other spins are fixed*. In other words, we are interested in probability measures P on $(\Omega, \mathcal{F})$ having prescribed conditional probabilities

$$G_\Lambda(\xi_\Lambda | \xi_{\Lambda^c}) \tag{1}$$

for a configuration $\xi_\Lambda \in S^\Lambda$ within a finite set $\Lambda \subset \mathbb{Z}^d$ given a fixed configuration $\xi_{\Lambda^c} \in S^{\Lambda^c}$ off Λ. In the following we write $\Lambda \subset\subset \mathbb{Z}^d$ when Λ is a *finite* subset of $\mathbb{Z}^d$. The specific form of these conditional distributions does not matter at the moment. Two special cases are

- *the Markovian case:* the conditional distribution (1) only depends on the value of the spins along the *boundary* $\partial \Lambda = \{i \notin \Lambda : |i - j| = 1 \text{ for some } j \in \Lambda\}$ of Λ, i.e.,

$$G_\Lambda(\xi_\Lambda | \xi_{\Lambda^c}) = G_\Lambda(\xi_\Lambda | \xi_{\partial \Lambda}) \tag{2}$$

(with a slight abuse of notation); $| \cdot |$ stands for the Euclidean norm.

- *the Gibbsian case:* the conditional distribution (1) is defined in terms of a *Hamilton function* H_Λ by the Boltzmann–Gibbs formula

$$G_\Lambda(\xi_\Lambda | \xi_{\Lambda^c}) = Z_{\Lambda | \xi_{\Lambda^c}}^{-1} \exp[-H_\Lambda(\xi)] , \tag{3}$$

where $Z_{\Lambda | \xi_{\Lambda^c}} = \sum_{\xi' \in \Omega : \xi' \equiv \xi \text{ off } \Lambda} \exp[-H_\Lambda(\xi')]$ is a normalizing constant. Physically, $H_\Lambda(\xi)$ describes the energy excess of the total configuration ξ over the energy of the outer configuration ξ_{Λ^c}. (Physicists will miss here the factor β, the inverse temperature; we will assume that β is subsumed into H_Λ or, equivalently, that the units are chosen in such a way that $\beta = 1$.)

In the following, $G_\Lambda(\cdot | \xi_{\Lambda^c})$ will be viewed as a probability measure on Ω for which the configuration outside Λ is almost surely equal to ξ_{Λ^c}.

The above idea of prescribing conditional probabilities leads to the following concept introduced in the late 1960's independently by R.L. Dobrushin, and O.E. Lanford and D. Ruelle.

Definition 1 *A probability measure P on $(\Omega, \mathcal{F})$ is called a Gibbs measure, or DLR-state, for a family $\mathbf{G} = (G_\Lambda)_{\Lambda \subset \subset \mathbb{Z}^d}$ of conditional probabilities (satisfying the natural consistency condition) if*

$$P(\xi_\Lambda \text{ occurs in } \Lambda \,|\, \xi_{\Lambda^c} \text{ occurs off } \Lambda) = G_\Lambda(\xi_\Lambda \,|\, \xi_{\Lambda^c})$$

for P-almost all ξ_{Λ^c} and all $\Lambda \subset \subset \mathbb{Z}^d$.

If $\mathbf{G}$ is Gibbsian for a Hamiltonian H as in (3), each Gibbs measure can be interpreted as an equilibrium state for a physical system with state space S and interaction H. This is because the Boltzmann–Gibbs distribution maximizes the entropy when the mean energy is fixed; we will discuss this point in more detail in the continuum setting in Sect. 4.2 below.

A general account of the theory of Gibbs measures can be found in the monograph [5]; here we will only present the principal ideas. In contrast to the situation for Markov chains, Gibbs measures do not exist automatically. However, in the present case of a finite state space S, Gibbs measures do exist whenever $\mathbf{G}$ is Markovian in the sense of (2), or almost Markovian in the sense that the conditional probabilities (1) are continuous functions of the outer configuration ξ_{Λ^c}. In this case one can show that any weak limit of $G_\Lambda(\cdot \,|\, \xi_{\Lambda^c})$ for fixed $\xi \in \Omega$ as $\Lambda \uparrow \mathbb{Z}^d$ is a Gibbs measure.

The basic observation is that the Gibbs measures for a given consistent family $\mathbf{G}$ of conditional probabilities form a convex set $\mathcal{G}$. Therefore one is interested in its extremal points. These can be characterized as follows.

Theorem 1 *Let $\mathcal{T} = \bigcap \sigma(\xi_{\Lambda^c} : \Lambda \subset \subset \mathbb{Z}^d)$ the tail σ-algebra, i.e., the σ-algebra of all macroscopic events not depending on the values of any finite set of spins. Then the following statements hold:*

(a) A Gibbs measure $P \in \mathcal{G}$ is extremal in $\mathcal{G}$ if and only if P is trivial on $\mathcal{T}$, i.e., if and only if any tail measurable real function is P-almost surely constant.

(b) Any two distinct extremal Gibbs measure are mutually singular on $\mathcal{T}$.

(c) Any non-extremal Gibbs measure is the barycenter of a unique probability weight on the set of extremal Gibbs measures.

A proof can be found in [5], Theorems (7.7) and (7.26). Statement (a) means that the extremal Gibbs measures are *macroscopically deterministic*: on the macroscopic level all randomness disappears, and an experimenter will get non-fluctuating measurements of macroscopic quantities like magnetization or energy per lattice site. Statement (b) asserts that *distinct extremal Gibbs measures show different macroscopic behavior.* So, they can be distinguished by looking at typical realizations of the spin configuration through macroscopic glasses. Finally, statement (c) implies that any realization which is typical for a non-extremal Gibbs measure is in fact typical for a suitable extremal Gibbs measure. In physical terms: *any configuration which can be seen in nature is governed by an extremal Gibbs measure*, and the non-extremal Gibbs measures can only be interpreted in a Bayesian way as measures describing the uncertainty of the experimenter. These observations lead us to the following definition.

Definition 2 *Any extremal Gibbs measure is called a phase of the corresponding physical system. If distinct phases exist, one says that a phase transition occurs.*

So, in terms of this definition the existence of phase transition is equivalent to the non-uniqueness of the Gibbs measure. In the light of the preceding theorem, this corresponds to a "macroscopic ambivalence" of the system's behavior. We should add that not all critical phenomena in nature can be described in this way: even when the Gibbs measure is unique it may occur that it changes its qualitative behavior when some parameters of the interaction are changed. However, we will not discuss these possibilities here and stick to the definition above for definiteness. In this contribution we will ask:

> What are the driving forces giving rise to a phase transition? Is there any mechanism relating microscopic and macroscopic behavior of spins?

As we will see, in a number of cases one can give the following geometric answer:

> One such mechanism is the formation of infinite clusters in suitably defined random graphs. Such infinite clusters serve as a link between the local and global behavior of spins, and make visible how the individual spins unite to form a specific collective behavior.

In the next section we will discuss two lattice models for which this answer is correct. In Sect. 4 we will show that a similar answer can also be given in a continuum set-up. A useful technical tool is the stochastic comparison of probability measures.

Suppose the state space S is a subset of $\mathbb{R}$ and thus linearly ordered. Then the configuration space Ω has a natural partial order, and we can speak of increasing real functions. Let P, P' be two probability measures on Ω. We say that P is *stochastically smaller* than P', and write $P \preceq P'$, if $\int f\, dP \leq \int f\, dP'$ for all local increasing functions (or, equivalently, for all measurable bounded increasing functions) f on Ω. A sufficient condition for stochastic monotonicity is given in the proposition below. Although this condition refers to the case of finite products (for which stochastic monotonicity is similarly defined), it is also useful in the case of infinite product spaces. This is because (by the very definition) the relation $\preceq$ is preserved under weak limits.

Proposition 1 (Holley's inequality) *Let S be a finite subset of $\mathbb{R}$, Λ a finite index set, and P, P' two probability measures on the finite product space S^Λ giving positive weight to each element of S^Λ. Suppose the single-site conditional probabilities at any $i \in \Lambda$ satisfy*

$$P(\,\cdot\,|\xi_{\Lambda\backslash\{i\}} \text{ occurs off } i) \preceq P'(\,\cdot\,|\xi'_{\Lambda\backslash\{i\}} \text{ occurs off } i) \quad \text{whenever } \xi \leq \xi'.$$

Then $P \preceq P'$. If this condition holds with $P' = P$ then P has positive correlations in the sense that any two bounded increasing functions are positively correlated.

For a proof (and a slight extension) we refer to Theorems 4.8 and 4.11 of [10].

2 Phase Transition and Percolation: Two Lattice Models

To provide the necessary background for our results on continuum particle systems let us still stick to the lattice case. We will consider two classical models which allow an understanding of phase transition in random-geometric terms. Many further examples for the relation between random geometry and phase transition can be found in [10].

Let us start recalling some basic facts on *Bernoulli percolation* on $\mathbb{Z}^d$ for $d \geq 2$. Consider $\mathbb{Z}^d$ as a graph with vertex set $\mathbb{Z}^d$ and edge set $E(\mathbb{Z}^d) = \{e = \{i, j\} \subset \mathbb{Z}^d : |i - j| = 1\}$. We fix two parameters $0 \leq p_s, p_b \leq 1$, the site and bond probabilities, and construct a random subgraph $\Gamma = (X, E)$ of $(\mathbb{Z}^d, E(\mathbb{Z}^d))$ by setting

$$X = \{i \in \mathbb{Z}^d : \xi_i = 1\}, \quad E = \{e \in E(X) : \eta_e = 1\},$$

where $E(X) = \{e \in E(\mathbb{Z}^d) : e \subset X\}$ is the set of all edges between the sites of X, and ξ_i, $i \in \mathbb{Z}^d$, and η_e, $e \in E(\mathbb{Z}^d)$, are independent Bernoulli variables satisfying $P(\xi_i = 1) = p_s$, $P(\eta_e = 1) = p_b$. This construction is called the *Bernoulli mixed site-bond percolation model*; setting $p_b = 1$ we obtain pure site percolation, and the case $p_s = 1$ corresponds to pure bond percolation.

Let $\{0 \leftrightarrow \infty\}$ denote the event that Γ contains an infinite path starting from 0, and

$$\theta(p_s, p_b; \mathbb{Z}^d) = \mathrm{Prob}(0 \leftrightarrow \infty)$$

be its probability. By Kolmogorov's zero-one law, we have $\theta(p_s, p_b; \mathbb{Z}^d) > 0$ if and only if Γ contains an infinite cluster with probability 1. In this case one says that percolation occurs. The following proposition asserts that this happens in a nontrivial region of the parameter square, which is separated by the so-called critical line from the region where all clusters of Γ are almost surely finite. The change of behavior at the critical line is the simplest example of a critical phenomenon.

Proposition 2 *The function $\theta(p_s, p_b; \mathbb{Z}^d)$ is increasing in p_s, p_b and d. Moreover, $\theta(p_s, p_b; \mathbb{Z}^d) = 0$ when $p_s p_b$ is small enough, while $\theta(p_s, p_b; \mathbb{Z}^d) > 0$ when $d \geq 2$ and $p_s p_b$ is sufficiently close to 1.*

Sketch proof: The monotonicity in p_s and p_b follows from Proposition 1, and the one in d from an obvious embedding argument. To show that $\theta = 0$ when $p_s p_b$ is small, we note that the expected number of neighbors in Γ of a given lattice site is $2d p_s p_b$. Comparison with a branching process thus shows that $\theta = 0$ when $2d p_s p_b < 1$.

Next, let $d = 2$ and suppose $0 \in X$ but the cluster C_0 of Γ containing 0 is finite. Consider $\partial_{ext} C_0$, the part of ∂C_0 belonging to the infinite component of C_0^c. For each site $i \in \partial_{ext} C_0$, either this site or all bonds leading from i to C_0 do not belong to Γ. This occurs with probability at most $1 - p_s p_b$. So, the probability that $\partial_{ext} C_0$ has a fixed location is at most $(1 - p_s p_b)^\ell$ with $\ell = \#\partial_{ext} C_0$. Counting all possibilities for this location one finds that $1 - \theta < 1$ when $1 - p_s p_b$ is small

enough. By the monotonicity in d, the same holds a fortiori in higher dimensions.
□

The above proposition is all what we need here on Bernoulli percolation; an excellent source for a wealth of further rigorous results is the book of [13].

We now ask for the role of percolation for Gibbs measures, and in particular for the existence of phase transitions. Of course, in contrast to the above this will involve *dependent*, i.e., non-Bernoulli percolation. We consider here two specific examples. In these examples, the family $\mathbf{G}$ of conditional probabilities will be Gibbsian for a nearest-neighbor interaction; this means that both (2) and (3) are valid.

2.1   The Ising Model

This is by far the most famous model of Statistical Mechanics, named after E. Ising who studied this model in the early 1920s in his thesis suggested by W. Lenz. It is the simplest model of a ferromagnet in equilibrium. One assumes that the spins have only two possible orientations, and therefore defines $S = \{-1, 1\}$. The family $\mathbf{G}$ is defined by (3) with

$$H_\Lambda(\xi) = J \sum_{\{i,j\} \cap \Lambda \neq \emptyset : |i-j|=1} 1_{\{\xi_i \neq \xi_j\}} , \tag{4}$$

where $J > 0$ is a coupling constant which is inversely proportional to the absolute temperature. This means that neighboring spins of different sign have to pay an energy cost J. There exist two configurations of minimal energy, so-called ground states, namely the configuration '+' which is identically equal to $+1$, and the configuration '−' identically equal to -1. The behavior of the model is governed by these two ground states. To see this we begin with some useful consequences of the ferromagnetic character of the interaction. First, it is intuitively obvious that the measures $G_\Lambda^+ = G_\Lambda(\cdot \,|+)$ decrease stochastically when Λ increases (since then the effect of the + boundary decreases). This follows easily from Holley's inequality, Proposition 1. Since the local increasing functions are a convergence determining class, it follows that the weak infinite-volume limit $P^+ = \lim_{\Lambda \uparrow \mathbb{Z}^d} G_\Lambda^+$ exists. Likewise, the weak limit $P^- = \lim_{\Lambda \uparrow \mathbb{Z}^d} G_\Lambda(\cdot \,|-)$ exists (and by symmetry is the image of P^+ under simultaneous spin flip). These limits are Gibbs measures and invariant under translations. Holley's inequality also implies that they are stochastically maximal resp. minimal in $\mathcal{G}$, and in particular extremal. This gives us the following criterion for phase transition in the Ising model.

Proposition 3 *For the Ising model on $\mathbb{Z}^d$ with any coupling constant $J > 0$ we have $\#\mathcal{G} > 1$ if and only if $P^- \neq P^+$ if and only if $\int \xi_0 \, dP^+ > 0$.*

The last equivalence follows from the relation $P^- \preceq P^+$, the translation invariance of these Gibbs measures, and the spin-flip symmetry. A detailed proof of the proposition and the previous statements can be found in Sect. 4.3 of [10].

How can we use this criterion? This is where random geometry enters the scenery. The key is the following geometric construction tracing back to [4] and in this form to [3]. It is called the *random-cluster representation of the Ising model*.

Let $\mathcal{E}_\Lambda^+ = \{E \subset E(\mathbb{Z}^d) : E \supset E(\Lambda^c)\}$ be the set of all edge configurations in $\mathbb{Z}^d$ which include all edges outside Λ, and define a probability measure ϕ_Λ on $\mathcal{E}_\Lambda^+$ by setting

$$\phi_\Lambda(E) = Z_\Lambda^{-1} \, 2^{k(E)} \, p^{\#E \setminus E(\Lambda^c)} \, (1-p)^{\#E(\mathbb{Z}^d) \setminus E} \quad \text{when } E \supset E(\Lambda^c), \quad (5)$$

where $p = 1 - e^{-J}$, $k(E)$ is the number of clusters of the graph $(\mathbb{Z}^d, E)$, and Z_Λ is a normalizing constant. ϕ_Λ is called the *random-cluster distribution in Λ with wired boundary condition*. This measure turns out to be related to $G_\Lambda^+ = G_\Lambda(\,\cdot\,|+)$. It will be convenient to identify a configuration $\xi \in \Omega$ with the pair (X^+, X^-), where X^+ and X^- are the sets of all lattice sites i for which $\xi_i = +1$ resp. -1.

Proposition 4 *For any hypercube Λ in $\mathbb{Z}^d$ there exists the following correspondence between the the Gibbs distribution G_Λ^+ for the Ising model and the random-cluster distribution ϕ_Λ in (5).*

$(G_\Lambda^+ \rightsquigarrow \phi_\Lambda)$ *Take a spin configuration $\xi = (X^+, X^-) \in \Omega$ with distribution G_Λ^+, and define an edge configuration $E \in \mathcal{E}_\Lambda^+$ as follows: Independently for all $e \in E(\mathbb{Z}^d)$ let $e \in E$ with probability*

$$p_\Lambda(e) = \begin{cases} 1 - e^{-J} & \text{if } e \subset X^+ \text{ or } e \subset X^-, \text{ and } e \cap \Lambda \neq \emptyset, \\ 1 & \text{if } e \subset \Lambda^c, \\ 0 & \text{otherwise,} \end{cases}$$

and $e \notin E$ otherwise. Then E has distribution ϕ_Λ.

$(\phi_\Lambda \rightsquigarrow G_\Lambda^+)$ *Pick an edge configuration $E \in \mathcal{E}_\Lambda^+$ according to ϕ_Λ, and define a spin configuration $\xi = (X^+, X^-) \in \Omega$ as follows: For each finite cluster C of $(\mathbb{Z}^d, E)$ let $C \subset X^+$ or $C \subset X^-$ according to independent flips of a fair coin; the unique infinite cluster of $(\mathbb{Z}^d, E)$ containing Λ^c is included into X^+. Then ξ has distribution G_Λ^+.*

Proof. A joint description of a spin configuration $\xi \in \Omega$ with distribution G_Λ^+ and an edge configuration $E \in \mathcal{E}_\Lambda^+$ with distribution ϕ_Λ can be obtained as follows. Any edge $e \in E(\mathbb{Z}^d)$ is independently included into E with probability $p = 1 - e^{-J}$ resp. 1 according to whether $e \cap \Lambda \neq \emptyset$ or not; each spin in Λ is equal to $+1$ or -1 according to independent flips of a fair coin; the spins off Λ are set equal to $+1$. The measure P thus described is then conditioned on the event A that no spins of different sign are connected by an edge. Relative to $P(\,\cdot\,|A)$, ξ has distribution G_Λ^+ and E has distribution ϕ_Λ. This is because $\exp[-H_\Lambda(\xi)]$ is equal to the conditional P-probability of A given ξ, and $2^{k(E)-1}$ is proportional to the conditional P-probability of A given E. Now, it is easy to see that the two constructions in the proposition simply correspond to the conditional distributions of E given ξ resp. of ξ given E relative to $P(\,\cdot\,|A)$. $\square$

Intuitively, the edges in the random-cluster representation indicate which pairs of spins "realize" their interaction, in that they decide to take the same orientation to avoid the dealignment costs. On the one hand, this representation is the basis of an efficient simulation procedure, the algorithm of Swendsen-Wang [20], which together with its continuous counterpart will be discussed at the end of Sect. 4.4. On the other hand, it is the key for a geometric approach to the phenomenon of phase transition, as we will now show.

The construction $\phi_\Lambda \rightsquigarrow G_\Lambda^+$ implies that, for $0 \in \Lambda$, the conditional expectation of ξ_0 given E is 1 when 0 is connected to $\partial \Lambda$ by edges in E, and 0 otherwise. Hence

$$\int \xi_0 \, dP_\Lambda^+ = \phi_\Lambda(0 \leftrightarrow \partial \Lambda) . \tag{6}$$

By Proposition 1, the measures ϕ_Λ decrease stochastically when Λ increases, so that the infinite-volume random-cluster distribution $\phi = \lim_{\Lambda \uparrow \mathbb{Z}^d} \phi_\Lambda$ exists. Letting $\Lambda \uparrow \mathbb{Z}^d$ in (6) we thus find that $\int \xi_0 \, dP^+ = \phi(0 \leftrightarrow \infty)$. Combining this with Proposition 3 we obtain the first statement of the following theorem, the equivalence of percolation and phase transition. This gives us detailed information on the existence of phase transition.

Theorem 2 *Consider the Ising model on $\mathbb{Z}^d$ with Hamiltonian (4) for any coupling constant $J > 0$. Then $\#\mathcal{G} > 1$ if and only if $\phi(0 \leftrightarrow \infty) > 0$. Consequently, there exists a coupling threshold $0 < J_c < \infty$ (corresponding to a critical inverse temperature) such that $\#\mathcal{G} = 1$ when $J < J_c$ and $\#\mathcal{G} > 1$ when $J > J_c$.*

Sketch proof. It only remains to show the second statement. This follows from Holley's inequality, Proposition 1. First, this inequality implies that ϕ_Λ is stochastically increasing in the parameter $p = 1 - e^{-J}$. Hence $\phi(0 \leftrightarrow \infty)$ is an increasing function of p. Moreover, one finds that ϕ is stochastically dominated by the Bernoulli bond percolation measure, whence $\phi(0 \leftrightarrow \infty) = 0$ when p is so small that $\theta(1, p; \mathbb{Z}^d) = 0$. Finally, ϕ is stochastically larger than the Bernoulli bond percolation measure with parameter $\tilde{p} = p/(p + 2(1 - p))$. Hence $\phi(0 \leftrightarrow \infty) > 0$ when p is so large that $\theta(1, \tilde{p}; \mathbb{Z}^d) > 0$. Details of this computation can be found in Sect. 6 of [10], which deals in fact with the extension of these results to the Potts model in which each spin has q different values. □

One may ask whether the connection between percolation and phase transition can be seen more directly in the behavior of spins. The following corollary gives an answer to this question. Let $\{0 \xleftrightarrow{+} \infty\}$ denote the event that 0 belongs to an infinite cluster of the graph $(X^+, E(X^+))$ induced by the set of plus-spins.

Corollary 1 *For the Ising model on $\mathbb{Z}^d$ with arbitrary coupling constant $J > 0$ we have $P^+(0 \xleftrightarrow{+} \infty) > 0$ whenever $\#\mathcal{G} > 1$. The converse holds only when $d = 2$.*

The first part follows readily from the construction in Proposition 4 which shows that $P^+(0 \xleftrightarrow{+} \infty) \geq \phi(0 \leftrightarrow \infty)$. For its second part see [10]. Pursuing the idea of plus-percolation further one can obtain the following result independently obtained in the late 1970s by Aizenman and Higuchi on the basis of previous work of L. Russo; a simpler proof has recently been given by [11].

Theorem 3 *For the Ising model on $\mathbb{Z}^2$ with $J > J_c$, there exist no other phases than P^+ and P^-.*

A celebrated result of Dobrushin asserts that in three or more dimensions there exist non-translation invariant Gibbs measures which look like P^+ in one half-space and like P^- in the other half-space.

2.2   The Widom–Rowlinson Lattice Gas

The Widom–Rowlinson lattice gas is a discrete analog of a continuum model to be considered in Sect. 4. It describes the random configurations of particles of two different types, plus or minus, which can only sit at the sites of the lattice $\mathbb{Z}^d$. Multiple occupations are excluded. So, at each site i of the lattice there are three possibilities: either i is occupied by a plus-particle, or by a minus-particle, or i is empty. The configuration space is thus $\Omega = S^{\mathbb{Z}^d}$ with $S = \{-1, 0, 1\}$. The basic assumption is that there is a hard-core repulsion between plus- and minus-particles, which means that particles of distinct type are not allowed to sit next to each other. In addition, there exists a chemical "activity" $z > 0$ which governs the overall-density of particles. The Hamiltonian thus takes the form

$$H_\Lambda(\xi) = \sum_{\{i,j\} \cap \Lambda \neq \emptyset: |i-j|=1} U(\xi_i, \xi_j) - \log z \sum_{i \in \Lambda} |\xi_i| , \tag{7}$$

where $U(\xi_i, \xi_j) = \infty$ if $\xi_i \xi_j = -1$, and $U(\xi_i, \xi_j) = 0$ otherwise. The associated family **G** of conditional probabilities is again given by (3). Just as in the Ising model, for $z > 1$ there exist two distinguished configurations of minimal energy, namely the constant configurations '+' and '−' for which all sites are occupied by particles of the same type. Moreover, one can again apply Holley's inequality to show that the Gibbs distributions $G_\Lambda^+ = G_\Lambda(\cdot|+)$ and $G_\Lambda^- = G_\Lambda(\cdot|-)$ converge to translation invariant limits $P^+, P^- \in \mathcal{G}$ which are stochastically maximal resp. minimal in $\mathcal{G}$, and therefore extremal. This implies that Proposition 3 holds verbatim also in the present case.

Is there also a geometric representation of the model, just as for the Ising model? The answer is yes, with interesting analogies and differences. There exists again a random-cluster distribution with an appearance very similar to (5), but this involves site percolation rather than bond percolation. Namely, consider the probability measure ψ_Λ on the set $\mathcal{X}_\Lambda^+ = \{Y \subset \mathbb{Z}^d : Y \supset \Lambda^c)\}$ which is given by

$$\psi_\Lambda(Y) = Z_\Lambda^{-1} 2^{k(Y)} p^{\#Y \setminus \Lambda^c} (1-p)^{\#\mathbb{Z}^d \setminus Y} \quad \text{for } Y \supset \Lambda^c; \tag{8}$$

here $p = \frac{z}{1+z}$, $k(Y)$ is the number of clusters of the graph $(Y, E(Y))$, and Z_Λ is a normalizing constant. ψ_Λ is called the *site random-cluster distribution in Λ with* *parameter p and wired boundary condition.* We will again identify a configuration $\xi \in \Omega$ with a pair (X^+, X^-), where X^+ and X^- are the sets of all lattice points i such that $\xi_i = +1$ resp. -1; thus $X^+ \cup X^-$ is the set of occupied sites, and $\xi \equiv 0$ on its complement. Here is the *random-cluster representation of the Widom–Rowlinson lattice gas* which is analogous to Proposition 4.

Proposition 5 *For any hypercube Λ in $\mathbb{Z}^d$ there exists the following correspondence between the the Gibbs distribution G_Λ^+ for the Widom–Rowlinson model and the site random-cluster distribution ψ_Λ in (8).*

$(G_\Lambda^+ \rightsquigarrow \psi_\Lambda)$ For a random spin configuration $\xi = (X^+, X^-) \in \Omega$ with distribution G_Λ^+, the random set $Y = X^+ \cup X^-$ has distribution ψ_Λ.

$(\psi_\Lambda \rightsquigarrow G_\Lambda^+)$ Pick a random set $Y \in \mathcal{X}_\Lambda^+$ according to ψ_Λ, and define a spin configuration $\xi = (X^+, X^-) \in \Omega$ with $X^+ \cup X^- = Y$ as follows: For each finite cluster C of $(Y, E(Y))$ let $C \subset X^+$ or $C \subset X^-$ according to independent flips of a fair coin; the unique infinite cluster of $(Y, E(Y))$ containing Λ^c is included into X^+. Then ξ has distribution G_Λ^+.

The random-cluster representation of the Widom-Rowlinson model is simpler than that of the Ising model because the randomness involves only the sites of the lattice. (This is a consequence of the hard-core interaction; in the case of a soft repulsion the situation would be different, as we will see in the continuum setting in Sect. 4.) On the other hand, there is a serious drawback of the site random-cluster distribution ψ_Λ: it does not satisfy the conditions of Proposition 1 for positive correlations. This is because the conditional probabilities in (10) below are not increasing in Y. So, we still have a counterpart to (6), viz.

$$\int \xi_0 \, dG_\Lambda^+ = \psi_\Lambda(0 \leftrightarrow \partial\Lambda) , \qquad (9)$$

but we do not know if the measures ψ_Λ are stochastically increasing in z and stochastically decreasing in Λ. So we obtain a somewhat weaker theorem.

Theorem 4 *Consider the Widom–Rowlinson model on $\mathbb{Z}^d$, $d \geq 2$, defined by (7) with activity $z > 0$. Then $\#\mathcal{G} > 1$ if and only if $\lim_{\Lambda \uparrow \mathbb{Z}^d} \psi_\Lambda(0 \leftrightarrow \partial\Lambda) > 0$. In particular, we have $\#\mathcal{G} = 1$ when z is sufficiently small, and $\#\mathcal{G} > 1$ when z is large enough.*

Sketch proof. The first statement follows immediately from (9) and the analog of Proposition 3. To prove the second statement we note that ψ_Λ has single-site conditional probabilities of the form

$$\psi_\Lambda(Y \ni i \,|\, Y \setminus \{i\}) = p \Big/ \big[p + (1 - p)\, 2^{\kappa(i,Y)-1}\big] , \qquad (10)$$

where $\kappa(i, Y)$ is the number of clusters of $Y \setminus \{i\}$ that intersect a neighbor of i. Since $0 \leq \kappa(i, Y) \leq 2d$, it follows from Proposition 1 that ψ_Λ is stochastically dominated by the site-Bernoulli measure with parameter $p^* = p/(p + (1-p)2^{-1})$, and dominates the site-Bernoulli measure with parameter $p_* = p/(p + (1 - $

$p)2^{2d-1}$). Combining this with Proposition 2 we thus find that $\#\mathcal{G} = 1$ when z is so small that $\theta(p^*, 1; \mathbb{Z}^d) = 0$, and $\#\mathcal{G} > 1$ when z is so large that $\theta(p_*, 1; \mathbb{Z}^d) > 0$. $\square$

Let us note that, in contrast to Theorem 2, the preceding result does not extend to the case when there are more than two different types of particles; this is related to the lack of stochastic monotonicity in this model. However, due to its simpler random-cluster representation the Widom–Rowlinson model has one advantage over the Ising model, in that it satisfies a much stronger counterpart to Corollary 1. In analogy to the notation there we write $\{0 \overset{+}{\longleftrightarrow} \infty\}$ for the event that the origin belongs to an infinite cluster of plus-particles.

Corollary 2 *For the Widom–Rowlinson lattice gas on $\mathbb{Z}^d$ for arbitrary dimension $d \geq 2$ and with any activity $z > 0$, we have $\#\mathcal{G} > 1$ if and only if $P^+(0 \overset{+}{\longleftrightarrow} \infty) > 0$.*

Sketch proof. The construction in Proposition 5 readily implies that

$$\psi_\Lambda(0 \leftrightarrow \partial\Lambda) = G_\Lambda^+(0 \overset{+}{\longleftrightarrow} \partial\Lambda) .$$

Combining this with (9) and letting $\Lambda \uparrow \mathbb{Z}^d$ one obtains the result. $\square$

The above equivalence of phase transition and percolation even holds when $\mathbb{Z}^d$ is replaced by an arbitrary graph.

As noticed before Theorem 4, we have no stochastic monotonicity in the activity z, and therefore no activity threshold for the existence of a phase transition. We are thus led to ask if, at least, the particle density is an increasing function of z. This can be deduced from general thermodynamic principles relying on convexity of thermodynamic functions rather than stochastic monotonicity. This will be described in Sect. 4.2 in the continuum set-up.

3   Continuum Percolation

In the rest of this contribution we will show that quite a lot of the preceding results and techniques carry over to models of point particles in Euclidean space. In this section we deal with a simple model of continuum percolation. Roughly speaking, this model consists of Poisson points which are connected by Bernoulli edges. To be precise, let $\mathcal{X}$ denote the set of all locally finite subsets X of $\mathbb{R}^d$. $\mathcal{X}$ is the set of all point configurations in $\mathbb{R}^d$, and is equipped with the usual σ-algebra generated by the counting variables $X \to \#X_\Lambda$ for bounded Borel sets $\Lambda \subset \mathbb{R}^d$; here we use the abbreviation $X_\Lambda = X \cap \Lambda$. Next, let $\mathcal{E}$ be the set of all locally finite subsets of $E(\mathbb{R}^d) = \{\{x, y\} \subset \mathbb{R}^d : x \neq y\}$. $\mathcal{E}$ is the set of all possible edge configurations and is equipped with an analogous σ-algebra. For $X \in \mathcal{X}$ let $E(X) = \{e \in E(\mathbb{R}^d) : e \subset X\}$ the set of all possible edges between the points of X, and $\mathcal{E}_X = \{E \in \mathcal{E} : E \subset E(X)\}$ the set of edge configurations between the points of X. We construct a random graph $\Gamma = (X, E)$ in $\mathbb{R}^d$ as follows.

- Pick a random point configuration $X \in \mathcal{X}$ according to the Poisson point process π^z on $\mathbb{R}^d$ with intensity $z > 0$.
- For given $X \in \mathcal{X}$, pick a random edge configuration $E \in \mathcal{E}_X$ according to the Bernoulli measure μ_X^p on $\mathcal{E}_X$ for which the events $\{E \ni e\}$, $e = \{x, y\} \in E(X)$, are independent with probability $\mu_X^p(E \ni e) = p(x - y)$; here $p : \mathbb{R}^d \to [0, 1]$ is a given even measurable function.

The distribution of our random graph Γ is thus determined by the probability measure

$$P^{z,p}(dX, dE) = \pi^z(dX)\,\mu_X^p(dE) \tag{11}$$

on $\mathcal{X} \times \mathcal{E}$. It is called the *Poisson random-edge model*, or *Poisson random-connection model*, and has been introduced and studied first by M. Penrose [19]; a detailed account of its properties is given in [18].

A special case of particular interest is when $p(x - y)$ is equal to 1 when $|x - y| \leq 2r$ for some $r > 0$, and 0 otherwise. This means that any two points x and y are connected by an edge if and only if the balls $B_r(x)$ and $B_r(x)$ with radius r and center x resp. y overlap. The connectivity properties of the corresponding Poisson random-edge model are thus the same as those of the random set $\Xi = \bigcup_{x \in X} B_r(x)$, for random X with distribution π^z. This special case is therefore called the *Boolean model*, or the *Poisson blob model*.

Returning to the general case, we consider the *percolation probability* of a typical point. Writing $x \leftrightarrow \infty$ when x belongs to an infinite cluster of $\Gamma = (X, E)$, this is given by the expression

$$\theta(z, p; \mathbb{R}^d) = \int \frac{\#\{x \in X_\Lambda : x \leftrightarrow \infty\}}{|\Lambda|}\, P^{z,p}(dX, dE) \tag{12}$$

for an arbitrary bounded box Λ with volume $|\Lambda|$. By translation invariance, $\theta(z, p; \mathbb{R}^d)$ does not depend on Λ. In fact, in terms of the Palm measure $\hat{\pi}^z$ of π^z and the associated measure $\hat{P}^{z,p}(dX, dE) = \hat{\pi}^z(dX)\,\mu_X(dE)$ we can write

$$\theta(z, p; \mathbb{R}^d) = \hat{P}^{z,p}(0 \leftrightarrow \infty)\,.$$

The following result of M. Penrose [19] is the continuum analog of Proposition 2.

Theorem 5 $\theta(z, p; \mathbb{R}^d)$ *is an increasing function of the intensity z and the edge probability function $p(\cdot)$. Moreover, $\theta(z, p; \mathbb{R}^d) = 0$ when $z \int p(x)\,dx$ is sufficiently small, while $\theta(z, p; \mathbb{R}^d) > 0$ when $z \int p(x)\,dx$ is large enough.*

Sketch proof: The monotonicity follows from an obvious stochastic comparison argument. Since $z \int p(x)\,dx$ is the expected number of edges emanating from a given point, a branching argument shows that $\theta(z, p; \mathbb{R}^d) = 0$ when $z \int p(x)\,dx < 1$. It remains to show that $\theta(z, p; \mathbb{R}^d) > 0$ when $z \int p(x)\,dx$ is large enough. By scaling we can assume that $\int p(x)\,dx = 1$. For simplicity we will in fact suppose that p is bounded away from 0 in a neighborhood of the origin, i.e.,

$p(x - y) \geq \delta > 0$ whenever $|x - y| \leq 2r$. (The following is a special case of an argument of [10].)

We divide the space $\mathbb{R}^d$ into cubic cells $\Delta(i)$, $i \in \mathbb{Z}^d$, with diameter at most r. We also pick a sufficiently large number n and introduce the following two concepts.

- Call a cell $\Delta(i)$ *good* if it contains at least n points which form a connected set relative to the edges of Γ in between them. This event does not depend on the configurations in all other cells and has probability at least

$$\pi^z(N_i \geq n) \left[1 - (n-1)(1 - \delta^2)^{n-2}\right] \equiv p_s ;$$

here, N_i is the random number of points in cell $\Delta(i)$, and the second term in the square bracket is an estimate for the probability that one of the n points is not connected to the first point by a sequence of two edges. The essential fact is that p_s is arbitrarily close to 1 when n and z are large enough.

- Call two adjacent cells $\Delta(i), \Delta(j)$ *linked* if there exists an edge from some point in $\Delta(i)$ to some point in $\Delta(j)$. Conditionally on the event that $\Delta(i)$ and $\Delta(j)$ are good, this has probability at least $1 - (1 - \delta)^{n^2} = p_b$, which is also close to 1 when n is large enough.

Now the point is the following: whenever there exists an infinite cluster of linked good cells (i.e., an infinite cluster in the countable graph with vertices at the good cells and with edges between pairs of linked cells) then there exists an infinite cluster in the original Poisson random-edge model. Hence $\theta(z, p; \mathbb{R}^d) \geq \frac{n}{v} \theta(p_s, p_b; \mathbb{Z}^d)$, where v is the cell volume and $\theta(p_s, p_b; \mathbb{Z}^d)$ is as in Proposition 2. Hence $\theta(z, p; \mathbb{R}^d) > 0$ when z is large enough. $\square$

How can one extend a percolation result as above from the Poisson case to point processes with spatial dependencies? Just as in the lattice gas, one can take advantage of stochastic comparison techniques. To this end we need a continuum analog of Holley's theorem.

A simple point process P on a bounded Borel subset Λ of $\mathbb{R}^d$ (i.e., a probability measure on $\mathcal{X}_\Lambda = \{X \in \mathcal{X} : X \subset \Lambda\}$) is said to have Papangelou (conditional) intensity $\gamma : \Lambda \times \mathcal{X}_\Lambda \to [0, \infty[$ if P satisfies the identity

$$\int P(dX) \sum_{x \in X} f(x, X \setminus \{x\}) = \int dx \int P(dX) \gamma(x|X) f(x, X) \qquad (13)$$

for any measurable function $f : \Lambda \times \mathcal{X}_\Lambda \to [0, \infty[$. (This is a non-stationary analog of the Georgii–Nguyen–Zessin equality discussed in the contribution of D. Stoyan to this volume.) This equation roughly means that $\gamma(x|X)\, dx$ is proportional to the conditional probability for the existence of a particle in an infinitesimal volume dx when the remaining configuration is X. Formally, it is not difficult to see that (13) is equivalent to the statement that P is absolutely continuous with respect to the intensity-1 Poisson point process $\pi_\Lambda = \pi_\Lambda^1$ in Λ with Radon–Nikodym density g satisfying $g(X \cup \{x\}) = \gamma(x|X)\, g(X)$, see e.g. [12]. In particular, the Poisson process π_Λ^z of intensity $z > 0$ on Λ has Papangelou intensity $\gamma(x|X) = z$.

Proposition 6 (Holley-Preston inequality) *Let $\Lambda \subset \mathbb{R}^d$ be a bounded Borel set and P, P' two probability measures on $\mathfrak{X}_\Lambda$ with Papangelou intensities γ resp. γ'. Suppose $\gamma(x|X) \leq \gamma'(x|X')$ whenever $X \subset X'$ and $x \notin X' \setminus X$. Then $P \preceq P'$. If this condition holds with $P' = P$ then P has positive correlations.*

Of course, the stochastic partial order $P \preceq P'$ is defined by means of the inclusion relation on $\mathfrak{X}_\Lambda$. Under additional technical assumptions the preceding proposition was first derived by Preston in 1975; in the present form it is due to Georgii and Küneth [12]. In the next section we will see how this result can be used to establish percolation in certain continuum random-cluster models, and thereby the existence of phase transitions in certain continuum particle systems.

4   The Continuum Ising Model

The continuum Ising model is a model of point particles in $\mathbb{R}^d$ of two different types, plus and minus. Rather than of particles of different types, one may also think of particles with a ferromagnetic spin with two possible orientations. The latter would be suitable for modelling ferrofluids such as the Au-Co alloy, which have recently found some physical attention. Much of what follows can also be extended to systems with more than two types, but we stick here to the simplest case. A configuration of particles is then described by a pair $\xi = (X^+, X^-)$, where X^+ and X^- are the configurations of plus- resp. minus-particles. The configuration space is thus $\Omega = \mathfrak{X}^2$.

We assume that the particles interact via a repulsive interspecies pair potential of finite range, which is given by an even measurable function $J : \mathbb{R}^d \to [0, \infty]$ of bounded support. The Hamiltonian in a bounded Borel set $\Lambda \subset \mathbb{R}^d$ of a configuration $\xi = (X^+, X^-)$ is thus given by

$$H_\Lambda(\xi) = \sum_{x \in X^+, y \in X^- : \{x,y\} \cap \Lambda \neq \emptyset} J(x - y) . \tag{14}$$

In view of its analogy to (4) this model is called the *continuum Ising model*. Setting $J(x - y) = \infty$ when $|x - y| \leq 2r$ and $J(x - y) = 0$ otherwise, we obtain the classical *Widom–Rowlinson model* [22] with a hard-core interspecies repulsion (which in spatial statistics is occasionally referred to as the *penetrable spheres mixture model*). Of course, this case corresponds to the Widom–Rowlinson lattice gas considered above. Here we make only the much weaker assumption that J is bounded away from zero on a neighborhood of the origin. That is, there exist constants $\delta, r > 0$ such that

$$J(x - y) \geq \delta \text{ when } |x - y| \leq 2r. \tag{15}$$

An interesting generalization of the Hamiltonian (14) can be obtained by adding an interaction term which is independent of the types of the particles. In a ferrofluid model this would mean that in addition to the ferromagnetic interaction of particle spins there is also a molecular interaction which is spin-independent. Such an extension is considered in [10].

Given the Hamiltonian (14), the associated *Gibbs distribution in Λ with activity $z > 0$ and boundary condition* $\xi_{\Lambda^c} = (X^+_{\Lambda^c}, X^-_{\Lambda^c}) \in \mathcal{X}^2_{\Lambda^c}$ is defined by the formula

$$G_\Lambda(d\xi_\Lambda | \xi_{\Lambda^c}) = Z^{-1}_{\Lambda | \xi_{\Lambda^c}} \exp[-H_\Lambda(\xi)] \, \pi^z_\Lambda(dX^+_\Lambda) \, \pi^z_\Lambda(dX^-_\Lambda) \qquad (16)$$

which is completely analogous to (3). The corresponding set $\mathcal{G} = \mathcal{G}(z)$ of Gibbs measures is then defined as in Definition 1 (with Λ running through the bounded Borel sets in $\mathbb{R}^d$ instead of the finite subsets of $\mathbb{Z}^d$).

In general, the existence of Gibbs measures in continuum models is not easy to establish. In the present case, however, it is simple: Thinking of $\mathcal{X}^2_\Lambda$ as the space of configurations on two disjoint copies Λ^+ and Λ^- of Λ, we see that $G_\Lambda(\cdot | \xi_{\Lambda^c})$ has the Papangelou intensity

$$\gamma(x | X^+, X^-) = \begin{cases} z \exp[-\sum_{y \in X^-} J(x-y)] \text{ if } x \in \Lambda^+ \\ z \exp[-\sum_{y \in X^+} J(x-y)] \text{ if } x \in \Lambda^- \end{cases} \le z. \qquad (17)$$

Proposition 6 therefore implies that $G_\Lambda(\cdot | \xi_{\Lambda^c}) \preceq \pi^z_\Lambda \times \pi^z_\Lambda$. Standard compactness theorems for point processes now show that for each $\xi \in \mathcal{X}^2$ the sequence $G_\Lambda(\cdot | \xi_{\Lambda^c})$ has an accumulation point P as $\Lambda \uparrow \mathbb{R}^d$, and it is easy to see that $P \in \mathcal{G}$.

4.1   Uniqueness and Phase Transition

We will now show that the Gibbs measure is unique when z is small, whereas a phase transition occurs when z is large enough. Both results rely on percolation techniques. As in the Widom–Rowlinson lattice gas, it remains open whether there is a sharp activity threshold separating intervals of uniqueness and non-uniqueness.

Proposition 7 *For the continuum Ising model we have* $\#\mathcal{G}(z) = 1$ *when z is sufficiently small.*

Sketch proof: Let $P, P' \in \mathcal{G}(z)$. We show that $P = P'$ when z is small enough. Let R be the range of J, i.e., $J(x) = 0$ when $|x| \le R$, and divide $\mathbb{R}^d$ into cubic cells $\Delta(i)$, $i \in \mathbb{Z}^d$, of linear size R. Let p^*_c be the Bernoulli site percolation threshold of the graph with vertex set $\mathbb{Z}^d$ and edges between all points having distance 1 *in the max-norm*. Consider the Poisson measure $Q^z = \pi^z \times \pi^z$ on the configuration space $\Omega = \mathcal{X}^2$.

Let ξ, ξ' be two independent realizations of Q^z, and suppose z is so small that $Q^z \times Q^z(N_i + N'_i \ge 1) < p^*_c$, where N_i and N'_i are the numbers of particles (plus or minus) in ξ resp. ξ'. Then for any finite union Λ of cells we have $Q^z \times Q^z(\Lambda \xrightarrow{\ge 1} \infty) = 0$, where $\{\Lambda \xrightarrow{\ge 1} \infty\}$ denotes the event that a cell in Λ belongs to an infinite connected set of cells $\Delta(i)$ containing at least one particle in either ξ or ξ'. Proposition 6 together with (17) imply that $P \times P' \preceq Q^z \times Q^z$. Hence $P \times P'(\Lambda \xrightarrow{\ge 1} \infty) = 0$. In other words, given two independent realizations ξ and ξ' of P and P' there exists a random corridor of width R around Λ which is completely free of particles. In particular, this means that ξ and ξ' coincide on this random corridor. By a spatial strong Markov property of Gibbs measures, it follows that P and P' coincide on the σ-algebra of events in Λ. As Λ can be chosen arbitrarily large, this proves the proposition. $\square$

After this result on the absence of phase transition (following from the absence of some kind of percolation) we turn to the existence of phase transition. This will follow from the existence of percolation in a suitable random-cluster model. In analogy to Propositions 4 and 5, we will derive a random-cluster representation of the Gibbs distribution

$$G^+_\Lambda = \int \pi^z_{\Lambda^c}(dY^+_{\Lambda^c}) \, G_\Lambda(\cdot \,|Y^+_{\Lambda^c}, \emptyset) \tag{18}$$

with a Poisson boundary condition of plus-particles and no minus-particle off Λ. Its random-cluster counterpart is the following probability measure χ_Λ on $\mathcal{X} \times \mathcal{E}$ describing random graphs (Y, E) in $\mathbb{R}^d$:

$$\chi_\Lambda(dY, dE) = Z^{-1}_{\Lambda|Y_{\Lambda^c}} \, 2^{k(Y,E)} \, \pi^z(dY) \, \mu^{p,\Lambda}_Y(dE) \,. \tag{19}$$

In the above, $k(Y, E)$ is the number of clusters of the graph (Y, E), $Z_{\Lambda|Y_{\Lambda^c}} = \int 2^{k(Y,E)} \pi^z_\Lambda(dY_\Lambda)$ normalizes the conditional probability of χ_Λ given $Y_{\Lambda^c} = Y \cap \Lambda^c$ (so that Y_{Λ^c} still has the Poisson distribution $\pi^z_{\Lambda^c}$), and $\mu^{p,\Lambda}_Y$ is the probability measure on $\mathcal{E}$ for which the edges $e = \{x, y\} \subset Y$ are drawn independently with probability $p(x - y) = 1 - e^{-J(x-y)}$ if $e \not\subset Y_{\Lambda^c}$, and probability 1 otherwise. The probability measure χ_Λ in (19) is called the *continuum random-cluster distribution in Λ with connection probability function p and wired boundary condition*. Note that (in contrast to (5) and (8)) this distribution describes random configurations of both points and edges. In the Widom–Rowlinson case of a hard-core interspecies repulsion the randomness of the edges disappears, and χ_Λ describes a dependent Boolean percolation model which is the direct continuum analog of (8). The *random-cluster representation of the continuum Ising model* now reads as follows.

Proposition 8 *For any bounded box Λ in $\mathbb{R}^d$ there is the following correspondence between the Gibbs distribution G^+_Λ in (18) for the continuum Ising model and the random-cluster distribution χ_Λ in (19).*

$(G^+_\Lambda \rightsquigarrow \chi_\Lambda)$ *Take a particle configuration $\xi = (X^+, X^-) \in \Omega$ with distribution G^+_Λ and define a random graph $(Y, E) \in \mathcal{X} \times \mathcal{E}$ as follows: Let $Y = X^+ \cup X^-$, and independently for all $e = \{x, y\} \in E(Y)$ let $e \in E$ with probability*

$$p_\Lambda(e) = \begin{cases} 1 - e^{-J(x-y)} & \text{if } e \subset X^+ \text{ or } e \subset X^-, \text{ and } e \cap \Lambda \neq \emptyset, \\ 1 & \text{if } e \subset \Lambda^c, \\ 0 & \text{otherwise.} \end{cases}$$

Then (Y, E) has distribution χ_Λ.

$(\chi_\Lambda \rightsquigarrow G^+_\Lambda)$ *Pick a random graph $(Y, E) \in \mathcal{X} \times \mathcal{E}$ according to χ_Λ. Define a particle configuration $\xi = (X^+, X^-) \in \Omega$ with $X^+ \cup X^- = Y$ as follows: For* *each finite cluster C of (Y, E) let $C \subset X^+$ or $C \subset X^-$ according to independent flips of a fair coin; the unique infinite cluster of (Y, E) containing Y_{Λ^c} is included into X^+. Then ξ has distribution G_Λ^+.*

Just as in the lattice case, the random-cluster representation above gives the following key identity: for any finite box $\Delta \subset \Lambda$,

$$\int [\#X_\Delta^+ - \#X_\Delta^-]\, G_\Lambda^+(dX^+, dX^-)$$

$$= \int \#\{x \in Y_\Delta : x \leftrightarrow Y_{\Lambda^c}\}\, \chi_\Lambda(dY, dE)\,; \qquad (20)$$

in the above, the notation $x \leftrightarrow Y_{\Lambda^c}$ means that x is connected to a point of Y_{Λ^c} in the graph (Y, E). In other words, the difference between the mean number of plus- and minus-particles in Δ corresponds to the percolation probability in χ_Λ. How can one check that the latter is positive for large z? The idea is again a stochastic comparison.

Let $\nu_\Lambda = \chi_\Lambda(\cdot \times \mathcal{E})$ the point marginal of χ_Λ. Then $\chi_\Lambda(dY, dE) = \nu_\Lambda(dY)\,\phi_{\Lambda,Y}(dE)$ with an obvious analog $\phi_{\Lambda,Y}$ of (5). An application of Proposition 1 shows that $\phi_{\Lambda,Y}$ is stochastically larger than the Bernoulli edge measure $\mu_Y^{\tilde{p}}$ for which edges are drawn independently between points $x, y \in Y$ with probability $\tilde{p}(x - y) = (1 - e^{-\delta})/[(1 - e^{-\delta}) + 2e^{-\delta}]$ when $|x - y| \leq 2r$, and with probability 0 otherwise. Here, δ and r are as in assumption (15). Moreover, ν_Λ has the Papangelou intensity

$$\gamma(x|Y) = z \int 2^{k(Y \cup \{x\},\, \cdot)}\, d\phi_{\Lambda, Y \cup \{x\}} \Big/ \int 2^{k(Y,\, \cdot)}\, d\phi_{\Lambda, Y}\,.$$

To get a lower estimate for $\gamma(x|Y)$ one has to compare the effect on the number of clusters in (Y, E) when a particle at x and corresponding edges are added. In principle, this procedure could connect a large number of distinct clusters lying close to x, so that $k(Y \cup \{x\}, \cdot)$ was much smaller than $k(Y, \cdot)$. However, one can show that this occurs only with small probability, so that $\gamma(x|Y) \geq \alpha z$ for some $\alpha > 0$. By Proposition 6, we can conclude that χ_Λ is stochastically larger than the Poisson random-edge measure $P^{\alpha z, \tilde{p}}$ defined in (11). The right-hand side of (20) is therefore not smaller than $\theta(\alpha z, \tilde{p}; \mathbb{R}^d)$. Finally, since $G_\Lambda^+ \preceq \pi^z \times \pi^z$ by (17), the Gibbs distributions G_Λ^+ have a cluster point $P^+ \in \mathcal{G}(z)$ satisfying

$$\int [\#X_\Delta^+ - \#X_\Delta^-]\, P^+(dX^+, dX^-) \geq \theta(\alpha z, \tilde{p}; \mathbb{R}^d)\,.$$

By spatial averaging one can achieve that P^+ is in addition translation invariant. Together with Theorem 5 this leads to the following theorem.

Theorem 6 *For the continuum Ising model on $\mathbb{R}^d$, $d \geq 2$, with Hamiltonian (14) and sufficiently large activity z there exist two translation invariant Gibbs measures P^+ and P^- having a majority of plus- resp. minus-particles and related to each other by the plus-minus interchange.*

This result is due to [10]. In the special case of the Widom-Rowlinson model it has been derived independently in the same way by Chayes, Chayes, and Kotecký [2]. The first proof of phase transition in the Widom-Rowlinson model was found by Ruelle in 1971, and for a soft but strong repulsion by Lebowitz and Lieb in 1972. Gruber and Griffiths [14] used a direct comparison with the lattice Ising model in the case of a species-independent background hard core.

As a matter of fact, one can make further use of stochastic monotonicity. (In contrast to the preceding theorem, this only works in the present case of two particle types.) Introduce a partial order '$\leq$' on $\Omega = \mathcal{X}^2$ by writing

$$(X^+, X^-) \leq (Y^+, Y^-) \text{ when } X^+ \subset Y^+ \text{ and } X^- \supset Y^-. \tag{21}$$

A straightforward extension of Proposition 6 then shows that the measures G_Λ^+ in (18) decrease stochastically relative to this order when Λ increases. (This can be also deduced from the couplings obtained by perfect simulation, see Sect. 4.4 below.) It follows that P^+ is in fact the limit of these measures, and is in particular translation invariant. Moreover, one can see that P^+ is stochastically maximal in $\mathcal{G}$ in this order. This gives us the following counterpart to Corollary 2.

Corollary 3 *For the continuum Ising model with any activity $z > 0$, a phase transition occurs if and only if*

$$\int \hat{P}^+(dX^+, dX^-)\, \mu_{X^+}^p (0 \xleftrightarrow{+} \infty) > 0 \, ;$$

here $\hat{P}^+$ is the Palm measure of P^+, and the relation $0 \xleftrightarrow{+} \infty$ means that the origin belongs to an infinite cluster in the graph with vertex set X^+ and random edges drawn according to the probability function $p = 1 - e^{-J}$.

It is not known whether P^+ and P^- are the only extremal elements of $\mathcal{G}(z)$ when $d = 2$, as it is the case in the lattice Ising model. However, using a technique known in physics as the Mermin–Wagner theorem one can show the following.

Theorem 7 *If J is twice continuously differentiable then each $P \in \mathcal{G}(z)$ is translation invariant.*

A proof can be found in [8]. The existence of non-translation invariant Gibbs measures in dimensions $d \geq 3$ is an open problem.

4.2   Thermodynamic Aspects

Although we were able to take some advantage of stochastic comparison techniques in the continuum Ising model, the use of Proposition 6 is much more limited than that of its lattice analog. The reason is that its condition requires some kind of attractivity, which is in conflict with stability (preventing the existence of infinitely many particles in a bounded region). This implies that a continuum Gibbs distribution G_Λ with a pair interaction cannot satisfy the conditions of Theorem 6 with $P = P' = G_\Lambda$, which would imply that G_Λ has positive correlations and is stochastically increasing with the activity z. Fortunately, this gap can be closed to some extent by the use of classical convexity techniques of Statistical Mechanics. These will allow us to conclude that at least the particle density of Gibbs measures is an increasing function of the activity z.

It should be noted that these ideas are standard in Statistical Physics; they are included here because they might be less known among spatial statisticians, and because we need to check that the general principles really work in the model at hand. One should also note that this technique does not depend on the specific features of the model; in particular, it applies also to the gas of hard balls discussed in H. Löwen's contribution to this volume.

Let us begin recalling the thermodynamic justification of Gibbs measures. Let $\Lambda \subset \mathbb{R}^d$ be a finite box. For any translation invariant probability measure P on $\Omega = \mathcal{X}^2$ consider the *entropy per volume*

$$s(P) = \lim_{|\Lambda| \to \infty} |\Lambda|^{-1} S(P_\Lambda).$$

Here we write P_Λ for the restriction of P to $\mathcal{X}_\Lambda^2$, the set of particle configurations in Λ, and

$$S(P_\Lambda) = \begin{cases} -\int \log f \, dP_\Lambda & \text{if } P_\Lambda \ll \pi_\Lambda^1 \times \pi_\Lambda^1 \text{ with density } f, \\ -\infty & \text{otherwise} \end{cases}$$

for the entropy of P_Λ relative two the two-species Poisson process $\pi_\Lambda^1 \times \pi_\Lambda^1$ on $\mathcal{X}_\Lambda^2$ with intensity 1. The notation $|\Lambda| \to \infty$ means that Λ runs through a specified increasing sequence of cubic boxes with integer sidelength. The existence of $s(P)$ is a multidimensional version of Shannon's theorem; see [5] for the lattice case to which the present case can be reduced by identifying Ω with $(\mathcal{X}_C^2)^{\mathbb{Z}^d}$ for a unit cube C.

Next consider the *interaction energy per volume*

$$u(P) = \int \hat{P}(dX^+, dX^-) \left[1_{\{0 \in X^+\}} \sum_{x \in X^-} J(x) + 1_{\{0 \in X^-\}} \sum_{x \in X^+} J(x) \right] \qquad (22)$$

defined in terms of the Palm measure $\hat{P}$ of P. $u(P)$ can also be defined as a per-volume limit, cf. [6], Sect. 3. Also, consider the *particle density*

$$\varrho(P) = \hat{P}(\Omega) = |\Lambda|^{-1} \int [\#X_\Lambda^+ + \#X_\Lambda^-] \, dP_\Lambda$$

of P; by translation invariance the last term does not depend on Λ. The term $-\varrho(P) \log z$ is then equal to the chemical energy per volume.

Finally, consider the *pressure*

$$p(z) = -\min_P \left[u(P) - \varrho(P) \log z - s(P) \right]; \qquad (23)$$

the minimum extends over all translation invariant probability measures P on Ω. The large deviation techniques of Georgii [6] show that $\mathsf{p}(z) = \lim_{|\Lambda| \to \infty} |\Lambda|^{-1} \log Z_{\Lambda|\xi_{\Lambda^c}}$ for each $\xi \in \Omega$. (The paper [6] deals only with particles of a single type and superstable interaction, but the extension to the present case is straightforward because J is nonnegative and has finite range.) The variational principle for Gibbs measures then reads as follows.

Theorem 8 *Let P be a translation invariant probability measure on $\Omega = X^2$. Then $P \in \mathcal{G}(z)$ if and only if $\mathsf{u}(P) - \varrho(P) \log z - \mathsf{s}(P)$, the free energy per volume, is equal to its minimum $-\mathsf{p}(z)$.*

The "only if" part can be derived along the lines of [6] and Proposition 7.7 of [7]. The "if" part follows from the analogous lattice result (see Sect. 15.4 of [5]) by the identification of Ω and $(X_C^2)^{\mathbb{Z}^d}$ mentioned above.

What does the theorem tell us about the particle densities of Gibbs measures? Let us look at the pressure $\mathsf{p}(z)$. First, it follows straight from the definition (23) that $\mathsf{p}(z)$ is a convex function of $\log z$. In other words, the function $\tilde{\mathsf{p}}(t) = \mathsf{p}(e^t)$ is convex. Next, inserting $P = \pi^z \times \delta_\emptyset$ into the right-hand side of (23) we see that $\tilde{\mathsf{p}} > -\infty$, and that the slope of $\tilde{\mathsf{p}}$ at t tends to infinity as $t \to \infty$.

Now, suppose $P \in \mathcal{G}(z)$ is translation invariant. The variational principle above then implies that the function $t \to (t - \log z)\varrho(P) + \mathsf{p}(z)$ is a tangent to $\tilde{\mathsf{p}}$ at $\log z$. For, on the one hand we have

$$\mathsf{u}(P) - \varrho(P) \log z - \mathsf{s}(P) = -\mathsf{p}(z) < \infty$$

and thus $\mathsf{u}(P) - \mathsf{s}(P) < \infty$, and on the other hand

$$\mathsf{u}(P) - \varrho(P) t - \mathsf{s}(P) \geq -\tilde{\mathsf{p}}(t) \quad \text{for all } t.$$

Inserting the former identity into the last inequality we get the result. As a consequence, the particle density $\varrho(P)$ lies in the interval between the left and right derivative of $\tilde{\mathsf{p}}$ at $\log z$. By convexity, these derivatives are increasing and almost everywhere identical. In fact, they are strictly increasing. For, if they were constant on some non-empty open interval I then for each $t_0 \in I$ and $P \in \mathcal{G}(e^{t_0})$ the function $t \to \varrho(P) t - \tilde{\mathsf{p}}(t)$ would be constant on I, and thus by the variational principle $P \in \mathcal{G}(e^t)$ for all $t \in I$. This is impossible because the conditional Gibbs distributions depend non-trivially on the activity. We thus arrive at the following conclusion.

Corollary 4 *Let $0 < z < z'$ and $P \in \mathcal{G}(z)$, $P' \in \mathcal{G}(z')$ be translation invariant. Then $\varrho(P) < \varrho(P')$, and $\varrho(P) \to \infty$ as $z \to \infty$.*

In the present two-species model it is natural to consider also the case when each particle species has its own activity, i.e., the plus-particles have activity z^+ and the minus-particles have activity z^-. It then follows in the same way that the pressure $\mathsf{p}(z^+, z^-)$ is a strictly convex function of the pair $(\log z^+, \log z^-)$, and therefore that the density of plus-particles is a strictly increasing function of z^+, and the density of minus-particles is a strictly increasing function of z^-; these densities tend to infinity as z^+ resp. z^- tends to infinity. Moreover, Theorem 6 implies that $\mathrm{p}(z^+, z^-)$ has a kink at (z, z) when z is large enough; this means that the convex function $t \to \mathrm{p}(ze^t, ze^{-t})$ is not differentiable at $t = 0$.

4.3 Projection on Plus-Particles

As the continuum Ising model is a two-species model, it is natural to ask what kind of system appears if we forget all minus-particles and only retain the plus-particles. The answer is that their distribution is again Gibbsian for a suitable Hamiltonian. This holds also in the case of different activities z^+ and z^- of plus- and minus-particles, which is the natural context here. To check this, take any box $\Delta \subset \mathbb{R}^d$ and let $\Lambda \supset \Delta$ be so large that the distance of Δ from Λ^c exceeds the range R of J. Integrating over X_Λ^- in (16) and conditioning on $X_{\Lambda \setminus \Delta}^+$ one finds that the conditional distribution of X_Δ^+ for given $X_{\Lambda \setminus \Delta}^+$ under $G_\Lambda(\cdot | \xi_{\Lambda^c})$ does not depend on Λ and ξ_{Λ^c} and has the Gibbsian form

$$\tilde{G}_\Delta(dX_\Delta^+ | X_{\Delta^c}^+) = \tilde{Z}_{\Delta | X_{\Delta^c}^+}^{-1} \exp[-z^- \tilde{H}_\Delta(X^+)] \, \pi_\Delta^{z^+}(dX_\Delta^+) \qquad (24)$$

with the Hamiltonian

$$\tilde{H}_\Delta(X^+) = \int_\Delta \left(1 - \exp\left[-\sum_{x \in X^+} J(x - y)\right]\right) dy \, . \qquad (25)$$

Thus, writing $\mathcal{G}(z^+, z^-)$ for the set of all continuum-Ising Gibbs measures on $\Omega = \mathcal{X}^2$ with Hamiltonian (14) and activities z^+ and z^-, and $\tilde{\mathcal{G}}(z^+, z^-)$ for the set of all Gibbs measures on $\mathcal{X}$ with conditional distributions (24), one obtains the following corollary.

Corollary 5 *Let $P \in \mathcal{G}(z^+, z^-)$ and $\tilde{P}$ be the distribution of the configuration of plus-particles. Then $\tilde{P} \in \tilde{\mathcal{G}}(z^+, z^-)$. In particular, $\#\tilde{\mathcal{G}}(z, z) > 1$ when z is large enough.*

The last statement follows from Theorem 6.

In the Widom–Rowlinson case when $J = \infty 1_{\{|\cdot| \le 2r\}}$, the relationship between $\mathcal{G}(z^+, z^-)$ and $\tilde{\mathcal{G}}(z^+, z^-)$ has already been observed in the original paper by Widom and Rowlinson [22]. The plus-Hamiltonian (25) then takes the simple form

$$\tilde{H}_\Delta(X^+) = \left|\Delta \cap \bigcup_{x \in X^+} B_{2r}(x)\right|$$

that is, $\tilde{H}_\Delta(X^+)$ is the volume in Δ of the Boolean model with radius $2r$ induced by X^+. In this case, the model with distribution $\tilde{G}_\Delta(dX_\Delta^+ | X_{\Delta^c}^+)$ was reinvented by Baddeley and van Lieshout [1]. Having the two-dimensional case in mind, they coined the suggestive term *area-interaction process*. From here one can go one step further to Hamiltonians which use not only the volume but also the other Minkowski functionals. This has been initiated by Likos et al. [16] and Mecke [17]; see also Mecke's contribution to this volume.

One particularly nice feature of the area-interaction process and its generalization (25) is that it seems to be the only known (non-Poisson) model to which Proposition 6 can be applied for establishing positive correlations of increasing functions. This attractiveness property makes the model quite attractive for statistical modelling. (By way of contrast, repulsive point systems can be modelled quite easily, for example by a nonnegative pair interaction.)

However, some caution is necessary due to the phase transition when $z^+ = z^- = z$ is large: The typical configurations of $\tilde{G}_\Delta(\cdot\,|\,\emptyset)$ for a large finite window Δ then can be typical for either phase, $\tilde{P}^+$ or $\tilde{P}^-$, and thus can have different particle densities. Due to finite size effects, this phenomenon already appears when z^+ and z^- are sufficiently close to each other. So, the spatial statistician should be aware of such an instability of observations and should examine whether this is realistic or not in the situation to be modelled.

Finally, one can use Proposition 6 to show that the Gibbs measures $\tilde{P} \in \tilde{G}(z^+, z^-)$ are stochastically increasing in z^+ and decreasing in z^-. In particular, the density of plus-particles for any $P \in \mathcal{G}(z^+, z^-)$ increases when z^+ increases or z^- decreases, as can also be seen using the partial order (21). The monotonicity results in the last paragraph of Sect. 4.2 thus follow also from stochastic comparison techniques, but Corollary 4 cannot be derived in this way.

4.4 Simulation

There are various reasons for performing Monte–Carlo simulations of physical or statistical systems, as discussed in a number of other contributions to this volume. In the present context, the primary reason is to sharpen the intuition on the system's behavior, so that one can see which properties can be expected to hold. This can lead to conjectures which then hopefully can be checked rigorously.

Here we will show briefly how one can obtain simulation pictures of the continuum Ising model. We start with a continuum Gibbs sampler which is suggested by Proposition 8; in the Widom–Rowlinson case it has been proposed by Häggström, van Lieshout and Møller [15].

Consider a fixed window $\Lambda \subset \mathbb{R}^d$ and the Gibbs distribution $G_\Lambda^{\text{free}} = G_\Lambda(\cdot\,|\,\emptyset, \emptyset)$ for the continuum Ising model in Λ with activity z and free (i.e., empty) boundary condition off Λ. We define a random map $F : \mathcal{X}_\Lambda \to \mathcal{X}_\Lambda$ by the following algorithm:

- take an input configuration $X \in \mathcal{X}_\Lambda$,
- select a Poisson configuration $Y \in \mathcal{X}_\Lambda$ with distribution π_Λ^z,
- define a random edge configuration $E \subset \{\{x,y\} : x \in X, y \in Y\}$ by independently drawing an edge from $x \in X$ to $y \in Y$ with probability $p(x - y) = 1 - e^{-J(x-y)}$,
- set $F(X) = \left\{ y \in Y : \{x, y\} \notin E \text{ for all } x \in X \right\}$.

That is, $F(X)$ is a random thinning of π_Λ^z obtained by removing all points which are connected to X by a random edge. Its distribution is nothing other than the Poisson point process $\pi_{\Lambda|X}^{z,J}$ on Λ with inhomogeneous intensity measure $\rho_X^J(dy) = z \, 1_\Lambda(y) \exp\left[-\sum_{x\in X} J(y-x)\right]\} \, dy$. (Of course, this could also be achieved by setting $F(X) = \{y \in Y : U_y \leq \exp\left[-\sum_{x\in X} J(y-x)\right]\}$ for independent $U(0,1)$-variables U_y, $y \in Y$. Although this was simpler in the case of the MCMC below, it would considerably increase the running time of the perfect algorithm of Theorem 9, as will be explained there.) Now the point is that $\pi_{\Lambda|X^+}^{z,J}$ is the conditional distribution of X^- given X^+ relative to the Gibbs distribution G_Λ^{free}, and similarly with $+$ and $-$ interchanged. So, if $\xi = (X^+, X^-)$ has distribution G_Λ^{free} and F^+, F^- are independent realizations of F then $(F^+(X^-), F^- \circ F^+(X^-))$ has again distribution G_Λ^{free}. This observation gives rise to the following *Markov chain Monte Carlo algorithm* (MCMC).

Proposition 9 *Let F_n^+, F_n^-, $n \geq 0$, be independent realizations of F, and $X_0^+ \in \mathcal{X}_\Lambda$ any initial configuration. Define recursively*

$$X_0^- = F_0^-(X_0^+), \quad X_n^+ = F_n^+(X_{n-1}^-), \quad X_n^- = F_n^-(X_n^+) \text{ for } n \geq 1.$$

Then the distribution of (X_n^+, X_n^-) converges to G_Λ^{free} in total variation norm at a geometric rate.

Proof. It suffices to observe that $F \equiv \emptyset$ with probability $\delta = e^{-z|\Lambda|}$. This shows that for any two configurations $X, X' \in \mathcal{X}_\Lambda$ and any $A \subset \mathcal{X}_\Lambda$

$$|\text{Prob}(F(X) \in A) - \text{Prob}(F(X') \in A)| \leq 1 - \delta.$$

So, if one looks at the process (X_n^+, X_n^-) for two different starting configurations then each application of F reduces the total variation distance by a factor of $1 - \delta$. $\square$

A nice property of the random mapping F is its monotonicity: if $X \subset X'$ then $F(X) \supset F(X')$ almost surely. This allows to modify the preceding algorithm to obtain *perfect simulation* in the spirit of Propp and Wilson, as described in the contribution of E. Thönnes to this volume. According to (17), G_Λ^{free} is stochastically dominated by independent Poisson processes of plus- and minus-particles. So one can use the idea of *dominated perfect simulation* in her terminology. We describe here only the algorithm and refer to her contribution for more details.

Roughly speaking, the perfect algorithm consists of repeated simultaneous runs of the preceding MCMC, starting from two particular initial conditions at some time $N_k < 0$ until time 0. The two initial conditions are chosen extremal relative to the ordering (21), namely with no initial plus-particle (the minimal case), and with a Poisson crowd of intensity z of plus-particles (which is maximal by stochastic domination). Since the same realizations of F are used in both cases, the two parallel MCMC's have a positive chance of coalescing during the time interval from N_k to 0. If this occurs, one stops. Otherwise one performs a further run which starts at some time $N_{k+1} < N_k$.

Theorem 9 *Let* F_n^+, F_n^-, $n \leq 0$, *be independent realizations of* F, *and* $(N_k)_{k \geq 1}$ *a strictly decreasing sequence of negative run starting times. For each run indexed by* $k \geq 1$ *let*

$$\Phi_k = F_0^+ \circ F_{-1}^- \circ F_{-1}^+ \circ \cdots \circ F_{N_k+1}^- \circ F_{N_k+1}^+ \circ F_{N_k}^-$$

be the random mapping corresponding to the MCMC of Proposition 9 for the time interval from N_k *to* 0, *and consider the processes*

$$X_{k,\min}^+ = \Phi_k(\emptyset), \quad X_{k,\max}^+ = \Phi_k \circ F_{N_k}^+(\emptyset) .$$

Then there exists a smallest (random) $K < \infty$ *such that* $X_{K,\min}^+ = X_{K,\max}^+$, *and the random particle configuration* $\xi_K = (X_{K,\min}^+, F_0^-(X_{K,\min}^+))$ *has distribution* G_Λ^{free}.

Since the random mapping F can be simulated by simple standard procedures, the implementation of the preceding algorithm is quite easy; a Macintosh application can be found at `http://www.mathematik.uni-muenchen.de/~georgii/CIsing.html` . The main task is to store the random edge configuration E in each application of F during a time interval $\{N_k, \ldots, N_{k-1} - 1\}$ for use in the later runs (which should be done in a file on the hard disk when $z|\Lambda|$ is large). As a matter of fact, once the set E is determined one can forget the positions of the particles of Y and only keep their indices. In this sense, E contains all essential information of the mapping F. As a consequence, knowing E one needs almost no time to apply the same realization of F in later runs. This is not the case for the alternative definition of F mentioned above.

However, there are some difficulties coming from the phase transition of the model. Running the perfect algorithm for small z is fine and raises no problem. But if z is large then the algorithm requires a considerably longer time to terminate. This is because for each run k the distribution of $\xi_{k,\max} = (X_{k,\max}^+, F_0^-(X_{k,\max}^+))$ will be close to P^+ and thus show a large crowd of plus-particles giving the minus-particles only a minimal chance to spread out. Likewise, the distribution of $\xi_{k,\min} = (X_{k,\min}^+, F_0^-(X_{k,\min}^+))$ will be close to P^-, so that the minus-particles are in the great majority. The bottleneck between these two types of configurations is so small that K will typically be much too large for practical purposes, at least for windows Λ of satisfactory size. In order to reduce this difficulty, one should not simulate the Gibbs distribution G_Λ^{free} with free boundary condition (as we have done above for simplicity), as this distributes most of its mass on two quite opposed events. Rather one should simulate one of the phases, say P^+. For a finite window, this can be achieved by imposing a random boundary condition of Poisson plus-particles outside Λ as in (18). Such a boundary condition helps $X_{k,\max}^+$ and $X_{k,\min}^+$ quite a lot to coalesce within reasonable time. (If one is willing to accept long running times, one should impose periodic boundary conditions to reduce the finite-size effects.)

The pictures shown are obtained in this way. The underlying interaction potential is $J(x) = 3(1 - |x|)^2$ for $|x| \leq 1$, $J(x) = 0$ otherwise. The size of

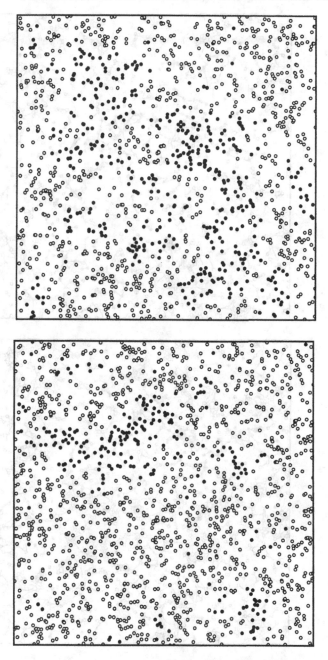

Fig. 1. Perfect samples of the continuum Ising model.
Top: $z = 4$, with particle density $\rho = 4.22$. Bottom: $z = 4.5$, with $\rho = 4.89$.

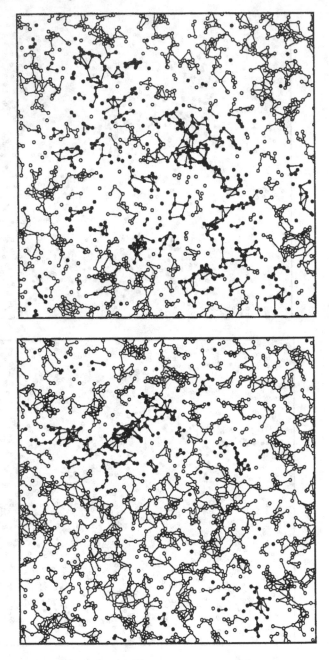

Fig. 2. Random-cluster representation.
Top: the subcritical case $z = 4$. Bottom: the supercritical case $z = 4.5$.

the window is 20×20. Outside of the window there is an invisible boundary condition of white Poisson particles. The activities are $z = 4.0$ (top) and $z = 4.5$ (bottom), and the corresponding coalescence times are $-N_K = 300$ resp. 400.

In the subcritical case $z = 4$, the particles in the bulk do not feel the boundary condition: there is no dominance of white over black. In the supercritical case $z = 4.5$ however, the influence of the white boundary condition is strong enough to dominate the whole window, and the phase transition becomes manifest. This is nicely illustrated by the random-cluster representation, which according to Proposition 8 is obtained from the point configuration (X^+, X^-) by adding random edges within X^+ and X^- separately. Here one sees that in the subcritical case the influence of the plus-boundary condition is only felt by a the particles near the boundary, while in the supercritical case the global behavior is dominated by a macroscopic cluster reaching from the boundary far into the interior of Λ. This visualizes the equivalence of phase transition and percolation derived in Corollary 3.

To conclude we mention two other algorithms. First, there is another perfect algorithm using a rejection scheme due to Fill, which has been studied in detail by Thönnes [21] in the Widom-Rowlinson case; its extension to the present case is straightforward. A further possibility, which is particularly useful in the supercritical case, is to use a continuum analog of the Swendsen–Wang algorithm [20]. In its classical version for the lattice Ising model, this algorithm consists in alternating applications of the two procedures in Proposition 4. In its continuum version, one can again alternate between the two procedures of Proposition 8, but one has to combine this with applications of the random mapping F in order to obtain a resampling of particle positions. Unfortunately, this algorithm does not seem to admit a perfect version because of its lack of monotonicity, but it has the advantage of working also for the many-species extension of the model.

References

1. Baddeley, A.J., M.N.M. van Lieshout (1995): 'Area-interaction point processes', *Ann. Inst. Statist. Math.* **46**, pp. 601–619
2. Chayes, J.T., L. Chayes, R. Kotecký (1995): 'The analysis of the Widom-Rowlinson model by stochastic geometric methods', *Commun. Math. Phys.* **172**, pp. 551–569
3. Edwards, R.G., A.D. Sokal (1988): 'Generalization of the Fortuin–Kasteleyn–Swendsen–Wang representation and Monte Carlo algorithm', *Phys. Rev.* D **38**, pp. 2009–2012
4. Fortuin, C.M., P.W. Kasteleyn (1972): 'On the random-cluster model. I. Introduction and relation to other models', *Physica* **57**, pp. 536–564
5. Georgii, H.-O. (1988): *Gibbs Measures and Phase Transitions* (de Gruyter, Berlin New York)
6. Georgii, H.-O. (1994): 'Large Deviations and the Equivalence of Ensembles for Gibbsian Particle Systems with Superstable Interaction', *Probab. Th. Rel. Fields* **99**, pp. 171–195
7. Georgii, H.-O. (1995): 'The Equivalence of Ensembles for Classical Systems of Particles', *J. Statist. Phys.* **80**, pp. 1341–1378

8. Georgii, H.-O. (1999): 'Translation invariance and continuous symmetries in two-dimensional continuum systems'. In: *Mathematical results in Statistical Mechanics*, ed. by S. Miracle-Sole, J. Ruiz, V. Zagrebnov (World Scientific, Singapore etc.), pp. 53–69
9. Georgii, H.-O., O. Häggström (1996): 'Phase transition in continuum Potts models', *Commun. Math. Phys.* **181**, pp. 507–528
10. Georgii, H.-O., O. Häggström, C. Maes (1999): 'The random geometry of equilibrium phases'. In: *Critical phenomena*, ed. by Domb and J.L. Lebowitz (Academic Press), forthcoming
11. Georgii, H.-O., Y. Higuchi (1999): 'Percolation and number of phases in the 2D Ising model', submitted to *J. Math. Phys.*
12. Georgii, H.-O., T. Küneth (1997): 'Stochastic comparison of point random fields', *J. Appl. Probab.* **34**, pp. 868–881
13. Grimmett, G.R. (1999): *Percolation*, 2nd ed. (Springer, New York)
14. Gruber, Ch., R.B. Griffiths (1986): 'Phase transition in a ferromagnetic fluid', *Physica A* **138**, pp. 220–230
15. Häggström, O., M.N.M. van Lieshout, J. Møller (1997): 'Characterization results and Markov chain Monte Carlo algorithms including exact simulation for some spatial point processes', *Bernoulli*, to appear
16. Likos, C.N, K.R. Mecke, H. Wagner (1995): 'Statistical morphology of random interfaces in microemulsions', *J. Chem. Phys.* **102**, pp. 9350–9361
17. Mecke, K.R. (1996): 'A morphological model for complex fluids', *J. Phys.: Condens. Matter* **8**, pp. 9663–9668
18. Meester, R., R. Roy (1996): *Continuum Percolation* (Cambridge University Press)
19. Penrose, M.D. (1991): 'On a continuum percolation model', *Adv. Appl. Probab.* **23**, pp. 536–556
20. Swendsen, R.H., J.-S. Wang (1987): 'Nonuniversal critical dynamics in Monte Carlo simulations', *Phys. Rev. Lett.* **58**, pp. 86–88
21. Thönnes, E. (1999): 'Perfect simulation of some point processes for the impatient user', *Adv. Appl. Probab.* **31**, pp. 69–87
22. Widom, B., J.S. Rowlinson (1970): 'New model for the study of liquid-vapor phase transition', *J. Chem. Phys.* **52**, pp. 1670–1684

Fun with Hard Spheres

Hartmut Löwen

Institut für Theoretische Physik II, Heinrich-Heine-Universität Düsseldorf,
Universitätsstraße 1
D-40225 Düsseldorf, Germany

Abstract. Thermostatistical properties of hard sphere and hard disk systems are discussed. In particular we focus on phase transitions such as freezing in the thermodynamic limit. Results based on theory and computer simulations are given. It is emphasized that suspensions of sterically-stabilized colloids represent excellent realizations of the hard sphere model. Finally a survey of current research activities for hard sphere systems is presented and some recent results are summarized.

1 Motivation

This article aims at several points: First, it is a brief introduction to classical statistical physics of hard sphere-like systems ranging from elementary definitions to recent research activities. In this respect it represents both a tutorial and a review. Second, it is written by a physicist, not by a mathematician. This implies that emphasis is put on simple physical pictures omitting any mathematical rigour. However, it is tried to link to the literature of mathematical physics and to establish thereby a connection between physics and mathematics. Third, if possible, relations between statistical physics and geometry are discussed.

2 Introduction: The Model

The hard sphere model is defined by a pair interaction between two classical particles that only involves a non-overlap condition. The potential energy of a pair of hard spheres is

$$V(r) = \begin{cases} \infty & \text{if } r < \sigma \\ 0 & \text{else} \end{cases} \qquad (1)$$

where σ is the diameter of the spheres and r is the distance between the two centers of the spheres, see Fig. 1 on page 315. The potential $V(r)$ is sketched in Fig. 2. It is very "steep" for touching spheres. More formally, the peculiarity of the hard sphere (or any other hard body) potential is that it sets a length scale (namely σ) but it does not set any energy scale. Clearly, a configuration of two overlapping spheres is punished by an infinite energy. Having a Boltzmann factor in mind, this implies that such overlapping configurations do not occur, i.e., they have no statistical weight in a thermal average. If one sphere is fixed, a second sphere can possess any center position except for a sphere around the first sphere with a *radius* σ. This is the reason why one calls the potential (1) an "excluded volume" interaction.

Interesting questions and non-trivial effects arise when many hard spheres are interacting at high density. To be specific we consider N hard spheres in a system volume Ω at a temperature T. The particle number density is then

$$\rho = N/\Omega \tag{2}$$

An equivalent dimensionless measure of the number density is provided by the so-called *packing fraction* (or volume fraction) η which is the ratio of the volume of the N spheres and the total accessible volume Ω:

$$\eta = N\Omega_s/\Omega = \pi\rho\sigma^3/6 \tag{3}$$

where $\Omega_s = \pi\sigma^3/6$ is the volume of a single sphere.

Let the set of three-dimensional vectors $\{R_1, R_2, ..., R_N\}$ denote an arbitrary configuration of N spheres, see Fig. 3. Then the total potential energy associated to this configuration is

$$V(R_1, R_2, ..., R_N) = \sum_{i,j=1;i<j}^{N} V(|R_i - R_j)|) \tag{4}$$

where we distinguish different functions by giving them different arguments. Obviously, as $V(R_1, R_2, ..., R_N)$ is a sum over hard sphere potentials, it only takes the two values 0 and ∞. Consequently, for the Boltzmann factor, we get

$$\exp(-\beta V(R_1, R_2, ..., R_N)) \equiv \exp(-V(R_1, R_2, ..., R_N)) \tag{5}$$

with $\beta = 1/k_B T$, k_B denoting Boltzmann's constant. This implies that the temperature scales out trivially. Or, in other words, hard objects are *athermal*; all their structural and thermodynamical properties do not depend on temperature so that the density ρ (or the packing fraction η) is the only relevant thermodynamical variable. The only relevance of temperature is that $k_B T$ sets the natural energy scale. This can directly be seen by defining the classical canonical partition function

$$Z = \frac{1}{\Lambda^{3N} N!} \int_\Omega d^3 R_1 ... \int_\Omega d^3 R_N \exp(-V(R_1, R_2, ..., R_N)) \tag{6}$$

Here, the de Broglie thermal wave-length Λ is just an arbitrary length scale to make the partition function dimensionless. Λ is irrelevant since multiplying Λ by a scaling factor simply means that the chemical potential μ is shifted by an (irrelevant) constant (the definition of μ is given later, see (27) and [18]). In (6), the factor $1/N!$ avoids multiple counting of configurations that arise simply from interchanging particle labels. Then the classical canonical (Helmholtz) free energy is

$$F = -k_B T \ln Z \tag{7}$$

from which we can extract all information required for equilibrium thermodynamics. Here, it becomes again evident that the thermal energy $k_B T$ simply sets the energy scale of the Helmholtz free energy but the reduced quantity $F/k_B T$ is independent of temperature.

Interesting collective phenomena are conveniently studied in the so-called thermodynamic limit (TDL) where the number of particles, N, and the system volume, Ω, become infinite such that the particle number density $\rho = N/\Omega$ stays finite. A phase transition is signalled by a nonanalytical dependence of the Helmholtz free energy on the thermodynamic parameters such as density ρ and temperature T. The non-analyticities determine the phase diagram of the system. At this stage we shall not study the existence of the thermodynamic limit but simply take it for granted. Unfortunately the thermodynamic limit implies many integrations in (6) for the partition function Z. Hence one has either to rely on Monte Carlo techniques to evaluate this high-dimensional integral or to perform approximation methods. As far as I know, there is no exact solution for a phase transition in D-dimensional hard sphere systems if D is larger than 1, for a more detailed discussion see below.

One can also add an *external potential* $V_{ext}(r)$ to the system. If $V_{ext}(r)$ vanishes, one speaks about a bulk system. The presence of an external potential means that the total potential energy now reads

$$V(R_1, R_2, ..., R_N) = \sum_{i=1}^{N} V_{ext}(R_i) +$$

$$\sum_{i,j=1;i<j}^{N} V(|R_i - R_j|) \tag{8}$$

Typical examples for sources of an external potential are system walls and system boundaries, a gravitational field, external laser-optical fields etc. For a fixed external potential one can also perform the thermodynamic limit. In cases of a symmetry-breaking phase transition, one can force the system to be in a symmetry-broken phase by imposing a suitable external potential $V_{ext}(r) = \epsilon f(r)$. Now the sequence of the TDL and the limit $\epsilon \to \infty$ is crucial and interchanging both leads to different results. While one always gets a homogeneous bulk system if the limit $\epsilon \to \infty$ is performed first, a symmetry-broken state may be reached if the limit $\epsilon \to \infty$ is performed after the TDL.

One basic question concerns the problem of close-packing in the TDL: A first part of the problem addresses the maximum of the packing fraction which defines the socalled close-packed fraction η_{cp}. The second part is the corresponding close-packed configuration which leads to this close-packed density. Physicists have assumed over centuries that the close-packed situation of spheres is a stack of intersecting two-dimensional triangular lattices. This gives

$$\eta_{cp} = \pi/3\sqrt{2} = 0.740... \tag{9}$$

In fact, a rigorous mathematical proof for this was lacking until 1998 when Hales discovered one (see e.g. [19]). The problem was that locally one can achieve closer packings by icosahedral structures but these packings cannot be joint together to fill the whole space. In fact, it is much simpler to prove that $\eta_{cp} = \pi/3\sqrt{2}$ among all periodic lattice structures which was done in 1831 by Gauss. See the contribution of J. Wills in this volume for different problems of close-packing. It is clear that there are many configurations leading to the same close-packed fraction η_{cp}, or, in other words, there is a high *degeneracy* of close-packed configurations. In one of them, the spheres are sitting on face-centered-cubic (fcc) lattice positions which corresponds to a stacking sequence $ABCABC....$ Another structure is the hexagonal-close-packed (hcp) lattice corresponding to a stacking sequence $ABAB....$ A more exotic structure is the double hcp lattice with a stacking sequence $ABACABAC...$, but also a random stacking sequence as $ABCBACBABCA...$ is conceivable.

One may now look for a phase transition in the range of intermediate densities $0 < \eta < \eta_{cp}$. In fact, it is by now well-established and accepted that hard spheres exhibit a freezing transition which we shall discuss in detail in chapter 3 and 4.

Let us finish with two more general remarks on the hard sphere model: First, due to its temperature-independence, it is the simplest non-trivial model for an interaction. In this sense, it is useful as a reference system for systems with more complicated interactions and particle shapes. Since a theoretical physicist typically first tries to incorporate the essential ingredients in a simple model in order to study the principle mechanisms, the hard sphere model is the first choice of a reasonable approximation. This is illustrated in the cartoon of Fig. 4: if a theoretician studies a herd of elephants, his first thought is to approximate them by hard spheres neglecting any details (trunks, tails, etc.). On a length scale compatible with the overall size of an eliphant this approximation is not completely ridiculous!

The second important fact is that the equilibrium thermodynamical properties of the hard sphere model can actually be probed in nature by examining suspensions of spherical sterically-stabilized colloidal particles (for a review, see [75]). Such particles have a mesoscopic size between $1nm$ and $1\mu m$. They are coated by polymer brushes and suspended in a microscopic solvent. A schematic picture is given in Fig. 5. A typical colloidal material is polymethylmethacrylate (PMMA). The omnipresent van-der-Waals attraction between two colloidal particles can be tuned to be extremely small by "index-matching" the particles. If the typical length ℓ of the adsorbed polymer brushes is much smaller than the diameter σ of the colloidal spheres, the total interaction between the particles is dominated by excluded-volume effects. This enables one to directly compare experimental data with predictions from the hard-sphere model. An electron-micrograph of colloidal particles is shown in Fig. 6. Indeed one sees that all the spheres have the same diameter, i.e. the socalled size-polydispersity is small. Still, in an actual quantitative comparison, there are three caveats: i) Is the size-polydisperity really small in the samples? ii) Are the colloidal particles really spherical i.e. isotropic? iii) Are the interactions stiff or is there still a penetrability of spheres? In recent experiments it has been proved that all these possible problems can be avoided by carefully "cooking" the suspensions [75].

3 The Hard Sphere Model in Arbitrary Spatial Dimension

3.1 General

It is instructive to generalize the hard sphere model to an arbitrary spatial dimension D. The reason to do so is twofold: First, one can formally embed the three-dimensional case in a general context. Second, in some special spatial dimensions, exact results are available.

The interaction between two hard "hypersphere" in spatial dimension D is

$$V(|r|) = \begin{cases} \infty & \text{if } |r| < \sigma \\ 0 & \text{else} \end{cases} \tag{10}$$

Here, $r = (r_1, r_2, ..., r_D)$ is a D-dimensional position vector and $|r| = \sqrt{\sum_{i=1}^{D} r_i^2}$ is the distance between the centers of two hyperspheres. It is straightforward to generalize the formalism developed in the last chapter to arbitrary D. The canonical partition function now is

$$Z = \frac{1}{\Lambda^{DN} N!} \int_\Omega d^D R_1 ... \int_\Omega d^D R_N \cdot$$
$$\exp(-V(R_1, R_2, ..., R_N)) \tag{11}$$

where R_i now is the D-dimensional position vector of the ith particle. Let us subsequently discuss some special cases.

3.2 One-Dimensional Case

In the *one-dimensional case* $(D = 1)$ we are dealing with N hard rods of width σ on a line of length L, see Fig. 7. Here the number "line" density is $\rho = N/L$ and the packing fraction is simply $\eta = \rho\sigma$. The close-packed situation is trivial in this case leading to $\eta_{cp} = 1$. This case is remarkable insofar as the partition function Z and the Helmholtz free energy F can be calculated analytically even in the thermodynamic limit. This was done by Tonks [83] in the early days of statistical mechanics. The final result for the reduced Helmholtz free energy per particle is

$$\frac{F}{k_B T N} = \ln(\rho\Lambda) - 1 - \ln(1 - \rho\sigma) \tag{12}$$

As a result, there is no phase transition as F is analytic in the particle density ρ in the domain $0 < \rho < 1/\sigma$. The only nonanalyticity occurs at the boundaries for $\rho \to 1/\sigma$ (close-packing) where F diverges to infinity as all rods are forced to touch each other. But this is not a true phase transition.

3.3   Two-Dimensional Case

In the *two-dimensional case* $(D = 2)$ of hard disks the close-packed area fraction is $\eta_{cp} \equiv \pi \rho_{cp} \sigma^2 / 4 = \pi/2\sqrt{3} = 0.907...$ corresponding to a perfect triangular lattice with long-ranged translational order. The proof is attributed to Lagrange who did it as early as in 1773. Note that apart from trivial translations, there is no degeneracy as in the three-dimensional case. As far as phase transitions are concerned, no rigorous result is known so far. The Fröhlich-Pfister argument for the absence of long-ranged translational order [35] is not possible for hard disks as some smoothness conditions are required for the mathematical proof which are not fulfilled for hard disks. The nature of the freezing transition for hard disks is still debated in recent literature [49,60] but at least there is evidence from computer simulations for a freezing transition into a triangular lattice occuring for a density well-separated from close-packing. Most probably the transition is in accordance with the Kosterlitz-Thouless scenario [49].

3.4   Arbitrary Dimension

Situations of hyperspheres with $D > 3$ are also conceivable, at least formally. If one restricts the consideration within periodic lattice types, the close-packed density is known for $D = 4, 5, 6, 7, 8$ and only for peculiar higher dimensions but not in general, see [55] for a compilation of recent data. Apart from very recent work [30], I am not aware of any investigation of phase transitions in higher dimensions.

3.5   Degenerate Cases

Finally let us discuss two "degenerate" situations, namely $D = 0$ and the limit $D \to \infty$. The zero-dimensional case can be viewed as a sphere in a cavity that holds only one particle. In this case one can compute the partition function exactly, of course. This limiting case is important to check the validity of different density functional approximations. The limit of infinite dimension is more tricky. There are bounds for the close-packed density proved at the beginning of this century by Minkowski and by Blichfeld (in [55]). In fact, one knows

$$\frac{\zeta(D)}{2^{D-1}} \leq \eta_{cp} \leq \frac{D+2}{2} 2^{-D/2} \tag{13}$$

where $\zeta(D)$ denotes the Riemannian zeta-function. This implies that the close-packed density vanishes in the limit $D \to \infty$. A virial analysis shows that only the first and the second virial coefficients survive in the limit $D \to \infty$ such that one can use an Onsager-type analysis to extract the instability of the fluid, see [32, 33] and [89]. The rigorous location of the freezing which should also depend on the structure of the crystalline phase is still an open question. However there are recent investigations using diagrammatic expansions [34]. The question might be easier to answer for parallel hard hypercubes where the problem of close-packing is trivial and an instability analysis suggest a second-order transition from a fluid phase into a hypercubic lattice [50].

4   Hard Spheres and Phase Transitions: Theory

In this chapter we review some popular theories for the many-body hard sphere system. Most theories are constructed in such a way that they only work in a certain phase. The easiest theory which applies to the solid phase is the cell or "free volume" approach. We also mention the Percus-Yevick and scaled-particle theory which describes the fluid state. Finally a unifying theory which works in both phases can be obtained by exploiting the density functional technique.

4.1   Intuitive Arguments

Why is there a fluid-solid transition in the hard sphere system? This is not obvious at all from intuition and it is still not accepted by everybody in the physics community. In order to discuss this further, we stress that the averaged potential energy vanishes, i.e.

$$< V(\boldsymbol{R}_1, \boldsymbol{R}_2, ..., \boldsymbol{R}_N) >= 0 \qquad (14)$$

as configuration where spheres are at contact have zero statistical weight. Here $< A >$ denotes the canonical average of the quantity A

$$< A > = \frac{1}{Z \Lambda^{3N} N!} \cdot$$
$$\int_\Omega d^3 R_1 ... \int_\Omega d^3 R_N \ A \ \exp(-V(\boldsymbol{R}_1, \boldsymbol{R}_2, ..., \boldsymbol{R}_N)) \qquad (15)$$

This immediately implies that the total free energy can be written as

$$F = \bar{H}_{kin} - TS = \frac{3}{2} N k_B T - TS \qquad (16)$$

where $\bar{H}_{kin} = 3Nk_BT/2$ is the averaged kinetic energy of the spheres and S is the entropy. Hence, apart from the trivial constant $\frac{3}{2}Nk_BT$, there is only entropy. This is the reason why one says that hard spheres are an *entropic* system. The intuitive feeling is that high entropy means low order. According to this intuition, an ordered phase should have a lower entropy or a higher free energy than a disordered phase. Hence a disordered phase has minimal free energy and should be the thermodynamically stable phase. This simple argument, however, is wrong. In fact, at high densities, the entropy of hard spheres is smaller in the ordered (solid) phase than in the disordered (fluid) phase! A more refined argument splits the entropy into two different parts which is visualized for hard disks in Fig. 8. For any random or disordered situation, one has many possible configurations but as the density grows more and more of these configurations are blocked by touching nearest neighbours. In an ordered solid-like phase, on the other hand, there is only one basic lattice configuration possible, but one can generate further non-overlapping configurations by moving the disks slightly away form their lattice position. Hence, a fluid phase has a high configurational entropy but a small "free volume" entropy, while a solid phase has a low configurational entropy but a high correlational (or "free volume") entropy. Both kind of entropies depend on the density. The configurational entropy is dominating for small densities whereas the "free volume" entropy is dominating near close-packing. Consequently there has to be a phase transition between these two situations for intermediate densities. It has to be emphasized that the freezing transition is not driven by competition between potential energy and entropy but by competition between these two different kinds of entropies. This is even the generic mechanism for freezing which also works for soft repulsive potentials. We finally remark that these two different kinds of entropies can be properly defined and calculated by density functional theory [9].

4.2   The Cell Model

The cell theory or free volume approach (for a general introduction and historical remarks see [61]) starts from a given hard sphere solid. We now draw the Wigner-Seitz (or Voronoi) cells of this lattice, see Fig. 9. Let us assume that each sphere can move freely only within its own Wigner-Seitz cell. Obviously we are neglecting some further configurations by this restriction, therefore this theory clearly is an approximation. But this approximation should be justified near close-packing. Equivalently, this means that any center of the spheres can move within a small "free volume" Ω_f which has the same form as the Wigner-Seitz cell, see again Fig. 9. By counting the configurations and considering trivial particle exchanges we estimate the partition function as follows:

$$Z \geq Z_{CT} = \frac{1}{\Lambda^{3N}} \Omega_f^N \tag{17}$$

and the free energy is

$$\frac{F}{k_B T N} \leq \frac{F_{CT}}{k_B T N}$$

$$= \ln \sqrt{2} + 3 \ln(\Lambda/\sigma) - 3 \ln[(\frac{\pi\sqrt{2}}{6\eta})^{1/3} - 1] \tag{18}$$

The cell theory thus establishes a rigorous upper bound to the free energy. Clearly, the free energy diverges as $\eta \to \eta_{cp}$. The leading logarithmic divergence becomes in fact asymptotically exact as $\eta \to \eta_{cp}$ [61]. In the one-dimensional case ($D = 1$) the cell theory gives the exact equation of state which is the pressure

$$P = -\frac{\partial F}{\partial \Omega}|_{N,T} \tag{19}$$

as a function of density, but the free energy itself is not exact.

4.3   Percus-Yevick Theory

Another obvious approach is to start from very low densities where the system is an ideal gas and calculate perturbatively the next leading corrections. The equation of state can be expressed in powers of the density. Clearly the starting point here is the fluid phase. It is known that the virial expansion has a finite radius of convergence in the density for any spatial dimension D [53] but the actual convergence radius might be much larger. One could surmise that all the virial coefficients are positive but a rigorous mathematical proof is still lacking.

The virial expansion can be improved by solving socalled liquid-integral equations [41]. In fact the following Percus-Yevick closure relation has been found to give good result even for intermediate packing fractions up to $\eta \approx 0.3$. The closure is expressed in terms of two correlation functions. The first is the *pair distribution function* $g(r)$ defined as

$$g(r) = \frac{1}{\rho N} < \sum_{i,j=1; i<j}^{N} \delta(r - (R_i - R_j)) > \qquad (20)$$

This function gives the probability of finding a particle at distance r from a given fixed particle. A typical $g(r)$ for hard spheres is shown in Fig. 10. For $r < \sigma$, $g(r)$ vanishes which is just the non-overlap condition:

$$g(r) = 0 \quad \text{for} \quad r < \sigma \qquad (21)$$

For $r \to \infty$, $g(r)$ is normalized to 1. For very small densities, $g(r) = \Theta(r - \sigma)$ where $\Theta(x)$ denotes the unit step function while for large densities, $g(r)$ exhibits a structure from neighbouring shells of particles. At very high densities $\eta \approx 0.5$, the contact value $g(r \to \sigma^+) \equiv g(\sigma^+)$ increases to large values and the second neighbour shell becomes split exhibiting a shoulder [84], see Fig. 10, which is in accordance with confocal microscopy measurements on sterically-stabilized colloidal suspensions [85]. The pair correlation function is also discussed in the contribution of Döge et al. in this volume. The equation of state can exactly be related [41] to the contact value of $g(r)$ by using the virial expression:

$$\frac{P}{\rho k_B T} = 1 + 4\eta g(\sigma^+) \qquad (22)$$

So once one knows $g(\sigma^+)$ for any density one gets F by integrating (22).

The second correlation function is the Ornstein-Zernike or *direct correlation function*. It is implicitly defined via the Ornstein-Zernike relation

$$g(r) - 1 = c(r) + \rho \int d^3 r' (g(r') - 1) c(|r - r'|) \qquad (23)$$

The Percus-Yevick closure combines the exact relation (21) with the approximation

$$c(r) = 0 \quad \text{for} \quad r > \sigma \qquad (24)$$

The advantage of the Percus-Yevick theory is that it can be solved analytically for $c(r)$. Using the Ornstein-Zernike relation and the virial expression one can deduce an analytical form for the Helmholtz free energy as follows

$$\frac{F}{k_BTN} \approx \frac{F_{PY}}{k_BTN} = 3 \ln(\Lambda/\sigma) - 1 + \ln(6\eta/\pi)$$

$$- \ln(1 - \eta) + \frac{3}{2}[\frac{1}{(1 - \eta)^2} - 1] \tag{25}$$

This expression clearly diverges when $\eta \to 1$ being an artifact of the approximation which is meant only for small η. Finally we remark that the Percus-Yevick direct correlation function and the free energy are exact in one spatial dimension.

4.4   Estimation of the Freezing Transition

Knowing analytical expressions (18) and (25) for the free energy in the solid and fluid state, we can estimate the location of the freezing transition. There are three conditions for phase coexistence. The first concerns thermal equilibrium, i.e. the temperature in the two coexisting phases has to be equal, $T_1 = T_2$. Due to the trivial temperature dependence of the free energy for hard sphere, this condition is fulfilled. Second, the pressure in the two coexisting phases has to coincide, $P_1 = P_2$, (mechanical equilibrium). Third, chemical equilibrium requires the same chemical potential in the two coexisting phases, $\mu_1 = \mu_2$. The latter two conditions are equivalent to Maxwell's common-tangent construction. This is easily explained in terms of the free energy per volume $f_i = F_i/V = f_i(T, \rho)$ of the two phases ($i = 1, 2$). The two pressures and chemical potentials can be written as

$$P_i = -\frac{\partial F_i}{\partial \Omega}|_{T,N} = f_i - \rho_i \frac{\partial f_i(T, \rho = \rho_i)}{\partial \rho}|_T \tag{26}$$

and

$$\mu_i = \frac{\partial F_i}{\partial N}|_{T,\Omega} = \frac{\partial f_i(T, \rho = \rho_i)}{\partial \rho}|_T \tag{27}$$

where $i = 1, 2$ labels the two different phases. The two conditions $P_1 = P_2$ and $\mu_1 = \mu_2$ hence are expressed as

$$f_1'(\rho_1) = f_2'(\rho_2) \tag{28}$$

and

$$f_2(\rho_2) = f_1(\rho_1) + f_1'(\rho_1)(\rho_2 - \rho_1) \tag{29}$$

with $f_1'(\rho) \equiv \partial f_i/\partial \rho|_T$. These equations mean that one finds the two coexisting densities ρ_1 and ρ_2 by a common tangent construction for the two free energy densities plotted as a function of density. In our case of hard sphere freezing this is visualized in Fig. 11. Assuming that the cell theory (solid line in Fig. 11) and the Percus-Yevick expression (dashed line in Fig. 11) are valid for any density the Maxwell common tangent construction leads to coexisting packing fractions of $\eta_f = 0.57$ for the fluid phase and $\eta_s = 0.65$ for the solid phase. Of course, actual data for free energies are required for intermediate densities where the two theoretical expressions are expected to fail. Nevertheless, we shall see later that the coexisting densities and the relatively large density jump across the transition are in fairly good agreement with "exact" computer simulations.

These considerations also give a clue of how to prove rigorously the existence of the freezing transition. The solid cell model gives an *upper bound* to the free energy in the solid phase. If one would know a *lower bound* of the free energy in the fluid phase and could show that this lower bound hits the upper bound of the cell model, then the existence of the phase transition would be proven. The construction of a lower bound in the fluid phase is strongly linked to the virial coefficients which determine the convergence of the virial expansions in powers of the density. If all the virial coefficients would be positive, then there has to be a freezing transition. Even if only the virial expansion truncated after the second coefficient would be a lower bound, then it hits the solid cell theory for spatial dimensions $D > 8$ [30]. However, although all these assumptions seem to be plausible for physicists, they need to be proved mathematically. Therefore, to establish rigorously the exostence of the freezing transition is still an open unsolved problem.

4.5   Scaled Particle Theory

The scaled particle theory ([76]; for a review see [8]) considers the reversible work to create a spherical cavity of radius R_0 in a hard sphere fluid. Formally the cavity can be regarded as a further "scaled" particle. One knows the relation of this work to the bulk pressure for the special case $R_0 = 0$. For $R_0 \to \infty$, this work is connected to the interfacial free energy γ between a planar hard walls and a hard sphere fluid. Interpolating between these special cases one gets the work for $R_0 = \sigma$ from which one deduces the contact value $g(\sigma^+)$. Using the exact virial expression (22), one gets the free energy. Remarkably, though a completely different phyical picture is used, the scaled-particle results coincides with the Percus-Yevick virial expression (25). The scaled-particle theory cannot be applied to the solid but it has the advantage that it can be generalized to hard convex bodies with non-spherical shapes as e.g. hard spherical-capped cylinders.

4.6   Density Functional Theory

Density functional theory (DFT) provides a unified picture of the solid and fluid phase. In fact as we shall show below, it is a way to combine the cell theory for the solid with the Percus-Yevick (or scaled particle) theory of the liquid. As for general reviews, see e.g. [57] and [29]. The cornerstone of DFT is the Hohenberg-Kohn-Sham theorem which was generalized to finite temperatures by Mermin. It guarantees the existence of a functional for the excess free energy $\mathcal{F}_{exc}[\rho]$ of the (in general inhomogeneous) one-particle density $\rho(r)$. This functional has the unique property that the functional for the grandcanonical free energy

$$\bar{\Omega}[\rho] := \mathcal{F}_{exc}[\rho] + \int_{\Omega} d^3r \rho(r)\{V_{ext}(r) - \mu$$
$$- 1 + k_B T \ln(\Lambda^3 \rho(r))\} \tag{30}$$

is minimized by the equilibrium one-particle density

$$\rho_0(r) = < \sum_{i=1}^{N} \delta(r - R_i) > \tag{31}$$

and the minimum $\bar{\Omega}[\rho_0(r)]$ is the actual grandcanonical free energy which is equal to $-P\Omega$ in the bulk case. The problem, however, is that nobody knows the actual form of the functional $\mathcal{F}_{exc}[\rho]$. It is only for the trivial case of an (non-interacting) ideal gas that $\mathcal{F}_{exc}[\rho]$ is known to vanish.

Different approximations for $\mathcal{F}_{exc}[\rho]$ designed for strongly interacting systems (in particular for hard spheres) are on the market. Most of them make use of the fact that the direct (Ornstein-Zernike) correlation function introduced in chapter 3.3 is the second functional derivative of $\mathcal{F}_{exc}[\rho]$ in the homogenous bulk fluid [41]:

$$c(|r_1 - r_2|) = \frac{1}{k_B T} \frac{\delta^2 \mathcal{F}_{exc}}{\delta\rho(r_1)\delta\rho(r_2)}|_{hom} \tag{32}$$

The most elaborated and reliable functional for hard spheres is that recently developed by [79]. It is fixed by approximating

$$\mathcal{F}_{exc}[\rho] \approx k_B T \int_{\Omega} d^3 r \Phi[\{n_\alpha(r)\}] \tag{33}$$

where one introduced a set of weighted densities

$$n_\alpha(r) = \int_{\Omega} d^3 r' \rho(r') w_\alpha(r - r') \tag{34}$$

Here, the index $\alpha = 0, 1, 2, 3, V1, V2$ labels six different weighted densities and six different associated weight functions. Explicitly these six weight functions are given by

$$w_0(r) = \frac{w_2(r)}{\pi\sigma^2} \tag{35}$$

$$w_1(r) = \frac{w_2(r)}{2\pi\sigma} \tag{36}$$

$$w_2(r) = \delta(\frac{\sigma}{2} - r) \tag{37}$$

$$w_3(r) = \Theta(\frac{\sigma}{2} - r) \tag{38}$$

$$w_{V1}(r) = \frac{w_{V2}(r)}{2\pi\sigma} \tag{39}$$

and

$$w_{V2}(r) = \frac{r}{r}\delta(\frac{\sigma}{2} - r) \tag{40}$$

Note that the index V denotes a vector weight function. We can express this fact by writing $w_{V1} \equiv w_{V1}, n_{V1} \equiv n_{V1},...$ Finally the function Φ is given by

$$\Phi = \Phi_1 + \Phi_2 + \Phi_3 \tag{41}$$

with

$$\Phi_1 = -n_0 \ln(1 - n_3) \tag{42}$$

$$\Phi_2 = \frac{n_1 n_2 - \boldsymbol{n}_{V1} \cdot \boldsymbol{n}_{V2}}{1 - n_3} \tag{43}$$

and

$$\Phi_3 = \frac{n_2^3 (1 - (\boldsymbol{n}_{V2}/n_2)^2)^3}{24\pi(1 - n_3)^2} \tag{44}$$

Let us emphasize few points: First, the six weight functions are connected to the geometrical (fundamental) Minkowski measures. In fact, the derivation of the Rosenfeld functional requires a convolution property which can nicely be evaluated by using the linear decomposition into the four Minkowski measures for an arbitrary additive measure. One might therefore conjecture that there is a deeper connection between the geometry and density functional theory which, however, still has to be discovered and worked out! Second, Rosenfeld's functional gives the analytical Percus-Yevick direct correlation function as an output by using the relation (32). Consequently the Percus-Yevick theory is included in this density functional approach. Also the cell model is included near close-packing [77]. Hence the density functional approach provides a unifying theory of fluid and crystal. In particular, a configuration of overlapping spheres (which implies $n_3 \to 1$) is avoided as the functional gives an infinite energy penalty to such densities, see again (42) and the denominator in (43). From this respect, the Rosenfeld functional is superior to weighted density approximations proposed earlier where $c(r)$ is taken as an input and overlapping configurations of hard spheres are not excluded, for a more detailed discussion see e.g. [70]. Third, one can test the quality of any density functional by subjecting the three-dimensional functional to a strongly confining external potential such that the resulting system lives in a reduced spatial dimension. For instance, by applying a hard tube of diameter σ one can squeeze the three dimensional hard sphere system into a system of hard rods. As the density functional for hard rods is exactly known, one can test whether the resulting projected three-dimensional functional respects this dimensional crossover [78]. The ultimate reduction occurs for an external hard cavity potential that can hold only a single particle. For this trivial situation, the exact functional is known. It was shown that this limit requires some conditions which can be exploited to fix some freedom in the original functional [78]. Finally, the freezing transition can be calculated by plugging in a constant density field for the fluid phase and a lattice sum of Gaussian peaks in the solid phase. If the width of the Gaussians and the prefactor are taken as variational parameters one gets a first-order freezing transition with coexisting packing fractions of $\eta_f = 0.491$ and $\eta_s = 0.540$ which are very close to "exact" simulation data.

5 Hard Spheres and Phase Transitions: Computer Simulations

Most of our knowledge for hard sphere systems is based on "exact" results obtained by computer simulations, see e.g. [1]. In a bulk computer simulation the system is typically confined to a finite cubic box with periodic boundary conditions in all three directions to minimize finite-size effects. The typical number of particles is in the range from $N = 100$ to $N = 1000000$.

The recipe is as follows: one starts from a given overlap-free configuration of spheres. Then one generates a new configuration by using either a Molecular Dynamics code or a Monte Carlo technique. Using Molecular Dynamics means that Newton's equation of motion are solved. The hard spheres are then moving along the classical trajectories which are straight lines interrupted by elastic collisons. In Monte Carlo one randomly displaces a randomly chosen particle and checks for particle overlap: if the new configuration is free of any overlaps the move is accepted, if not it is rejected. Then one carefully has to equilibrate the system. Finally statistics is gathered to perform the canonical averages. We remark that Monte Carlo techniques are also possible in different ensembles where the pressure is fixed instead of the system volume, or the chemical potential is fixed instead of the particle number. For an example, see the method described in the contribution of Döge et al. in this volume.

A problem is that only averages are readily calculated by a simulation. The key quantity for phase transitions, however, is the Helmholtz free energy, which cannot be written as an average. One possible solution of this problem is to calculate the contact value $g(\sigma^+)$ of the pair distribution function which can clearly be written as an average, see (20). One thereby gains the pressure (or the reduced equation of state) by using the virial expression (22). In doing so for arbitrary densities, one can plot directly $P(\rho)$ and look for van-der-Waals loops indicating a first order phase transition, see Fig. 12. Typically the hysteresis is small decreasing with increasing system size and therefore it is difficult to see whether really a phase transition takes place. A more accurate alternative is to obtain the Helmholtz free energy F by integration as follows

$$\frac{F}{N} = k_B T[\ln(\rho_r \Lambda^3) - 1] + \int_{\rho_r}^{\rho} d\rho' \frac{P(\rho')}{\rho'^2} \qquad (45)$$

Here, the reference density ρ_r is so small that the system can be considered to be an ideal gas where the free energy is known. This is the simplest way of so-called *thermodynamic integration* starting from a well-known reference system. This strategy readily applies to the fluid phase.

As a remark, the virial expression also works in the solid phase if the contact value of the spherically averaged pair distribution function is inserted into (22). However there are technical problems in applying this recipe to the solid phase as the pair distribution function strongly piles up near contact and extrapolation of $g(r)$ to contact bears a large extrapolation error. A smarter way of thermodynamic integration in the hard sphere solid is to start from an Einstein solid

[31]. Here, all particles are harmonically bound to a lattice position of a given lattice as described by an external potential H_{ext}

$$H_{ext} = \sum_{i=1}^{N} \frac{K}{2} (R_i - R_i^{(0)})^2 \tag{46}$$

where $\{R_i^{(0)}\}$ are the positions of the given lattice. If the spring constant K which confines the particles to their lattice positions is very large, then the particles do not feel the hard sphere interaction. Hence the system is practically a set a decoupled harmonic oscillators for which the reference free energy F_0 can readily be calculated. Now the harmonic external potential is switched off continuously, i.e. we consider the total Hamiltonian

$$H_{tot} = H_{kin} + (1 - \lambda) H_{ext} + V_{int} \tag{47}$$

where H_{kin} is the total kinetic energy, V_{int} is the pairwise hard core interaction, and the parameter λ is a formal coupling parameter by which we can switch off continuously the external harmonic potential and turn on the hard-core interaction.. It is readily calculated that the derivative $\partial F/\partial \lambda|_{T,\Omega,N}$ can be written as an average:

$$\frac{\partial F}{\partial \lambda} = -k_B T \frac{\frac{\partial Z}{\partial \lambda}}{Z} = - < H_{ext} >_\lambda$$

$$= -\frac{NK}{2} < (R_i - R_i^{(0)})^2 >_\lambda \tag{48}$$

Here the canonical average $< ... >_\lambda$ means that an external potential of strength $1 - \lambda$ is present. (48) implies that one has to calculate the Lindemann parameter (or the mean-square-displacement) of the solid in order to access $\partial F/\partial \lambda$. Finally integration with respect to λ yields the desired free energy:

$$F = F_0 + \int_0^1 d\lambda \frac{\partial F}{\partial \lambda} \tag{49}$$

It is important to remark that one needs a whole set of simulations (for different λ) to access a single free energy. In practice typically 10-30 integration points are needed to get a good accuracy. Apart from numerical integration errors and statistical and finite-system-size errors, this methods leads in principle to exact results for the free energy. The only requirement is that one should not cross a phase boundary during the integration. Also the lattice structure is not known a priori but different lattice types have to be tried and the resulting free energy which is minimal corresponds to the realized structure.

Computer simulations of the hard sphere system have given a coherent picture of what is going on as far as phase transformations of the system are concerned. By a careful study of finite system size effects it has been established from the early days of computer simulation [45] that the hard sphere system freezes indeed from a fluid into an ordered solid with a strongly first-order transition, i.e. the density jump across the transition is pretty large. The data of the coexisting packing fractions are $\eta_f = 0.494$ and $\eta_s = 0.545$. The phase diagram is sketched in Fig. 13. The combination of Percus-Yevick and cell theory gives coexistence densities that are too high while there is perfect agreement with density functional theory.

There is another interesting non-equilibrium phase transition for $\eta = \eta_G \approx 0.58$ where a rapidly compressed hard sphere fluid freezes into an amorphous glass as signalled by a very slow decay of dynamical correlations. However, if one waits for a long time, the system will recrystallize in its thermodynamically stable solid [27]. Above a certain threshold density $\eta_{RCP} \approx 0.64$ called random-closed packing there is no glass transition possible and the system is forced to freeze into a regular solid. We finally mention that the whole phase diagram including the glass transition was confirmed in detail by experiments on sterically-stabilized colloidal suspensions [75].

What is the stable crystal lattice away from close-packing? This question has attracted some attention in the past years. A simple cubic and body-centered-cubic lattice can be ruled out from the very beginning, since these lattices are mechanically unstable with respect to shear. A tricky competition arises between the possible close-packed structures fcc, hcp, double hcp, and random stacking, see again chapter 1. It was shown by computer simulation [14,17,58,74] that for $\eta_s < \eta < \eta_{cp}$ an fcc solid has a slightly lower free energy than all other stacking sequences, but the relative difference in the free energy per particle is smaller than $10^{-3} k_B T$.

As already mentioned in chapter 2, the freezing transition in the hard disk system is much more difficult to compute by simulation and is still controversial. The reason is that the transition is not strongly first order as for the hard sphere system. Therefore the free energy differences are tiny and also finite system size effects are much more pronounced in two spatial dimensions.

6   A Selection of Recent Research Activities on Hard-Sphere-Like Systems

6.1   Binary Mixtures

It is straightforward to generalize the one-component hard-sphere model to two species. The three parameters determining the system are now the diameter ratio $q = \sigma_1/\sigma_2 \leq 1$ and the two partial packing fractions, η_s and η_l of the small and big spheres. Obviously, the one-component model is obtained as the special case $q = 1$.

Depending on the ratio q, one might expect quite different phase diagrams. If q is not much different from 1, then an fcc crystal is stable which is randomly occupied by the two species. Computer simulations [52] and numerous density functional calculations (see e.g.[22,23] have been performed here. The results for the phase diagram are in accordance with measurements on sterically-stabilized colloidal suspensions.

For intermediate q, there are more exotic crystalline phases. For certain values of q there are crystalline solids with an AB AB_2 and AB_{13} (superlattice) structure. These structures were obtained by experiments [4], computer simulation [28], density functional [90] and cell theory [5] studies and demonstrate nicely the fruitful interaction between these different approaches. The existence of such solid lattices crucially depends on the close-packed structure. Even more complicated lattice structures can be expected upon further reducing q. Also it has been speculated about the existence of stable quasicrystals for certain ratios q although they can most probably be ruled out for $q > 0.85$ [59]. The stability of quasicrystals is closely related to the question whether the close-packed structure is a periodic lattice or not. There is no mathematical proof known for general q.

Another interesting case is the limit of small q. Here there has been some debate about possible phase separation over the last decade. An analytical Percus-Yevick solution is possible predicting no fluid-fluid phase separation for hard sphere mixtures but the theory fails in the limit of $q \to 0$ if η_s and η_l is kept finite [10]. The phase diagram of strongly asymmetric hard sphere mixtures was recently obtained by computer simulations by Dijkstra, van Roij and Evans [25,26] which answered the story after all. For three different ratios $q = 0.2, 0.1, 0.05$ the phase diagrams are shown in Fig. 14. In fact a fluid-fluid phase separation is preempted by the fluid-solid transition but an isostructural solid-solid transition shows up for high packing fractions of the large particle due to the strong and short-ranged depletion attraction induced by the small particles [38].

Finally we remark that the kinetic glass transition is different for large and for small q. While both particle species freeze-in simultaneously for $q \approx 1$, there is a crossover at $q_c \approx 0.15$ where the big spheres are frozen-in on a lattice and the small spheres are still liquid-like. This was found theoretically [16,64,65] and confirmed experimentally for colloidal suspensions [46,47].

6.2 Size Polydispersity

A size-polydisperse hard-sphere system can be understood as a mixture with an infinite number of different species whose diameter is distributed according to a normalized probability function $p(\sigma)$. A relative small distribution is characterized by its first two moments, or equivalently by its mean diameter $\bar{\sigma} = \int d\sigma p(\sigma)$ and the relative polydispersity $s^2 = \overline{\sigma^2}/\bar{\sigma}^2 - 1$. A study of effects induced by polydispersity is important if one has a quantitative comparison with a real colloidal sample in mind. It has been established by density functional theory [3] and computer simulation [12,13,51] that above a certain critical polydispersity of roughly 6% a solid lattice is no longer stable. The corresponding phase diagram is shown in Fig. 15. A regular random occupied solid lattice structure coexists with a fluid that has a higher polydispersity as the solid as indicated by the tie-lines in Fig. 15. For high densities the solid exhibits reentrant melting into an amorphous phase [7].

At higher polydispersities, the phase behaviour depends more and more on the details of the diameter distribution $p(\sigma)$. A randomly occupied solid is expected to separate into two or more solids with different lattice constants [5,82]. Also fluid-fluid phase separations are probable to occur 21,87]. We finally mention that the Percus-Yevick direct correelation fucntion can be explicitly calculated involving only the first three moments of the diameter distribution [11] and that the cell model is again a reliable description of the solid for high densities and small polydispersities [72].

6.3 Hard Spheres near Hard Plates

A planar hard wall can be described as an external potential

$$V_{ext}(z) = \begin{cases} \infty \text{ if } z < \sigma/2 \\ 0 \quad \text{else} \end{cases} \tag{50}$$

where z is the coordiante perpendicular to the wall. The insertion of such a planar hard plate costs free energy as there are less configurations possible. This additional free energy scales with the plate surface and gives rise to a positive surface free energy γ. For a fluid phase in contact with a wall, scaled particle theory makes a theoretical prediction for γ. In Fig. 16, γ is plotted versus the bulk packing fraction η. The agreement between scaled particle theory, density functional theory [37] and computer simulation [42,43] is convincing. If a solid is in contact with a hard wall, γ depends on the orientation. It has recently be shown that the cell model provides a reasonable analytical theory for γ which agrees perfectly with the computer simulation data [43].

For a fluid in contact with a hard wall, there is an interesting wetting effect if the bulk density is slightly below bulk freezing. Precrystallization on the hard walls [20] occurs, i.e. few layers on top of the wall have an in-plane long-ranged order corresponding to a intersecting triangular lattice sheets.

Other interesting phase transitions occur for two parallel hard plates with a slit distance H. The phase diagram depends solely on two parameters, namely the reduced density $\rho_H = N\sigma^3/(AH)$ (where A is the system area) and the reduced plate distance $h = H/\sigma - 1$. Clearly one can continuously interpolate between two and three spatial dimensions by tuning the plate distance: For $H = \sigma$, our model reduces to that of two-dimensional hard discs while for $H \rightarrow \infty$ the three-dimensional bulk case is recovered.

The equilibrium phase diagram as obtained by Monte-Carlo computer simulation in the $\rho_H - h$-plane [80,81] is shown in Fig. 17 for moderate plate distances h. The phase behaviour is very rich and much more complicated than in the bulk. Cascades of different solid-solid transitions are found. For low densities the stable phase is an inhomogeneous fluid. All possible stable solid phases are also realized as close-packed configurations [71,73] for a certain plate distance. Accordingly one finds stable layered structures involving intersecting triangular lattices ($1\triangle$, $2\triangle$) and intersecting square lattices ($2\square$). Also a buckled phase (b) and a phase with a rhombic elementary cell (rhombic phase (r)) are stable. All transitions are first-order. Results of the cell model together with a simple fluid state theory are given in Fig. 18. Clearly the simple cell theory gives the correct topology of the phase diagram.

Similar phases were found in experiments of highly salted charged colloids between glass plates [62,63,68,86,88]. Here even higher reduced plate distances were studied. There is compelling evidence that a prism-phase consisting of alternating prisms built up by spheres is the close-packed configuration in certain domains of h [68]. Still a full quantitative mapping of the experimental data onto the theoretical phase digram of Fig. 17 has to be performed.

Let us comment on further related aspects: First it would be nice to perform a full theoretical calculation for the phase diagram of hard spheres between hard plates using a density functional calculation with Rosenfeld's functional. Second, one should investigate different confining shapes. Intriguing examples are circular and polyhedral boundaries in two dimensions. Studies have been made for confined hard discs [66] and confined hard spheres within spherical cavities [67]. Finally it is an unsolved mathematical problem to rigorously establish the close-packed structure for different h.

6.4 Hard Spherocylinders

Finally let us discuss phase transitions for hard convex bodies that are non-spherical, for a recent review see [2]. In particular, if these bodies are rotational invariant around one axis they may serve as a model for colloidal liquid crystals [54]. In particular hard spherocylinders with an additional orientational degree of fredom have been studied. These are spherical capped cylinders of cylindrical length L and diameter σ whose anisotropy is characterized by the aspect ratio $p = L/\sigma$. For $p = 0$ one recovered the case of hard spheres. The phase diagram of hard spherocylinders depends on the aspect ratio p as well as on the particle density ρ. It has recently been explored by computer simulations [14] and is shown in Fig. 19. A number of mesophases or liquid-crystalline phase are stable in the plane spanned by p and ρ. There is a plastic (or rotator) crystals for small p. For larger p, a nematic and a smectic A phase become stable for intermediate densities. Possible stacking sequences in the solid phase are AAA where all triangular sheets are put directly on top of each other and ABC which is the close-packed structure. Note that the AAA stacking sequence is not a close-packed situation but is still stable for intermediate densities. Cell theory combined with scaled particle theory can reproduce this diagram satisfactorily [39]. Also density functional theory studies have been performed [40,44]. However, it is not easy to generalize Rosenfeld's theory to the case of anisotropic particles.

7 Conclusions

To summarize: Systems of hard spheres and its variants show interesting phase transitions. Although they are purely entropically driven, they exhibit ordering transitions. These transitions are seen in theory, computer simulation and in real matter, namely in sterically-stabilized colloidal suspensions.

A few final remarks are in order: First, there are further fascinating phenomena occurring for dynamial correlations and non-equilibrium situations of hard sphere systems [36] which have not been addressed at all in this article. Another field of physics where hard sphere system play an important role are simulations of granular matter [56]. Second, most stable crystalline phases observed in phase diagrams of hard sphere problems are close-packed ones. Therefore it would be very helpful to provide mathematical proofs for the close-packed structures in confining geometry.

As a final perspective, such simple intuitive systems as hard spheres are nontrivial enough to be studied also over the next decades. One might surmise that further interesting unexpected transitions will be discovered in the near future. Hence the final conclusion is that hard spheres are fun both for physicists and for mathematicians.

Acknowledgments:
I thank Matthias Schmidt, Martin Watzlawek, Arben Jusufi, Martin Heni, Yasha Rosenfeld, Thomas Palberg, Reimar Finken, and Siegfried Dietrich for helpful remarks. I am grateful to D. Stoyan, Joachim Dzubiella, and Matthias Schmidt for a critical reading of the manuscript. Financial support by the Deutsche Forschungsgemeinschaft via the Schwerpunktsprogramm Benetzung und Strukturbildung an Grenzflächen is gratefully acknowledged.

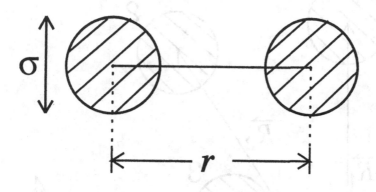

Fig. 1. Two hard spheres of diameter σ at center-of-mass distance r.

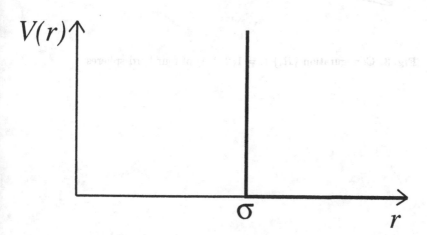

Fig. 2. Pair potential of hard spheres $V(r)$ as a function of their center-of-mass distance r.

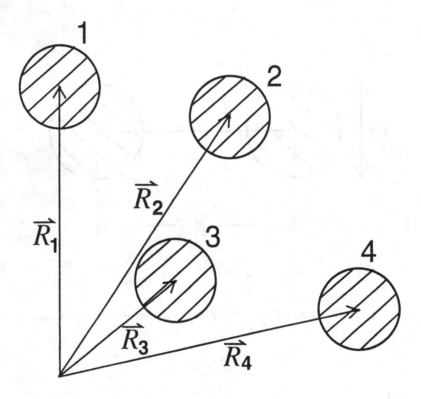

Fig. 3. Configuration $\{\vec{R}_i\}$ $(i = 1, 2, 3, 4)$ of four hard spheres.

Fig. 4. The hard sphere model at work: zeroth approximation for almost any problem in the brain of the theoretician.

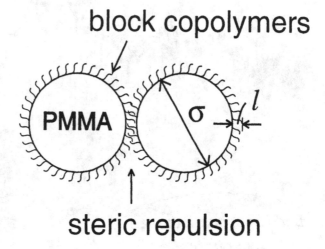

Fig. 5. Sterically-stabilized colloidal suspensions of PMMA spheres with coated block copolymer brushes of length ℓ.

Fig. 6. Electron micrograph of colloidal microspheres. From [69].

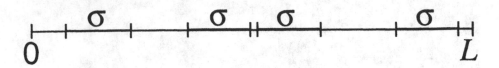

Fig. 7. Hard sphere model in one spatial dimensions: hard rods along a line of length L.

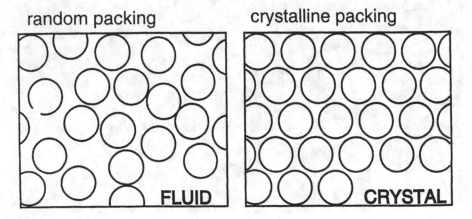

Fig. 8. Random configuration in the fluid phase and regular packed configuration in the solid phase of a hard disk system.

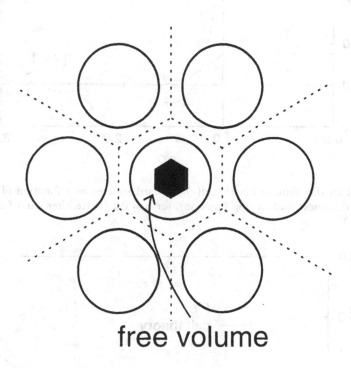

free volume

Fig. 9. Cell theory for the hard sphere crystal. The Wigner-Seitz cell (dashed line) and the free-volume cell is shown schematically.

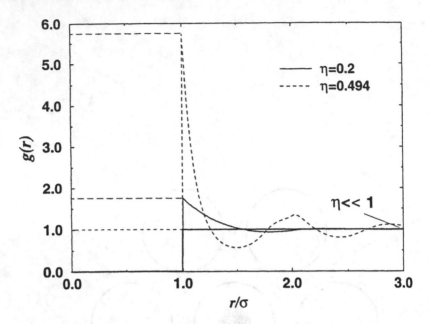

Fig. 10. Pair distribution function $g(r)$ for hard spheres as a function of distance r for small densities (dashed step function), for $\eta = 0.2$ (dotted line) and for $\eta = 0.494$ (solid line).

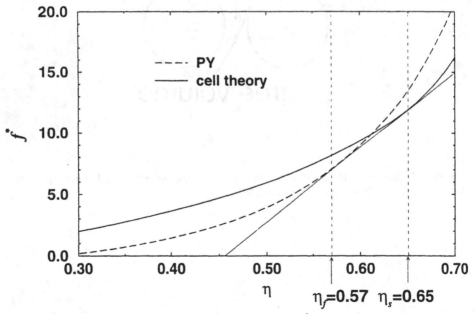

Fig. 11. Reduced free energy per unit volume $f^* = f\sigma^3/k_B T$ versus packing fraction η. The dashed line is the Percus-Yevick virial expression, the solid line is from solid cell theory. The Maxwell common tangent is also shown.

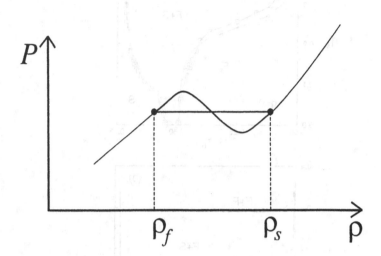

Fig. 12. Isothermal equation of state P as a function of density. The Maxwell equal-area construction is shown.

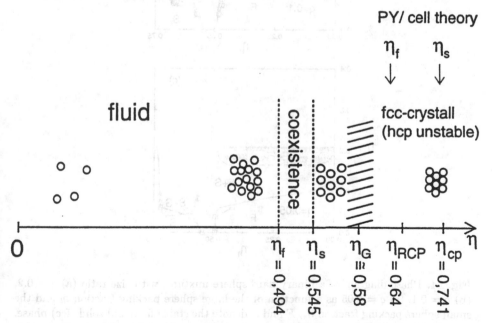

Fig. 13. Phase diagram of hard spheres versus packing fraction η. The freezing transition together with the two coexisting packing fractions η_f and η_s are shown. Also the glass transition is indicated.

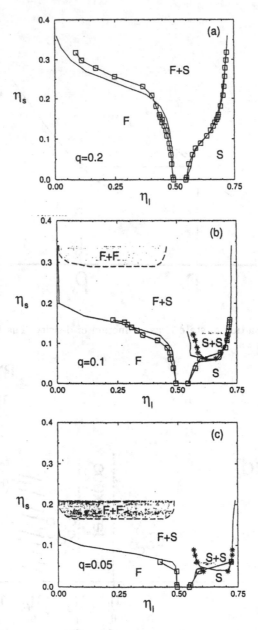

Fig. 14. Phase diagram of a binary hard-sphere mixtures with size ratio (a) $q = 0.2$, (b) $q = 0.1$, (c) $q = 0.05$ as a function of the large sphere packing fraction η_l and the small sphere packing fraction η_s. F and S denote the stable fluid and solid (fcc) phase. $F + S$, $F + F$, and $S + S$ denote, respectively, the stable fluid-solid, the metastable fluid-fluid and (meta) stable solid-solid coexistence region. The solid and dashed lines are from one effective one-component depletion potential. The symbols joined by lines to guide the eye are from computer simulations of the full binary system. From [26].

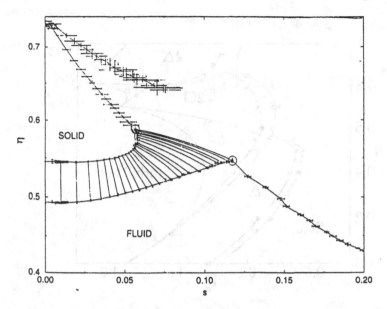

Fig. 15. Phase diagram of polydisperse hard spheres in the η, s plane. The fluid and solid phase together with their tie-lines are shown. The terminating polydispersity for a solid phase is roughly 6%. From [12,13].

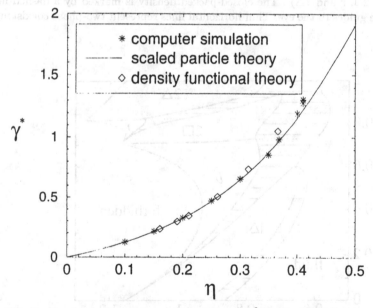

Fig. 16. Reduced interfacial free energy $\gamma^* = \gamma\sigma^2/k_BT$ of the hard sphere fluid in contact with a hard wall versus bulk packing fraction η. Solid line: scaled-particle theory; *: simulation data from [42] and [43]; diamonds: density functional results from [37].

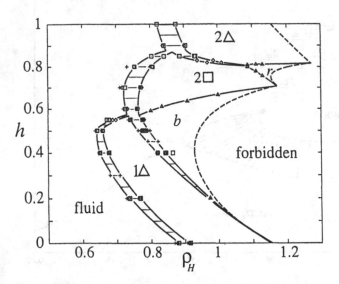

Fig. 17. Phase diagram for hard spheres of reduced density ρ_H between parallel plates with effective reduced distance h. Symbols indicate different system sizes: $N = 192(+); N = 384, 512(\diamond); N = 576(\triangle); N = 1024, 1156(\square)$. Six phases occur (fluid, $1\triangle$, b, $2\square$, r and $2\triangle$) . The closed-packed density is marked by a dashed line. Solid lines are guides to the eye. Thin horizontal lines represent two-phase coexistence. From [80].

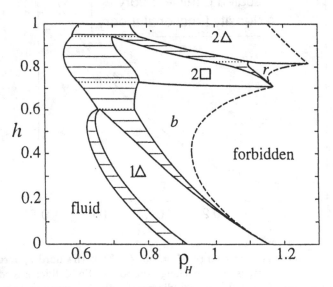

Fig. 18. Same as Fig. 17, but now obtained within the cell model for the solid phases and a simple mapping theory for the fluid phase. From [81].

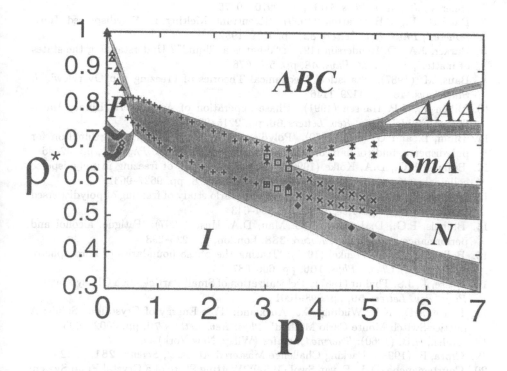

Fig. 19. Phase diagram of hard spherocylinders obtained by cell and scaled-particle theory in the $\rho^* - p$ plane where $\rho^* \equiv \eta/\eta_{cp}$. The coexistence regions are shown as shaded areas. Simulation results from [15] are shown as dots. There is an isotropic fluid (I), an ABC-stacked solid, an AAA stacked solid, a plastic crystal (P), a nematic (N) and a smectic-A (SmA) phase. The meaning of the symbols for the simulational data are: (+) I-ABC transition, ($\Diamond$) I-P transition, ($\Box$) I-SmA transition, ($\blacklozenge$) I-N transition, ($\times$) N-SmA transition, (*) SmA-ABC transition, ($\blacktriangle$) P-ABC transition. From [39].

References

1. Allen, M.P., D.J. Tildesley (1989): *Computer Simulation of Liquids* (Clarendon Press, Oxford)
2. Allen, M.P., G.T. Evans, D. Frenkel, B. M. Mulder (1993): 'Hard Convex Body Fluids', *Advances in Chemical Physics*, Vol. **LXXXVI**, pp. 1–166
3. Barrat, J.L., J.P. Hansen (1986): 'On the stability of polydispersed colloidal crystals', *J. Physique* **47**, Paris, pp. 1547–1553
4. Bartlett, P., R.H. Ottewill, P.N. Pusey (1992): 'Superlattice Formation in Binary Mixtures of Hard-Sphere Colloids', *Phys. Rev. Letters* **68**, pp. 3801–3804
5. Bartlett, P. (1997): 'A geometrically-based mean-field theory of polydisperse hard-sphere mixture', *J. Chem. Phys.* **107**, pp. 188–196

6. Bartlett, P. (1998): 'Fractionated crystallization in a polydisperse mixture of hard spheres', *J. Chem. Phys.* **109**, pp. 10970–10975
7. Bartlett, P., P.B. Warren (1999): 'Reentrant Melting in Polydispersed Hard Spheres', *Phys. Rev. Letters* **82**, pp. 1979–1982
8. Barker, J.A., D. Henderson (1976): 'What is a "liquid"? Understanding the states of matter', *Rev. Mod. Phys.* **48**, pp. 587–676
9. Baus, M. (1987): 'Statistical Mechanical Theories of Freezing: An Overview', *J. Stat. Phys.* **48**, pp. 1129–1146
10. Biben T., J.P. Hansen (1991): 'Phase Seperation of Asymmetric Binary Hard-Sphere Fluids', *Phys. Rev. Letters* **66**, pp. 2215–2218
11. Blum, L. and G. Stell (1979): 'Polydisperse systems. I. Scattering function for polydisperse fluids of hard or permeable spheres', *J. Chem. Phys.* **71**, pp. 42–46
12. Bolhuis, P.G., D.A. Kofke (1996a): 'Numerical study of freezing in polydisperse colloidal suspensions', *J. Phys. Condensed Matter* **8**, pp. 9627–9631
13. Bolhuis, P.G., D.A. Kofke (1996b): 'Monte Carlo study of freezing of polydispersed hard spheres', *Phys. Rev. E* **54**, pp. 634–643
14. Bolhuis, P.G., D. Frenkel, S.-C. Mau, D.A. Huse (1997): 'Fatigue, alcohol and performance impairment', *Nature* **388**, London, pp. 235–238
15. Bolhuis, P.G., D. Frenkel (1997): 'Tracing the phase boundaries of hard spherocylinders', *J. Chem. Phys.* **106**, pp. 666–687
16. Bosse, J., J.S. Thakur (1987): 'Delocalization of Small Particles in a Glassy Matrix', *Phys. Rev. Letters* **59**, pp. 998–1001
17. Bruce, A.D., N.B. Wilding, G.J. Auckland: 'Free Energy of Crystalline Solids: A Lattice-Switch Monte Carlo Method', *Phys. Rev. Letters* **79**, pp. 3002–3005
18. Callen, H.B. (1960): *Thermodynamics* (Wiley, New York)
19. Cipra, B. (1998): 'Packing Challenge Mastered At Last', *Science* **281**, p. 1267
20. Courtemanche, D.J., F. van Swol (1992): 'Wetting State of a Crystal Fluid System of Hard Spheres', *Phys. Rev. Letters* **69**, pp. 2078–2081
21. Cuesta, J.A. (1999): 'Demixing in a single-peak distributed polydisperse mixture of hard spheres', *Europhysics Letters* **46**, pp. 197–203
22. Denton, A.R., N.W. Ashcroft (1990): 'Weighted-density-functional theory of nonuniform fluid mixtures: Application to freezing of binary hard-sphere mixtures', *Phys. Rev. A* **42**, pp. 7312–7329
23. Denton, A.R., N.W. Ashcroft (1991): 'Vegard's law', *Phys. Rev. A* **43**, pp. 3161–3164
24. Dietrich, S. (1988):'Wetting Phenomena'. In: *Phase transitions and Critical Phenomena*, ed. by C. Domb, J.L. Lebowitz, Vol. 12 (Academic Press, London), pp. 1–128
25. Dijkstra, M., R. van Roij, R. Evans (1999a): 'Direct Simulation of the Phase Behaviour of Binary Hard-Sphere Mixtures: Test of the Depletion Potential Description', *Phys. Rev. Letters* **82**, pp. 117–120
26. Dijkstra, M., R. van Roij, R. Evans (1999b): 'Phase diagramm of highly asymmetric binary hard-sphere mixtures', *Phys. Rev. E* **59**, pp. 5744–5771
27. Doliwa, B., A. Heuer (1998): 'Cage Effect, Local Anisotropies and Dynamic Heterogeneities at the Glass Transiton: A Computer Study of Hard Spheres', *Phys. Rev. Letters* **80**, pp. 4915–4918
28. Eldridge, M.D., P.A. Madden, D. Frenkel (1993): 'The stability of the AB_{13} crystal in a binary hard sphere system', *Mol. Phys.* **79**, pp. 105–120
29. Evans, R. (1992): 'Density Functionals in the Theory of Nonuniform Fluids'. In: *Fundamentals of Inhomogeneous Fluids*, ed. by D. Henderson (Marcel Dekker, New York), pp. 85–176

30. Finken, R., M. Schmidt, H. Löwen: to be published
31. Frenkel, D., B. Smit (1996): *Understanding Molecular Simulation* (Academic, San Diego)
32. Frisch, H.L., N. Rivier and D. Wyler (1985): 'Classical Hard-Sphere Fluid in Infinitely Many Dimensions', *Phys. Rev. Letters* **54**, pp. 2061–2063
33. Frisch, H.L., J.K. Percus (1987): 'Nonuniform classical fluid at high dimensionality', *Phys. Rev. A* **35**, pp. 4696–4702
34. Frisch, H.L., J.K. Percus (1999): 'High dimensionality as an organizing device for classical fluids', *Phys. Rev. E* **60**, pp. 2942–2948
35. Fröhlich, J., C. Pfister (1981): 'On the Absence of Spontaneous Symmetry Breaking and of Crystalline Ordering in Two-Dimensional Systems', *Commun. Math. Phys.* **81**, pp. 277–298
36. Garzo, V., J.W. Dufty (1999): 'Dense fluid transport for inelastic hard spheres', *Phys. Rev. E* **59**, pp. 5895–5911
37. Götzelmann, B., A. Haase, S. Dietrich (1996): Structure factor of hard spheres near a wall', *Phys. Rev. E* **53**, pp. 3456–3467
38. Götzelmann, B., R. Evans, S. Dietrich (1998): 'Depletion forces in fluids', *Phys. Rev. E* **57**, pp. 6785–6800
39. Graf, H., H. Löwen, M. Schmidt (1997): 'Cell theory for the phase diagram of hard spherocylinders', *Prog. Colloid Polymer Science* **104**, pp. 177–179
40. Graf, H., H. Löwen (1999): 'Density functional theory for hard spherocylinders: phase transitions in the bulk in the presence of external fields', *J. Phys. Condensed Matter* **11**, pp. 1435–1452
41. Hansen, J.P., I.R. McDonald (1986): *Theory of Simple Liquids*, 2nd ed. (Academic Press, London)
42. Henderson, J.R., F. van Swol (1984): 'On the interface between fluid and a planar wall: Theory and simulations of a hard sphere fluid at a hard wall', *Mol. Phys.* **51**, pp. 991-1010
43. Heni, M., H. Löwen (1999): 'Interfacial free energy of hard sphere fluids and solids near a hard wall', *Phys. Rev. E* (in press)
44. Holyst, R., A. Poniewierski (1989): 'Nematic-smectic-A transition for perfectly aligned hard spherocylinders: Application of the smoothed-density approximation', *Phys. Rev. A* **39**, pp. 2742–2744
45. Hoover, W.G., F.H. Ree (1968): 'Melting Transitions and Communal Entropy for Hard Spheres', *J. Chem. Phys.* **49**, pp. 3609–3617
46. Imhof, A., J.K.G. Dhont (1995): 'Long-time self-Diffusion in binary colloidal hard-sphere dispersions', *Phys. Rev. E* **52**, pp. 6344–6357
47. Imhof, A., J.K.G. Dhont (1997): 'Phase behaviour and long-time self-diffusion in a binary hard sphere dispersion', *Colloids and Surfaces* **122**, pp. 53–61
48. Jaster, A. (1998): 'Orientational order of the two-dimensional hard-disk system', *Europhys. Letters* **42**, pp. 277–282
49. Jaster, A. (1999): 'Computer simulations of the two-dimensional melting transition using hard disks', *Phys. Rev. E* **59**, pp. 2594–2602
50. Kirkpatrick, T.R. (1986): 'Ordering in the parallel hard hypercube gas', *J. Chem. Phys.* **85**, pp. 3515–3519
51. Kofke, D.A., P.G. Bolhuis (1999): 'Freezing of polydisperse hard spheres', *Phys. Rev. E* **59**, pp. 618–622
52. Kranendonk, W.G.T., D. Frenkel (1991): 'Thermodynamic properties of binary hard sphere mixtures', *Mol. Phys.* **72**, pp. 715–733
53. Lebowitz, J.L., O. Penrose (1964): 'Convergence of virial expansions', *J. Math. Phys.* **5**, pp. 841–847

54. Lekkerkerker, H.N.W., P. Buining, J. Buitenhuis, G.J. Vroege, A. Stroobants (1995): 'Liquid Crystal Phase Transition in Dispersions of Rodlike Colloidal Particles'. In: *Observation, Prediction and Simulation of Phase Transitions in Complex Fluids*, ed. by M. Baus et al. (Kluwer, Holland), pp. 53–112

55. Leppmeier, M. (1997): *Kugelpackungen: von Kepler bis heute* (Vieweg, Wiesbaden)

56. Luding, S. (1995): 'Granular materials under vibration: simulations of rotating spheres', *Phys. Rev. E* **52**, pp. 4442–4457

57. Löwen, H. (1994): 'Melting, Freezing and Colloidal Suspensions', *Phys. Rep.* **237**, pp. 249–324

58. Mau, S.-C., D.A. Huse (1999): 'Stacking entropy of hard-sphere crystals', *Phys. Rev. E* **59**, pp. 4396–4401

59. McCarley, J.S., N.W. Ashcroft (1994): 'Hard-Sphere quasicrystals', *Phys. Rev. B* **49**, pp. 15600–15606

60. Mitus A.C., H. Weber, D. Marx (1997): 'Local structure analysis of the hard-disk fluid near melting', *Phys. Rev. E* **55**, pp. 6855–6857

61. Münster, A. (1974): *Statistical Thermodynamics.* Vol. **II** (Springer-Verlag, Berlin), pp. 337–346

62. Murray, C.A., W.O. Sprenger, R.A. Wenk (1990): 'Comparison of melting in three and two dimensions: Microscopy of colloidal spheres', *Phys. Rev. B* **42**, pp. 688–703

63. Murray, C.A. (1992): 'Experimental Studies of Melting and Hexatic Order in Two-Dimensional Colloidal Suspensions'. In: *Bond-orientational Order in Condensed Matter Systems*, ed. by K.J. Strandburg (Springer, New York), pp. 137–215

64. Nägele, G., J.K.G. Dhont (1998): 'Tracer-diffusion in colloidal mixtures: A mode-coupling scheme with hydrodynamic interactions', *J. Chem. Phys.* **108**, pp. 9566–9576

65. Nägele, G., J. Bergenholtz (1998): 'Linear viscoelasticity of colloidal mixtures', *J. Chem. Phys.* **108**, pp. 9893–9904

66. Németh, Z.T., H. Löwen (1998): 'Freezing in finite systems: hard discs in circular cavities', *J. Phys. Condensed Matter* **10**, pp. 6189–6203

67. Németh, Z.T., H. Löwen (1999): 'Freezing and glass transition of hard spheres in cavities', *Phys. Rev E* **59**, pp. 6824–6829

68. Neser, S., C. Bechinger, P. Leiderer, T. Palberg (1997a): 'Finite-Size Effects on the Closest Packing of Hard Spheres', *Phys. Rev. Letters.* **79**, pp. 2348–2351

69. Neser, S.T. Palberg, C. Bechinger, P. Leiderer (1997b): 'Direct observation of a buckling transition during the formation of thin colloidal crystals', *Progr Colloid Polymer Science* **104**, pp. 194–197

70. Ohnesorge, R., H. Löwen, H. Wagner (1991): 'Density-funcional theory of surface melting', *Phys. Rev. A* **43**, pp. 2870–2878

71. Pansu, B., P. Pieranski (1984): 'Structures of thin layers of hard spheres: high pressure limit', *J. Physique* **45**, pp. 331–339

72. Phan, S.-E., W.B. Russel, J. Zhu, P.M. Chaikin (1999): 'Effects of polydispersity on hard sphere crystals', *J. Chem. Phys.* **108**, pp. 9789–9795

73. Pieranski, P., L. Strzelecki, B. Pansu (1983): 'Thin Colloidal Crystals', *Phys. Rev. Lett.* **50**, pp. 900–902

74. Pronk, S., D. Frenkel (1999): 'Can stacking faults in hard-sphere crystals anneal out spontaneously?' *J. Chem. Phys.* **110**, pp. 4589–4592

75. Pusey, P.N. (1991): 'Colloidal Suspensions'. In: *Liquids, Freezing and the Glass Transition*, ed. by J.P. Hansen, D. Levesque, J. Zinn-Justin (North Holland, Amsterdam), pp. 763–942

76. Reiss, H., H.L. Frisch, E. Helfand, J.L. Lebowitz (1960): 'Aspects of the Statistical Thermodynamics of Real Fluids', *J. Chem. Phys.* **32**, pp. 119–124

77. Rosenfeld, Y. (1996): 'Close-packed configurations, 'symmetry breaking', and the freezing transition in density functional theory', *J. Phys. Condensed Matter* **8**, pp. L795–L801

78. Rosenfeld, Y. , M. Schmidt, H. Löwen, P. Tarazona (1997): 'Fundamental-measure free-energy density functional for hard spheres: Dimensional crossover and freezing', *Physical Review E* **55**, pp. 4245–4263

79. Rosenfeld, Y. (1998): 'Self-consistent density functional theory and the equation of state for simple liquids', *Mol. Phys.* **94**, pp. 929–936

80. Schmidt, M., H. Löwen (1996): 'Freezing between Two and Three Dimensions', *Phys. Rev. Letters* **76**, pp. 4552–4555

81. Schmidt, M., H. Löwen (1997): 'Phase diagram of hard spheres confined between two parallel plates', *Phys. Rev. E* **55**, pp. 7228–7241

82. Sear, R.P. (1998): 'Phase seperation and crystallisation of polydisperse hard spheres', *Europhys. Letters* **44**, pp. 531–535

83. Tonks, L. (1936): 'The Complete Equation of State of One, Two and Three-Dimensional Gases of Hard Elastic Spheres', *Phys. Rev.* **50**, pp. 955–963

84. Truskett, T.M., S. Torquato, S. Sastry, P.G. Debenedetti, F.H. Stillinger (1998): 'Structural precursor to freezing in the hard-disk and hard-sphere system', *Phys. Rev. E* **58**, pp. 3083–3088

85. Van Blaaderen, A. and P. Wiltzius (1995): 'Real-Space Structure of Colloidal Hard-Sphere Glasses', *Science* **270**, pp. 1177–1179

86. Van Winkle, D.H., C.A. Murray (1986): 'Layering transitions in colloidal crystals as observed by diffraction and direct lattice imaging', *Phys. Rev. A* **34** (1986), pp. 562–573

87. Warren, P.B. (1999): 'Fluid-fluid phase seperation in hard spheres with a bimodal size distribution', *Europhys. Letters* **46**, 295–300

88. Weiss, J., D.W. Oxtoby, D.G. Grier, C.A. Murray (1995): 'Martensitic transition in a confined colloidal suspension', *J. Chem. Phys.* **103**, pp. 1180–1190

89. Wyler, D., N. Rivier, H. Frisch (1987): 'Hard-Sphere fluid in infinite dimensions', *Phys. Rev. A* **36**, pp. 2422–2431

90. Xu, H., M. Baus (1992): 'A density functional study of superlattice formation in binary hard-sphere mixtures', *J. Phys. Condensed Matter* **4**, pp. L663–L668

Finite Packings and Parametric Density

Jörg M. Wills

Math. Department, Universität Siegen,
D-57068 Siegen

Abstract. Finite Packings of circles, spheres or other convex bodies are investigated in various fields. We give a broad survey of planar results based on parametric density, because there is no such survey yet. For 3–dimensional packings we describe the atomistic approach to Wulff shape for general crystals, i.e. via periodic packings of balls of different size. The corresponding approach for quasicrystals is in the survey by [7].

1 Introduction

Packings of atoms, molecules or particles are investigated in various branches of physics (see [10,21]), chemistry (see [17]), crystallography (see next section), geology (see [23]), biology and biotechnology (see [18]), astronomy (see [19] and [33]) and ceramics and polymers (see [16]). Of course all these reference lists are far from being complete.

The packings can be ordered in a lattice or periodically as in ideal crystals (see [20,26,35,46]) or quasiperiodically as in quasicrystals (see [1,4,5,6,8,27,31,37,38]) or they can be partially or totally disordered as e.g. in liquids, colloids or in gases (see [17,18,19,25]). The packing objects can be hard or soft, randomly distributed or not (see [23]), they can be assumed as balls, or ellipsoids, or as convex or nonconvex objects.

Usually one considers packings in 3–space or 3D, but in many cases, e.g. for observations by the microscope, the plane, i.e. 2D is appropriate. Clearly all packings in real world are finite, but in many cases (e.g. crystal structure classifications) one can simplify and assume the packing as infinite. In this case boundary effects and their difficulties can be avoided, which is an essential advantage as formulated in the wellknown saying: "The bulk was created by god, and the boundary by the devil".

But here finite packings are considered and in the 3D case they are divided into: small packings (cluster, microcluster) (see [10,15,26]), local packings (growth in facets, whiskers) and global packings (crystals and quasicrystals) (see [5,6,37,38]), in particular for spheres of different size. It is the goal of this article to consider just finite packings and coverings, and we will use methods from geometry, in particular discrete and convex geometry, and asymptotic methods for the Wulff shape. In fact we use only one method (parametric density), but this one is very flexible and uses classical tools of convex and discrete geometry, in particular mixed volumes and Minkowski functionals (see e.g. [32]). Of course finite packings include also infinite packings as limit case. There are two principal types of packings (see [13] and Fig. 1 in Sect. 2):

a) Bin Packings, i.e. packings of objects in a given Container or Bin, or in other words: with prescribed external boundaries and

b) Free Packings, i.e. packings without external boundaries, which are generated by inner forces, e.g. Coulomb, van der Waal, or by no forces at all.

For the following reasons we consider Free Packings; and mention Bin Packings only briefly in Sect. 2:

a) Packings in nature (cluster, crystals, etc) usually are Free Packings.

b) For Bin Packings there is no general mathematical theory, although various algorithms exist.

c) For Free Packings there is the well developed theory of mixed volumes and Minkowski functionals, which are used for the parametric density.

After the general introduction the next section deals with basic definitions. As 2D packing theory is much better developed than 3D (and higher–dimensional) theory, we consider 2D and 3D separately, namely 2D in Sects. 3 and 4 and 3D in Sects. 6 – 8. In many cases physical objects can also overlap, partially or completely. So we add Sect. 5 on finite coverings in 2D (see [2]). In all cases, except in the last section, we consider translates of a given 0–symmetric convex body. So we do not allow rotations (except for the trivial case of circles or balls), which is of course a strong restriction. For rotations one needs an individual approach.

In 2D we give a rather comprehensive approach, as the general theorems in Sects. 4 and 5 are not so wellknown, and the literature is scattered (but see [29]). In particular packings and coverings of nonconvex cell complexes are considered, which are useful for some applications.

In 3D we describe global packings, i.e. the Wulff shape for the general case of periodic packings in detail. For the Wulff shape of quasicrystals we refer to the detailed description by Böröczky and Schnell in 1998 and 1999 [5,6]. We further refrain from describing our method for online packings, which model whiskers (see [43]) and for microclusters (see [10,15,26]), because these are still in progress. In particular there is a current project supported by DFG on modelling of microclusters.

2   Density Definitions

Throughout this paper we consider packings und coverings in ordinary (Euclidean) 2–space E^2 or 3–space E^3 which we also denote by $2D$ and plane or by $3D$ and space. In many cases definitions and results hold for all dimensions $d \geq 2$; we then write E^d, but one can read E^2 or E^3. We always understand packings or coverings as a static situation and not as a dynamic process. The packing (or covering) objects are always convex bodies and in most cases these convex bodies are translates of a given convex body, e.g. of a ball. If one wants to measure or to compare two packings, one needs an appropriate density measure. Most physical density measures are based on energy, as e.g. the Lennard–Jones potential and Morse potential, but of course also geometric tools, methods and intuition is used. Geometric densities are based on volume V (Jordan, Riemann or Lebesgue), which means the area in the plane. For simplicity we also use V in the planar case.

By a packing we always mean that any two objects do not overlap, i.e. have at most boundary points in common. As mentioned in Sect. 2, finite packings can be classified into Bin Packings (with a given container or bin) and Free Packings without a bin.

So, if nonoverlapping convex bodies $K_i, i = 1, \ldots, n$ and a container or bin C with $\bigcup_{i=1}^{n} K_i \subset C$ are given, then the packing density is

$$\delta_{\text{Bin}} = (\Sigma_{i=1}^{n} V(K_i))/V(C) \tag{2.1}$$

This notion traces back to Archimedes who first observed that the ratio of volumes of a ball packed into its circumscribed cylinder is $2/3$.

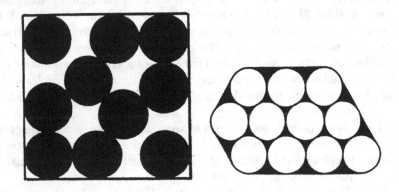

Fig. 1. Bin packing of 10 circles in a square (left) and free packing of 11 circles in their convex hull.

If no container (bin) is given, then (the Norwegian mathematician) A. Thue proposed in 1892 to use instead the convex hull $\text{conv}(K_1, \ldots, K_n)$, i.e. the smallest convex body containing the K_i. So

$$\delta_{\text{Thue}} = (\Sigma_{i=1}^{n} V(K_i))/V(\text{conv}(\bigcup_{i=1}^{n} K_i)) \tag{2.2}$$

Both definitions are useful for explicit density calculation in $2D$ and $3D$, but they are too general to permit a theory with nontrivial general results.

So we restrict ourselves to translates of a given convex body and moreover we assume that all convex bodies are centrally symmetric, or briefly 0-symmetric. This includes balls, ellipsoids, parallelotopes, cylinders, but exlcudes e.g. simplices. The same restriction holds for coverings. Parametric density is a generalization of Thue's density and will be introduced in Sects. 4 and 6.

A main goal for estimating finite packings and coverings it to compare them with optimal lattice packing or covering. For this we remind the reader that a lattice $L \subset E^d$ is a packing lattice for a convex body K, if $L + K$ is a packing (see [14,28] or [29]). The packing density V of $L + K$ is $\delta_L(K) = V(K)/\det L$.

The minimum of all determinants of packing lattices of K exists and is called the critical packing determinant of K and denoted by $\Delta_p(K)$.

A packing lattice of K with critical determinant is called a critical lattice and it provides the densest lattice packing of K:

$$\delta_e(K) = V(K)/\Delta_p(K)$$

Similarly for coverings (see [14]): Any lattice L with the property that $L + K$ covers E^d (or equivalently: $L + K = E^d$), is called a covering lattice and the covering density is $\vartheta_L(K) = V(K)/\det L$.

Again the extremum, here the maximum of all determinants of covering lattices exists and is called the critical covering determinant $\Delta_c(K)$. The correspoding lattice is the thinnest or most economic covering lattice of K and its density is

$$\vartheta_e(K) = V(K)/\Delta_c(K)$$

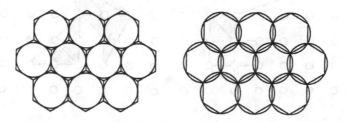

Fig. 2. Parts of the densest lattice packing (left) and the thinnest lattice covering (right) of unit circles. The circumscribed hexagons (left) have area $\Delta_p(B^2) = 2\sqrt{3}$; the inscribed hexagons (right) have area $\Delta_c(B^2) = 3\sqrt{3}/2$.

Examples in 3–space: For the ball B^3 we have $\delta_e(B^3) = \frac{4\pi}{3}/4\sqrt{2} = 0,74048...$ (Gauss 1831) and $\vartheta_e(B^3) = 1,464...$ (Bambah 1954).

3   Cell Complexes in the Plane

Packings and coverings in the plane can be considered in cell complexes, in particular in simplicial complexes and hence in a very general setting, so that most planar packing and covering problems of translates of convex bodies can be subsumed. Further the proofs for the most general results in the following sections use simplicial complexes.

As the notion of density only makes sense for convex sets, but not for general simplicial complexes and cell complexes, the inequalities in Sects. 4 and 5 are given in terms of volumes, and only in the convexity case also for densities. The best known simplicial complexes are Delone (or Delauney, same author) triangulations, see [12].

A planar simplicial complex $\mathcal{C}$ is a collection of points, segments and triangles with the following 3 properties:

1. Every segment (or edge or 1–face) of $\mathcal{C}$ connects 2 vertices (or 0–faces) of $\mathcal{C}$, and contains no other vertex of $\mathcal{C}$.
2. The intersection of any 2 edges of $\mathcal{C}$ is either empty or a vertex of $\mathcal{C}$,
3. Every triangle (or 2–face) T of $\mathcal{C}$ is the convex hull of 3 vertices of $\mathcal{C}$, and its 3 edges are also edges of $\mathcal{C}$, and T contains no other vertices or edges of $\mathcal{C}$.

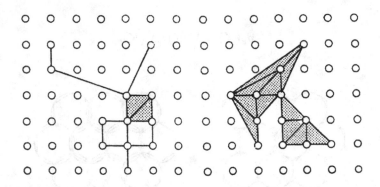

Fig. 3. Two simplicial complexes in $\mathbb{Z}^2$

If the number of vertices of $\mathcal{C}$ is finite, $\mathcal{C}$ is called finite; otherwise infinite. In the first case the number of vertices, edges and triangles is denoted by $f_i(\mathcal{C}), i = 0, 1, 2$. For $f_0(\mathcal{C})$ we also write n, the number of packing or covering objects. The alternating sum

$$\chi(\mathcal{C}) = f_0(\mathcal{C}) - f_1(\mathcal{C}) + f_2(\mathcal{C})$$

is called the Euler characteristic of $\mathcal{C}$.

Examples: If $\mathcal{C}$ is a full triangle, then $f_0 = f_1 = 3$ and $f_2 = 1$, hence $\chi(\mathcal{C}) = 1$. If $\mathcal{C}$ is a triangle without its interior, then $f_0 = f_1 = 3$, and $f_2 = 0$, hence $\chi(\mathcal{C}) = 0$. If $\mathcal{C}$ consists of n isolated points, then $\chi(\mathcal{C}) = n$.

The volume or area of $\mathcal{C}$ is the sum of areas of its triangles and denoted by $V(\mathcal{C})$. Here we underline that the cell complex $\mathcal{C}$ has to be distinguished from its underlying point set, the so–called support of $\mathcal{C}$ or simply set $\mathcal{C}$. So $V(\mathcal{C}) = V(\text{set}\mathcal{C})$. The elements of $\mathcal{C}$ are the vertices, edges and triangles: the elements of set $\mathcal{C}$ are points of the plane.

Example: If $\mathcal{C}$ is an edge with its endpoints, then $\mathcal{C}$ has 3 elements, but set $\mathcal{C}$ has infinitely many elements.

The boundary $bd\mathcal{C}$ of a cell complex $\mathcal{C}$ is a cell complex itself, namely the union of the boundary edges and boundary vertices, which form finitely many Jordan-polygons which do not cross each other and which have at most vertices of $\mathcal{C}$ in common. The number of vertices of $\mathcal{C}$ on the boundary is denoted by $\dot{n}(\mathcal{C})$ or briefly by $\dot{n}$.

Finally we introduce the normed perimeter $P(\mathcal{C})$, which is less obvious than the other functions on $\mathcal{C}$. For this let $u \in S^1$ be a unit vector and $l_u(K)$ the length of the intersection of K with a line through 0 in direction of u.

For any edge of $bd\mathcal{C}$ let e' be the parallel line through 0 and $l(e' \cap K)$ the length of the segment. Then

$$P(\mathcal{C}) = \sum l(e)/l(e' \cap K)$$

where the summation is over all edges of $bd\mathcal{C}$.

4   Finite Packings in the Plane

A finite packing $C_n + K$ is called a lattice packing, if there is a packing lattice L of K (i.e. $L + K$ is a packing) with $C_n \subset L$. In this case one has (for the next sections see also the survey by [13] or the book by [29]):

Theorem 4.1 (Pick 1899) Let $K \in \mathcal{K}_0^2, n \in \mathbb{N}$ and $C_n + K$ be a finite lattice packing. a) Then for any simplicial complex $\mathcal{C}$ with $\text{vert}\mathcal{C} = C_n$ is

$$n(\mathcal{C}) \le \frac{V(\mathcal{C})}{\Delta_p(K)} + \frac{1}{2}\dot{n}(\mathcal{C}) + \chi(\mathcal{C}) \tag{4.1}$$

b) Equality holds for all simplicial complexes of the critical lattice.

The result is a trivial consequence of Pick's identity. With the weighted lattice point enumerator $\hat{n}(\mathcal{C}) = n(\mathcal{C}) - \frac{1}{2}\dot{n}(\mathcal{C}) - \chi(\mathcal{C})$ the theorem simplifies to

$$\hat{n}(\mathcal{C}) \le V(\mathcal{C})/\Delta_p(K) \tag{4.2}$$

Unfortunately this elegant inequality does not hold for nonlattice packings, as the following simple example shows:

Example: Let $K = B^2$ be the unit circle and $C_n = \{(2i,0)|i = 1,\ldots,n-2\} \cup \{(0,\varepsilon),(2n-2,-\varepsilon)\}$. Then $C_n + B^2$ is a packing and $\dot{n} = 4, V(\mathrm{conv}C_n) = 2\varepsilon(n-1), \chi(\mathrm{conv}C_n) = 1, \Delta_p(B^2) = \sqrt{3}/2$ and (5.1) becomes

$$n \leq (4/\sqrt{3})\varepsilon(n-1) + 2 + 1$$

which is false for each $\varepsilon < \sqrt{3}/4$, if n is sufficiently large.

The search for the appropriate packing inequality in the nonlattice case began with A. Thue 1892 and ended with the following general result:

Theorem 4.2 (Folkman, Graham, Witsenhausen and Zassenhaus 1969 / 72): Let $K \in \mathcal{K}_0^2, n \in \mathbb{N}$ and $C_n + K$ be a packing. a) Then for any simplicial complex $\mathcal{C}$ with $\mathrm{vert}\mathcal{C} = C_n$ is

$$n(\mathcal{C}) \leq \frac{V(\mathcal{C})}{\Delta_p(K)} + \frac{1}{2}P(\mathcal{C}) + \chi(\mathcal{C}) \qquad (4.3)$$

b) Equality holds for all simplicial complexes of the critical lattice of K, whose edges are all shortest edges.

Remarks:
a) The equality case is Pick's identity for the critical packing lattice of K.
b) Theorems 4.1 and 4.2 are closely related, but theorem 4.2 is much harder to prove. Clearly $P(\mathcal{C})$ is harder to determine than $\dot{n}(\mathcal{C})$.
c) For any given $C_n, n \geq 2$ there are finitely many simplicial complexes $\mathcal{C}$ with $\mathrm{vert}\mathcal{C} = C_n$. Among them those with $\mathrm{set}\mathcal{C} = \mathrm{conv}C_n$ are distinguished, because their $\mathrm{set}\mathcal{C}$ is maximal with respect to the volume. Besides this maximality property $\mathrm{conv}C_n$ has by definition the minimality property to be the smallest convex body containing C_n. It turns out that for this relevant case one obtains an elegant formula, which can be generalized to higher dimensions.

Theorem 4.3 (Groemer, Oler 1960/61)
a) For a finite packing $C_n + K$ and $P_n = \mathrm{conv}\, C_n$ is

$$(n-1)\Delta_p(K) + V(\rho_p K) \quad \leq V(P_n + \rho_p K) \qquad (4.4)$$

Here $\rho_p = \rho_p(K) = \Delta_p(K)/Z_p(K)$, where $Z_p(K)$ is the area of the smallest parallelogram circumscribed at K.

b) Equality holds for all convex lattice polygons of the critical lattice of K, whose edges are all shortest edges.

Remarks: One can show that $3/4 \leq \rho_p(K) \leq 1$. In particular, $\rho_p = 3/4$ for the regular hexagon, $\rho_p = 1$ for a parallelogram and $\rho_p = \sqrt{3}/2$ for B^2. The definition of ρ_p permits another version of (5.4):

$$V(S_n + \rho_p K) \leq V(P_n + \rho_p K) \qquad (4.5)$$

where S_n is a segment such that $C_n + K$ is a linear packing of translates $K_i = s_i + K$ and $S_n + K$ a "sausage" (this name was created by L. Fejes Toth). Inequality (4.5) was the motivation to introduce in [40] and in [2] the

Definition 4.1 Let $K \in \mathcal{K}_0^2, C_n + K$ be a packing and $\rho > 0$. Then

$$\delta(K, C_n, \rho) = nV(K)/V(\text{conv}C_n + \rho K)$$

is the parametric density.

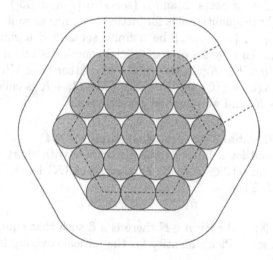

Fig. 4. A packing of 19 circles, with $\rho = 1$ and $\rho = 3$.

From this definition and (4.5) follows with $S'_n = \{s_1, \ldots, s_n\}$, i.e. $\text{conv}S'_n = S_n$:

$$\delta(K, C_n, \rho) \le \delta(K, S'_n, \rho) \quad \text{for all } \rho \ge \rho_p.$$

From (4.4) one gets with $\delta_e(K) = V(K)/\Delta_p(K)$ and the parametric density $\delta(K, C_n, \rho) = nV(K)/V(P_n + \rho K)$:

$$\delta(K, C_n, \rho_p) \le \delta_e(K) + \varepsilon_n$$

for all C_n. With $\varepsilon_n \to 0$ for $n \to \infty$ follows by simple standard arguments:

Corollary 4.1 (Rogers 1952) Let $K \in \mathcal{K}_0^2$. Then

$$\delta(K) = \delta_e(K)$$

and for the special case of $K = B^2$ unit circle:

Corollary 4.2 (Thue, L. Fejes Toth, Segre, Mahler 1892/1940)

$$\delta(B^2) = \delta_e(B^2)$$

These results complete the theory of packings by translates of 0–symmetric convex bodies in the plane.

5   Finite Coverings in the Plane

We use the notation of Sects. 2 and 3 (see also [2] and [13]). In the covering case the same tight inequality holds for lattice coverings as well as for nonlattice coverings. Let $C_n = \{c_1, \ldots, c_n\}$ be a finite set and $\mathcal{C}$ a simplicial complex with vert $\mathcal{C} = C_n$. An edge $\overline{xy}$ of a simplicial complex is called a short edge, if $\overline{xy} \subset (x + K) \cup (y + K)$. Again we write $V(\mathcal{C})$ rather than $V(\text{set } \mathcal{C})$. Again let $\hat{n}(\mathcal{C}) = n(\mathcal{C}) - \frac{1}{2}\hat{n}(\mathcal{C}) - \chi(\mathcal{C})$ and $K \in \mathcal{K}_0^2$. Then $C_n + K$ is called a covering of $\mathcal{C}$ if set $\mathcal{C} \subset C_n + K$, and we have the general result:

Theorem 5.1 (Bambah, Rogers, Zassenhaus 1964)
a) Let $K \in \mathcal{K}_0^2$ and let $\mathcal{C}$ be a simplicial complex with short boundary edges and vert$\mathcal{C} = C_n$ and set$\mathcal{C} \subset C_n + K$. Further let $\Delta_c(K)$ be the critical covering determinant of K. Then

$$\hat{n}(\mathcal{C}) \geq V(\mathcal{C})/\Delta_c(K) \qquad (5.1)$$

b) For each $K \in \mathcal{K}_0^2$ and each $n \in \mathbb{N}$ there is a $\mathcal{C}$ such that equality holds.
c) The equality case is Pick's identity for the critical covering lattice of K.

Remarks. a) In (5.1) low dimensional parts of $\mathcal{C}$ (2–sided edges or isolated points) can be cancelled as they count 0 on both sides of (5.1)
b) Equality holds for all lattice polygons of the critical lattice of K, which are covered by the corresponding lattice translates of K. There are also nonlattice complexes, for which (5.1) holds with equality, e.g. two disjoint triangles, which do not belong to one lattice.
c) The condition of short boundary edges is necessary, because Theorem 5.1 does not hold for general simplicial complexes as the following simple example shows:
Example: Let $K = B^2$. Then $\Delta_c(B^2) = 3\sqrt{3}/2$. Let $C_n = \{0, X, Y, Z\}$ with $0 = (0,0), X = (2,0), Y = (0, 2 - \varepsilon), Z = (1,1)$ and $0 < \varepsilon < 1/2$. Let $\mathcal{C}$ be the simplicial complex with vert$\mathcal{C} = C_n$ and set $\mathcal{C} = \text{conv}\{0, X, Y\} \cup \overline{XZ}$. Then $\chi(\mathcal{C}) = 1, \hat{n}(\mathcal{C}) = \frac{1}{2}, V(\mathcal{C}) = 2 - \varepsilon$ and set $\mathcal{C} \subset C_n + B^2$. So $\hat{n}(\mathcal{C})\Delta_c(B^2) = 3\sqrt{3}/4 < V(\mathcal{C})$, which contradicts (5.1).
d) One can avoid counterexamples of this like the previous one either by the requirement of short boundary edges as in theorem 5.1 or by the condition that the intersection of any translate $K_i = K + c_i$ with set C is starshaped.
e) Restriction to the simplicial complex $\mathcal{C}$ with set$\mathcal{C} = \text{conv } C_n$ leads to the most important special case of the convex hull, where no further restriction is needed:

Theorem 5.2 (Bambah, Woods 1972)
a) Let $K \in \mathcal{K}_0^2$ and $P_n = \text{conv} C_n \subset C_n + K$. Then

$$\hat{n}(P_n) \geq V(P_n)/\Delta_c(K) \tag{5.2}$$

b) For any $K \in \mathcal{K}_0^2$ and any $n \in \mathbb{N}$ there is a C_n such that equality holds.

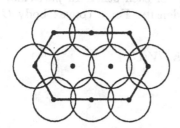 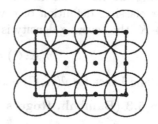

Fig. 5. Two circle coverings with $n = 10, \hat{n} = 5$ and $n = 12, \hat{n} = 6$

Remarks. (5.2) is equivalent to the density inequality

$$\hat{\vartheta}(K, P_n) = (\hat{n}(P_n)V(K))/V(P_n) \geq V(K)/\Delta_c(K) = \vartheta_e(K) \tag{5.3}$$

Now $\liminf\limits_{n \to \infty} \hat{\vartheta}(K, P_n) = \vartheta(K)$ and one gets with $\vartheta_e(K) \geq \vartheta(K)$:

Corollary 5.1 (L. Fejes Toth 1949) Let $K \in \mathcal{K}_0^2$. Then

$$\vartheta(K) = \vartheta_e(K)$$

and for the special case $K = B^2$:

Corollary 5.2 (Kershner 1939).

$$\vartheta(B^2) = \vartheta_e(B^2)$$

Examples:
1) Let $K = B^2, n = 3$ and conv C_n a regular triangle with edgelength $\sqrt{3}$. Then conv $C_n \subset C_n + B^2, V(\text{conv } C_n) = \frac{3}{4}\sqrt{3}$, $\hat{n} = 3 - 3/2 - 1 = \frac{1}{2}, \Delta_c(B^2) = \frac{3}{2}\sqrt{3}$ and (6.1) holds with equality.
2) Let $K = B^2, n = 3$ and convC_n be a triangle with 2 edges of length $\sqrt{3}$ and one of length 3. Again conv$C_n \subset C_n + B^2, V(\text{conv} C_n) = \frac{3}{4}\sqrt{3}, \hat{n} = \frac{1}{2} \Delta_c(B^2) = \frac{3}{2}\sqrt{3}$ and (6.1) holds again with equility.
3) Any union of triangles from 1) and 2), which is edge–to–edge, gives a simplicial complex, such that (5.1) holds with equality.

The previous results complete the theory of coverings by translates of 0–symmetric convex bodies in the plane. Different from the packing case here one has several independent approaches cf. theorems 5.1, 5.2 and 5.3.

The covering density in (5.3) differs from the parametric density in Sect. 4. This is due to the fact that the theorems 4.3 by Groemer and Oler and 5.2 by Bambah and Woods do not correspond directly to each other and that packing and covering are no direct dual operations of each other. In particular there is another classical definition of covering density: For a convex body C with $C \subset C_n + K$, the covering density is

$$\vartheta(K,C) = nV(K)/V(C)$$

For $V(C)$ and hence for $\vartheta(K,C)$ one has:

Theorem 5.3 (Bambah, Rogers 1961)

$$V(C) \leq (n-1)\Delta_c(K) + V(K) \tag{5.4}$$

Remarks: Obviously (5.1) and (5.4) are closely related and (5.4) can be derivated from (5.1). In (5.4) equality holds for $n = 1$ and for parallelograms and all n. (5.4) can not be generalized to simplicial complexes. But corollary 5.1 and 5.2 can also be easily deduced from (5.4).

In general it is hard to characterize the optimal convex bodies C in (5.4), whereas the characterization in (5.1) is much simpler.

6   Large Packings of Spheres

In the following three sections we describe large periodic packings of spheres which model crystals or quasicrystals. The structure is for physical and chemical reasons prescribed, and with the help of parametric density and its first "derivative", the density deviation, the Wulff shape is modeled. We describe the general case of crystals (see [35,45]). The simplest case was described in [41] and the quasicrystal case in [5,6]. We start with the basic definitions:

Let B^d denote the unit ball in Euclidean d–space $E^d, d \geq 2$. Let V denote the volume and $V(B^d) = \kappa_d$. For $x, y \in E^d$ let $\langle x, y \rangle$ denote their scalar product and $\langle x, x \rangle = \|x\|^2$. Let $L \subset E^d$ be a lattice with $\det L > 0$ and L^* its dual or polar lattice. For $t^j \in E^d$ let $L_j = L + t^j$, $j = 1, \ldots, n$ translates of L or briefly grids. Let $C_j \subset E^d$, $j = 1, \ldots, n$ be convex bodies such that $(L_1 + C_1) \cup \ldots \cup (L_n + C_n)$ is a packing. Then $M = L_1 \cup \ldots \cup L_n$ is a periodic set and also called a periodic packing for $C_1, \ldots, C_n$. With $V(C_j) = v_j$, $j = 1, \ldots, n$, $\mathfrak{V} = (v_1, \ldots, v_n)$ and $v = \sum\limits_{j=1}^{n} v_j$ the classical density of an infinite periodic packing is

$$\delta_M(\mathfrak{V}) = v/\det L$$

For the applications we always assume $C_j = r_j B^d$, $j = 1, \ldots, n$, i.e. the C_j are balls of radius r_j. This is no essential restriction, because the density depends on the volume of the packings objects, not on their shape.

Concerning M we need some simple notations: An $x \in M$ is called M–point, a finite subset $M' \subset M$ is called a finite M–set. The C_j are called the decoration of M or M'. Any finite M' with its decoration is called a finite M–packing. We are interested in dense finite M–packings. So we consider only finite $M' \subset M$, which are saturated with respect to M, i.e. with

$$M' = M \cap (\text{conv } M').$$

For a finite saturated M–set M' we call $P = \text{conv } M'$ its M–polytope. For any convex body $K \subset E^d$ we define its lattice point enumerator $L(K) = \text{card}(K \cap L)$ and for the grids:

$$L_j(K) = \text{card}(K \cap L_j) \qquad j = 1, \ldots, n.$$

We are now ready for the generalized parametric density:

Definition 6.1 Let $M, \mathfrak{V}$ and $\varrho > 0$ be given. Then for an M–polytope P its parametric density is

$$\delta_M(\mathfrak{V}, P, \varrho) = \frac{\sum_{j=1}^{n} v_j L_j(P)}{V(P + \varrho B^d)}.$$

In general P is "large" , namely a crystal with many atoms, i.e. points. So rthe quotient in definition 6.1 can be written as a Taylor expansion. The $L_j(P)$ can be written as polynomials (Ehrhart polynomials, see e.g. [14] and $V(P + \rho B^d)$ can be written as the Steiner polynomial, see e.g. [32]. This makes the Taylor expansion simple and this is the idea for the central theorem 8.1.

For any polytope P let $F_i(P), i = 1, \ldots, k$ denote its facets. For simplicity F_i denotes the point set as well as its $(d-1)$–volume. Hence $\sum_{i=1}^{k} F_i(P) = F(P)$, where F is the surface area. Further

$$f_i(P) = F_i(P)/(V(P))^{1-1/d}, i = 1, \ldots, k$$

denotes the relative surface area of a facet. With $r(P)$ and $R(P)$ we denote the inradius and the circumradius of P.

7   Facet Densities

Any $u \in L^*(u \neq 0)$ determines an array of lattice–hyperplanes in L with normal vector u. If $M = L$ is a lattice, then there is only one type of lattice–hyperplane and so the packing density of a facet with (outer) normal u is easy to determine. In the general case, if M is the superposition of translates L_i of L equipped with spheres of different size, the determination of the facet density is more complicated. One has to investigate the packing density for each of the L_i and has to take into consideration the influence of the packing density of the parallel layers of balls coming from the $L_j, j \neq i$. This needs a careful, but elementary analysis which is given in this section.

For any M–polytope P its parametric density can easily be calculated. Clearly one cannot expect general results for densest packings. But for suitable 'large' M–polytopes one obtains good asymptotic results. For this we need a rather technical but useful definition: For any $u^i \in L^*$ and $L_j = L + t^j$, $j = 1, \ldots, n$ let $\epsilon_{ij} = \{\langle u^i, t^j \rangle\}$, where $\{x\} = x - [x]$ and $[x]$ denotes as usual the largest integer $\leq x$. Now we define the appropriate class of polytopes:

Definition 7.1 Let L and M be given and $U = (u^1, \ldots, u^k) \subset L^*$ with pos $U = E^d$ (positive hull). Further let $z_i \in \mathbb{N}$, $i = 1, \ldots, k$, $j_i \in \{1, \ldots, n\}$ and $P' = \{x \in E^d | \langle x, u^i \rangle \leq z_i + \epsilon_{ij_i}, i = 1, \ldots, k\}$. Then $P = \text{conv}\,(P' \cap M)$ is called an (M, U)–polytope.

Remarks: (M, U)–polytopes are M–polytopes. By Definition 7.1 a facet with normal u^i intersects an L_{j_i} in a $(d-1)$–grid with determinant $\|u^i\| \det L$.

If the z_i are 'large', the other facets with normals $\notin U$ are 'small' and lie 'near' the $(d-2)$–faces of P'. In general P' is not the convex hull of M–points, hence no M–polytope. In [35] and [45] it is shown that $V(P) - V(P') < c(V(P))^{1-2/d}$, where c depends only on M and U. For any i one can choose $j_i \in [1, \ldots, n]$. It turns out that the choice of j_i is essential for the local packing density at the surface of P.

The following local density functions were introduced in [35]:

Definition 7.2 For $m \in \{1, \ldots, n\}$ let $\tau(i, m) = \frac{1}{2} - \sum\limits_{j=1}^{n} \frac{v_j}{v} \{\varepsilon_{im} - \varepsilon_{ij}\}$ and $\tau_i = \max\limits_m \tau(i, m)$. $\tau(i, m)$ is called the local density distribution. $\tau(i, m)/\|u^i\|$ is called the facet density and $\tau_i/\|u^i\|$ the optimal facet density in direction of u^i.

Remarks: Clearly $\tau(i, m) \in (-1/2, 1/2]$. If $M = L$ is a lattice, then $\tau_i = 1/2$, $i = 1, \ldots, k$. The facet density measures the volume of the balls per unit of the surface area. So it has homogeneity degree 1 rather than 0 as densities usually have. We use this notation, because it gives the appropriate density description.

8   Wulff Shapes

The first result is a Taylor expansion of parametric density, already mentioned after definition 6.1:

Theorem 8.1. Let M, U and $\mathfrak{V}$ be given, let P be an (M, U)–polytope with $r(P)/R(P) > \varepsilon_0$, $\mu = (V(P))^{1/d}$, $j_i \in [1, \ldots, n]$ and $\varrho > \max(\tau(i, j_i)/\|u^i\|)$. Then (for $\mu \to \infty$)

$$\delta_M(\mathfrak{V}, P, \varrho) = \delta_M - \mu^{-1} \Delta_M(P, \varrho) + O(\mu^{-2}) \tag{8.1}$$

and

$$\Delta_M(P, \varrho) = \delta_M \sum_{i=j}^{k} f_i(P)(\varrho - \tau(i, j_i)/\|u^i\|) \tag{8.2}$$

Definition 8.1 $\Delta_M(P, \varrho)$ is called the Density Deviation of P and ϱ.

Remarks: The first error term in (8.1) $\Delta_M(P, \varrho)$ measures the average difference between δ_M and parametric density. The second error term (or edge term) $O(\mu^{-2})$ depends on $M, U, \mathfrak{V}, \varepsilon_0$ and ϱ, but not on the size of P or of the f_i.

If $V(P)$ and hence μ is large, then $\Delta_M(P, \varrho)$ is the essential error term. The choice of ϱ guarantees that all summands are > 0 and hence $\Delta_M(P, \varrho) > 0$.

So among all (M, U)-polytopes of same volume, i.e. of same μ, those with minimal $\Delta_M(P, \varrho)$ are best packings and it is the central problem to find the shapes which minimize $\Delta_M(P, \varrho)$. We minimize $\Delta_M(P, \varrho)$ in two steps:

First we minimize the coefficients of the $f_i(P)$ in $\Delta_M(P, \varrho)$, i.e. we replace $\tau(i, m)/\|u^i\|$ by $\tau_i / \|u^i\|$, for which we have:

Proposition: $\frac{1}{2n} \leq \tau_i \leq \frac{1}{2}$, and both bounds are tight.

A short proof is in [35]. Second we observe that the density deviation in (8.2) can be written as

$$\Delta_M(P, \varrho) = \sum_{i=1}^{k} \sigma_i f_i(P)$$

with some $\sigma_i > 0$. If the σ_i are energies, then the problem of minimizing this sum for all polytopes with given unit normals $U = \{u^1, \ldots, u^k\}$ and fixed volume $V(P)$, hence for fixed μ, is the famous Gibbs–Curie energy problem and solved in Wulff's theorem (1901): The optimal polytope is the Wulff shape

$$\{x \in E^d | \langle x, u^i \rangle \leq \sigma_i, i = 1, \ldots, k\}.$$

Now $\sigma_i = \rho - \tau_i/\|u^i\|$ implies that ρ can be interpreted as an energy. But from the mathematical point of view the meaning of the numbers $\sigma_i > 0$ is irrelevant. In fact, if we consider the facet density (Def. 7.2) as weighted sum with weights v_j/v, the v_j can be volumes as well as weights, potentials or energies. So we get, independent of the physical meaning of the σ_i, in our case with the unit vectors $u^i/\|u^i\|$ the classical result:

Theorem 8.2. (Wulff 1909) Let $M, U, \mathfrak{V}$ and $\varrho > \max_i(\tau_i/\|u^i\|)$ be given. Then $\Delta_M(P, \varrho)$ is minimal for the Wulff shape

$$W_M(U, \varrho) = \{x \in E^d | \langle x, u^i \rangle / \|u^i\| \leq \varrho - \tau_i / \|u^i\|, i = 1, \ldots, k\}.$$

Remarks: The Wulff shape is a polytope with at most k facets; but as it is a normalized asymptotic shape, it is no (M, U)-polytope. The choice of ϱ guarantees that $W_M(U, \varrho)$ is nonempty. There are many proofs of Wulff's theorem so that we do not give one. The most famous one, based on energetic arguments, is due to [20]. The first geometric proof is by [9]; the shortest one by P.M. Gruber (unpubl.). In fact we got the first hint on the relation between theorem 8.1 and Wulff shape by P.M. Gruber in 1994.

If M and $\mathfrak{V}$ and ϱ are prescribed, the Wulff shape of course depends strongly on U. So there is the natural question, if the artificial restriction of prescribing $U \subset L^*$ can be replaced by admitting all infinitely many $u^i \in L^*$. It is one of the most surprising results, that even then $W_M(L^*, \varrho)$ is a polytope, i.e. only finitely many u^i contribute to W_M.

Theorem 8.3. The Wulff shape $W_M(L^*, \varrho)$ is a polytope.

Remarks: a) The first proof of Theorem 8.3 for the special case $M = L$ was given in [41,43] and it can easily be generalized to general M. In [35] the general case was proved with a different method, and moreover an upper bound for the number of facets of $W_M(L^*, \varrho)$ was given, if M is rational with respect to L. So for a proof of Theorem 8.3 we refer to these papers. An interesting application of theorem 8.3 is in [36], cf. also [46].

b) Theorems 8.1, 8.2 and 8.3 have been checked for various examples and in all cases the obtained Wulff shapes coincide for a certain interval of the parameter ρ with the shapes of the corresponding ideal crystals. Increasing parameter increases the number of facets. U. Schnell has developed an algorithm, which generates Wulff shapes from the given crytallographic data for any parameter.

References

1. Baake, M., R.V. Moody (1998): 'Diffractive Point Sets with Entropy', J. Phys. **A31** pp. 9023–9039
2. Betke, U., M. Henk, J.M. Wills (1994): 'Finite and infinite packings', J. reine angew. Math. **453**, pp. 165–191
3. Betke, U., M. Henk, J.M. Wills (1995): 'A new approach to covering', Mathematika **42**, pp. 251–263
4. de Boissieu, M., J.L. Verger–Gaugry, R. Currat (Eds.): Aperiodic '97, *Proc. Intern. Conf. of Aperiodic Crystals* (World Scientific, Singapore)
5. Böröczky, K., Jr., U. Schnell (1998): 'Wulff shape for non–periodic arrangements', Lett. Math. Phys. **45**, pp. 81–94
6. Böröczky, K., Jr., U. Schnell (1999): 'Quasicrystals and Wulff-shape', Disc. Comp. Geom. **21**, pp. 421–436
7. Böröczky, K., Jr., U. Schnell, J.M. Wills (1999): 'Quasicrystals, Parametric Density and Wulff shape'. In: *Directions in mathematical Quasicrystals*, ed. by M. Baake, R.V. Moody, to appear
8. Cockayne, E. (1995): 'The quasicrystalline sphere packing problem', *Proc. 5th. Int. Conf. Quasicr.*, ed. by C. Janot, R. Mosseri (World Scient.)

9. Dinghas, A. (1943): 'Über einen geometrischen Satz von Wulff über die Gleichge-wichtsform von Kristallen', Z. Kristallogr. **105**, pp. 304–314

10. Duncan, M.A., D.H. Rouvray (1989): 'Microclusters', Scient. Amer., Dec. 1989, pp. 60–65

11. Edelsbrunner, H., E.P. Mücke (1994): 'Three–dimensional alpha–shapes', ACM Transact. Graph **13**, pp. 43–72

12. Graham, R.L., H.S. Witsenhausen, H.J. Zassenhaus (1972): 'On tightest packings in the Minkowski plane', Pacific J. Math. **41**, pp. 699–715

13. Gritzmann, P., J.M. Wills (1993): 'Finite Packing and Covering'. In: *Handbook of Convex Geometry*, pp. 863–897 (North Holland, Amsterdam)

14. Gruber, P.M., C.G. Lekkerkerker (1987): *Geometry of Numbers*, 2nd edn. (North Holland, Amsterdam)

15. Hoare, M.R., J. McInness (1983): 'Morphology and statistical statics of simple microclusters', Adv. Phys. **32**, pp. 791–821

16. Jacobs, K., St. Herminghaus, K.R. Mecke (1998): 'Thin Liquid Polymer Films Rupture via defects', Langmuir **14**, pp. 965–969

17. Jacobs, K., St. Herminghaus, K.R. Mecke, J. Bischof, A. Fery, M. Ibn–Elhaj, St. Schlagowski (1998): 'Spinodal dewetting in Liquid Crystal and Liquid Metal Films', Science **282**, pp. 916–919

18. Janin, J., F. Rodier (1995): 'Protein Interaction at Crystal Contacts', Proteins: Structure, Function and Genetics **23**, pp. 580–587

19. Kerscher, M., J. Schmalzing et al. (1997): 'Minkowski funcionals of Abell/ACO clusters', Mon. Not. R. Astron. Soc. **284**, pp. 73–84

20. v. Laue, M. (1943): 'Der Wulffsche Satz für die Gleichgewichtsform von Kristallen', Z. Kristallogr. **105**, pp. 124–133

21. Löwen, H. (1990): 'Equilibrium shapes of crystals near the triple point', Surface Science **234**, pp. 315–323

22. Löwen, H., M. Schmidt (1997): 'Freezing in confined supensions', Progr. Colloid Polym. Sci. **104**, pp. 81–89

23. Mecke, J., R. Schneider, D. Stoyan, W.R.R. Weil (1990): *Stochastische Geometrie* (Birkhäuser, Basel)

24. Mosseri, R., J.F. Sadoc (1989): 'Description of metallic and covalent clusters with icosahedral symmetry: the polytope model', Z. Phys. D–Atoms, Molecules and Clusters **12**, pp. 89–92.

25. Nemeth, Z.T., H. Löwen (1998): 'Freezing in Finite: Systems: Hard discs in circular cavities', J. Phys. Condens. Matter **10**, pp. 6189–6204

26. Northby, J.A. (1987): 'Structure and binding of Lennard–Jones clusters', J. Chem. Phys. **87**, pp. 6166–6177

27. Olami, Z., S. Alexander (1988): 'Quasiperiodic packing densities', Phys. Rev. **37**, pp. 3973–3978

28. Pach, J. (Ed.) (1991): *New Trends in Discr. and Compr. Geom.* (Springer, Berlin 1991)

29. Pach, J., P.K. Agarwal (1995): *Combinatorial Geometry* (John Wiley, New York)

30. McPherson, A. (1989): 'Macromolecular Crystals', Scient. American, March 1989, pp. 42–49

31. Rivier, Ch., J.F. Sadoc (1988): 'Polymorphism and disorder in a close–packed–structure', Europhys. Lett. **7**, pp. 523–528

32. Sangwine–Yager, J.R. (1993): 'Mixed volumes'. Ch. 1.2 in: *Handbook of Convex Geometry*, ed. by P.M. Gruber, J.M. Wills (North Holland, Amsterdam)

348   Jörg M. Wills

33. Schmalzing, J., K.M. Gorski (1998): 'Minkowski functionals used in the morphological analysis of cosmic microwave background anisotropy maps', Mon. Not. R.
Astron. Soc. **297**, pp. 355–365

34. Schmidt, M., H. Löwen (1997): 'Phase Diagram of hard spheres confined between
two parallel plates', Phys. Review E **55**, pp. 7228–7241

35. Schnell, U. (1999): 'Periodic sphere packings and the Wulff–shape', Beitr. Allg.
Geom. **40**, pp. 125–140

36. Schnell, U. (2000): 'FCC versus HCP via Parametric Density', to appear in Discrete
Math

37. Smith, A.P. (1993): 'The sphere packing problem in quasicrystals', J. of Non-Cryst.
Solids **153/154**, pp. 258–263

38. Verger–Gaugry, J.L. (1988): 'Approximate icosahedral periodic tilings with
pseudo–icosahedral symmetry in reciprocal space', J. Phys. (France) **49**, pp. 1867–
1874

39. Wills, J.M. (1990): 'A quasicrystalline sphere–packing with unexpected high density', J. Phys. (France) **51**, pp. 1061–1064

40. Wills, J.M. (1993): 'Finite sphere packings and sphere coverings', Rend. Semin.
Mat., Messina, Ser. II 2, pp. 91–97

41. Wills, J.M. (1996): 'Lattice packings of spheres and Wulff–shape', Mathematika
86, pp. 229–236

42. Wills, J.M. (1997a): 'On large lattice packings of spheres', Geom. Dedicata **65**, pp.
117–126

43. Wills, J.M. (1997b): 'Parametric density, online packings and crystal growth', Ren-
dic.di Palermo **50**, pp. 413–424

44. Wills, J.M. (1998): 'Crystals and Quasicrystals, Sphere Packings and Wulff–Shape',
Proc. Aperiodic '97, ed. by J.L. Verger–Gaugry (World Scientific, Singapore)

45. Wills, J.M. (1999): 'The Wulff–shape of large periodic packings', *Proc. DIMACS
workshop Discrete Math. Chem.*, to appear

46. Woodcock, L.V. (1997): 'Entropy difference between the face centered cubic and
the hexagonal close–packed crystal structures', Nature **385**, pp. 141–142

A Primer on Perfect Simulation

Elke Thönnes

Mathematical Statistics, Chalmers University of Technology

Abstract. Markov Chain Monte Carlo has long become a very useful, established tool in statistical physics and spatial statistics. Recent years have seen the development of a new and exciting generation of Markov Chain Monte Carlo methods: perfect simulation algorithms. In contrast to conventional Markov Chain Monte Carlo, perfect simulation produces samples which are guaranteed to have the exact equilibrium distribution. In the following we provide an example-based introduction into perfect simulation focussed on the method called Coupling From The Past.

1 Introduction

A model that is sufficiently realistic and flexible often leads to a distribution over a high-dimensional or even infinite-dimensional space. Examples for such complex distributions include Markov random fields in statistical physics and Markov point processes in stochastic geometry. For many of these complex distributions direct sampling is not feasible. However, there is a very useful tool which may produce (approximate) samples, Markov Chain Monte Carlo (MCMC).

MCMC methods base the sampling of a distribution on a Markov chain. An ergodic Markov chain whose equilibrium distribution is the target distribution is sampled after it has run for a long time. There are many standard methods, like the Metropolis-Hastings algorithm or the Gibbs Sampler, see [10], which allow the construction of Markov chains whose distribution, under regularity conditions, converges to the target distribution. A notoriously difficult problem however remains: when has the chain run for long enough to be sufficiently close to equilibrium? The MCMC literature refers to the initial time the Markov chain is run until it is assumed to be close enough to stationarity as the *burn-in period*.

In the last years a new variant of MCMC methods have been developed, so-called *perfect simulation algorithms*. These are algorithms which automatically ensure that the Markov chain is only sampled after equilibrium has been reached. Thus they produce samples which are guaranteed to have the target distribution and solve the problem of choosing an adequate burn-in period.

The aim of this paper is to give the reader a detailed introduction to the ideas of perfect simulation. We concentrate on one particular method called *Coupling From The Past* (CFTP) and its extensions. This algorithm was developed by [27] and, at the moment, is the more widely used method. However, we would like to point out that there is an alternative general perfect simulation method, Fill's perfect rejection sampling algorithm, see [5]. In contrast to Coupling From The

Past, this method is interruptible, that is the state sampled by the algorithm and its runtime are independent. Thus the algorithm is also known as Fill's interruptible algorithm. The interested reader is referred to [5,6] as well as [25] for a presentation of the method. Further applications and extensions of Fill's algorithm may be found in [7,22] and [35].

In the first section of this paper we will motivate the problem of choosing a burn-in period using the example of a random walk. The next section then discusses couplings for Markov chains which are a basic tool in perfect simulation. In Sect. 4 we present Coupling From The Past as developed in [27]. This is followed by the discussion of two very useful extensions of the method, Dominated Coupling From The Past and perfect simulation in space.

Before we embark on our journey into the world of perfect simulation, let us introduce some assumptions which we make throughout the paper. We consider Markov chains which live on a state space E which is equipped with a separable σ-algebra $\mathcal{E}$. We assume that the Markov chain of interest is ergodic, that is irreducible, aperiodic and positive recurrent. For a general introduction into Markov chain theory the reader may consult [19] or [26]. Standard Markov chain theory tells us that the distribution of an ergodic Markov chain converges towards the limit distribution, see for example [19]. This distribution is called the equilibrium or stationary distribution and is denoted by π throughout this paper. Our aim is to produce an exact sample from the distribution π.

2   Conventional Markov Chain Monte Carlo

As a simple introductory example let us consider the following urn model which leads to a random walk on the four integers $\{0, 1, 2, 3\}$. Urn models, like for example the Ehrenfest urn model, are useful tools as they provide simple models which may describe the movements of molecules. Because of their simplicity and amenability to the method, simple random walks are also often used to introduce the ideas of perfect simulation, see [14,17].

Example 1 A random walk: *Suppose we have three balls which are distributed over two urns. With probability 1/2 we pick a ball from the left urn and put it into the right urn. Alternatively, we take a ball from the right urn and put it into the left urn. If we find a chosen urn empty we do nothing. What is the long-run average number M of balls in the right urn?*

We may describe the number of balls in the right urn as a Markov chain X whose state-flow diagram is shown in Fig. 1. Suppose P denotes the transition matrix of X then, in this example, it is straightforward to compute the equilibrium distribution π by solving the linear equation system $\pi P = \pi$. Then we can determine M as the mean of π. However, let us assume that we would like to estimate M using simulation. We can do so by simulating the chain X for s steps and by estimating M as the average $\frac{1}{s} \sum_{n=1}^{s} X_n$. The chain X may be simulated by flipping a fair coin. Everytime the coin comes up heads we go a step upwards

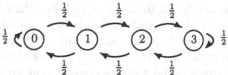

Fig. 1. State-flow diagram for the Markov chain in Example 1.

or if we are in state 3 we stay in state 3. If the coin comes up tails we go a step downwards or if we are in state 0 we stay in state 0.

But how do we choose the initial configuration X_0? Suppose we start in 3, that is we assume that in the beginning all 3 balls are in the right urn. Then the first samples $X_1, X_2, \ldots$ will be slightly higher than we expect in the long run. This is called the initialisation bias and it is due to choosing an initial state which is not sampled from the equilibrium distribution. Nevertheless, for any initial state the distribution of the Markov chain converges towards π. So a common procedure is to simulate the chain for a while without using the initial samples in the estimate. We choose a time $m \in \mathbb{N}$ and estimate M by $\frac{1}{s} \sum_{n=m}^{m+s} X_n$. Thus the samples we produce before time m, during the burn-in period, are ignored. But how long should we choose the burn-in period? We would like the effect of the initial state to wear off, but when can we assume this?

The choice of an appropriate burn-in period is a difficult problem which may be approached in different ways. One possibility is to try to examine analytically the convergence properties of the chain and thus to assess how fast the chain approaches equilibrium. It is usually a hard task to find bounds on the convergence rate and often the resulting bounds are not tight enough to be of any practical value. There is a vast literature on convergence rate computations and the interested reader is referred to a very incomplete selection: [29,30] and [31].

Another approach to determine an adequate burn-in period is to use convergence diagnostics. These are methods which observe the output of the MCMC algorithm and warn if convergence has not been reached yet. However, although these diagnostics may increase our confidence in that the Markov chain has converged, they do not guarantee convergence. For an overview on the large variety of convergence diagnostics see for example the reviews in [1,2].

A recent development are a new variant of MCMC methods which automatically decide whether the chain has reached equilibrium. These methods have become known as *perfect simulation algorithms* and have been particularly successful for models in statistical physics and stochastic geometry. The basis of perfect simulation are couplings and the next section is devoted to a detailed introduction into the coupling method.

3   Coupling

In the last section we encountered the problem of determining the length of the burn-in period. During this burn-in we would like the effect of choosing an initial state, which was not drawn from the equilibrium distribution, to wear off. A reasonable idea seems to start a path of the chain from each possible state and then to wait until they all produce the same results. The intuition is that then the results are no longer influenced by the starting value of the chain. To give the paths started from different initial states a chance to agree we adopt a method which is called the *coupling method*. A coupling specifies a joint distribution for given marginals. The couplings considered in our setting are of a more restrictive nature: two stochastic processes are coupled if their paths coincide after a random time, the coupling time. Couplings are an extremely useful tool in probability theory and are often used to determine convergence properties of Markov chains. For an introduction into the coupling method we refer the interested reader to the book by Lindvall [18]. As we will see at the end of this section, "forward" couplings as described here are not sufficient to produce a sample with the exact equilibrium distribution. Nevertheless, couplings are an essential tool for perfect simulation and thus we will discuss in greater detail how we may couple paths of a Markov chain which are started in different initial states.

3.1   Random Walks

Let us first consider the random walk from the previous section. Here we would like to start a path of the chain in each of the possible initial states $\{0, 1, 2, 3\}$. Recall that we can use a fair coin to produce paths of the chain. This also provides us with a simple way of coupling paths from different initial states. Whenever the coin comes up heads all paths go a step upwards or stay in 3 if in 3. Alternatively, if the coin comes up tails then all paths move a step downwards or stay in 0 if in 0. Figure 2 illustrates the procedure. Each of the resulting paths behaves like a path of the random walk started in the corresponding initial state. The coupling is such that once the state of two paths coincides subsequent states of the two paths also coincide. In other words, if paths meet then they merge, we say they *coalesce*. As we continue evolving the paths they all merge eventually and we reach complete coalescence. At this point the current state of the chain is the same regardless in which state it was started. Note from Fig. 2 that the paths started from the intermediate states 1 and 2 always lie between the path started in state 0 and the path started in state 3. This is due to the fact that we use a *monotone transition rule* to make the updates. A transition rule is a random map which specifies a transition for each state according to the transition kernel of the chain. If all realisations of the random map are monotone functions then we call it a monotone transition rule. In our example the transition rule is given by

$$f(n, C) \quad = \quad \begin{cases} \min(n+1, 3) & \text{if } C = H \\ \max(n-1, 0) & \text{if } C = T \end{cases} \quad n \in \{0, 1, 2, 3\} \quad (1)$$

where C describes whether the coin comes up heads (H) or tails (T). The transition rule is thus a random map whose realisations are specified by the realisations of the coin toss. We achieve the coupling of paths by applying the same realisation of the random map f to all paths.

Notice that $\mathbb{P}(f(n, C) = m) = p_{n,m}$ for $n, m \in \{0, 1, 2, 3\}$ where $p_{n,m}$ are the transition probabilities of the target chain X. Furthermore, observe that

$$\mathbb{P}\Big(f(n, C) = f(m, C)\Big) \geq \sum_{j=0}^{3} p_{n,j}\, p_{m,j} \qquad \text{for all } n, m \in \{0, 1, 2, 3\}. \quad (2)$$

This means that at each step of our coupling the probability of two paths merging when using the transition rule f is greater or equal than the probability of two paths merging in an independent coupling. An independent coupling is achieved if we use an independent coin for each path. This does not hold for every coupling. For example consider the simple symmetric random walk on the vertices of a square as given in Fig. 3. If we take a fair coin and move from each state clockwise if it comes up heads and anti-clockwise if it comes up tails, then paths started from different initial states will never meet. The perfect simulation algorithm, which is presented in the next section, assumes that we use a transition rule for which the analogue of (2) holds. This can always be satisfied as we can choose an independent coupling of paths. However, the speed of the algorithm is greatly increased if we choose a coupling such that paths coalesce quickly.

The realisations of the random map f in (1) are monotone and so f is a monotone transition rule. Thus the use of f leads to paths which maintain the initial order between the starting states. A necessary requirement for the existence of such a monotone transition rule is the stochastic monotonicity of the transition kernel of the Markov chain, for a definition see [18] or [32]. Due to the

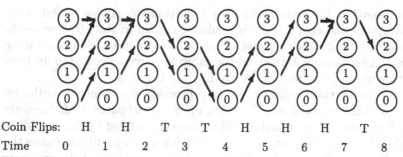

Coin Flips:	H	H	T	T	H	H	H	T	
Time	0	1	2	3	4	5	6	7	8

Fig. 2. Coupled paths of the random walk in Example 1 produced by applying the same outcomes of coin flips to all paths. Note how paths coalesce as they meet. Complete coalescence is achieved after 7 steps.

Fig. 3. For this random walk we may define a coupling such that paths started in different states do not meet.

monotonicity of f we can determine the time of complete coalescence simply by monitoring the path started in state 3 and the path started in state 0. Complete coalescence occurs if and only if these two paths merge and this occurs in finite time almost surely. For our random walk this may seem not such a big improvement. However, for many Markov chains on large state spaces the determination of complete coalescence is not practical if we have to monitor the paths from all initial states. One example for such a chain is the Gibbs Sampler for the Ising model, which will be presented in Sect. 3.2. But first let us discuss another urn model.

Example 2 Another random walk: *As before, we assume we have three balls distributed over two urns. However, in this example if we choose the left urn and it is empty, then we take a ball from the right urn and put into the left one. If we find the right urn empty we do nothing. Below is the state flow diagram of the resulting random walk. It only differs from the previous example in the type of moves the chain can make from state 3.*

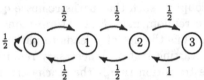

Fig. 4. State-flow diagram for the Markov chain in Example 2.

Again we can simulate the chain by flipping a fair coin. We choose the following strategy. Everytime the coin comes up heads we remain in state 0 if we are in 0, we move a step up if we are in state 1 or 2 and we move a step down if we are in state 3. Alternatively, if the coin comes up tails, we move a step up from state 0 and we move a step down if we are in state 1, 2 or 3.

Similar to the previous example we may produce a coupling of paths by using the same realisation of a coin flip when updating the paths. Unfortunately the resultant coupling is not monotone. (This is easily verified by drawing some sample paths). However, by using a cross-over trick we may still determine the time of complete coalescence by keeping track of two paths only. This cross-over technique was first used in [15] and is further examined for Markov random fields in [12].

As before, we may describe the coupling using a transition rule given by

$$f(n,C) = \begin{cases} n+1 & \text{if } C = H \text{ and } n \in \{1,2\} \\ n-1 & \text{if } C = T \text{ and } n \in \{1,2\} \\ 0 & \text{if } C = H \text{ and } n = 0 \\ 1 & \text{if } C = T \text{ and } n = 0 \\ 2 & \text{if } n = 3 \end{cases} \tag{3}$$

where C is the realisation of the coin flip. Suppose we impose the following partial order on the state space: $2 \preceq 0 \preceq 1 \preceq 3$. Then for fixed C and $n \preceq m$

the above transition rule satisfies $f(n, C) \succeq f(m, C)$. For example if $C = H$, then $0 = f(0, H) \succeq f(1, H) = 2$. Thus the transition rule is *anti-monotone*. In Figure 5 we have ordered the states according to $\preceq$ and drawn some coupled sample paths of the chain. The anti-monotonicity of our coupling can easily be seen in the figure. For example in the first update, the highest state 3 moves to the lowest state 2 and the lowest state 2 moves to the highest state 3.

The anti-monotonicity of the transition rule allows us to monitor complete coalescence by evolving two paths only. We denote the two paths by $X^{\min}$ and $X^{\max}$. We start the minimal path in the minimal state and the maximal path in the maximal state, that is

$$X_0^{\min} = 2 \quad \text{and} \quad X_0^{\max} = 3.$$

We then evolve the two paths as follows

$$X_{k+1}^{\min} = f\left(X_k^{\max}, C_k\right) \qquad X_{k+1}^{\max} = f\left(X_k^{\min}, C_k\right),$$

where C_k is the kth coin toss. Hence the two paths evolve as a two-component Markov chain in which the update of one component is made according to the current state of the other component. The two components are not individually Markov and, as long as they differ, they do not evolve according to the transition probabilities of our random walk. However, once the two components coincide they do evolve like our random walk. Most importantly, the minimal and maximal path sandwich between them all paths of our random walk if evolved using the same coin flip realisations. Thus coalescence of the minimal and maximal path implies complete coalescence of the paths started from all initial states. In Fig. 5 we have drawn the maximal and the minimal path as dotted lines. In this realisation the minimal and the maximal path $(X^{\min}, X^{\max})$ start in $(2, 3)$ respectively and then evolve as $(2, 3), (2, 3), (2, 3), (2, 1), (0, 1), (2, 0)$ and finally coalesce after 6 steps in state 1. In the seventh step they jointly reach state 2.

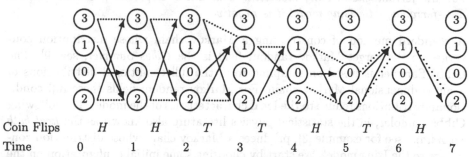

Coin Flips		H		H		T		T		H		T		H
Time	0		1		2		3		4		5		6	7

Fig. 5. Coupled sample paths of the random walk in Example 2. Note that we reordered the initial states and that the coupling is anti-monotone with respect to the new ordering. The dotted lines show the maximal and the minimal path.

3.2   The Ising Model

The next example, the *Ising model* as described for example in [37], is taken from statistical physics and was one of the first models to be considered for perfect simulation, see [5,27]. The Ising model is probably the simplest form of a *Markov random field*. Markov random fields are defined on a discrete lattice Λ of sites where each site may take a value from a finite set of states S. The distribution of a Markov random field is given by the expression

$$\pi(x) \;=\; \frac{1}{Z}\,\exp\Big(-H(x)\Big),$$

where Z is the normalizing constant known as the *partition function* and $H(x)$ is the *energy function*. For most Markov random fields there is no closed form expression for the partition function and therefore direct sampling of these models is not feasible. However, we may produce (approximate) samples of these models using MCMC.

Example 3 Ising model:
The Ising model has energy function

$$H(x) \;=\; -\frac{1}{KT}\left[\,J\sum_{j\sim k} x_j x_k - Bm \sum_k x_k \right]$$

where j, k are sites on a square lattice Λ. Here $j \sim k$, that is site j is a neighbour of site k, if j and k are sites at Euclidean distance one. We may imagine the Ising model as a lattice which at each site has a small dipole or spin which is directed either upwards or downwards. Thus each site j may take a value $x_j \in \{-1, +1\}$ representing a downward respectively an upward spin. The constant K is the Boltzmann factor and T is the absolute temperature in the system. The external field has intensity B and m describes a property of the material. For the ferromagnetic Ising model the constant J is positive, whereas for the anti-ferromagnetic Ising model it is negative.

A standard method of constructing a Markov chain whose distribution converges to the target Markov random field is the *Gibbs Sampler*, see [9]. The Gibbs Sampler is based on sampling from the full conditional distributions of a multi-dimensional Markov chain. For Markov random fields these full conditional distributions reduce to the local characteristics of the model. The following Gibbs Sampler, in the statistical physics literature also known as the *heat bath algorithm*, see for example [3], produces a Markov chain whose distribution converges to the Ising model. We start by choosing some initial configuration on the sites of a finite lattice Λ. Then, step by step we go from one site to the next and update its spin. At site n we assign an upward spin with probability

$$\mathbb{P}\Big(x_n = +1 \,\Big|\, x_{-n}\Big) \;=\; \frac{\pi(x_n = +1,\ x_{-n})}{\pi(x_n = +1,\ x_{-n}) + \pi(x_n = -1,\ x_{-n})}.$$

Here x_{-n} denotes the configuration x on Λ excluding the site n and so the above is the conditional probability of the Ising model having an upwards spin at site n given the current spin configuration on all other sites. The reader may verify that

$$\mathbb{P}\Big(x_n = +1 \,\Big|\, x_{-n}\Big) \;\;=\;\; \Big(1 + \exp\Big(-\frac{2}{KT}\Big[mB + J\sum_{j\sim n} x_j\Big]\Big)\Big)^{-1},$$

which, notably, does not depend on the partition function Z.

More specifically, at each step k we independently draw a random number U_k which is uniform on the interval $(0,1)$ and a random number N_k which is uniform on the lattice Λ. We then assign an upward spin to the site N_k if

$$U_k \;\leq\; \mathbb{P}\Big(x_{N_k} = +1 \,\Big|\, x_{-N_k}\Big),$$

otherwise, we assign a downward spin.

As the lattice Λ is finite we need to specify how we treat sites which are on the boundary of Λ. One possibility is to impose periodic boundary conditions (also called the torus condition). Here the lattice is mapped onto a torus by identifying opposite boundaries. However, edge-effects may occur, that is the sample we draw may not behave exactly like a finite lattice sample of an Ising model defined on an infinite grid. In Sect. 6 we discuss how these edge-effects may be avoided but for now let us assume the torus condition.

We can couple paths of the Gibbs Sampler started from different initial states by reusing the sampled random variates N_k and $U_k, k \in \mathbb{N}$. At time k we update the same site N_k in each path using the same realisation of U_k for all paths. As in the previous example we may describe our updating procedure using a transition rule. We set

$$f(x, U, N) \;=\; \begin{cases} \{x_N = +1, x_{-N}\} \text{ if } U \leq \mathbb{P}(x_N = +1 \mid x_{-N}) \\ \{x_N = -1, x_{-N}\} \text{ otherwise} \end{cases} \tag{4}$$

where $\{x_N = +1, x_{-N}\}$ is the configuration which we obtain by setting $x_N = +1$ and leaving the spins of all other sites in x unchanged.

One problem we encounter is that the set of all initial states of the Ising model is usually very large. Thus it may be prohibitively expensive to monitor all paths. However, as for the random walk examples, we may exploit the monotonicity or anti-monotonicity of the transition rule to determine efficiently the time of complete coalescence. Let us have a closer look at the update rule which we are using. We assign an upward spin to site N if

$$U \;\leq\; \mathbb{P}\Big(x_N = +1 \,\Big|\, x_{-N}\Big) \;=\; \Big(1 + \exp\Big(-\frac{2}{KT}\Big[mB + J\sum_{j\sim N} x_j\Big]\Big)\Big)^{-1}.$$

First consider the case when $J > 0$, that is the ferromagnetic Ising model. Then the probability $\mathbb{P}(x_N = +1 \mid x_{-N})$ is the greater the more neighbours of N have an upward spin. We may exploit this fact by equipping the state space of the

Ising model with an appropriate partial order $\preceq$. We say the spin configuration x is smaller than y, that is $x \preceq y$ if

$$x_j \;\leq\; y_j \qquad \text{for all } j \in \Lambda.$$

This partial order, which was used in [27], may seem counter-intuitive from a physical point of view as a larger state may not necessarily have smaller energy. However, we do not attempt to attach any physical meaning but simply define a partial order for which our transition rule is monotone. Figure 6 shows a triple of configurations which are ordered with respect to $\preceq$.

Now, if $x \preceq y$ then

$$\mathbb{P}\Big(x_N = +1 \,\Big|\, x_{-N}\Big) \;\leq\; \mathbb{P}\Big(y_N = +1 \,\Big|\, y_{-N}\Big) \quad \text{for any } N \in \Lambda.$$

It follows that, for fixed U and N, whenever f assigns an upwards spin to x_N then it also assigns an upwards spin to y_N as

$$U \;\leq\; \mathbb{P}\Big(x_N = +1 \,\Big|\, x_{-N}\Big) \;\leq\; \mathbb{P}\Big(y_N = +1 \,\Big|\, y_{-N}\Big).$$

Analogously, for fixed U and N, whenever f assigns a downward spin to y_N then it assigns a downward spin to x_N. Thus if $x \preceq y$ then the updated configurations maintain their partial ordering, that is $f(x, U, N) \preceq f(y, U, N)$ for fixed U and N. It follows that our transition rule is monotone with respect to $\preceq$ and so the partial ordering between paths is preserved. The state space has a maximal state $x_{\max}$ with respect to $\preceq$ which is the configuration consisting of upward spins only. Similarly, the minimal state $x_{\min}$ is given by the configuration consisting of downward spins only. Due to the monotonicity of the transition rule f the paths are coupled in such a way that all paths lie between the path started in the maximal state and the path started in the minimal state, see also Fig. 7. Complete coalescence occurs if and only if these two paths coalesce which will occur in almost surely finite time.

Now let us discuss the anti-ferromagnetic case, that is if $J < 0$. Careful inspection of the transition rule leads to the observation that f is anti-monotone because for $x \preceq y$

$$\mathbb{P}\Big(x_N = +1 \,|\, x_{-N}\Big) \;\geq\; \mathbb{P}\Big(y_N = +1 \,|\, y_{-N}\Big).$$

Moreover, if we start two paths in two states which are comparable with respect to $\preceq$ then after some updates the states of the two paths may no longer be

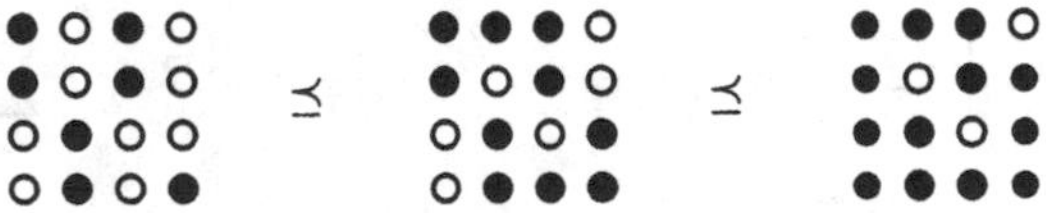

Fig. 6. Three spin configuration on a 4×4 lattice. Upward spins are represented as black sites and downward spins as white sites. We ordered the three configurations with respect to $\preceq$.

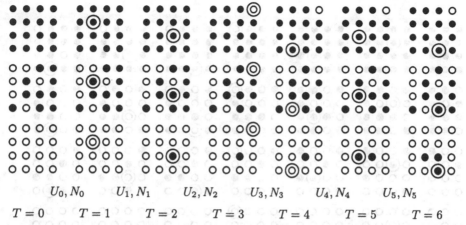

$$U_0, N_0 \qquad U_1, N_1 \qquad U_2, N_2 \qquad U_3, N_3 \qquad U_4, N_4 \qquad U_5, N_5$$

$$T = 0 \qquad T = 1 \qquad T = 2 \qquad T = 3 \qquad T = 4 \qquad T = 5 \qquad T = 6$$

Fig. 7. Coupled paths of the Gibbs Sampler for the ferromagnetic Ising model on a 4×4 lattice. Black sites have an upward spin, white sites a downward spin. The site that has just been updated is encircled. The uppermost path is started in $x_{\max}$, the lowermost path in $x_{\min}$. Observe how the path started from the intermediate configuration is sandwiched between the two paths started in the maximal and minimal state.

comparable. However, we can still monitor complete coalescence by monitoring a minimal and a maximal path which are evolved according to a cross-over. We start a path $X^{\max}$ in $x_{\max}$ and another path $X^{\min}$ in $x_{\min}$. We then update the two paths according to

$$X_{k+1}^{\min} = f\left(X_k^{\max}, U_k, N_k\right) \quad \text{and} \quad X_{k+1}^{\max} = f\left(X_k^{\min}, U_k, N_k\right).$$

This leads to a maximal path $X^{\max}$ and a minimal path $X^{\min}$ which sandwich between them the paths which are started from all initial states and evolved according to the transition rule f and same realisations of U_k and N_k, $k \in \mathbb{N}$. Thus we may determine complete coalescence by monitoring whether $X^{\max}$ and $X^{\min}$ coalesce. This can be shown to occur in almost surely finite time. Figure 8 illustrates the coupling and cross-over procedure.

3.3 Immigration-Death Process

Our final example is an immigration-death process on the natural numbers and thus a Markov chain on an infinite state space.

Example 4 Immigration-Death process:
Consider the number of dust particles contained in a given small volume. If there are N particles in the system, then new particles enter at a rate $\lambda (N+1)/(N+2)$, where λ is some positive constant. Thus the immigration rate increases the more particles are already in the system. Particles stay in the system for an exponential amount of time with unit mean.

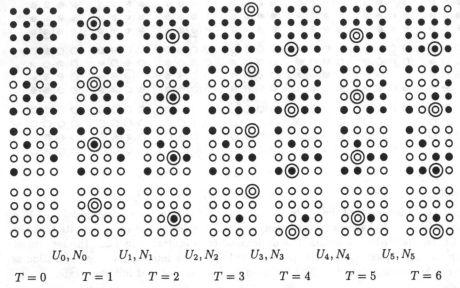

$$U_0, N_0 \qquad U_1, N_1 \qquad U_2, N_2 \qquad U_3, N_3 \qquad U_4, N_4 \qquad U_5, N_5$$

$$T = 0 \qquad T = 1 \qquad T = 2 \qquad T = 3 \qquad T = 4 \qquad T = 5 \qquad T = 6$$

Fig. 8. Coupled paths of the Gibbs Sampler for the anti-ferromagnetic Ising model on a 4×4 lattice. The uppermost and lowermost path are the maximal and minimal path respectively and evolved according to a cross-over. The two intermediate paths in the middle are evolved according to the standard coupling construction. Note that although these two paths are comparable at time $T = 0$, they are no longer comparable at time $T = 6$. However they are still comparable to the states of the minimal and the maximal path.

The number of particles N in the given volume is an immigration-death process with transition rates

$$\begin{aligned}
N &\to N + 1 & &\text{at rate } \lambda (N + 1)/(N + 2), \\
N &\to N - 1 & &\text{at rate } N & &\text{for } N \in \mathbb{N}.
\end{aligned} \tag{5}$$

We may simulate N as follows. We start by simulating an immigration-death process D with transition rates

$$\begin{aligned}
D &\to D + 1 & &\text{at rate } \lambda \\
D &\to D - 1 & &\text{at rate } D & &\text{for } D \in \mathbb{N}_0.
\end{aligned} \tag{6}$$

Observe that D and N have the same death rate, but the immigration rate of D is larger than for N. In mathematical terms, D stochastically dominates N. A description on how to simulate a constant rate immigration-death process may be found in [28].

Given a realisation of D we may derive a realisation of the process N. As the initial configuration N_0 at time 0 we choose a number from $\{0, 1, \ldots, D_0\}$. Now, whenever a particle arrives in D say at time t, it enters the given volume with probability $(N_{t-} + 1)/(N_{t-} + 2)$ where N_{t-} is the number of particles in the system immediately before the arrival of the new particle. We may achieve this by marking every immigration time t of D with an independent random variable U_t which is uniform on the unit interval. The new particle enters the system if and only if

$$U_t \ \leq \ \frac{N_{t-} + 1}{N_{t-} + 2}.$$

Finally consider a death in D at time t which leads to a death in N at time t with probability N_{t-}/D_{t-}. Again we may achieve this by marking any death time t in D with a mark U_t which is uniform on the unit interval. A death occurs in N at time t if

$$U_t \ \leq \ \frac{N_{t-}}{D_{t-}}.$$

In the following whenever we speak of the dominating process D we implicitly mean the process D and its jump time marks. By coupling N to D we not only define a simulation procedure, but, as we will see in Sect. 5, we also make the process N amenable to a perfect simulation algorithm. Perfect simulation of birth-and-death processes on the natural numbers like the one above was first described in [14].

The coupling procedure may be illustrated using a Hasse diagram as in Fig. 9. It consists of a sequence of horizontal levels which stand for the states $\{0, 1, 2, \ldots\}$. On each level we have arrows representing the jump times of D. An arrow pointing upwards indicates an immigration time and an arrow pointing downwards a death time. The arrow corresponding to the jump time t is marked with U_t. Now, for each level we may delete arrows according to the rules described above. For example, on level 3 corresponding to state 3 we delete any upwards arrow whose associated mark U_t exceeds $(N_{t-} + 1)/(N_{t-} + 2) = 4/5$. We delete any downwards arrow whose associated mark U_t exceeds $N_{t-}/D_{t-} = 3/D_{t-}$. The process N started in some state $j \in \{0, \ldots, D_0\}$ may now be constructed as follows. We start on level j and move from left to right. Whenever we come across an upward arrow we go a level upwards. Alternatively, whenever we come across a downward arrow then we go a level downwards, see Fig. 9.

For the immigration-death process in Example 4 we may produce a coupling by using the same realisation of D and the associated jump time marks $U_t, t \geq 0$, and applying the above procedure to all paths started in a state $j \in \{0, \ldots, D_0\}$. From Fig. 9 we may see that two paths can never cross each other but can only meet and then merge. Thus the paths maintain the partial ordering of their initial states and so the coupling is monotone. It follows that complete coalescence of paths started from every state in $\{0, \ldots, D_0\}$ occurs if and only if the path started in state 0 and the path started in state D_0 coalesce. The state space of the target process N are the natural numbers $\mathbb{N}_0$ which, of course, is much larger than the finite set $\{0, \ldots, D_0\}$. However, as we will see in Sect. 5, to produce a perfect sample we only need complete coalescence for the set bounded above by D. Complete coalescence occurs in almost surely finite time as a sufficient event is that D hits zero, in which case all relevant paths of N also hit zero.

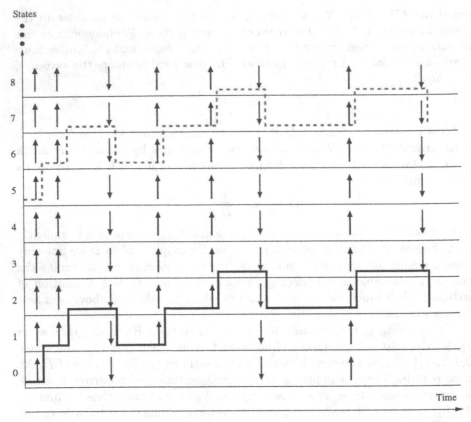

Fig. 9. The Hasse diagram for the immigration-death process in Example 4. The dashed line shows D which was started at time 0 in state 5. Some of the jump times were deleted according to our decision rules. Notice that the more jump times of D are deleted the lower the level. The solid line shows a path of N started in state 0 and evolved coupled to D.

3.4   Forward Coupling and Exact Sampling

We started this section with the motivation of finding a time when the effect of the initial state of the chain has worn off. We may argue now that this has happened when complete coalescence occurs as in this case the chain is in the same state regardless of its starting value. Let S be the time of complete coalescence which is a random stopping time. We may think that X_S has the equilibrium distribution, however, this intuition is flawed! Even if the chain had been started in equilibrium we cannot conclude that X_S has the equilibrium distribution. This is due to the fact that S is not a fixed but a random time. As an illustration let us consider again our random walk from Example 1. From Figure 2 we can see that at the time of complete coalescence, that is at time S, the chain is necessarily either in state 0 or in state 3. Clearly, this is not a sample of the equilibrium distribution which is uniform on the integers $\{0, 1, 2, 3\}$.

Fortunately, a rather simple but very effective modification enables us to sample X in equilibrium. This modification is called Coupling From The Past and is due to [27].

4   Coupling from the Past

The last section showed how to couple paths of a Markov chain from different initial states such that after a random time S, the time of complete coalescence, all paths have merged into one. Although the state of any of these paths at time S does not depend on its starting value we have seen that sampling the chain at time S may give a biased sample which is caused by the fact that S is not a fixed time but a random stopping time. In the following we discuss an alternative approach called *Coupling From The Past* (CFTP) which was introduced by Propp and Wilson [27]. It is also based on coupling and complete coalescence but samples the chain at a fixed time, namely time 0.

Recall from the previous section that we produced coupled sample paths of the chain X started from every initial state by sampling transition rules. At each time step $k \in \mathbb{N}$ we independently sampled a transition rule f_k and produced a sample path starting in $x \in E$ by setting

$$X_k(x) \quad = \quad f_k(X_{k-1}(x)) \quad = \quad f_k \circ f_{k-1} \circ \cdots \circ f_1(x).$$

For instance, for our random walk in Example 1 we sampled an independent coin C_{k-1} and set

$$X_k(x) \quad = \quad f(X_{k-1}(x), C_{k-1})$$

where f is defined as in (1). We have also seen in the previous section that an adequate choice of transition rules eventually leads to complete coalescence of the sample paths. Or, in other words, the image of the composite map defined as

$$F_{0,k} \quad = \quad f_k \circ f_{k-1} \circ \cdots \circ f_1 \tag{7}$$

eventually becomes a singleton as k approaches infinity. Let S be the time of complete coalescence then, unfortunately, the unique image of $F_{0,S}$ does not have the equilibrium distribution in general.

In 1995, Propp and Wilson made a simple but ingenious observation: if we *reverse* the order in which we compose the sampled transition rules and continue sampling until the image of the composite map becomes a singleton then this unique image has the equilibrium distribution!

Let us look at this in more detail. At time k we sample the transition rule f_k but now we define a composite map by

$$\tilde{F}_{0,k} \quad = \quad f_1 \circ \cdots \circ f_k.$$

Thus we have reversed the order of composition compared to (7). The above is equivalent to the following procedure. We go backwards in time and sample at time $-k$ the transition rule f_{-k} and now define a composite map by

$$F_{-k,0} \quad = \quad f_{-1} \circ \cdots \circ f_{-k} \qquad k \in \mathbb{N}. \tag{8}$$

A nice illustration of the difference between forward coupling as in (7) and backward coupling as in (8) on the example of Matheron's dead leaves model can be found in [17]. It is also illustrated in an animated simulation on http://www.warwick.ac.uk/statsdept/Staff/WSK/dead.html.

How can we interpret the composite map in (8)? If we set

$$X_j^{-k}(x) \;=\; f_{-j-1} \circ \cdots \circ f_{-k}(x)$$

then $\{X_j^{-k}(x), -k \leq j \leq 0\}$ behaves like a path of X started at time $-k$ in state x. Thus $F_{-k,0}(x)$ is the state at time 0 of a path of X started at time $-k$ in state x. It follows that $F_{-k,0}$ has a singleton image if and only if the corresponding coupled paths of X started at time $-k$ in all initial states achieve complete coalescence by time 0. Thus the above procedure produces coupled paths started in all initial states at earlier and earlier times until they achieve complete coalescence by time 0. The time when all paths coalesce is not necessarily time 0 but can occur earlier. Nevertheless, we only ever sample at the fixed time 0. If complete coalescence is achieved then the common state at time 0 is an exact sample from π.

We may describe the procedure using the pseudo-code notation from computer science. Suppose we have an algorithm RandomMap$(-k)$ which samples the transition rule f_{-k}. Then we can describe the CFTP algorithm as follows:

CFTP:
   $k \leftarrow 0$
   $F_0 \leftarrow$ identity map
   Repeat
      $k \leftarrow k - 1$
      $f \leftarrow$ RandomMap(k)
      $F_k \leftarrow F_{k+1} \circ f$
      until image of F_k is a singleton
   return image of F_k

We will first give a heuristical argument which provides an intuitive explanation why this method produces an exact sample. This is followed by some examples. A rigorous proof for the correctness of the procedure is given at the end of the section.

Let $-T$ be the first time when the image of $F_{-k,0}$ becomes a singleton, that is $T = \min\{k : F_{-k,0}$ has a singleton image$\}$. Suppose we could start the chain X at time $-\infty$ and run it up to time $-T$. As the chain is ergodic and has run for an infinite amount of time, the heuristic suggests that the chain is in equilibrium at time $-T$. Suppose now that the value of the infinite time simulation at time $-T$ is x, then x is a sample from π. The transition rules $f_{-k}, k \in \mathbb{N}$ describe transitions according to the transition kernel P of X. As $\pi P = \pi$ the transition rules preserve the equilibrium distribution and so it follows that $F_{-T,0}(x)$ is also a sample from π. Of course, we do not know x but, as the image of the composite map is unique, it does not matter which value x the infinite time chain takes at time $-T$. Hence, if the image of the composite map becomes a singleton then we may deduce the state of the infinite time simulation at time 0. Thus, by extending backwards in time until the image of the composite map becomes a singleton, we reconstruct the path of the infinite time simulation in the recent past. In other words, we create a "virtual simulation from time $-\infty$".

Let us now apply the above algorithm to our examples. For the random walk in Example 1 the procedure runs as follows. Recall that we may simulate the random walk using fair coin flips. We go step by step backwards in time and perform the following routine.

1. At time $-k, k \in \mathbb{N}$, we independently flip a fair coin C_{-k}.
2. We then start a path from all initial states $\{0, 1, 2, 3\}$ and evolve them from time $-k$ till time 0 according to the coin flips $C_{-k}, C_{-k+1}, \ldots, C_{-1}$ (note the order of the coins!).
3. If all paths coalesce at time 0 then we return their common state as a sample from π.

If the paths do not coalesce, then we go a step further backwards in time and repeat the above steps. Thus we independently flip another biased coin C_{-k-1}, and again evolve the paths started in all initial states from time $-k-1$ to time 0 using the coin flips $C_{-k-1}, C_{-k}, \ldots, C_{-1}$. We continue going successively further backwards in time until we finally reach complete coalescence at time 0.

Remark 1 *It is essential that in the kth iteration the coins are used in the order $C_{-k}, C_{-k+1}, \ldots C_{-1}$ and that we reuse all previously sampled coin flips in the appropriate order. Only then is the sample guaranteed to have the equilibrium distribution.*

We can make the above procedure more efficient by noting the following. It is not necessary to evolve paths started in *all* initial states till time 0. As discussed in the previous section, due to the monotonicity of the transition rule, complete coalescence occurs if and only if the path started in state 0 and the path started in state 3 coalesce. Thus we only need to monitor these two paths for coalescence.

Neither is it necessary to check for coalescence at each time step. Recall that complete coalescence of paths started from time $-T$ means that the map $F_{-T,0}$ has a singleton image. But then the composite map $F_{-T-1,0} = F_{-T,0} \circ f_{-T-1}$ has exactly the same singleton image and, by induction, so has any $F_{-S,0}$ with $S > T$. Thus we may proceed as follows. Let $0 = T_0 < T_1 < T_2 \ldots$ be an increasing sequence of time points. Then for $k = 1, 2, \ldots$, we perform the following steps until we reach complete coalescence:

1. Sample independent coin flips $C_{-T_k}, C_{-T_k+1}, \ldots, C_{-T_{k-1}-1}$.
2. Evolve one path started in state 0 and one started in state 3 from time $-T_k$ to time 0 using the coin flips $C_{-T_k}, C_{-T_k+1}, \ldots, C_{-1}$.
3. Check for coalescence at time 0.

A recommended choice for the sequence of time points is $T_k = 2^{k-1}$, which in [27] is shown to be close to optimal.

For simulation purposes we do not need to store all coin toss realisations. Instead we can reproduce them by resetting the seed of a seeded pseudo-random number generator.

Figure 10 illustrates the CFTP algorithm for the random walk from Example 1.

Iteration 1: $T_1 = 3$

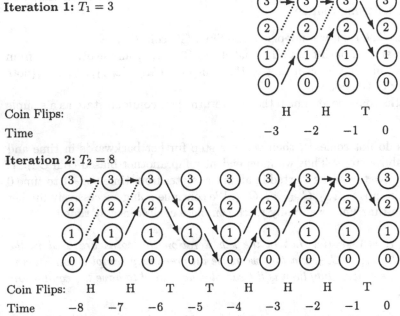

Coin Flips: H   H   T

Time -3 -2 -1 0

Iteration 2: $T_2 = 8$

Coin Flips: H   H   T   T   H   H   H   T

Time -8 -7 -6 -5 -4 -3 -2 -1 0

Fig. 10. CFTP for Example 1. The paths started in state 0 and in state 3 are shown as solid lines. The dotted lines are the paths started from intermediate states. However we do not need to monitor these to determine complete coalescence. Note how the coin toss realisations of the previous iteration are reused! Complete coalescence occurs at time -1, however we continue till time 0 and sample state 2.

For the random walk in Example 2 we may proceed in a similar fashion.

As discussed earlier, in this setting we may use a cross-over to detect complete coalescence of all paths by monitoring two paths only, a minimal and a maximal path. At time $-T_k$ we start a path in the minimal state 2 and a path in the maximal state 3. (Recall that we chose a partial ordering $\preceq$ which differs from the natural partial ordering on the integers.) We then evolve the minimal and the maximal path by updating the minimal path according to the current configuration of the maximal path and vice versa. If coalescence of these two paths occurs by time 0, then their common state at time 0 has the equilibrium distribution.

For the Ising model we may proceed as follows. In iteration k we sample independent random variables $U_{-T_k}, U_{-T_k+1}, \ldots, U_{-T_{k-1}-1}$ which are uniform on the unit interval and random variables $N_{-T_k}, N_{-T_k+1}, \ldots, N_{-T_{k-1}-1}$ which are uniform on the lattice Λ. In the ferromagnetic case, we start a path in $x_{\max}$, that is the configuration consisting of only upwards spins, and a path in $x_{\min}$, that is the configuration with only downward spins. The two paths are evolved from time $-T_k$ to time 0 using the transition rule f as defined in (4) together with the realisations of $U_{-T_k}, U_{-T_k+1}, \ldots U_{-1}$ and $N_{-T_k}, N_{-T_k+1}, \ldots N_{-1}$. If the two paths coalesce then we output their common state at time 0 as a sample from the equilibrium distribution. If coalescence has not been achieved yet then we extend further backwards in time.

In the anti-ferromagnetic case we adopt the following procedure. In iteration k we also sample independent random variables $U_{-T_k}, U_{-T_k+1}, \ldots, U_{-T_{k-1}-1}$ and $N_{-T_k}, N_{-T_k+1}, \ldots N_{-T_{k-1}-1}$. We again start two paths at time $-T_k$, one in $x_{\max}$ and one in $x_{\min}$. However, we now evolve the two paths according to a cross-over as described in the previous section. As before, if coalescence of these two paths occurs then we may deduce complete coalescence of the paths started in all initial values. The unique state at time 0 is then a perfect sample. If coalescence has not occurred yet then we extend further backwards in time.

Remark 2 *The heat bath algorithm is known to mix slowly for temperatures close to criticality. To produce samples of the Ising model close to the critical temperature [27] apply Coupling From The Past to a Gibbs Sampler for the random cluster model. (This type of Gibbs Sampler is also known as single bond heat bath.) By assigning random colours to the obtained clusters a realisation of a random cluster model is turned into a realisation of an Ising or Potts model. Propp and Wilson [27] produced samples of the Ising model at critical temperature on a 512×512 toroidal grid in about 20 seconds on a Sparcstation. CFTP needed to go back only to about time -30 to produce such a sample. The authors also show how to produce samples from the Ising model simultaneously for a range of temperature values.*

We now give a rigorous proof for the correctness of the CFTP algorithm for finite state spaces, see also [27].

Theorem 1 *Let X be an ergodic Markov chain with transition matrix P and stationary distribution π. Coupling From The Past as presented above produces an exact sample of the target equilibrium distribution π.*

Proof: For $k \in \mathbb{N}$ consider the composite map

$$F_{-k,-j} \;=\; f_{-j-1} \circ \cdots \circ f_{-k} \qquad \text{where } j \leq k.$$

We will first show that the image of $F_{-k,0}$ almost surely becomes a singleton as k approaches infinity. We then proceed to show that this unique image has the distribution π.

Let $z \in E$ be an arbitrary state in the state space of X. As the chain X is irreducible and aperiodic there is a finite $N > 0$ such that $P^N(y, z) > 0$

for all $y \in E$. Here P^N denotes the N–step transition matrix of X, that is $P^N(y, z) = \mathbb{P}(X_N = z | X_0 = y)$. It follows the existence of a constant $\epsilon > 0$ such that for any $k \in \mathbb{N}$ we have

$$\mathbb{P}\Big(\text{image of } F_{-kN,-(k-1)N} \text{ is a singleton}\Big) > \epsilon.$$

Now, the events

$$\Big\{\text{image of } F_{-kN,-(k-1)N} \text{ is a singleton}\Big\}, \qquad k \in \mathbb{N}$$

are independent and have a probability of at least ϵ. Thus by Borel-Cantelli

$$S \;=\; \min\Big\{k \in \mathbb{N}: \text{ image of } F_{-kN,-(k-1)N} \text{ is a singleton}\Big\}$$

is almost surely finite. But then the composite map

$$F_{-SN,0} \;=\; F_{-SN,-(S-1)N} \circ F_{-(S-1)N,0}$$

has also a unique image.

Like the "forwards" composite map

$$F_{0,k} \;=\; f_k \circ \cdots \circ f_1$$

the map $F_{-k,0}$ is composed of k independent transition rules and thus $X_0^{-k}(x) = F_{-k,0}(x)$ has the same distribution as $X_k(x) = F_{0,k}(x)$. Now, $X_k(x)$ for $k \in \mathbb{N}_0$ is a path of X started at time 0 in x and so, due to the ergodicity of X, the distribution of $X_k(x)$ converges to π as k approaches infinity. As $X_0^{-k}(x)$ has the same distribution as $X_k(x)$ it follows that its distribution also converges to π. Moreover, by the definition of X_0^{-k} we have

$$\lim_{k \to \infty} X_0^{-k}(x) \;=\; \lim_{k \to \infty} F_{-k,0}(x) \;=\; \text{unique image of } F_{-SN,0}$$

where S is defined as above. As $X_0^{-k}(x)$ tends in distribution to π it follows that the unique image of $F_{-SN,0}$ must also have the equilibrium distribution π. $\qquad \square$

5   Dominated Coupling from the Past

In this section we discuss an extension of the original CFTP algorithm as in [27]. This extension is called *Dominated Coupling From The Past*, or Coupling Into And From The Past, and is due to [14], see also [15]. Suppose $T \geq 0$ is the smallest random time such that coupled paths of the target chain started in *all* initial states at time $-T$ have coalesced by time 0. Foss and Tweedie [8] showed that T is almost surely finite if and only if the chain is uniformly ergodic. Thus CFTP as described in the previous section only applies to uniformly ergodic Markov chains. However, many Markov chains of interest, in particular chains which converge to point process distributions, are not uniformly, but only geo-metrically ergodic. Fortunately, Dominated Coupling From The Past may enable us to sample the stationary distribution of these Markov chains.

Dominated CFTP essentially specifies a time-evolving bounded random set $\Theta_t, t \in \mathbb{R}$, such that

1. there exists an almost surely finite time T such that paths started at time $-T$ from all initial states in Θ_{-T} coalesce at time 0,
2. if coalescence as in 1. occurs then the unique state at time 0 has equilibrium distribution.

Heuristically we may think of Θ_t as a random set which provides a stochastically varying upper and lower bound on the values at time t of an infinite time simula-tion. This is best explained using an example, so consider our immigration-death process from Example 4. In Sect. 3 we showed how to couple paths of the process started from different initial states. We have seen that this coupling is monotone with respect to the natural ordering on the integers. The state space has a min-imal state with respect to this partial order, the state 0, but it does not have a maximal state. However, due to our coupling construction, we do know that the process N is bounded above at any time by the process D. Thus, although we do not have a fixed bound on N_t for any $t \in \mathbb{R}$, we do have a random bound given by D_t and we may set $\Theta_t = \{0, \ldots, D_t\}$.

How can we exploit this in a CFTP-type algorithm? Let us first use the heuristic of an infinite time simulation to provide the intuition; a formal proof follows later. Consider an infinite time simulation of the target Markov chain N started at time $-\infty$. We denote the infinite time simulation by $N^{-\infty}$. Our aim is to reconstruct the path of $N^{-\infty}$ in the recent past. Clearly, $N^{-\infty}$ is bounded below by 0. But how about an upper bound? If we assume that the infinite time simulation $N^{-\infty}$ is started in state 0 and coupled to an infinite time simulation of D then $N^{-\infty}$ is bounded above by $D^{-\infty}$. So our first task is to reconstruct the path of $D^{-\infty}$ in a finite interval $[-T, 0]$. A little thought shows that this is easily done. If D was started at time $-\infty$ then heuristics suggest that it is in equilibrium at time $-T$. Now, the stationary distribution of D is a Poisson distribution which is easy to sample. Thus, if we start D at time $-T$ in its equilibrium and simulate it till time 0 then we may interpret this realisation as the path of $D^{-\infty}$ on $[-T, 0]$.

Because $N^{-\infty}$ is coupled to and thus bounded above by D, we may deduce that the path of $N^{-\infty}$ on $[-T, 0]$ lies below the given realisation of D on the same time interval. In particular we have that $N_{-T}^{-\infty} \leq D_{-T}$. Suppose that the paths of N started at time $-T$ from all initial states in $\{0, 1, \ldots, D_{-T}\}$ and coupled to the realisation of D on $[-T, 0]$ coalesce by time 0. Then, according to our heuristic, their common state at time 0 is also the state of the infinite time simulation $N^{-\infty}$ at time 0. Therefore this state is a sample from the equilibrium distribution. Recall that the coalescence of the path of N started in state 0 and the path of N started in state D_{-T} implies the coalescence of all paths started in $\{0, \ldots, D_{-T}\}$. Thus it is sufficient to monitor only these two paths for coalescence.

If the paths started from $\{0, 1, \ldots, D_{-T}\}$ do not coalesce by time 0, then we need to extend backwards in time and repeat the above procedure. To do this we need to extend the realisation of D on $[-T, 0]$ backwards in time, that is we need to produce a realisation of D on $[-T - S, 0]$ which coincides with the previous realisation on $[-T, 0]$. We can do this by exploiting the time-reversibility of D. We start D at time 0 in equilibrium and simulate it up to time T. Then we set

$$\widetilde{D}_{-t} \;=\; D_t \qquad \text{for } t \in [0, T]$$

that is we reverse the path of D in time. This produces a path of D on $[-T, 0]$. If we extend backwards then we just continue our simulation of D from time T to time $T + S$ and again reverse the resulting path in time.

Here is a pseudo-code description of the algorithm, which is also illustrated in Fig. 11. The algorithm $\texttt{Extend}(D, -T)$ extends a given path of D to a path on $[-T, 0]$ and assigns marks to any new jump times. The algorithm $\texttt{Evolve}(D, -T)$ then starts a path of N in 0 and one in D_{-T} and evolves them coupled to the path D till time 0.

Dominated CFTP:
```
  T ← 0
  D ← ∅
  Repeat
     T ← T - 1
     D ← Extend(D, -T)
     N₀ ← Evolve(D, -T)
     until N₀ is a singleton
  return Y₀
```

What are the characteristics of the above procedure? Firstly, we started with a time-homogeneous process D which stochastically dominated N. This process D had a standard stationary distribution which is easy to sample. Furthermore, we made use of the fact that D was time-reversible.

Paths of the target process N were derived as an adapted functional of D. The coupling between D and N ensured that if we started N at time $t \in \mathbb{R}$ in a configuration bounded above by D then the path of N was bounded by the path of D at any later time.

We determined coalescence by starting a path of N at time $-T$ in D_{-T} and one in state 0. Let the "upper" path started at time $-T$ in D_{-T} be denoted by U^{-T} and the "lower" path started at time $-T$ in state 0 by L^{-T}. The two processes U and L have the following properties, some of which can also been seen in Fig. 11.

1. Conditional on a realisation of D on $[-T - S, 0]$ we have a *funneling* property, that is

$$L_t^{-T} \leq L_t^{-T-S} \leq U_t^{-T-S} \leq U_t^{-T} \qquad \text{for } t \in [-T, 0]. \tag{9}$$

Thus the earlier we start U and L the closer the two paths get.

2. If $L_t^{-T} = U_t^{-T}$ for $t \in [-T, 0]$ then $L_u^{-T} = U_u^{-T}$ for $u \in [t, 0]$. Hence once the upper and lower path coalesce they remain coalesced. We call this the *coalescence* property.

3. Suppose we start a path of N at time $-T$ in some $j \in \{0, \ldots, D_{-T}\}$ and evolve it coupled to the same realisation D as the lower and upper path, then we have the following *sandwiching property*:

$$L_t^{-T} \quad \leq \quad N_t^{-T} \quad \leq \quad U_t^{-T} \qquad \text{for } t \in [-T, 0]. \tag{10}$$

Therefore a path of N started at time $-T$ in some state bounded above by D_{-T} and evolved according to D lies between the upper and lower path started at time $-T$ and coupled to the same realisation of D. Together with the funneling property it follows that any path of N started at time $-T - S$ in a state bounded above by D_{-T-S} lies on $[-T, 0]$ between the lower and upper path started at time $-T$ and evolved according to the same realisation of D.

We now prove rigorously that our Dominated CFTP algorithm does in fact sample the desired equilibrium distribution. The proof is a special case of the proof in [16] for general Dominated CFTP algorithms.

Theorem 2 *Suppose U and L are defined as above and satisfy the funneling, coalescence and sandwiching properties. If*

$$T_C \quad = \quad \inf \left\{ T \geq 0 : U_0^{-T} = L_0^{-T} \right\}$$

is almost surely finite then $U_0^{-T_C}$ has the distribution π.

Proof: As T_C is almost surely finite the funneling property implies that the limit $\lim_{T \to \infty} U_0^{-T}$ exists and that

$$\lim_{T \to \infty} U_0^{-T} \quad = \quad \lim_{T \to \infty} L_0^{-T} \quad = \quad U_0^{-T_C}.$$

Now, let N_0^{-T} be the state at time 0 of a path of N started at time $-T$ in state 0. Then the distribution of N_0^{-T} is the same as of N_T, that is a path of N started at time 0 in state 0 and run up to time T. Due to the ergodicity of N, it follows that the distribution of N_0^{-T} converges to π as $T \to \infty$. The limit of N_0^{-T} may be interpreted as our infinite time simulation. The sandwiching property ensures that

$$\lim_{T \to \infty} L_0^{-T} \quad \leq \quad \lim_{T \to \infty} N_0^{-T} \quad \leq \quad \lim_{T \to \infty} U_0^{-T}$$

and so

$$\lim_{T \to \infty} N_0^{-T} \quad = \quad U_0^{-T_C}$$

which implies that $U_0^{-T_C}$ has the equilibrium distribution π. $\qquad\square$

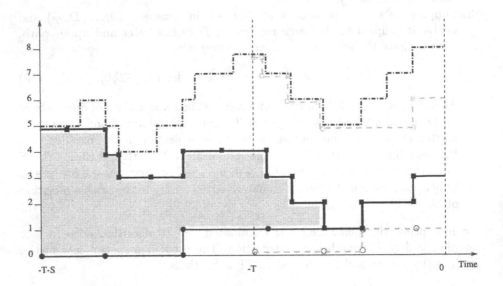

Fig. 11. The immigration-death process example 4. The dashed-dotted line is the immigration-death process D. The solid line marked with squares shows the maximal path and the solid line marked with disks the minimal path started at time $-T - S$. The two paths have coalesced by time 0. The shaded area is the area in which any path of N started at time $-T - S$ in some state smaller than D_{-T-S} lies. The dashed lines show an earlier CFTP iteration started from time $-T$ in which coalescence of the lower and upper path did not occur. Note how the process D has been extended backwards in time. Observe also that the lower process lies below the upper process and how both processes satisfy the funneling property.

Dominated Coupling From The Past was originally developed for locally stable Markov point processes, see [14,15,16]. Markov point processes are usually specified by a density π with respect to a unit rate homogeneous Poisson point process on a bounded window W. For example the Strauss process, which is described in [34] and which models repulsive point patterns, is given by

$$\pi(x) \;=\; \alpha \, \beta^{n(x)} \, \gamma^{t(x)} \qquad x \subset W,$$

where $\beta > 0$ and $0 < \gamma < 1$. Here $n(x)$ counts the number of points in x and $t(x)$ the number of pairs of neighbour points, that is points which are less than the interaction range R apart. Like for many other point process models, the normalizing constant α cannot be computed in closed form. From the density of a Markov point process we may derive its Papangelou conditional intensity

$$\lambda(x, \xi) \quad = \quad \begin{cases} \frac{\pi(x \cup \{\xi\})}{\pi(x)} & \text{if } \pi(x) > 0 \\ 0 & \text{otherwise,} \end{cases}$$

where $x \subset W$ is a point pattern and $\xi \in W$ an individual point. For example, for the Strauss process the Papangelou conditional intensity is given by

$$\lambda(x, \xi) \quad = \quad \beta \, \gamma^{t(x, \xi)},$$

where $t(x, \xi)$ counts the number of neighbours of ξ in x. More information on Papangelou conditional intensities may be found in [4]. A Markov point process that is locally stable has a Papangelou conditional intensity which is uniformly bounded above by some constant λ^*. For example, for the Strauss model $\lambda^* = \beta$.

Spatial birth-death processes may converge in distribution to point processes. These birth-and-death processes are Markov jump processes whose states are point patterns and which evolve in time through births and deaths of individual points. For an introduction see [33] or [21]. A spatial birth-death process is specified by its birth rate and its death rate. If we choose a unit death rate and a birth rate which is equal to the Papangelou conditional intensity of a Markov point process then, under regularity conditions specified for example in [21], the resultant spatial birth-death process Y converges to the distribution of the point process.

Similar to the immigration-death process example, we can produce exact samples for locally stable Markov point processes using Dominated CFTP. Suppose the ergodic spatial birth-and-death process Y converges to the distribution of such a Markov point process with density π. Note that we may derive a realisation of Y from a realisation of a spatial birth-and-death process Z with the same death rate and a higher birth rate. In our setting we may choose Z to have unit death rate and birth rate λ^*. Then Z is time-reversible and has a Poisson point process as its equilibrium. We mark every birth time t of Z with a mark V_t which is uniform on $(0, 1)$. We then can derive a path of Y from a path of Z as follows. A birth of a point ξ in Z at time t leads to the birth of the same point at the same time in Y if

$$V_t \quad \leq \quad \lambda(Y_{t-}, \xi)/\lambda^*.$$

Thus the acceptance rule for births is very similar to our acceptance rule for births in the immigration-death process example. The acceptance rule for deaths differs from the procedure for the immigration-death process. For spatial birth-and-death processes we can distinguish the individual elements of a configuration. Thus we may adopt the following simple procedure. Whenever a point η dies in Z we check whether this point exists in Y and if so, let it die at the same time in Y. The reader may verify that this coupling construction leads to the correct birth and death rate for Y.

The above coupling is very similar to the coupling we chose for our immigration-death process. Careful inspection leads to the observation that Y_t is always a subset of Z_t if we start it in a configuration which is a subset of Z at the starting time. We can use the set-up for a Dominated CFTP algorithm as follows. We produce a stationary path of Z on $[-T, 0]$ and mark all birth times. Then we start a path of Y at time $-T$ in every point pattern which is a subset of Z_{-T}. We evolve the paths according to the above coupling till time 0 and check for complete coalescence. If all paths have coalesced at time 0 their common state has the distribution π. If they have not coalesced, then we need to extend backwards in a similar manner as for the immigration-death process.

We can detect complete coalescence more efficiently if $\lambda(x, \xi)$ is monotone, that is $\lambda(x, \xi) \leq \lambda(y, \xi)$ if $x \subseteq y$. Then we only need to monitor a path started at time $-T$ in the empty set and a path started at time $-T$ in the point pattern Z_{-T}. If the two paths coalesce by time 0 then their common state at time 0 is a perfect sample. If $\lambda(x, \xi)$ is anti-monotone, as for example for the Strauss process, then we may use a cross-over to monitor coalescence efficiently.

For a more detailed introduction into the perfect simulation of locally stable Markov point processes see [16]. Further examples may found in [14,15] and [21]. The method is extended to random set processes in [17]. It may also be applied to general distributions, see for example [20,23] or [24].

6   Perfection in Space

When introducing the Ising model we promised the reader a simulation method which avoids edge-effects. A perfect sample in space may be achieved by extending not only backwards in time but also in space. This idea was first presented in [14] and for Markov random fields is discussed in more detail in [13] or in [36]. We explain the method using the example of a ferromagnetic Ising model.

Suppose we would like to produce a perfect sample on the $m \times m$ lattice $\Lambda = \Lambda_0$. For $k \in \mathbb{N}$ let $\Lambda_{-k-1} = \Lambda_{-k} \cup \partial(\Lambda_{-k})$ be the lattice we achieve by adding the neighbours of the boundary sites of Λ_{-k} to the lattice.

In the kth iteration of the CFTP algorithm we now perform the following procedure.

1. Sample independent random variables $U_{-T_k}, U_{-T_k+1}, \ldots, U_{-T_{k-1}-1}$ which are uniform on the unit interval and random variables $N_{-T_k}, N_{-T_k+1}, \ldots, N_{-T_{k-1}-1}$ which are uniform on the lattice Λ_{-k+1}.
2. Start one path of the Gibbs Sampler in $x_{\max}$ and one in $x_{\min}$ on the lattice Λ_{-k}. Evolve the paths from time $-T_k$ to time 0 using the transition rule h as defined in (4) together with the realisations of $U_{-T_k}, U_{-T_k+1}, \ldots U_{-1}$ and $N_{-T_k}, N_{-T_k+1}, \ldots N_{-1}$.

If the two paths coalesce at time 0 on Λ_0, then we output their common state at time 0 as a sample from the equilibrium distribution. If coalescence has not been achieved yet we extend further backwards in time and space. Figure 12 further illustrates the procedure. The reader is invited to compare this algorithm to the standard CFTP procedure on page 366.

The algorithm is set up such that from time $-T_j$ until time $-T_{j-1} - 1$ only sites on the interior of the lattice Λ_{-j} are updated but taking into account the configuration on the boundary sites.

Iteration 1:

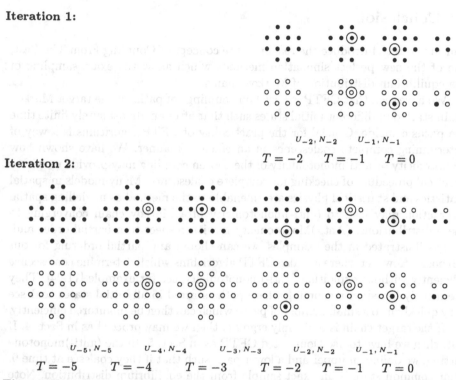

Iteration 2:

Fig. 12. Two iterations of the "perfect in space" CFTP algorithm for the ferromagnetic Ising model on a 4×4 lattice. Sites with upward spins are black and sites with downward spins white. The upper path is started in $x_{\max}$ and the lower in $x_{\min}$. The encircled site has just been updated. Only sites in the interior of each lattice are updated, taking into account the configuration on the boundary sites. We need to achieve coalescence on the centre 4×4 lattice, which for clarity we have drawn at the end of each path. In the second iteration, we reduce the site of the lattice after time $T = -2$ because from then onwards only sites of the centre 4×4 lattice are updated.

If the two paths coalesce at time 0 on Λ then their common state is not only independent from the starting configuration on Λ but also independent from any starting configuration on the infinite lattice $\mathbb{Z}^2$. For a formal proof see [13]. Of course, it can happen that we do not achieve coalescence in finite time. This crucially depends on the strength of interaction between sites. In [13] a bound on the strength of interaction is given such that coalescence in almost surely finite time is ensured.

Many refinements of the algorithm presented here are possible. If we use a cross-over then we may also apply the method to the anti-ferromagnetic Ising model. An application to general Markov random fields can be found in [13,36]. Furthermore, the point process setting is discussed in [14].

7   Conclusion

Our aim was to introduce the reader to the concept of Coupling From The Past, one of the new perfect simulation methods which allow the exact sampling of the equilibrium distribution of Markov chains.

We have seen that CFTP is based on couplings of paths of the target Markov chain started in different initial states such that after an almost surely finite time the paths coincide. Crucial for the practicality of CFTP algorithms is a way of determining complete coalescence in an efficient manner. We have shown how monotonicity or anti-monotonicity of the chosen coupling may provide us with a practical procedure of checking for complete coalescence. Many models in spatial statistics and statistical physics are amenable to the method as modelling spatial interaction may lead to (anti-)monotonicity of a Markov chain converging to these distributions. (Anti-)Monotonicity is with respect to a partial order and, as was illustrated in the examples, we can choose any partial ordering for our purposes. However, there are also CFTP algorithms which determine coalescence efficiently without exploiting (anti-)monotonicity, see for example [11,24]. They are usually based on a coupling where paths started in all initial states coalesce very quickly into a small number of paths which can then be monitored efficiently.

If the target chain is uniformly ergodic, then we may proceed as in Sect. 4. If not, then we have to use Dominated CFTP as in Sect. 5. In the (anti-)monotone setting, we define an upper and a lower path such that if they coalesce at time 0, their common state is an exact sample from the equilibrium distribution. Note that the lower and upper path do not need to evolve like the target chain if they have not coalesced yet. Before coalescence they may even live on an augmented state space, for an example see [17]. However, once they coalesce, they behave like the target chain.

All in all, there is a lot of freedom in setting up a CFTP algorithm. The challenge is to construct an algorithm which is efficient enough to be practical. An annotated bibliography which contains a multitude of examples of perfect simulation algorithms may be found on the perfect simulation website www.dimacs.rutgers.edu/~dbwilson/exact.

Acknowledgement
The author would like to acknowledge the support by the EU TMR network ERB-FMRX-CT96-0095 on "Computational and statistical methods for the analysis of spatial data".

References

1. Brooks, S.P., G.O. Roberts (1999): 'Assessing convergence of Markov Chain Monte Carlo algorithms', *Statist. Comput.* **8**, pp. 319–335
2. Cowles, M.K., B.P. Carlin (1996): 'Markov chain convergence diagnostics: a comparative review', *J. Amer. Statist. Assoc.* **91**, pp. 883–904
3. Creutz, M. (1979): 'Confinement and critical dimensionality of space-time', *Phys. Rev. Lett.* **43**, pp. 553–556
4. Daley, D.J., D. Vere-Jones (1988): *Introduction to the Theory of Point Processes* (Springer, New York)
5. Fill, J.A. (1998): 'An interruptible algorithm for exact sampling via Markov chains', *Ann. Appl. Probab.* **8**, pp. 131–162
6. Fill, J.A., M. Machida, D. J. Murdoch, J.S. Rosenthal (1999): 'Extension of Fill's perfect rejection sampling algorithm to general chains', preprint available on www.mts.jhu.edu/~fill/.
7. Fill, J.A. (1998): 'The Move-To-Front rule: A case study for two perfect sampling algorithms', *Probab. Engrg. Inform. Sci.* **12**, pp. 283–302
8. Foss, S.G., R.L. Tweedie (1998) 'Perfect simulation and backward coupling', *Stochastic Models* **14**, pp. 187–203
9. Geman, S., D. Geman (1984): 'Stochastic relaxation, Gibbs distributions and the Bayesian restoration of images', *IEEE Trans. PAMI* **6**, pp. 721–741
10. Gilks, W.R., S. Richardson, D.J. Spiegelhalter (1996): *Markov Chain Monte Carlo In Practice* (Chapman & Hall, London)
11. Green, P.J., D.J. Murdoch (1999): 'Exact sampling for Bayesian inference: towards general purpose algorithms'. In: *Bayesian Statistics 6*, ed. by Bernardo, J.M., J.O. Berger, A.P. Dawid, A.F.M. Smith (Oxford University Press, Oxford), pp. 301–321
12. Häggström, O., K. Nelander (1998): 'Exact sampling from anti-monotone systems', *Statist. Neerlandica* **52**, pp. 360–380
13. Häggström, O., J.E. Steif (1999): 'Propp-Wilson algorithms and finitary codings for high noise random fields', to appear in *Combin. Probab. Comput.*
14. Kendall, W.S. (1997): 'Perfect simulation for spatial point processes'. In: *Proc. ISI 51st session, Istanbul (August 1997)*, volume 3, pp. 163–166
15. Kendall, W.S. (1998): 'Perfect simulation for the area-interaction point process'. In: *Probability Towards 2000*, ed. by L. Accardi, C.C. Heyde (Springer, New York), pp. 218–234
16. Kendall, W.S., J. Møller (1999): 'Perfect Metropolis-Hastings simulation of locally stable point processes', Research report 347, University of Warwick
17. Kendall, W.S., E. Thönnes (1999): 'Perfect simulation in stochastic geometry', *Pattern Recognition* bf 32, pp. 1569–1586
18. Lindvall, T. (1992): *Lectures On The Coupling Method*. Wiley Series in Probability and Mathematical Statistics (John Wiley & Sons)
19. Meyn, S.P., R.L. Tweedie (1993): *Markov Chains and Stochastic Stability* (Springer Verlag, New York)
20. Mira, A., J. Møller, G. O. Roberts (1998): 'Perfect slice samplers', preprint, available on www.dimacs.rutgers.edu/~dbwilson/exact
21. Møller, J. (1999): 'Markov Chain Monte Carlo and spatial point processes'. In: *Stochastic Geometry: Likelihood And Computation*, ed. by W.S. Kendall, O.E. Barndorff-Nielsen, M.N.M. van Lieshout. Proceedings Séminaire Européen de Statistique (Chapman & Hall/CRC, Boca Raton), pp. 141–172

22. Møller, J., K. Schladitz (1998): 'Extensions of Fill's algorithm for perfect simulation', preprint, to appear in *J. Roy. Statist. Soc.* **B**

23. Møller, J. (1999): 'Perfect simulation of conditionally specified models', *J. Roy. Statist. Soc.* **B 61**, pp. 251–264

24. Murdoch, D.J., P.J. Green (1998): 'Exact sampling from a continuous state space', *Scand. J. Statist.* **25**,pp. 483–502

25. Murdoch, D.J., J.S. Rosenthal (1998): 'An extension of Fill's exact sampling algorithm to non-monotone chains', preprint

26. Norris, J.R. (1997): *Markov Chains*. Cambridge Series in Statistical and Probabilistic Mathematics (Cambridge University Press, Cambridge)

27. Propp, J.G., D.B. Wilson (1996) 'Exact sampling with coupled Markov chains and applications to statistical mechanics', *Random Structures Algorithms* **9**, pp. 223–252

28. Ripley, B.D. (1987): *Stochastic Simulation* (John Wiley & Sons, New York)

29. Roberts, G.O., N.G. Polson (1994): 'On the geometric convergence of the Gibbs sampler', *J. Roy. Statist. Soc.* **B 56**, pp. 377–384

30. Rosenthal, J.S. (1995): 'Minorization conditions and convergence rates for Markov chain Monte Carlo', *J. Amer. Statist. Assoc.* **90**, pp. 558–566

31. Saloff-Coste, L. (1999): 'Simple examples of the use of Nash inequalities for finite Markov chains'. In: *Stochastic Geometry: Likelihood and Computation*, ed. by O. Barndorff-Nielsen, W.S. Kendall, M.N.M van Lieshout (Chapman & Hall/CRC, Boca Raton), pp. 365–400

32. Stoyan, D. (1983): *Comparison Methods for Queues and Other Stochastic Models*. Wiley Series in Probability and Mathematical Statistics (John Wiley & Sons, Chichester)

33. Stoyan, D., W.S. Kendall, J. Mecke (1995) *Stochastic Geometry and its Applications*. Wiley Series in Probability and Mathematical Statistics (John Wiley & Sons, Chichester), 2nd ed.

34. Stoyan, D., H. Stoyan (1994): *Fractals, Random Shapes and Point Fields* (John Wiley & Sons, Chichester)

35. Thönnes, E. (1999): 'Perfect simulation of some point processes for the impatient user', *Adv. Appl. Probab.* **31**, pp. 69–87

36. van den Berg, J., J.E. Steif (1999): 'On the existence and non-existence of finitary codings for a class of random fields', to appear in *Ann. Probab.*

37. Winkler, G. (1991): *Image Analysis, Random Fields and Monte Carlo Methods*. Applications of Mathematics (Springer, Berlin)

Grand Canonical Simulations of Hard-Disk Systems by Simulated Tempering

Gunter Döge

Institute of Stochastics, Freiberg University of Mining and Technology,
Bernhard-von-Cotta-Str. 2, D-09596 Freiberg, Germany

Abstract. For the simulation of hard core Gibbs point processes simulated tempering is shown to be an efficient alternative to commonly used Markov chain Monte Carlo algorithms. The behaviour of the area fraction and various spatial characteristics of the hard core process is studied using simulated samples.

1 Introduction

The nature of the two-dimensional melting transition has been a matter of hot debate for the past three decades. Whereas a conventional first-order transition between the isotropic liquid and the solid was assumed since the pioneering work in 1962 by Alder and Wainwright [2], the scenario of dissociation of dislocations and disclinations in the solid phase was proposed 1979 by Nelson, Halperin and Young [22] which leads to two second order transitions according to the theory by Kosterlitz and Thouless [15]. The intermediate, so-called 'hexatic', phase displays an exponential decay of translational order but an algebraic decay of bond orientational order, in contrast to the two-dimensional solid phase, where the bond orientational order is long ranged and the translational order decays algebraicly. Even for a very simple system like hard disks, the issue of the order of the melting transition is not settled yet. For instance, hysteresis loops occurring at first order transitions depend strongly on systems size in any simulation at fixed particle numbers. Therefore, large scale computer simulations or novel algorithms are needed which circumvent the ambiguities of the results obtained by conventional techniques.

Simulation algorithms such as the usual Metropolis algorithm (M) [20] and molecular dynamics (MD) [3,25] play an important role in statistical physics (see [5,6,8]). These methods are also important tools in the study of canonical Gibbs hard-disk and hard-sphere systems with a fixed number of objects in a bounded set. Recent Monte Carlo simulations were done in the NVT ensemble (constant volume) [30,28,29] and in the NPT ensemble (constant pressure) [16,7] but none in the grand-canonical μVT ensemble (constant chemical potential), because of the obvious difficulty to add a hard core particle at high densities. Although numerical investigations of the two-dimensional melting could be done in several ways, these simulations do not give conclusive results neither on the location of the melting and freezing densities nor on the nature of the transition itself.

The aim of this paper is to point out that previous simulations performed at fixed particle number have several serious drawbacks and that grand canonical simulations may help to overcome these difficulties.

Two-dimensional solids differ from three-dimensional ones in lacking long-range translational order which arises from a divergence of long-wavelength fluctuations. Therefore, large system sizes are important for simulations of two-dimensional systems at fixed particle number in order to allow for significant fluctuation contributions to the entropy. It is found in many simulations that for small particle number the system exhibits significantly more order than for larger numbers with the same density. In order to explore the properties of these less ordered states one has either to simulate very large ensembles or to allow for fluctuations in the particle number by sampling a grand canonical system. Otherwise the solid structure is stabilized what eventually may suppress a transition into a less ordered phase completely until the solid structure becomes unstable at the melting density η_f. This results in a tie line between a liquid and a solid phase which moves to smaller chemical potentials μ and becomes shorter than in an equilibrium simulation.

In particular at high densities, simulations with a fixed number of particles suppress in a systematic way configurational fluctuations. Similar to the single-occupancy cell approximation, particles are almost surely constrained to motions within a single cell defined by their neighbours. This unrealistic restriction is due to the low acceptance rate of large moves in conventional Monte-Carlo simulations, i.e. of deleting and adding a particle at another position. Usually all particles are moved maximally only one tenth of the radius at each update in order to achive an acceptable rate, which nevertheless makes the convergence very slow and suppress large scale fluctuations. Thus in usual computer simulations with fixed particle number each particle almost never leaves its cell. This is obviously the case for fixed volume simulations (NVT ensemble) but even in fixed pressure simulations (NPT ensemble) only an overall scaling of distances, i.e. an isotropic change of volume is performed which does not change the configuration or the respective arrangement of the particles. Thus we expect the solid phase to be stabilized by all simulation techniques so far applied, making the need of another method obvious.

The second possibility, namely simulating a grand canonical ensemble with fluctuating particle number, is commonly rejected because of 'obvious reasons'. But we will show in Sect. 2 that simulated tempering (ST) makes such simulations possible even for hard disks at high densities, so that grand canonical simulations are not hampered in the high-density regime anymore.

Due to the irregular spatial structure of a fluid it is difficult to invent a simulation method (such as a grand canonical simulated tempering technique) which stabilizes a fluid in the transition regime. Such a technique would allow to determine η_s more precisely than η_f and would be a complementary approach to simulations at fixed particle number.

Since one can suspect that most simulations stabilize the solid phase at densities for which it would melt already (i.e., determining η_f is possibly more precise than η_s), it seems to be necessary to implement a simulation technique which stabilizes the fluid phase, i.e., which allows for metastable fluid configurations but decreases the stability of solid configurations. Large systems sizes are expected to be not relevant for such a simulation method since large scale fluctuations are introduced by particle exchange, i.e., deleting and adding hard disks.

There are two other important reasons for an implementation of grand canonical simulations, first, the direct determination of the entropy in the solid phase and, second, to overcome the constraints on the lattice structure induced by periodic boundary conditions.

In the fluid phase both the pressure and the free energy (or entropy) are known since one has a continuous path to the ideal gas limit which can serve as a reference system with known thermodynamic quantities. But in the solid phase the entropy is by integration techniques also known, except for an additive constant which cannot be determined directly. This difficulty arises because the entropy, a part of the chemical potential, is not a function of coordinates (dynamical variable) which can be averaged but is defined relative to a reference system by integrating along a reversible path to the considered state. In simulations with a fixed number of particles one has no possibility to determine the entropy and thus to compare the amount of disorder with analytical results on the solid side of the transition. This is only possible for grand canonical simulations. Additionally, a grand canonical simulation would not only allow the direct determination of the entropy and the additional constant in the solid phase but also the measurement of the van-der-Waals loop $\eta(\mu)$ in the transition regime.

In order to minimize the influence of the simulation box upon the spatial structure of the solid phase one can apply periodic boundary conditions on a rectangular box of aspect ratio $\sqrt{3} : 2$ allowing hexagonal lattice structures. Nevertheless, in canonical systems with fixed particle numbers such devices are by no means sufficient to avoid influences of the boundary conditions on the solid structure. For instance, the net number of vacancies, i.e., the difference of vacancies and interstitial particles is constant [26]. Since in most canonical simulations the net number of vacancies is set equal to zero, the configurations will probably never reach full equilibrium globally. The hope of applying large scale simulations is that local subsystems exhibit random boundaries and resemble a multicanonical system which is large enough to reach equilibrium values of thermodynamic quantities. In principle one could avoid such problems by applying grand canonical simulations.

The grand canonical case with a random number of objects cannot be attacked by MD and is difficult with M in the neighborhood of the melting and freezing point. In this situation and probably in many other cases a novel method called simulated tempering (ST) [10,18] may be helpful. It is able to improve greatly the mixing properties of M.

The idea of ST applied to a system with a high degree of order is to carry out simulations for a series of coupled systems which work under different 'temperatures'. The 'coldest' system is that system which has really to be simulated, where M has unsatisfactory mixing properties, while the 'hotest' is completely random. In the case of simulation of hard-disk systems, overlappings of disks in different extent in the stages above the cold system are possible. During the simulation, the process moves randomly between the various systems, producing samples of the cold systems during stays in that system. The price of this procedure is a more sophisticated simulation approach and a bit more computer storage space. Perhaps, the simulated complex system with the various stages may have an own physical meaning.

Note that ST is different from simulated annealing: during a ST simulation the temperature changes in both directions (cold and warm) even in the equilibrium state, while Simulated annealing is based on continuously decreasing temperature.

The aim of the present paper is to explain the application of ST to simulation of grand canonical Gibbs hard-disk systems. In Sect. 2 we introduce and compare the algorithms. The results of simulated tempering simulations are reported in Sect. 4. It represents, for instance, the functional relationship between packing fraction η and the chemical potential μ. For values of μ below the freezing point the Padé approximation for the fluid phase [12,13] and the simulation results are in very good agreement. Similarly, a cell model approximation and the simulation results for μ above the melting point are close together. Finally, the ST simulations produce the same pair correlation functions as those presented in Truskett et al. [27], which are obtained by MD.

2 Grand Canonical Simulations for Hard-Disk Systems

The hard core Gibbs process is a common model for systems of hard disks ($d = 2$) or hard spheres ($d = 3$). The local Gibbs process is defined on a bounded region W and has the unnormalized density

$$f(\varphi) = \exp\left(\mu \cdot \#\varphi - \sum_{1 \leq i < j \leq \#\varphi} V\left(|x_i - x_j|\right)\right),\tag{1}$$

for $\varphi = \{x_1, \ldots, x_{\#\varphi}\} \subset W$, where $\#\varphi$ is the number of disks. The parameter μ is called *chemical potential*, it is related to the *chemical activity* $z = e^\mu$. Larger values of μ mean closer packings of disks; the value $\mu = \infty$ corresponds to the maximum packing, a hexagonal pattern of disks in the two-dimensional case. The (normalized) probability density for a variable number of disks is given by $f(\varphi)/\Xi$, where

$$\Xi = \sum_{\substack{N=0 \\ (\#\varphi=N)}}^{\infty} \frac{1}{|W|^N N!} \int_{W^N} f(\varphi)\, d\varphi$$

is the well-known *grand partition function* on W, where $|W|$ is the d-dimensional volume of W. For a fixed number $\#\varphi = N$ of disks the *canonical partition function*

$$Z = \frac{1}{|W|^N N!} \int_{W^N} f(\varphi)\, d\varphi$$

is used as normalizing factor instead of Ξ, see the paper of H. Löwen in this volume. The pair potential V for the hard core Gibbs process with disk diameter σ is

$$V(r) = \begin{cases} \infty, & r < \sigma \\ 0, & r \geq \sigma. \end{cases}$$

The simulations are performed for case $d = 2$ on a square $W = [0, a]^2$ with periodic boundary conditions, i.e. our target density with respect to the unit rate Poisson process on W is proportional to

$$f(\varphi) = \begin{cases} \exp\left(\mu \cdot \#\varphi\right), & \text{if } |x_i - x_j| > \sigma \text{ for all } i \neq j \\ 0 & \text{else}. \end{cases} \tag{2}$$

Here and henceforth,

$$|x|^2 = \left(\min(x_1, a - x_1)\right)^2 + \left(\min(x_2, a - x_2)\right)^2, \quad x = (x_1, x_2) \in W,$$

denotes the geodesic distance when W is wrapped on a torus.

Samples of this local process are used as representatives of the global hard core Gibbs process Φ, ignoring edge effects. $\lambda = \lambda(\mu)$, denotes the intensity, the mean number of disks per area, as a function of μ. Equivalently we may consider the area fraction

$$\eta(\mu) = \lambda(\mu)\pi\sigma^2/4 \tag{3}$$

of the set of all disks; note that $\lambda(\mu)$ but not $\eta(\mu)$ depends on the diameter σ. $\eta = \eta(\mu)$ is a decreasing function, see Sect. 4.2 of the paper by H.-O. Georgii in this volume. Many other questions concerning the qualitative behaviour of $\eta(\mu)$ are still open, in particular the question concerning existence of discontinuities of $\eta(\mu)$ or its derivatives.

The simulation problem becomes difficult when μ increases. In order to obtain a simulation algorithm with good mixing properties we combine various Metropolis-Hastings algorithms with simulated tempering, as described in Sect. 3. The experimental results concerning area fraction, pair correlation function and some other characteristics are discussed in Sect. 4.

3   Algorithms

Our basic algorithm is the Metropolis-Hastings algorithm (MH) studied in [9,11] and [21]. Section 3.1 provides a short description of this algorithm; note that it is applicable for the canonical ensemble, i.e. fixed number of disks, as well as the grand canonical ensemble, i.e. when the number of points fluctuates. However, in this paper the grand canonical ensemble is mainly considered, as mentioned and explained in the introduction. The combination of the MH algorithm and simulated tempering [10,18] is introduced in Sect. 3.2.

3.1 Basic Algorithm

To understand the Simulated Tempering (ST) algorithm, it is convenient to describe the MH algorithm for the case when we want to simulate samples from any unnormalized density f of a point process on W (in particular (1)), i.e. f is a non-negative integrable function with respect to the unit rate Poisson process on W.

Assume that φ with $f(\varphi) > 0$ is the current state of the Markov chain generated by the MH algorithm. It is then proposed to either **(a)** add, **(b)** delete, or **(c)** move a point with probabilities $p_1(\varphi)$, $p_2(\varphi)$, and $1 - p_1(\varphi) - p_2(\varphi)$, respectively. The proposal φ' for the next state in the chain is given as follows:

(a) $\varphi' = \varphi \cup \{x\}$ where the new point $x \in W$ is sampled from a density $b(\varphi, \cdot)$ on W;

(b) $\varphi' = \varphi \setminus \{x\}$ where $x \in \varphi$ is chosen with probability $d(\varphi, x)$ (if $\varphi = \emptyset$ we set $\varphi' = \varphi$);

(c) $\varphi' = (\varphi \setminus \{x\}) \cup \{y\}$ where $x \in \varphi$ is chosen with probability $d(\varphi, x)$ and y is sampled from a density $m(\varphi \setminus \{x\}, x, \cdot)$ (if $\varphi = \emptyset$ we set $\varphi' = \emptyset$).

The probability resp. density functions b, d and m can be chosen arbitrarily under mild regularity conditions. The proposed state φ' is finally accepted with probabilities $\min\{1, r(\varphi, \varphi')\}$, where the Hastings ratio $r(\varphi, \varphi')$ depends on the type of transition and is given by

(a) $\dfrac{f(\varphi')p_2(\varphi')d(\varphi', x)}{f(\varphi)p_1(\varphi)b(\varphi, x)}$;

(b) $\dfrac{f(\varphi')p_1(\varphi')b(\varphi', x)}{f(\varphi)p_2(\varphi)d(\varphi, x)}$ (if $\varphi = \emptyset$ then $r(\varphi, \varphi') = r(\emptyset, \emptyset) = 1$);

(c) $\dfrac{f(\varphi')\left(1 - p_1(\varphi') - p_2(\varphi')\right)d(\varphi', y)m(\varphi \setminus \{x\}, y, x)}{f(\varphi)\left(1 - p_1(\varphi) - p_2(\varphi)\right)d(\varphi, x)m(\varphi \setminus \{x\}, x, y)}$.

If φ' is rejected, the Markov chain remains in φ. These acceptance probabilities are affected by the choice of b, d and m; a bad choice can cause inefficient low acceptance probabilities, i.e., a Markov chain with bad mixing properties.

In the simulations, $p_1(\varphi) = p_2(\varphi) = p$ are constant; the densities $d(\varphi, \cdot)$ and $b(\varphi, \cdot)$ are uniform on φ and W, respectively; and the density $m(\varphi \setminus \{x\}, x, \cdot)$ is uniform on a square of side length $2 \times \epsilon$ centered in x, where ϵ is small with respect to the edge length a of W. Note that the Metropolis algorithm [20] is the special case $p = 0$ where the number of points is fixed.

Theoretical properties of the MH algorithm are studied in [9,11] and [21]. By construction the Markov chain is reversible with invariant density specified by f with respect to the unit rate Poisson process on W if $p > 0$ or with respect to the Bernoulli process on W if $p = 0$. In particular, if f is the target density (2), the Markov chain is uniformly ergodic when $p > 0$, and also when $p = 0$ provided that σ is sufficiently small (this is needed to ensure irreducibility).

For large values of μ, however, the chain, despite the property of uniform ergodicity, converges very slowly and produces highly autocorrelated samples.

3.2   Simulated Tempering

Much better results are obtained when the MH algorithm is combined with simulated tempering as described in the following.

The equilibrium distribution of our implementation of simulated tempering is a mixture of distributions of repulsive point process models with unnormalized densities $f_1, \ldots, f_n$, $n \geq 2$, where the MH algorithm for f_i mixes well when i is small, while it produces highly autocorrelated samples when i increases towards n. Specifically, for $i = 1, \ldots, n$,

$$f_i(\varphi) = \exp\left(\mu_i \cdot \#\varphi - \frac{\gamma_i}{2} \sum_{\substack{x,y \in \varphi \\ x \neq y}} \left[\mathbf{1}(|x - y| \leq \sigma) + c \frac{|\mathcal{B}(x, \sigma/2) \cap \mathcal{B}(y, \sigma/2)|}{|\mathcal{B}(0, \sigma/2)|} \right] \right)$$

with $0 = \gamma_1 < \gamma_2 < \cdots < \gamma_{n-1} < \gamma_n = \infty$ and $c > 0$, and where we set $0 \times \infty = 0$. $\mathcal{B}(x, r)$ denotes the disk with centre x and radius r and $|\cdot|$ means the area. The terms $\gamma_i \mathbf{1}(|x - y| \leq \sigma)$ and $\gamma_i c |\mathcal{B}(x, \sigma/2) \cap \mathcal{B}(y, \sigma/2)|/|\mathcal{B}(0, \sigma/2)|$ both introduce a penalty whenever two disks overlap; the latter term enables us to distinguish between point patterns with the same number of overlapping pairs of disks, but where the degree of overlap differs. In particular, f_n is the target density with $\mu = \mu_n$, while f_1 specifies the Poisson process with rate $\exp(\mu_1)$. The penalizing parameter γ_i is, by analogy with physics, referred to as an inverse temperature, so that the Poisson process is the "hot" distribution and the target process is the "cold" distribution. For the simulations reported in this paper, the value $c = 10$ was chosen as a result of some pilot simulations. Below is discussed how to choose the other parameters in order to obtain an algorithm which inherits the good mixing properties of the MH algorithms for small i.

Simulated tempering generates a Metropolis-Hastings chain $(X_l, I_l)_{l \geq 0}$, where I is a 'auxiliary variable'; X describes the disk configuration and I the value of i. The equilibrium distribution of this chain is given by the (unnormalized) density

$$\tilde{f}(\varphi, i) = f_i(\varphi)\delta_i, \quad i = 1, \ldots n,$$

where the $\delta_i > 0$ are specified as follows. Suppose that $(X, I) \sim \tilde{f}$. The marginal distribution of X is then the mixture $\sum_{i=1}^{n} f_i \delta_i$; f_i is the (unnormalized) conditional density of $X|I = i$; and $P(I = i) \propto \delta_i c_i$, where c_i denotes the normalizing constant (grand partition function) of f_i. Estimates $\hat{c}_i$ of c_i can up to a constant of proportionality be obtained in different ways as described in [10]. One possibility is to use stochastic approximation, another is reverse logistic regression [11] where the normalizing constants are estimated from preliminary samples obtained with Metropolis-coupled Markov chains. Our experience is that stochastic approximation is not feasible for large n while reverse logistic regression is computationally demanding but secure. By choosing $\delta_i = 1/\hat{c}_i$ an approximate uniform mixture is obtained.

Now, for the simulated tempering algorithm a proposal kernel Q on $\{1, \ldots, n\}$ is defined by $Q(i, i + 1) = Q(i, i - 1) = 1/2$ for $1 < i < n$ and $Q(1, 2) = Q(n, n - 1) = 1$. Given a current state (φ, i), the two components are updated in turn using first the MH update $\varphi \to \varphi'$ using the density $f_i(\cdot)$, and secondly the kernel Q is used to propose an update $(\varphi', i) \to (\varphi', i')$: we return (φ', i') with probability $\min\{1, r(i, i'|\varphi')\}$ and retain (φ', i) otherwise, where $r(i, i'|\varphi') = \tilde{f}(\varphi', i')Q(i', i)/(\tilde{f}(\varphi', i)Q(i, i'))$. By construction the Markov chain (X_l, I_l) is reversible with invariant density $\tilde{f}$; in particular $(X_l)_{l \geq 1: I_l = n}$ has equilibrium density f_n.

Regeneration may be useful for estimation of Monte Carlo errors as explained in [10]. In the ST algorithm a regeneration step can be done at the hot temperature, i.e. if the simulated tempering chain reaches a state (X_l, I_l) with $I_l = 1$, the point pattern X_l is replaced by a completely new generated point pattern with the Poisson process density f_1.

Provided that the pairs of parameter values (μ_i, γ_i) and $(\mu_{i+1}, \gamma_{i+1})$ are chosen sufficiently close so that reasonable acceptance rates between 20% and 40% for transitions $(\varphi, i) \leftrightarrow (\varphi, i \pm 1)$ are obtained, $(X_l)_{l \geq 1: I_l = n}$ yields a well-mixed sample from the target model f_n. Let (p_i, ϵ_i) denote the parameter values of the MH algorithms combined in the simulated tempering algorithm, $i = 1, \ldots, n$. We choose the parameter ϵ_i to be decreasing as a function of i so that reasonable acceptance rates for proposed moves are obtained for each temperature. The values of p_i are also taken to be decreasing since insertion or deletion proposals have low acceptance probabilities for the low temperatures. The intensity of the Poisson process with density f_1 is chosen as $\exp(\mu_1) = 1/\sigma^2$. This value corresponds to the area fraction $\eta = \pi/4 = 0.785$ of a planar global pure hard core Gibbs process with the same intensity. The remaining parameters are chosen as

$$\mu_i = \mu_1 + t_i (\mu_n - \mu_1)$$

and

$$\gamma_i = \begin{cases} t_i \gamma^* & \text{for } 1 \leq i < n \\ \infty & \text{for } i = n \end{cases}$$

with n normalized 'temperatures' $0 = t_1 < t_2 < \ldots < t_n = 1$ and a value of γ^* such that there are almost no overlapping disks in the $(n - 1)$th chain $(X_l)_{l \geq 1: I_l = n-1}$. Finally, the adjustment of n and $(t_i)_{i=1,\ldots,n}$ to obtain reasonable acceptance rates for transitions $(\varphi, i) \leftrightarrow (\varphi, i \pm 1)$ are done similarly to [10, Sect. 2.3]. Typically the number n of temperatures lies between 20 and 50.

4   Results

In the following we report on some simulated results for the hard-core model (2):

- area fraction (Sect. 4.1),
- pair correlation function (Sect. 4.2),
- point process order characteristics: alignment function (Sect. 4.3) and hexagonality statistics (Sect. 4.4).

These characteristics seem to describe various aspects of the phase transition described in the introduction. Moreover, these characteristics (except area fraction) measure the degree of order in the point patterns compared to an equilateral triangular lattice.

For each considered value of μ we used the simulated tempering algorithm for the grand canonical ensemble (i.e. all $p_i > 0$) in the simulation study of the area fraction, while for the other characteristics we used the less computer intensive method of simulated tempering for the canonical ensemble (i.e. all $p_i = 0$).

4.1   Area Fraction

Because of the periodic boundary condition the area fraction $\eta(\mu)$ can be estimated by

$$\hat{\eta}(\mu) = \frac{\pi\sigma^2 \overline{N}(\mu)}{4|W|}$$

where $\overline{N}(\mu)$ is the empirical mean number of disks in the cold chain. For $W = [0,10]^2$, $\sigma = 1$ and μ between 3 and 14 the increasing curve of estimated area fraction $\hat{\eta}(\mu)$ shown in Fig. 1 is obtained. For each considered value of μ, the simulated tempering chain has a length between 5×10^8 and 10^{10}. The calculation of $\overline{N}(\mu)$ is based on point samples φ at the lowest temperature, i.e., when $I = n$. The numbers of such point samples are usually between 2×10^7 and 10^8; here we used an appropriate burn-in (about 10% of the sample).

Figure 1 shows that for a wide range of μ values the curve of $\hat{\eta}(\mu)$ nearly coincides with the curve obtained by a Padé approximation:

$$\mu = \log\frac{4\eta}{\pi\sigma^2} + \frac{4\eta - 6.04\eta^2 + 3.1936\eta^3 - 0.59616\eta^4 + 0.03456\eta^5}{(1 - 1.34\eta + 0.36\eta^2)^2}.$$

This approximation is derived from

$$\mu = \lambda\frac{\partial\beta F/N}{\partial\lambda} + \frac{\beta F}{N}$$

(Hansen and McDonald [12]) using a Padé approximation

$$\frac{\beta F}{N} = \log\lambda - 1 + b\lambda\frac{1 - 0.28b\lambda + 0.006b^2\lambda^2}{1 - 0.67b\lambda + 0.09b^2\lambda^2}$$

from Hoover and Ree [13]. λ is the intensity (see (3)) and $b = \pi\sigma^2/2$. The cell approximation is based on disk configurations in the solid phase, i.e., for high area fractions near the maximum. It is described in Sect. 3.2 of the paper by H. Löwen in this volume and is based on the fact, that a disk never leave its (Voronoi) cell in the solid phase, i.e., the topology of the neighbourhood relations between disks is stable over the time.

The graph indicates that for values of $\mu < 9$ (corresponding to $\eta < 0.65$) both the Padé formula and our simulations yield good approximations of $\eta(\mu)$. Notice the change in the $\hat{\eta}(\mu)$ curve at values close to the freezing point $\eta = 0.69$ and

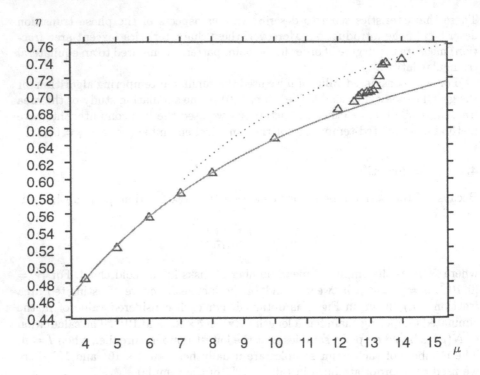

Fig. 1. Estimated values of $\eta(\mu)$ using simulated tempering ($\triangle$), Padé approximation (solid line) and cell approximation (dotted line).

the melting point point $\eta = 0.716$ mentioned after equation (3); in particular this may indicate a discontinuity in the curve at the melting point. The behaviour of the curve of $\hat{\eta}(\mu)$ for $\mu > 13.3$ raises doubt about if sufficiently long simulated tempering chains have been used for the values of $\mu > 13.3$. Particularly we believe that the curve should increase further and not show a flat behaviour as for the largest values of μ in Fig. 1.

4.2   Pair Correlation Function

For the results in this and subsequent sections we used $W = [0, 20]^2$, $\sigma = 1$ and determined the number N of points for every value of area fraction $\eta = 0.65$, 0.67, 0.69, 0.696, 0.701, 0.707, 0.71, 0.715, 0.721, 0.735, so that $\eta = \pi\sigma^2 N/(4|W|)$ in accordance with (3). Hence N is ranging from 331 to 374. For the estimation of each considered statistic (pair correlation function, hexagonality number, and so on) we used for various reasons subsamples of 100 point patterns of the cold chain obtained from mucher longer runs of the ST chain: The estimation was first done for each point pattern, then we averaged over the 100 estimates; but the cost of the estimation for one point pattern is about 500 times the cost of one step in the simulated tempering chain. Moreover, because of the small changes in the simulated tempering chain, estimates based on subsequent point patterns look almost the same. Therefore, for the 100 point patterns, we used a spacing of at least $10Nn$.

The pair correlation function g is a well-known characteristic for point processes, see e.g. [24,25] and [27]. In $\mathbf{R}^2$, assuming stationarity and isotropy of the hard core Gibbs point process, $\lambda(\mu)^2 g(r)(\pi\delta^2)^2$ can be interpreted as the probability of observing a point in each of two infinitesimally small disks of radius δ and with arbitrary but fixed centers located in distance $r > 0$ from each other. Under the target model (2), when the number of points is fixed and W is identified with a torus, $(N/|W|)^2 g(r)(\pi\delta^2)^2$ has a similar interpretation (using the periodic boundary condition when calculating inter point distances).

The pair correlation function can be estimated by non-parametric kernel methods as described in [25] apart from the following modifications. We replaced the intensity λ with $N/|W|$. Furthermore, because of the the high number of points per sample and since we averaged over 100 samples, we used a very small band width in the kernel (of value 0.03, see [25], page 285). Reducing the band width reduces the bias in the estimator at the cost of higher variances, by the averaging we still obtain a smooth curve. Furthermore, because of the averaging, the variance in the estimator is substantially reduced.

Estimated pair correlation functions with $\eta = 0.65$ and $\eta = 0.735$ are shown in Fig. 2. As expected, with increasing η the pair correlation function reflects more order. The peaks of the estimated pair correlation functions can be compared with the modes at $r = 1, 1.732, 2, \ldots$ for the pair correlation function of the limiting regular hexagonal pattern of hard disks with diameter $\sigma = 1$. Clearly the curve for $\eta = 0.735$ is in better agreement to the limiting case than the curve for $\eta = 0.65$. In particular, the second mode for the curve with $\eta = 0.65$ splits into two modes as η increases.

As mentioned at the end of Sect. 1, in statistical physics mostly the ordinary Metropolis algorithm (i.e. the MH algorithm for the canonical ensemble) and molecular dynamics have been used for simulations. Molecular dynamics (MD) is based on the equations of motion of N molecules described by a local Gibbs process (see e.g. [3,26]). The theoretical convergence properties of MD are not well understood, but numerical evidence, e.g. obtained by comparison with results produced by the ordinary Metropolis algorithm, supports that MD produces reliable results for pure hard core Gibbs processes (Torquato, 1998, personal communication). This is also supported by our results: Figure 2 is in agreement with the results in [28] obtained by MD.

4.3   Alignment Function

The alignment function $z_B(r)$ is a kind of third-order characteristic which is well adapted to show if there are linear chains of points as for lattice-like point patterns [25]. For $r > 0$, consider any $r \in \mathbf{R}^2$ with $\|r\| = r$ and let B_r be a square centered at $r/2$ and of side length ξr, where one side is parallel to r and $0 < \xi < 1$ is a user-specified parameter. In $\mathbf{R}^2$, assuming stationarity and isotropy of the hard core Gibbs point process, $\lambda(\mu)|B_r|z_B(r)$ can be interpreted

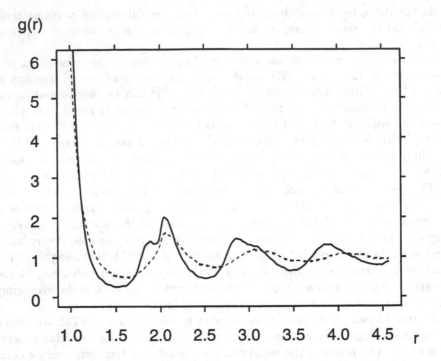

Fig. 2. Estimated pair correlation functions for $\eta = 0.65$ (dashed line) and $\eta = 0.735$ (solid line).

as the mean number of points in B_r under the condition that there is a point in each of the locations $\mathbf{o} = (0,0)$ and $\mathbf{r}$. For $r = 2$, this mean number is exactly 1 in the limiting case of a regular hexagonal pattern of disks with diameter 1. For a stationary Poisson point process we have that $z_B \equiv 1$, while if e.g. $z_B(r) > 1$, then B_r contains on the average more points than an arbitrarily placed rectangle of the same area. Large and small values of $z_B(r)$ for suitable r may thus indicate a tendency of alignment in the point pattern. In particular, if ξ is sufficiently small, one may expect $z_B(2)$ to be an increasing function of η with limit $0.2165/\xi^2$ obtained at the maximal area fraction $\eta = 0.907$.

The statistical estimation of $z_B(r)$ follows the same lines as in [25], page 294, except that we again replace $\lambda(\mu)$ with $N/|W|$ (since the number of points is fixed) and use the torus convention. After some experimentation we decided to use $\xi = 0.1$.

Simulations show as expected that $z_B(2)$ increases with increasing η; but it is $z_B(2) = 1.83$ for $\eta = 0.735$ and this is still far from the maximum value 21.65 obtained at $\eta = 0.907$. The alignment of the point patterns is more apparent for slightly increased r, e.g. $r = 2.2$. Figure 3 shows estimates of $z_B(2.2)$ and $z_B(3)$ as functions of η. Also $z_B(2.2)$ is an increasing function of η. Note that the curve of $z_B(2.2)$ is steepest for values of η between the freezing and melting points. The value 14.96 of $z_B(2.2)$ for $\eta = 0.735$ is not very far from the upper bound 17.89 obtained by assuming that $\lambda(\mu)|B_r|z_B(r) \leq 1$ (which holds as $\mu \to \infty$). However the curve of $z_B(3)$ decreases nearly linearly and slowly towards 0; perhaps surprisingly, this curve does not show any change at the freezing and melting points.

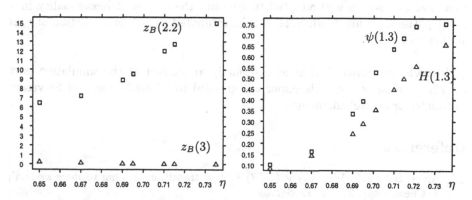

Fig. 3. The alignment functions $z_B(2.2)$ ($\square$) and $z_B(3)$ ($\triangle$) versus area fraction (left figure). Estimated hexagonality number $H(1.3)$ ($\triangle$) and hexagonality statistic $\psi(1.3)$ ($\square$) versus area fraction (right figure).

4.4 Hexagonality Number

The idea behind any hexagonality characteristic is to look for deviations from the hexagonal arrangement of neighbouring points to a point in an equilateral triangular lattice.

A first possibility is to use Ripley's K function (see e.g. [25]). In $\mathbf{R}^2$, assuming stationarity, $\lambda(\mu)K(r)$ is the mean number of points in a disk of radius r centred at the typical point (which is not counted). It vanishes for $r < 1$ and takes the value 6 for values of r a bit larger than 1 in the case of an equilateral triangular lattice with side length 2. Thus, for the hard core Gibbs point process when r is a bit larger than 1 and the number of points is fixed, one should expect an abrupt change of the values of $K(r)$ for η in the phase transition region. This, however, was not observed in our simulations, where we observed for $K(r)$ a continuous and nearly linear dependence of η.

Quite different is the behaviour of the hexagonality number $H(r)$, the *probability* that a disk of radius r centred at the typical point contains exactly 6 other points. Figure 3 shows the estimated $H(1.3)$ as an increasing function of η. The curve is steepest when η is between the freezing and melting points.

Weber *et al.* (see [30]) consider the bond-orientational number $\psi(r)$ defined as the norm of the mean of the following sum taken over all points of the hard core Gibbs point process contained in a disk of radius r centred at the typical point:

$$\sum_j e^{6i\phi_j}$$

where i denotes the imaginary unit and ϕ_j is the angle between the x-axis and the line through the typical point and the jth point contained in the disk. Clearly, this characteristic is well adapted to quantify the degree of hexagonality in a point pattern. Figure 3, which shows the estimated $\psi(1.3)$ as a function of η, is similar to Fig. 3.

Acknowledgement: This article is mainly an extract of the simulation part of [19]. The authors of this paper are grateful to Klaus Mecke and Salvatore Torquato for helpful comments.

References

1. Alder, B.J., T.E. Wainwright (1957): 'Phase transition of a hard sphere system', J. Chem. Phys. **27**, pp. 1208–1209
2. Alder, B.J., T.E. Wainwright (1962): 'Phase transition in elastic disks', Phys. Rev. **127**, pp. 359–361
3. Allen, M.P., D.J. Tildesley (1987): *Computer Simulation of Liquids* (Oxford University Press, Oxford)
4. Bagchi, K., H.C. Andersen, W. Swope (1996): 'Computer simulation study of the melting transition in two dimensions', Phys. Rev. Lett. **76**, pp. 255–258
5. Binder, K. (Ed.) (1995): *The Monte Carlo Method in Condensed Matter Physics* (Topics in Applied Physics Vol. 71, Springer, Berlin)
6. Ciccotti, G., D. Frenkel, I.R. McDonald (Eds.) (1987): *Simulation of liquids and solids. Molecular Dynamics and Monte Carlo Methods in Statistical Mechanics* (North-Holland, Amsterdam)
7. Fernández, J.F., J.J. Alonso, E. Stankiewicz (1995): 'One-stage continuous melting transition in two dimensions', Phys. Rev. Lett. **75**, pp. 3477–3480
8. Frenkel, D., B. Smit (1996): *Understanding molecular simulation. From algorithms to applications* (Academic Press, San Diego)
9. Geyer, C.J., J. Møller (1994): 'Simulation procedures and likelihood inference for spatial point processes', Scand. J. Statist. **21**, pp. 359–373
10. Geyer, C.J., E.A. Thompson (1995): 'Annealing Markov chain Monte Carlo with applications to pedigree analysis', J. Am. Statist. Ass. **90**, pp. 909–920
11. Geyer, C.J. (1999): 'Likelihood inference for spatial point processes'. In: *Stochastic Geometry: Likelihood and Computations*, ed. by O.E. Barndorff-Nielsen, W.S. Kendall, M.N.M. van Lieshout (Chapman and Hall/CRC, London) pp. 79–140
12. Hansen, J.-P., I.R. McDonald (1986): *Theory of Simple Liquids* (Academic Press, London)
13. Hoover, W.G., F.H. Ree (1969): 'Melting Transition and Communal Entropy for Hard Spheres', J. Chem. Phys. **49**, pp. 3609–3617
14. Jaster, A. (1999): 'An improved Metropolis algorithm for hard core systems', cond-mat/9810274 (21. Oct. 1998); Physica A **264**, p. 134
15. Kosterlitz, J.M., D.J. Thouless (1973): 'Ordering metastability and phase transformation in two-dimensional systems', J. Phys. C **6**, pp. 1181–1203; Kosterlitz, J.M. (1974): 'The critical properties of the two-dimensional xy model', J. Phys.

C 7, pp. 1046–1060; Berenzinskii, V.L. (1972): 'Destruction of long-range order in one-dimensional and two-dimensional systems possessing a continuous symmetry group. II.Quantum systems', Sov. Phys. JETP **34**, pp. 610–616

16. Lee, J., K.J. Strandburg (1992): 'First-order melting transition of the hard-disk system', Phys. Rev. B **46**, p. 11190—11193

17. Marcus, A.H., S.A. Rice (1996): 'Observations of First-Order Liquid-to-Hexatic and Hexatic-to-Solid Phase Transitions in a Confined Colloid Suspension', Phys. Rev. Lett. **77**, pp. 2577–2580

18. Marinari, E., G. Parisi (1992): 'Simulated tempering: A new Monte Carlo scheme', Europhysics Letters **19**, pp. 451–458

19. Mase, S., J. Møller, D. Stoyan, R.P. Waagepetersen, G. Döge (1999): 'Packing Densities and Simulated Tempering for Hard Core Gibbs Point Processes', (submitted 1999)

20. Metropolis, N., A.W. Rosenbluth, M.N. Rosenbluth, A.H. Teller, E. Teller (1953): 'Equation of state calculations by fast computing machines', J. Chemical Physics **21**, pp. 1087–1092

21. Møller, J. (1999): 'Markov chain Monte Carlo and spatial point processes'. In: *Stochastic Geometry: Likelihood and Computations*, ed. by O.E. Barndorff-Nielsen, W.S. Kendall, M.N.M. van Lieshout (Chapman and Hall/CRC, London) pp. 141–172

22. Nelson, D.R., B.I. Halperin (1979): 'Dislocation-mediated melting in two dimensions', Phys. Rev. B **19**, pp. 2457–2484; Young, A.P. (1979): 'Melting and the vector Coulomb gas in two dimensions', Phys. Rev. B **19**, p. 1855

23. Schmidt, M. (1997): Freezing in confined geometry. PhD-thesis, Düsseldorf (Shaker Verlag, Aachen)

24. Stoyan, D., W.S. Kendall, J. Mecke (1995): *Stochastic Geometry and its Applications*, 2nd edn. (Wiley & Sons, New York)

25. K.J. Strandburg (1988): 'Two-dimensional melting', Rev. Mod. Phys. **60**, pp. 161–207

26. Swope, W.C., H.C. Andersen (1992): 'Thermodynamics, statistical thermodynamics, and computer simulation of crystals with vacancies and interstitials', Phys. Rev. A **46**, pp. 4539–4548; 'A computer simulation method for the calculation of liquids and solids using the bicanonical ensemble', J. Chem. Phys. **102**, pp. 2851–2863 (1995)

27. Truskett, T.M., S. Torquato, S. Sastry, P.G. Debenetti, F.H. Stillinger (1998): 'A structural precursor to freezing in the hard-disk and hard-sphere systems', Phys. Rev. E **58**, pp. 3083–3088

28. Weber, H. D. Marx (1994): 'Two-dimensional melting approached via finite-size scaling of bond-orientational order', Europhys. Lett. **27**, pp. 593–598

29. Weber, H., D. Marx, K. Binder (1995): 'Melting transition in two dimensions: A finite-size scaling analysis of bond-orientational order in hard disks', Phys. Rev. B **51**, pp. 14636–14651

30. Zollweg, J.A., G.V. Chester, P.W. Leung (1992): 'Melting in two dimensions', Phys. Rev. B **46**, pp. 11186–11189

Dynamic Triangulations
for Granular Media Simulations

Jean-Albert Ferrez, Thomas M. Liebling, and Didier Müller

Chair of Operations Research, Department of Mathematics, Swiss Federal Institute of Technology, Lausanne (EPFL)

Abstract. We present an efficient method for the computer simulation of granular media based on dynamic triangulations. Round and polygonal grains are considered. In both cases, we explain the theoretical principles on which the methods are based. The simulation schemes are outlined and several examples are given.

1 Introduction

The distinct element method (DEM) is a natural tool for computer simulations of highly discontinuous phenomena. By considering each element, particle, or grain separately, it allows to take into account their local behavior while providing global measures. Classical examples include molecular dynamics and granular media simulations. The main drawback of this approach is traditionally associated with the high computational cost of detecting and resolving the interactions between the elements. To be more precise, there is a non-compressible cost associated with the implementation of the physical law governing those interactions, and an overhead associated with finding where those interactions occur. Efficient neighborhood functions can greatly reduce this overhead, thus allowing larger and longer simulations to be performed. The following sections present two such neighborhoods based on **dynamic triangulations**. One of them is restricted to round grains, in two or three dimensions, the other allows grains of any polygonal shape, but in 2D only. For both cases, the structures and algorithms are described, the underlying physics is briefly sketched, and several examples are given.

2 Simulation of Granular Media

Granular media cannot be assimilated to any of the three states of matter: solid, liquid, gas. The typical sandpile, for example, offers a wide range of behavior modes that violate some of the characteristics of each state. A granular medium is not a solid, because it does not resist stretching, but is not a gas either, because it resists compression. And unlike a liquid, its surface at rest may not be flat. The study of such behaviors conducts naturally to considering not the medium itself, but each grain independently, and to perform simulations with distinct element methods. Those methods only make sense if the number of grains is large enough to accurately replicate natural behaviors. This requires either very powerful computers, or extremely efficient simulation methods, or both! The main issues with a distinct element approach are:

1. The detection of the interactions between grains. A large number of grains means a very large number of potential collisions, especially if the medium is volatile. Various approaches exist to reduce this number: regular or adaptive grids (quadtrees, see [22]), distance-based neighborhood, etc. Coarse and fine detection are often combined: first decide which grains may collide, then locate the contact point (see [17]).

2. Applying a contact model that will ensure realistic behavior at the local as well as at the global level. Two approaches exist: The *hard body* model considers instantaneous impacts, with only one such impact happening at any given time and an immediate update of the grain trajectories. This works well for sparse, volatile media. The *soft body* model on the other hand allows a small continuous deformation of the grains at the contact point to absorb and restitute the energy of the shock. This works well for dense, static media.

Computer simulations are expected to bring a better understanding of qualitative phenomena like convection, segregation, arching effects, etc, as well as trustable values for quantitative parameters like local density, homogeneity, energy dissipation, etc.

One particular area of interest is the study of the packing process. The question is simple: "How to put as many elements as possible in a given container", but the answers are many and often difficult. Kepler's conjecture about the maximal density of balls of equal size, for example, was stated almost 400 years ago and only recently [10] proposed a proof. And if the packing of equal spheres finally seems to be mastered, optimal packings of spheres of unequal radii - or of general polyhedral grains - still cannot be determined analytically and require either experiments or simulations. They are, though, a key parameter of many industrial processes, and a better understanding of their principles would benefit both research and production in areas such as pharmacy, chemistry, civil engineering, etc.

3   When the Grains Are Round

In this section we describe a collision detection method that can be used when the grains are round. The idea is to identify a small subset of the pairs of grains where a collision may take place (Fig. 1), ignoring the vast majority of pairs for which we have an *a priori* criterion to exclude any contact.

3.1   Dynamic Triangulations

Let us first see how a static triangulation can be used to detect collisions. Given a set of n disjoint spheres where sphere S_i has center c_i and radius r_i (Fig. 2), we associate with each sphere its **Laguerre cell** L_i defined as:

$$L_i = \{x \mid P_i(x) \leq P_j(x) \quad \forall j \neq i\} \quad \text{where} \quad P_i(x) = ||c_i - x||^2 - r_i^2$$

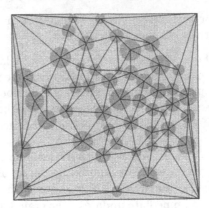

Fig. 1. The triangulation used to detect collisions among discs. Including the four corners in the triangulation is not mandatory, but has a nice side effect: the number of edges does not depend on the position of the discs in the box.

This definition is similar to that of the well known Voronoi cell. It essentially means that L_i contains all the points that are "closer" to S_i than to any other sphere, but using the power function to a sphere as a "distance". In particular, we have the following properties:

$$P_i(x) \leq 0 \ \forall x \in S_i \quad \text{and} \quad P_i(x) > 0 \ \forall x \notin S_i$$

So, if the spheres do not overlap, $S_i \subset L_i$. Therefore, there is a potential contact between S_i and S_j only if $L_i \cap L_j \neq \emptyset$, and the detection of the collisions between two spheres is reduced to the adjacency detection among the Laguerre cells. The **weighted Delaunay triangulation** is the dual structure of the Laguerre complex[1], that is: edge (i, j) exists if and only if $L_i \cap L_j \neq \emptyset$. Collision detection is now reduced to the enumeration of the edges in the weighted Delaunay triangulation.

It is possible to slightly relax one of the constraints: the spheres need not be disjoint. Overlaps are allowed and will appear as edges in the triangulation (Fig. 3) as long as they remain small, thus allowing soft body contact models to be used. Problems arise when four or more spheres intersect each other *at the same point*, but this never happens in granular media.

So far we have made no assumption about the dimension of the space, because these properties are valid in any dimension. Real life applications, though, are in dimensions two and three. In 2D, the number of edges in any triangulation is at most three times the number of points, yielding a very efficient ratio of edges to test for potential collision versus effective collisions. In 3D, there are nasty configurations, but for most practical uses, the Delaunay triangulation still has a linear number of edges (see [3]).

---

[1] The Laguerre complex is also known as *Power Diagram* (see [1] and references therein). The Delaunay triangulation of weighted sites used here is also known as *regular triangulation*. Boissonnat and Yvinec [3] made an extensive review of these fundamental computational geometry tools.

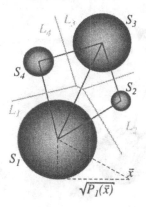

Fig. 2. The spheres S_i, the Laguerre cells L_i, and the weighted Delaunay triangulation on the center of the spheres. The power $P_1(x)$ of x to S_1 is shown, measured as the squared distance along the tangent from x to the contact point.

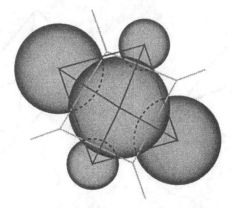

Fig. 3. The Laguerre cells and weighted Delaunay triangulation with overlapping spheres. Here the overlaps are exaggerated, but the detection still works. In practice, the overlaps are about 1% of the size of the balls.

This collision detection mechanism seems quite appealing, but what happens when the positions c_i of the spheres move? The distances $P_i(x)$ change and the minimum for a given x can be attained for a different i, yielding a different Laguerre complex and a different weighted Delaunay triangulation, as shown in Fig. 4. Recomputing the whole triangulation from scratch is not an option: it takes too much time and the expected benefit of the method vanishes.

Fortunately, another property of the Delaunay triangulation helps maintaining it efficiently. So far we have seen the Delaunay triangulation as the dual of the Laguerre complex. It is also among all possible triangulations the only one satisfying the *incircle* criterion: in 2D, every disc enclosing a triangle contains no other points, see Fig. 5.

It is therefore possible at any time to identify all the portions of the triangulation that have become invalid under the motion of the sites by checking whether

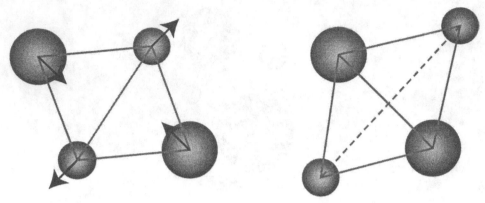

Fig. 4. As the spheres move, the original Delaunay triangulation may fail to detect contacts. Maintaining the triangulation is mandatory: a new edge appears while another one disapears.

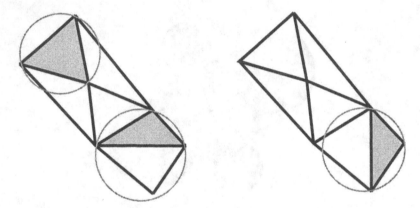

Fig. 5. The *incircle* test: in the right picture, the top grey triangle is valid while the bottom one is not. After flipping (right picture) the triangulation becomes the Delaunay triangulation.

the enclosing discs are empty. More precisely, for the configuration shown in Fig. 6 (left), the sign of the determinant

$$\Delta = \begin{vmatrix} x_1 \ y_1 \ r_1^2 - (x_1^2 + y_1^2) \ 1 \\ x_2 \ y_2 \ r_2^2 - (x_2^2 + y_2^2) \ 1 \\ x_3 \ y_3 \ r_3^2 - (x_3^2 + y_3^2) \ 1 \\ x_4 \ y_4 \ r_4^2 - (x_4^2 + y_4^2) \ 1 \end{vmatrix}$$

where x_i (resp. y_i) is the x (resp. y) coordinate of the center c_i and r_i the radius of the sphere i, tells whether the triangulation is locally valid or not.

Such a local Delaunay criterion is also valid higher dimensions. In 3D, the discs enclosing triangles are replaced by spheres enclosing tetrahedra. It also extends to the weighted generating sites used for our purpose.

Once illegal portions have been identified, local transformations - called flips - are then performed to re-enforce the Delaunay criteria in the triangulation. Flipping means alternating between the valid triangulations of a closed area defined by 4 (2D) or 5 (3D) points (see [6]). In 2D, only one operation is necessary, while in 3D two cases occur: the addition or the removal of an edge (Fig. 6).

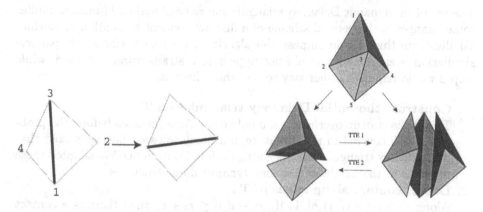

Fig. 6. The flips in 2D and in 3D.

Two common problems in Computational Geometry arise when computing the determinant Δ:

Degeneracy: If $\Delta = 0$, the four points are exactly on the same circle, and both possible triangulations can be seen as valid or invalid. This degenerate case is carefully avoided in theory. Since it clearly happens in practice, we just choose arbitrarily one of the two, knowing this does not break our collision detection. Some time later when the positions have evolved, chances are high that the degeneracy will have been lifted.

Numerical stability: The points may be very close to being on the same circle. In this case, the absolute value of Δ will be very small. So small that the limited precision of the computer is defeated, and it may give a false result to the *incircle* test. The problem here is twofold: again we could obtain a valid triangulation that is temporarily not the Delaunay triangulation, without jeopardizing our collision detection scheme, but we could also end up with inconsistencies in our data structures. This second case is dangerous for the whole simulation, and the workaround consists in detecting possibly false evaluations of Δ and repeating the computation with exact arithmetics.

At this stage, there only remains the construction of the initial triangulation, indeed a well known problem in numerical simulation. The theory and computational implementation closest to our needs is that of Joe [11,12]. We need a more elaborate version, though, because we are dealing with weighted generating sites and we want to transform the triangulation over time. While the former constraint is easily satisfied, the latter forces us to use elaborate data structures to represent the triangulation as well as proximity relations. This is achieved with *doubly connected edge lists* (DCEL) introduced by [8] in 2D, and *doubly connected face lists* (DCFL) introduced by [5] in 3D.

3.2   Distinct Element Modeling and Simulation

The use of the dynamic Delaunay triangulation as neighborhood function implies some changes in the general scheme of a distinct element simulation algorithm. To illustrate this, we decompose the algorithm used to perform soft particle simulation: steps 1 and 5 would not appear in a strait-forward method, while step 2 would require another way to find the collisions.

1. **Construct the initial Delaunay triangulation T**
   This is the startup overhead, paid only once. As mentioned before, the problem is well covered in the literature and many software packages exist that compute this triangulation efficiently, both in 2D or in 3D. We adapted these methods to the weighted case and dynamic data structures.
2. **Detect contact along edges of T**
   Along each edge (i, j) of T, if $\|c_i - c_j\| \leq r_i + r_j$ then there is a contact between spheres i and j. There may be edges with no contact, but there may not be a contact without an edge. This is where the use of the triangulation shows its strength, reducing the average detection complexity from $O(n^2)$ to $O(n)$.
3. **Apply some law at every contact point**
   Once all the contact points have been identified, proper action must be taken according to the nature of the simulation. This action can be based on values like the size of the overlap, relative normal and shear speeds, particle types, etc. Upon completion of this step, all the forces acting on the particles at every contact point should be known.
4. **Update the trajectories**
   After summing the forces contributed by the contacts in step 3 and external forces (gravitation, etc), it is now possible to integrate the motion equations over the duration of the iteration, resulting in new particle positions.
5. **Update the triangulation T with local operations**
   This is the running overhead of the triangulation. It involves
   5a. checking if every edge (2D) or triangle (3D) is locally legal and
   5b. performing the necessary flips.
   However, if the motions of the grains are relatively slow (as in a static, dense granular media), the triangulation changes seldom. Furthermore, the elapsed time between the appearance of an edge in the triangulation and the actual contact between the corresponding grains is such that this step may safely be skipped most of the time. The net result is a much faster algorithm that still detects all the collisions, even if the triangulation slightly breaks the Delaunay criterion.
6. **Go back to 2**

The iteration duration δt must be chosen carefully. Small values will require many more iterations for the same result, but too large values might introduce instabilities in the triangulation (step 5) and a loss of accuracy in the model (step 3). The soft body theory assumes a contact duration of about 10^{-5} seconds and requires at least 100 iterations for every contact to be on the safe side, thus setting $\delta t = 10^{-7}$. This is too small for today's computers, though, and most authors admit using a much larger value, in the 10^{-4} to 10^{-5} range. The triangulation usually needs to be updated 100 to 1000 times per second, depending of the volatility of the media, so step 5 is only executed every 100 iterations or so.

If a hard body model is used for the contacts, time is not sliced in iterations of equal duration δt anymore. The trajectories of the grains and the evolution of the triangulation give the exact time of each contact. A global scheduler is then used to deal with these collisions in chronological order. See [9] and references therein for a complete theoretical study of what they call a *kinetic data structure*.

3.3   The Physical Models

In the algorithm presented above, steps 1, 2 and 5 are "support" steps, in that they deal with creating, utilizing and maintaining the triangulation. On the other hand, steps 3 and 4 perform the actual simulation.

In step 4 we assume that all the forces acting on each grain i are known. These forces come from contacts with other grains - as computed in step 3 - or from contacts with the walls or floor, or from external factors like gravitation. From the sum of all these forces we compute the new speed and position of each grain by integrating the motion equations:

$$a_i(t) = \frac{1}{\text{mass}_i} \sum \text{forces}_i(t)$$
$$v_i(t) = v_i(t - \delta t) + \delta t a_i(t)$$
$$c_i(t) = c_i(t - \delta t) + \delta t v_i(t)$$

Step 3 holds most of the magic that turns a distinct element computer simulation into a realistic representation of real life phenomena. The complete study of the shock between two spheres is beyond the scope of this article. We give a brief description of the soft body models we have used. The interested reader will find extensive coverage of this topic for example in [19].

The shock can be seen as a two phase process: in the first phase, the kinetic energy is absorbed and transformed into an elastic deformation of the grains; in the second phase, a part of this deformation is given back as kinetic energy.

The force at the contact point can be decomposed into a normal and a tangential part. The normal force also has two distinct components: one elastic and one viscous. The elastic component acts like a spring. It is a function of the size of the overlap between the two spheres: the bigger this overlap, the stronger the repulsion force, although the relation may not be linear.

The viscous component acts like a dash. It is a function of the relative speed between the two spheres: in the first phase, it will add to the elastic part and further absorb the kinetic energy. In the second phase, however, it acts in the opposite direction and accounts for the permanent deformation of the grain, the heat dissipation, etc.

The pioneer model of [4] is based on linear expressions for both elastic and viscous components of the normal and tangential forces. Later additions include non-linear versions of the elastic component by [13] and hysteretic forces by [18].

The simplest form of tangential force is given by Coulomb's law that only consider a friction coefficient. More elaborate versions also take into account an elastic and a viscous component.

3.4   Parallelism

Besides efficient collision detection, an obvious way to tackle larger simulations is to use very powerful computers. Nowadays, this statement often involves some form of parallelism, either with shared memory parallel servers or with distributed clusters of machines.

The hard body models rely on a global scheduler to treat the degeneracies in chronological order. The amount of work required to locally update the triangulation is relatively small compared to the number of events. Therefore, the degree of parallelism inherent to the method is small, which means that the expected benefit only happens for small numbers of processors.

The soft body models, on the other hand, allow a spatial decomposition of the simulated area with a very high proportion of independent, local operations. Furthermore, clever design of the data structures avoids concurrent accesses throughout most of the simulation. Thus, that method has a high degree of parallelism and the performances still grow with large numbers of processors.

The algorithm given in Sect. 3.2 was implemented on a CRAY T3D with very good results (see [7]). Steps 2, 3, 4 and 5a were performed in parallel on up to 128 processors. The remaining parts of the code, step 1 and 5b, were left sequential because their impact on the overall running time is small and their efficient parallelization is much more difficult. Access and efficient use of this supercomputer was critical when performing some of the the the more demanding experiments (see below). For example, one run of the "falling rock" simulation that would take a whole week-end of computation on a standard workstation was typically completed in two hours on the T3D!

3.5   Examples

Many simulations were performed using 2D discs as grains, the reader is referred to [15] for a complete description and to [14] for more images and videos. Both hard and soft body models were used. Here are a few examples.

Figure 7 comes from the simulation of a rock fall on an embankment, a joint project with Prof. Descoeudres of the Civil Engineering Department of EPFL.

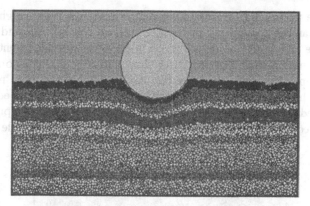

Fig. 7. Rock fall on an embankment.

Many mountain roads in Switzerland have to be protected against rock falls by galleries composed of a concrete slab topped with a granular bed. The absorption of falling rock impacts by this bed is the key issue. Physical experiments were performed and then duplicated by computer simulation. This allowed us to fit the model more precisely. The real experiment involved a 1000 Kg bloc of concrete fall with a 10 meters vertical drop. The simulation was performed with the same initial conditions. The picture shows the final position of the rock after absorption by a bed composed of approx. 20'000 grains. The colors identify the horizontal layers and their deformation due to the impact. No special technique was used to differentiate the falling rock from the other grains, proving the great versatility and robustness of the method.

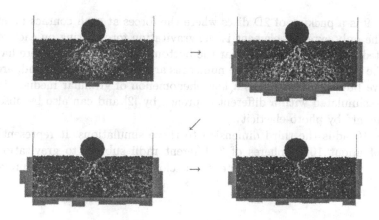

Fig. 8. Impact absorption by the granular bed and pressure distribution on the slab.

The four images in Fig. 8 were produced from the same simulation, at four different times immediately after the impact. They show the impact wave propagation inside the media (grains under higher pressure have a lighter shade of gray) as well as the repartition of the pressure on the bottom and sides of the bed. Only the soft body model allows such an aggregate representation of the forces in a dense media. The situation shown in Fig. 8 corresponds to a bed formed of grains of various sizes, the bed itself was obtained by a natural packing process. The experiment was repeated with regular beds formed of identical grains disposed on a grid, but the results were quite different: the pressure is not well distributed in the media, and transmitted mostly in a single spot on the slab.

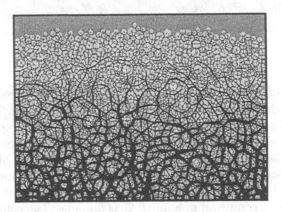

Fig. 9. Pressure distribution. All edges in the triangulation that actually correspond to a contact between two spheres are drawn with a thickness proportional to the force at that contact point. Forces against the walls are represented identically.

Figure 9 is a packing of 2D discs where the forces at each contact point are drawn. The only external element is the gravitation force acting on each grain. As expected, the forces increase near the bottom, since the grains there have to "carry" the grains above them. The numerous arches, on the other hand, are not so intuitive but are nevertheless a key phenomenon of granular media. Similar effects are simulated with a different approach by [2] and can also be observed in experiments by photo-elasticity.

Figure 10 adds the third dimension to those simulations. It represents the packing of about 1000 spheres of 3 different radii subject to gravitation (to initiate the packing) and vibration of the enclosing box (to further improve the packing).

4   When the Grains Are Polygonal

We describe in this section a collision detection method that can be used when the grains are represented by polygons. The idea is to decompose the interstices between the grains, maintaining a dynamic triangulation of this space, updating it according to the motion of the grains and rely on it to report any collision. 2D

Fig. 10. A 3D packing of about 1000 spheres in the unit cube. Left: 3D view. Right: bottom view.

simulations based on this technique have been performed by [16]. Unfortunately, it does not generalize trivially to 3D; we will list the problems that occur as well as possible solutions.

4.1   In 2D

In 2D, the grains may have any simple polygonal shape: convex or not, "round" or "flat", etc. (Fig. 11).

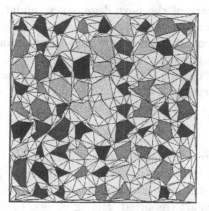

Fig. 11. The triangulation used for detecting collisions among 2D polygonal grains.

A triangulation is built on the vertices of the grains and constrained by the edges of the grains, resulting in a decomposition of the free space between the grains. Any movement - translation, rotation or both - of the grains will thus be transmitted to the triangulation, possibly causing a *degeneracy*, that is the disappearing of a triangle due to one of its vertices reaching the opposite edge.

- If that edge belongs to a grain, then there has been a contact, and an appropriate hard-body, instantaneous collision model is used to update the trajectories of those two grains.
- If that edge does not belong to a grain, a local transformation of the triangulation is performed to guarantee it remains valid (Fig. 12).

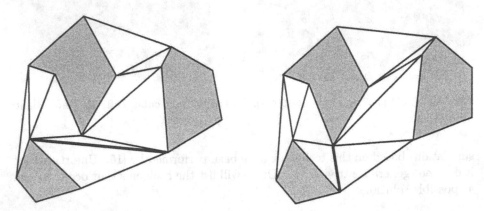

Fig. 12. A degeneracy in the triangulation and the local re-triangulation.

Given the positions and trajectories of all the grains, therefore of all vertices, it is possible to compute exactly for every triangle when it will become degenerate. All these events are inserted in a scheduler. The simulation then loops forever, taking care of the next degeneracy, updating the triangulation and adjusting future events.

Computer simulations of the flow of grains through an hourglass were performed. As long as enough grains remain in the upper part, the flow through the opening is independent of time (Fig. 13). Again, more examples and videos can be found on the web pages of [14].

4.2   In 3D

The technique of the preceding paragraph cannot be used in 3D as it is, because a 3D polyhedral shape may not admit a decomposition into tetrahedra. The smallest and simplest example of a non-triangulable polyhedron was already given in 1928 by Schönhart (reported, among others, by [3]). This means, if at any time the position of the grains is such that some part of the space between them cannot be triangulated, the whole detection process fails.

Here are two ways to possibly get around this limitation and still apply this method in 3D, both of which are still under theoretical study:

1. **Restrict the shapes of the grains**
   Even the restriction of the grains to regular tetrahedra of the same size may not be enough to guarantee the existence of a triangulation between

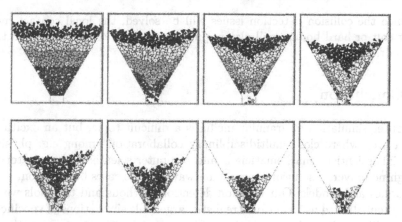

Fig. 13. The flow of grains in an hourglass.

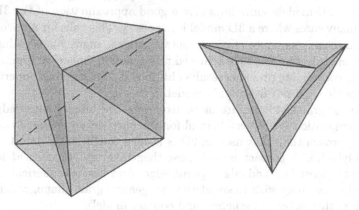

Fig. 14. Schönhart's polyhedron, 3D view (left) and deformed top view (right). It is basically a twisted, triangular-based cylinder.

them. Furthermore, such a restricted model would not be a big improvement over spheres. Using non-regular tetrahedra certainly leads to potentially non-triangulable area such as Schönhart's polyhedron.

2. **Use additional points**

   By adding enough auxiliary points (Steiner points, see Weil's paper in this volume), any space bouded by piecewise linear surfaces can be triangulated. The problem here is to balance between adding enough points to succeed, and not adding too many (or even remove some when they are not needed anymore) in order to remain efficient, especially in a dynamic context.

Since none of the two possibilities seems promising, our current plan is to encapsulate each polyhedral grain in a sphere, use the Delaunay triangulation as in Sect. 3 for a first stage, global detection scheme, and then perform local, pairwise contact detection where necessary. However, this will probably impose some restrictions on the shape and relative sizes of the grains.

Once the collision detection issues will be solved, the local contact model, either soft or hard bodies, will also require some attention to be applied to 3D polyhedral grains.

5 Conclusion

Computer simulation of granular media is a difficult topic, but an exciting research area where close multidisciplinary collaboration among e.g. physicists, materials scientists, mathematicians and computer scientists is mandatory. Although no universal solution seems underway, specific cases benefit from various techniques and models. Our collision detection methods and the tools we built using them allowed us to perform realistic large scale simulations of specific phenomena like grain flows, convection and segregation, pressure absorption and distribution in the special cases of round or 2D polygonal grains.

Although 2D models sometimes give a good approximation of the 3D world, there are many cases where a 3D model is necessary. This calls for more elaborate detection mechanisms, more efficient contact models, many more grains, better visualization tools, etc. We have achieved this goal with spherical grains and are now able to consolidate previous results obtained in 2D as well as undertake new simulations that did not fit in a 2D model.

Polyhedral grains still require more attention. The current knowledge in 3D space decomposition is not very helpful for our purpose, so maybe a completely different approach than that used in 2D is needed to efficiently detect collisions among polyhedra. A contact model must then take into account at least two cases: point against face and edge against edge. And besides spherical and polyhedral grains, one may wish to simulate more general grain shape, which in turn will require other detection schemes and contact models.

References

1. Aurenhammer, F. (1987): 'Power diagrams: properties, algorithms and applications', *SIAM J. Comput.*, **16**(1), pp. 78–96
2. Bakucz, P., G. Krause, D. Stoyan (1999): 'Force distribution in loaded planar disc systems'. In: *Computational Methods in Contact Mechanics IV*, ed. by L. Gaul, C. A. Brebbia (WIT Press, Southampton/Boston), pp. 273–282
3. Boissonnat, J.-D., M.Yvinec (1998): *Algorithmic Geometry* (Cambridge University Press)
4. Cundall, P.A., O.D.L. Strack. (1979): 'A discrete numerical model for granular assemblies', *Géotechnique*, **29**
5. Dobkin, D.P., M.J. Laszlo (1989): 'Primitives for the manipulation of three-dimensional subdivisions', *Algorithmica*, **4**, pp. 3–32
6. Edelsbrunner, H., N.R. Shah (1996): 'Incremental topological flipping works for regular triangulations', *Algorithmica*, **15**, pp. 223–241
7. Ferrez, J.-A., D. Müller, Th.M. Liebling (1996): 'Parallel implementation of a distinct element method for granular media simulation on the Cray T3D', *EPFL Supercomputing Review*, **8**

Online at http://sawww.epfl.ch/SIC/SA/publications/SCR96/scr8-page4.
html.

8. Guibas, L., J. Stolfi (1985): 'Primitives for the manipulation of general subdivisions
and the computation of voronoi diagrams', *ACM Transactions on Graphics*, **4**, pp.
74–123

9. Guibas, L.J., L. Zhang (1998) 'Euclidean proximity and power diagrams'. In: *Proc.
10th Canadian Conference on Computational Geometry*
http://graphics.stanford.EDU/~lizhang/interests.html.

10. Hales, Th.C. (1998): 'The kepler conjecture', http://www.math.lsa.umich.edu/
~hales/countdown/, 1998. Overview of the proof and related papers.

11. Joe, B. (1989): 'Three-dimensional triangulations from local transformations',
SIAM J. Sci. Stat. Comput., **10**, pp. 718–741

12. Joe, B. (1991): 'Construction of three-dimensional delaunay triangulations using
local transformation', *Computer Aided Geometric Design*, **8**, pp. 123–142

13. Kuwabara, G., K. Kono (1987): *Jap. J. Appl. Phys.*, **26**(1230)

14. Müller, D. (1996a): 'Simulations of granular media', http://rosowww.epfl.ch/
dm/sigma.html

15. : Müller, D. (1996b): Techniques informatiques efficaces pour la simulation de
mileux granulaires par des méthodes d'éléments distincts. PhD thesis, EPFL

16. Müller, D., Th.M. Liebling (1995): 'Detection of collisions of polygons by using a
triangulation'. In: *Contact Mechanics*, ed. by M. Raous et al. (Plenum Publishing
Corporation, New York), pp. 369–372

17. O'Connor, R.M. (1996): A distributed discrete element modeling environment -
Algorithms, implementation and applications. PhD thesis, MIT

18. Sadd, M.H., Q. Tai, A. Shukla (1993): 'Contact law effects on wave propagation
in particulate materials using distinct element modeling', *Int. J. Non-Linear Mechanics*, **28**(251)

19. Schäfer, J., S. Dippel, D.E. Wolf (1996): 'Force schemes in simulations of granular
materials', *J.Phys. I*, **6**(5)

20. Telley, H., Th. M. Liebling, A. Mocellin (1996a): 'The Laguerre model of grain
growth in two dimensions: Part I. Cellular structures viewed as dynamical Laguerre tesselations', *Phil. Mag. B*, **73**(3), pp. 395–408

21. Telley, H., Th.M. Liebling, A. Mocellin (1996b): 'The Laguerre model of grain
growth in two dimensions: Part II. Examples of coarsening simulations', *Phil.
Mag. B*, **73**(3), pp. 409–427

22. Wenzel, O., N. Bicanic (1993): 'A quad tree based contact detection algorithm'.
In: *Proceedings of the 2nd international conference on discrete element methods
(DEM), MIT*. IESL Publications

23. Xue, X., F. Righetti, H. Telley, Th. M. Liebling, A. Mocellin (1997): 'The Laguerre
model for grain growth in three dimensions', *Phil. Mag. B*, **75**(4), pp. 567–585

Index

Lecture Notes in Physics

For information about Vols. 1–519
please contact your bookseller or Springer-Verlag

Monographs

For information about Vols. 1–21
please contact your bookseller or Springer-Verlag